David Clements, PhD
Anil Shrestha, PhD
Editors

New Dimensions in Agroecology

New Dimensions in Agroecology has been co-published simultaneously as *Journal of Crop Improvement*, Volume 11, Numbers 1/2 (#21/22) and Volume 12, Numbers 1/2 (#23/24) 2004.

Pre-publication
REVIEWS,
COMMENTARIES,
EVALUATIONS . . .

"This book draws on outstanding leaders in agroecology to develop a broad basis for the need for sustainable agriculture and a secure global food supply. Confirming the need for agroecology and food security is the World Health Organization report that indicates that more than three billion people are malnourished in the world–the largest number in history."

David Pimentel, PhD
Professor of Ecology and Agricultural Sciences
College of Agriculture and Life Sciences
Cornell University

More pre-publication
REVIEWS, COMMENTARIES, EVALUATIONS . . .

"A VALUABLE ADDITION to the literature on making our agriculture and food systems more sustainable. . . Reviews the literature on the state of the art in agroecological research. . . . Shows how much progress has been made in the last decade in scientific thinking and research, especially on agroecology's environmental component."

Christoffel den Biggelaar, Dr. Ing.
Associate Professor of Agroecology
The Goodnight Family Program
in Sustainable Development
Department of Interdisciplinary Studies
Appalachian State University

"ENGAGING AND INFORMATIVE. . . . The chapters cover a wide range of expertise, including soil fertility, microbial ecology, weed management, pest control, and landscape and systems ecology, with contributions from prominent agroecologists such as Miguel Altieri, David Andow, Charles Francis, and Stephen Gliessman. New topics addressed in the book are research, education, and policy reforms for redesigning agricultural systems with an ecological focus, unintended effects of transgenic crops in ecological systems, and effects of global change on ecological processes. BACKGROUND INFORMATION MAKES EACH CHAPTER ACCESSIBLE TO STUDENTS AND RESEARCHERS across the various sub-fields of agroecology."

Deborah K. Letourneau, PhD
Professor
Department of Environmental Studies
University of California, Santa Cruz

Food Products Press
An Imprint of The Haworth Press, Inc.

New Dimensions in Agroecology

New Dimensions in Agroecology has been co-published simultaneously as *Journal of Crop Improvement*, Volume 11, Numbers 1/2 (#21/22) and Volume 12, Numbers 1/2 (#23/24) 2004.

The *Journal of Crop Improvement* Monographic "Separates"

(formerly the *Journal of Crop Production*)*

Below is a list of "separates," which in serials librarianship means a special issue simultaneously published as a special journal issue or double-issue *and* as a "separate" hardbound monograph. (This is a format which we also call a "DocuSerial.")

"Separates" are published because specialized libraries or professionals may wish to purchase a specific thematic issue by itself in a format which can be separately cataloged and shelved, as opposed to purchasing the journal on an on-going basis. Faculty members may also more easily consider a "separate" for classroom adoption.

"Separates" are carefully classified separately with the major book jobbers so that the journal tie-in can be noted on new book order slips to avoid duplicate purchasing.

You may wish to visit Haworth's website at . . .

http://www.HaworthPress.com

. . . to search our online catalog for complete tables of contents of these separates and related publications.

You may also call 1-800-HAWORTH (outside US/Canada: 607-722-5857), or Fax 1-800-895-0582 (outside US/Canada: 607-771-0012), or e-mail at:

docdelivery@haworthpress.com

New Dimensions in Agroecology, David Clements, PhD and Anil Shrestha, PhD (Vol. 11, No. 1/2 #21/22 and Vol. 12, No. 1/2 #23/24, 2004). *Provides extensive information on current innovative agroecological research and education as well as emerging issues in the field.*

Adaptations and Responses of Woody Plants to Environmental Stresses, edited by Rajeev Arora, PhD (Vol. 10, No. 1/2 #19/20, 2004). *Focuses on low-temperature stress biology of woody plants that are of horticultural importance.*

Cropping Systems: Trends and Advances, edited by Anil Shrestha, PhD* (Vol. 8, No. 1/2 #15/16 and Vol. 9, No. 1/2 #17/18, 2003). *"Useful for all agricultural scientists and especially crop and soil scientists. Students, professors, researchers, and administrators will all benefit. . . . THE CHAPTER AUTHORS INCLUDE PRESENT AND FUTURE LEADERS IN THE FIELD with broad international perspectives. I am planning to use this book in my own teaching" (Gary W. Fick, PhD, Professor of Agronomy, Cornell University)*

Crop Production in Saline Environments: Global and Integrative Perspectives, edited by Sham S. Goyal, PhD, Surinder K. Sharma, PhD, and D. William Rains, PhD* (Vol. 7, No. 1/2 #13/14, 2003). *"TIMELY. . . . COMPREHENSIVE. . . . The authors have considerable experience in this field. I hope this book will be read widely and used for promoting soil health and sustainable advances in crop production." (M. S. Swaminathan, PhD, UNESCO Chair in Ecotechnology, M. S. Swaminathan Research Foundation, Chennai, Tami Nadu, India)*

Food Systems for Improved Human Nutrition: Linking Agriculture, Nutrition, and Productivity, edited by Palit K. Kataki, PhD and Suresh Chandra Babu, PhD* (Vol. 6, No. 1/2 #11/12, 2002). *Discusses the concepts and analyzes the results of food based approaches designed to reduce malnutrition and to improve human nutrition.*

Quality Improvement in Field Crops, edited by A. S. Basra, PhD, and L. S. Randhawa, PhD* (Vol. 5, No. 1/2 #9/10, 2002). *Examines ways to increase nutritional quality as well as volume in field crops.*

Allelopathy in Agroecosystems, edited by Ravinder K. Kohli, PhD, Harminder Pal Singh, PhD, and Daizy R. Batish, PhD* (Vol. 4, No. 2 #8, 2001). *Explains how the natural biochemical interactions among plants and microbes can be used as an environmentally safe method of weed and pest management.*

The Rice-Wheat Cropping System of South Asia: Efficient Production Management, edited by Palit K. Kataki, PhD* (Vol. 4, No. 1 #7, 2001). *This book critically analyzes and discusses production issues for the rice-wheat cropping system of South Asia, focusing on the questions of soil depletion, pest control, and irrigation. It compiles information gathered from research institutions, government organizations, and farmer surveys to analyze the condition of this regional system, suggest policy changes, and predict directions for future growth.*

The Rice-Wheat Cropping System of South Asia: Trends, Constraints, Productivity and Policy, edited by Palit K. Kataki, PhD* (Vol. 3, No. 2 #6, 2001). *This book critically analyzes and discusses available options for all aspects of the rice-wheat cropping system of South Asia, addressing the question, "Are the sustainability and productivity of this system in a state of decline/stagnation?" This volume compiles information gathered from research institutions, government organizations, and farmer surveys to analyze the impact of this regional system.*

Nature Farming and Microbial Applications, edited by Hui-lian Xu, PhD, James F. Parr, PhD, and Hiroshi Umemura, PhD* (Vol. 3, No. 1 #5, 2000). *"Of great interest to agriculture specialists, plant physiologists, microbiologists, and entomologists as well as soil scientists and evnironmentalists. . . . very original and innovative data on organic farming." (Dr. André Gosselin, Professor, Department of Phytology, Center for Research in Horticulture, Université Laval, Quebec, Canada)*

Water Use in Crop Production, edited by M.B. Kirkham, BA, MS, PhD* (Vol. 2, No. 2 #4, 1999). *Provides scientists and graduate students with an understanding of the advancements in the understanding of water use in crop production around the world. You will discover that by utilizing good management, such as avoiding excessive deep percolation or reducing runoff by increased infiltration, that even under dryland or irrigated conditions you can achieve improved use of water for greater crop production. Through this informative book, you will discover how to make the most efficient use of water for crops to help feed the earth's expanding population.*

Expanding the Context of Weed Management, edited by Douglas D. Buhler, PhD* (Vol. 2, No. 1 #3, 1999). *Presents innovative approaches to weeds and weed management.*

Nutrient Use in Crop Production, edited by Zdenko Rengel, PhD* (Vol. 1, No. 2 #2, 1998). *"Raises immensely important issues and makes sensible suggestions about where research and agricultural extension work needs to be focused." (Professor David Clarkson, Department of Agricultural Sciences, AFRC Institute Arable Crops Research, University of Bristol, United Kingdom)*

Crop Sciences: Recent Advances, Amarjit S. Basra, PhD* (Vol. 1, No. 1 #1, 1997). *Presents relevant research findings and practical guidance to help improve crop yield and stability, product quality, and environmental sustainability.*

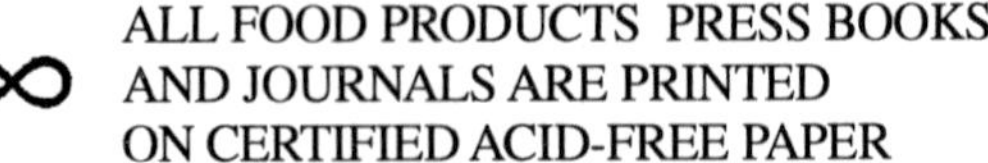

ALL FOOD PRODUCTS PRESS BOOKS
AND JOURNALS ARE PRINTED
ON CERTIFIED ACID-FREE PAPER

New Dimensions in Agroecology

David Clements, PhD
Anil Shrestha, PhD
Editors

New Dimensions in Agroecology has been co-published simultaneously as *Journal of Crop Improvement*, Volume 11, Numbers 1/2 (#21/22) and Volume 12, Numbers 1/2 (#23/24) 2004.

Food Products Press®
An Imprint of The Haworth Press, Inc.

Published by

Food Products Press®, 10 Alice Street, Binghamton, NY 13904-1580 USA

Food Products Press® is an imprint of The Haworth Press, Inc., 10 Alice Street, Binghamton, NY 13904-1580 USA.

New Dimensions in Agroecology has been co-published simultaneously as *Journal of Crop Improvement*, Volume 11, Numbers 1/2 (#21/22) and Volume 12, Numbers 1/2 (#23/24) 2004.

Cover design by Kerry E. Mack

Cover photo by Anil Shrestha, PhD

Library of Congress Cataloging-in-Publication Data

New dimensions in agroecology / David Clements, Anil Shrestha, editors.
p. cm.
"Simultaneously as Journal of crop improvement, volume 11, numbers 1/2 and volume 12, numbers 1/2, 2004."
Includes bibliographical references and index.
ISBN 1-56022-112-7 (hard cover : alk. paper) – ISBN 1-56022-113-5 (soft cover : alk. paper)
1. Agricultural ecology. I. Journal of crop improvement. II. Clements, David (David R.) III. Shrestha, Anil.
S589.7.N49 2004
577.5′5–dc22

2004011704

Indexing, Abstracting & Website/Internet Coverage

This section provides you with a list of major indexing & abstracting services. That is to say, each service began covering this periodical during the year noted in the right column. Most Websites which are listed below have indicated that they will either post, disseminate, compile, archive, cite or alert their own Website users with research-based content from this work. (This list is as current as the copyright date of this publication.)

Abstracting, Website/Indexing Coverage . . . . Year When Coverage Began

- ***AGRICOLA Database <www.natl.usda.gov/ag98>*** **1998**
- ***AGRIS*** **1998**
- ***BIOBASE (Current Awareness in Biological Science) <URL: http://www.elsevier.nl>*** **1998**
- ***Cambridge Scientific Abstracts (Water Resources Abstracts/ Agricultural & Environmental Biotechnology Abstracts) <www.csa.com>*** **2001**
- ***Chemical Abstracts Service <www.cas.org>*** **1998**
- ***Derwent Crop Production File*** **1998**
- ***Environment Abstracts. Available in print–CD-ROM– on Magnetic Tape. For more information check: <www.cispubs.com>*** **1998**
- ***Field Crop Abstracts (c/o CAB Intl/CAB ACCESS) <www.cabi.org/>*** **1998**
- ***Foods Adlibra*** **1998**

(continued)

- ***Food Science and Technology Abstracts (FSTA) Scanned, abstracted and indexed by the International Food Information Service (IFIS) for inclusion in Food Science and Technology Abstracts (FSTA) <www.ifis.org>*** **2002**
- ***HORT CABWEB (c/o CAB Intl/CAB ACCESS) <http://hort.cabweb.org>*** **1998**
- ***Plant Breeding Abstracts (c/o CAB Intl/CAB ACCESS) <www.cabi.org/>*** **1998**
- ***Referativnyi Zhurnal (Abstracts Journal of the All-Russian Institute of Scientific and Technical Information–in Russian)*** **1998**
- ***TROPAG & RURAL (Agriculture and Environment for Developing Regions), available from SilverPlatter, both via the Internet and on CD-ROM <www.kit.nl>*** **2003**

Special Bibliographic Notes related to special journal issues (separates) and indexing/abstracting:

- indexing/abstracting services in this list will also cover material in any "separate" that is co-published simultaneously with Haworth's special thematic journal issue or DocuSerial. Indexing/abstracting usually covers material at the article/chapter level.
- monographic co-editions are intended for either non-subscribers or libraries which intend to purchase a second copy for their circulating collections.
- monographic co-editions are reported to all jobbers/wholesalers/approval plans. The source journal is listed as the "series" to assist the prevention of duplicate purchasing in the same manner utilized for books-in-series.
- to facilitate user/access services all indexing/abstracting services are encouraged to utilize the co-indexing entry note indicated at the bottom of the first page of each article/chapter/contribution.
- this is intended to assist a library user of any reference tool (whether print, electronic, online, or CD-ROM) to locate the monographic version if the library has purchased this version but not a subscription to the source journal.
- individual articles/chapters in any Haworth publication are also available through the Haworth Document Delivery Service (HDDS).

New Dimensions in Agroecology

CONTENTS

Preface xiii

New Dimensions in Agroecology for Developing a Biological Approach to Crop Production 1
David R. Clements
Anil Shrestha

Education in Agroecology and Integrated Systems 21
Charles A. Francis

Redesigning Industrial Agroecosystems: Incorporating More Ecological Processes and Reducing Pollution 45
Eugene P. Odum
Gary W. Barrett

Integrating Agroecological Processes into Cropping Systems Research 61
Stephen R. Gliessman

An Agroecological Basis for Designing Diversified Cropping Systems in the Tropics 81
Miguel A. Altieri
Clara I. Nicholls

Soil Ecosystem Changes During the Transition to No-Till Cropping 105
Tami L. Stubbs
Ann C. Kennedy
William F. Schillinger

The Ecology of Crop-Weed Interactions: Towards a More Complete Model of Weed Communities in Agroecosystems 137
Bruce D. Maxwell
Edward Luschei

Research and Extension Supporting Ecologically Based IPM Systems 153
Fabián D. Menalled
Douglas A. Landis
Larry E. Dyer

Effects of Key Soil Organisms on Nutrient Dynamics in Temperate Agroecsosystems 175
Joann K. Whalen
Chantal Hamel

Nutrient Dynamics: Utilizing Biotic-Abiotic Interactions for Improved Management of Agricultural Soils 209
Chantal Hamel
Christine Landry
Abdirashid Elmi
Aiguo Liu
Timothy Spedding

Microbial and Genetic Diversity in Soil Environments 249
Ping Wang
Warren A. Dick

Impact of Global Change on Biological Processes in Soil: Implications for Agroecosystem Management 289
Shuijin Hu
Weijian Zhang

The Importance of Biodiversity in Agroecosystems 315
Lori Ann Thrupp

Biophysical and Ecological Interactions in a Temperate Tree-Based Intercropping System 339
N. V. Thevathasan
A. M. Gordon
J. A. Simpson
P. E. Reynolds
G. Price
P. Zhang

Agricultural Landscapes: Field Margin Habitats and Their Interaction with Crop Production 365
E. J. P. Marshall

Benefits of Re-Integrating Livestock and Forages in Crop Production Systems 405
E. Ann Clark

Nitrogen Efficiency in Mixed Farming Systems 437
Egbert A. Lantinga
Gerard J. M. Oomen
Johannes B. Schiere

Ecological Context for Examining the Effects of Transgenic Crops in Production Systems 457
Jennifer A. White
Jason P. Harmon
David A. Andow

Redesigning Pest Management: A Social Ecology Approach 491
Stuart B. Hill

Research Priorities in Natural Systems Agriculture 511
T. S. Cox
C. Picone
W. Jackson

Index 533

ABOUT THE EDITORS

David R. Clements, PhD, is Associate Professor of Biology and Environmental Studies at Trinity Western University in Langley, British Columbia. Dr. Clements completed his PhD in Kingston, Ontario, on integrated pest management in apple orchards. During his post-doctorate research at the University of Guelph, he continued to apply ecological principles to pest management, working in the area of weed management of row crops. He has taught ecology and botany courses at Trinity Western University since 1994, including field courses in plant ecology in the Gulf Islands off the coast of British Columbia and in the Hawaiian islands. He continues to research various aspects of integrated weed management in agriculture as well as study invasive weed biology in natural ecosystems.

Dr. Clements has published accounts for seven species in the Biology of Canadian Weed series of the Canadian Journal of Plant Science (CJPS) and was also recently named an Associate Editor of the CJPS, with responsibility for the Biology of Canadian Weed series. He has published more than 25 refereed articles. Dr. Clements is a member of several organizations, including the Canadian Weed Science Society, A Rocha Canada–Christians in Conservation, the Soil and Water Conservation Society, and the Langley Environmental Partners Society.

Anil Shrestha, PhD, is Weed Ecologist with the University of California's Statewide IPM Program. He is based at the University's Kearney Agricultural Center in Parlier. After completing his undergraduate degree at the Narendra Dev University of Agriculture and Technology in Faizabad, India, he worked several years with the Department of Agriculture and with the FAO Fertilizer Program in his homeland of Nepal. Dr. Shrestha is committed to the extension and research efforts in vegetation management, integrated pest management, cropping systems, and agroecology. He has authored more than 25 refereed papers and several extension articles in the field and has recently edited a book entitled *Cropping Systems: Trends and Advances.*

Preface

The term "agroecology" was proposed in the 1930s and has been in use for quite sometime. It became a discipline in its own right in the 1980s. Of course, the roots of an ecological approach to agriculture go back to the dawn of the agricultural revolution, when early farmers could contend with the elements *only* by understanding the ecological forces that could be wielded to design an agroecosystem. Recently, as traditional agronomic approaches have faced various economic and societal pressures, agroecology has been seen as a serious alternative and is now being introduced as a course or even a department at numerous colleges and universities. The idea of this book is to aid this process by incorporating contributions from researchers who have been working in agroecology for some time with that of other people making headway in the area (some of whom may not even consider themselves as "agroecologists"). The overall theme being developed is that ecological approaches offer numerous new dimensions to help meet the challenges of agriculture (or "the problem *of* agriculture" as Wes Jackson likes to put it).

Agroecology strives to bridge two fields, agronomy and ecology, and thus to bring holism to agriculture. This stretching of perspective is evident throughout this book, be it stretching from the field level to the landscape level, or from the level of bioengineering a gene to incorporating that gene within a complex ecological system–the agroecosystem. One of the authors, Eugene P. Odum, championed holistic thinking in the realm of ecology (which itself can become excessively reductionist), bringing the concept of the ecosystem to life. He passed away in August 2002, and it is in his memory that we dedicate this book.

[Haworth co-indexing entry note]: "Preface." Clements, David, Anil Shrestha, and Gary Barrett. Co-published simultaneously in *Journal of Crop Improvement* (Food Products Press, an imprint of The Haworth Press, Inc.) Vol. 12, No. 1/2 (#23/24), 2004, pp. xix-xxi; and: *New Dimensions in Agroecology* (ed: David Clements, and Anil Shrestha) Food Products Press, an imprint of The Haworth Press, Inc., 2004, pp. xiii-xv. Single or multiple copies of this article are available for a fee from The Haworth Document Delivery Service [1-800-HAWORTH, 9:00 a.m. - 5:00 p.m. (EST). E-mail address: docdelivery@haworthpress.com].

http://www.haworthpress.com/web/JCRIP

Although A. C. Tansley first proposed the term ecosystem in 1935, it was Eugene P. Odum who educated a generation of ecologists throughout the world with his classic text "Fundamentals of Ecology," first published in 1953; the fifth edition is scheduled for publication in 2004, co-authored with Gary W. Barrett. The clarity and enthusiasm regarding his holistic approach to terrestrial and aquatic systems were compelling. Many of his over 200 publications related to his research in agroecosystem ecology conducted at the HorseShoe Bend (HSB) experimental site located near campus at the University of Georgia.

In addition to HSB, he was the founder of the Sapelo Island Marine Biology Laboratory, the Savannah River Ecology Laboratory, and the world-recognized Institute of Ecology. Odum was a member of the National Academy of Sciences, and received numerous honors during his illustrious career, including receiving in 1975 the prestigious Tyler Prize for Environmental Achievement. In 1987, with his brother H. T. Odum, he received the Royal Swedish Academy's Crafoord Prize, considered to be equivalent to the Nobel Prize. For an overview of his career and honors, see an article entitled "Eugene P. Odum: Pioneer in Ecosystem Ecology" published in *BioScience*, 2002, 52:1047-1048.

We pay tribute to Dr. Odum in this special issue because it became recognized during the past half century that natural ecosystems, due to their evolutionary history, served as model systems regarding restoring ecological processes (e.g., nutrient cycling, improving soil quality, pest control, and biotic diversity) in agricultural systems. This model and underpinning also led to an array of emerging principles and fields of study such as agroecosystem ecology, agrolandscape ecology, and conservation biology to name a few. In fact, numerous research centers, such as the Land Institute in Salina, Kansas, base their research agenda on a natural systems approach. We are grateful to pioneers such as Eugene P. Odum for providing as early basis and understanding for this exciting and futuristic research.

The editors would like to thank various people who made this volume possible. We are thankful to Dr. Clarence Swanton, Chair of Plant Agriculture, University of Guelph, for mentoring and logistical support, and Trinity Western University for granting DRC a sabbatical leave. We are also grateful to Joyanna Lee who helped with technical editing. We also thank Dr. Amarjit Basra, Editor-in-Chief of the *Journal of Crop Improvement*, Dr. Shrikrishna Singh, Managing Editor, and all of the other staff at The Haworth Press that made this effort possible.

We also want to acknowledge the long hours committed to the project by the reviewers of the contributions. These were: M. A. Altieri, K.

Barker, L. Bastiaans, G. W. Barrett, B. Booth, C. Boutin, P. D. Brown, C. Butler Flora, D. C. Coleman, Sr. Augusta Collins, B. Dail, C. den Biggelaar, J. L. Flora, C. A. Francis, K. Giller, C. M. Ghersa, S. R. Gliessman, W. S. Harper, S. B. Hill, S. Jones, P. Jeranyama, N. Jordan, H. Karsten, P. C. Kerridge, D. K. Letourneau, W. M. Londsdale, J. M. Luna, T. L. Marsh, M. Marvier, J. A. E. Molina, P. F. Mooney, J. Paul Mueller, S. D. Murphy, P. Murray, E. P. Odum, C. Opio, M. C. Rillig, M. Sanderson, J. Sauerborn, M. M. Schoeneberger, P. Stahl, A. D. Tomlin, and B. Tracy.

David Clements
Anil Shrestha
Gary Barrett

New Dimensions in Agroecology for Developing a Biological Approach to Crop Production

David R. Clements
Anil Shrestha

SUMMARY. Agroecology is emerging from the conceptual realm to become a significant discipline in North America and many parts of the world. We explore 10 dimensions of agroecology that are important in developing a more biologically-based science of agriculture: (1) a new philosophy of agriculture, (2) systems thinking, (3) local adaptation, (4) the non-crop biota, (5) crop autecology, (6) encompassing the agricultural landscape, (7) closing the materials cycle: crops, livestock and local or global cycling, (8) technology and ecology, (9) human ecology, and (10) the natural dimension. Agroecology deals with the applications

---

David R. Clements is Professor of Biology and Environmental Studies, Trinity Western University, Langley, BC, Canada V2Y 1Y1 (E-mail: clements@twu.ca).

Anil Shrestha is IPM Weed Ecologist, University of California, Statewide IPM Program, Kearney Agricultural Center, 9240 South Riverbend Avenue, Parlier, CA 93648.

The authors acknowledge Eugene Odum, George Barrett, Charles Francis, and Stuart Hill for helpful comments on earlier versions of the manuscript. The authors also wish to thank all of the other authors in this volume for input and inspiration. The authors are thankful for the support of Clarence Swanton, in particular for funding A.S. to present this research in an invited paper addressed to a joint meeting of the American Society of Agronomy, the Soil Science Society of America and the Crop Science Society of America, November 2001.

[Haworth co-indexing entry note]: "New Dimensions in Agroecology for Developing a Biological Approach to Crop Production." Clements, David R., and Anil Shrestha. Co-published simultaneously in *Journal of Crop Improvement* (Food Products Press, an imprint of The Haworth Press, Inc.) Vol. 11, No. 1/2 (#21/22), 2004, pp. 1-20; and: *New Dimensions in Agroecology* (ed: David Clements, and Anil Shrestha) Food Products Press, an imprint of The Haworth Press, Inc., 2004, pp. 1-20. Single or multiple copies of this article are available for a fee from The Haworth Document Delivery Service [1-800-HAWORTH, 9:00 a.m. - 5:00 p.m. (EST). E-mail address: docdelivery@haworthpress.com].

http://www.haworthpress.com/web/JCRIP

Digital Object Identifer: 10.1300/J411v11n01_01

of ecological principles in agroecosystems and it represents a logical response to shortcomings of conventional agriculture. Current crop production approaches fail to account for biological complexities of agroecosytems and the need to feed the world without jeopardizing the sustainability of its life support systems. A key strategy employed by agroecologists is to compare agroecosystems and natural ecosystems systematically, and attempt to integrate knowledge of natural ecosystems into agricultural practice. Through this process, traditional agronomy is elevated to agroecology.

KEYWORDS. Sustainable agriculture, agroecosystems, landscape ecology, biotechnology, ecological integrity, philosophy

INTRODUCTION

The field of agroecology has been emerging with the increased need to understand the biological complexities of agroecosystems. The importance of this field has become more acute as a result of increased pressure to farm in an environmentally sound, economically viable, and socially acceptable manner. To many, agroecology may seem like a "buzz word," a more politically correct word for agronomy or merely a concept in agricultural science. In the past decade, however, 'agroecology' has been elevated from a conceptual realm and is now a major discipline in many educational and research institutions. The 'profit maximization approach' to agriculture or 'traditional agronomy' led to many ecological and ethical shortcomings which has now raised the need to assess the biological basis of sustainability (Shrestha and Clements, 2003). Several authors suggest that a logical response is required to address the shortcomings of conventional agriculture (Gliessman, 1998; Francis, 2004) and answer the dire call to feed the world without jeopardizing the sustainability of its life support systems (Barrett and Odum, 2000). There is a clear need for humility in our approach to solving agricultural problems (Zimdahl, 2002), and agroecology addresses the breadth of agriculture and food production beyond the simple application of new technologies in the field.

Agroecosystems are derived from manmade changes to the 'natural ecosystem' and encompass a complexity of biological interactions. We strongly believe that agriculture is a form of applied ecology and that many of the principles of ecology must be applied to better understand agriculture. Several authors have provided definitions for the term 'agroecology.' Gliessman (1998) defines agroecology as "the application of ecological concepts and principles to the design and management of sustainable agroecosystems." Francis (2003) proposed that agroecology be defined as "the ecology of food systems," reflecting both the importance of ecology and systems thinking in agroecology. Francis (2004) advocates communicating the issues of agriculture, within a food system perspective, looking at how ecological principles can be used to explain and improve not only on-farm production, but also the entire food-producing enterprise, including examination of issues such as global nutrient cycling.

In the above context, a key strategy used by agroecologists is to carefully and systematically compare 'agroecosystems' and 'natural ecosystems,' and attempt to integrate knowledge of natural ecosystems into agricultural practice (Odum, 1984; Conway, 1985; Altieri, 1995). Agroecology is not a recipe, rather it consists of several dimensions. The objective of this review is to outline the multidimensional nature of agroecology, and highlight promising dimensions that may help in elevating traditional agronomy to agroecology.

NEW DIMENSIONS: EXPANDING THE FRONTIERS OF AGRICULTURAL SCIENCE

Through the 20th century, a new paradigm evolved for studying agriculture in the developed and the developing world. The ability to increase crop yields through genetic improvements and improved agronomic techniques drove a system of agricultural research, based largely on small plot trials conducted by universities and government institutions (Stoskopf, 1981). One inescapable effect of this new paradigm was a 'reductionist' approach to agriculture, a strategy parallel to applications of science in other segments of society. This approach almost totally ignored the fact that agriculture operates as a system. Unintended consequences of this 'reductionist' strategy included environmental degradation (e.g., groundwater pollution, loss of soil or soil nutrients, deleterious effects of pesticides on non-target organisms), and cultural decline (e.g., disappearance of family farms, loss of vitality of farming communities) (Berry, 1977; Goldschmidt, 1998; Hill, 1998; Pearce, 1995;

Pimentel et al., 1995; Williams et al., 1998). As we have described elsewhere (Shrestha and Clements, 2003), partially in response to these unintended consequences, society is presently turning a corner in the learning cycle, coming back to the values associated with farming more closely in tune with nature. Agroecology, in a word, describes the ideal of a marriage between agriculture and ecology (Jackson and Piper, 1989) and thus is the natural and logical outcome of this turn of the learning cycle.

One might be tempted to say that agroecology is "re-expanding the frontiers of agricultural science" rather than simply expanding them. However, much of the context of agriculture has changed in the last 100 years. New agricultural technology and an industrial paradigm are entrenched in the system, and integral to and currently contribute to providing more food to the burgeoning population of the world than would be possible using 19th century technology (Evans, 1998). Our technology is driven by a simple cause and effect, single-factor and industrial mindset that ignores the complexity and potentials of whole systems and broader emergent properties. Although aspects of this technology has unintended consequences that partly drive the need for a "kinder, gentler" agriculture, much of our 'reductionist' research and results, can be skillfully woven into new agricultural systems.

Another major change in the past 100 years, is the advancement in our understanding of the way the world works–its ecology. Thus, now in addition to building on farmer intuition, we have the possibility of knowing in great detail the way natural ecosystems work. Extending this rationale, we can understand the way agroecosystems ought to work, if they are designed to mimic natural ecosystems as much as possible. Thus, the frontiers of agricultural science can be expanded in two ways: firstly by taking a less fragmented approach to development and application of agricultural technology, and secondly by incorporating principles of biology and ecology more explicitly within agriculture. This expansion involves a number of dimensions, ten of which are outlined in Table 1. This list is by no means exhaustive, but is reflective of the contents of the other contributions to this volume.

Dimension 1: A New Philosophy of Agriculture

Whereas many agricultural scientists might be content to simply design systems to maximize crop yield, a broader sustainability should involve economic, social and environmental sustainability (Gliessman, 1998). Traditional agronomic approaches, in maximizing yields, have

TABLE 1. New dimensions in agroecology in comparison to traditional agronomic approaches.

New dimensions added by agroecology	Transcendence of traditional agronomic approaches
1. a new philosophy of agriculture	integrates biological and cultural knowledge to strive for economic, social, and environmental sustainability
2. systems thinking	holistic, attempting to understand interactions among various agroecosystem components
3. local adaptation	accounts for local ecology and cultural knowledge
4. the non-crop biota	non-crop biota such as insects, non-crop vegetation recognized as part of the agroecosystem
5. crop autecology	complexity of cropping systems possible–e.g., polycultures, intercropping or agroforestry
6. encompassing the agricultural landscape	landscape scale considered vs. simply field scale or experimental plot
7. closing the materials cycle: crops, livestock and local or global cycling	systems developed to avoid excessive spatial separation of crops and the livestock they feed
8. technology and ecology	ecological implications of technology (including biotechnology) carefully considered
9. human ecology	cultural aspects of agriculture explicitly recognized in designing agroecosystems for sustainability
10. the natural dimension	natural ecosystems are mimicked more intentionally in attempts to improve ecological integrity and resilience of agroecosystems

sacrificed environmental and social sustainability, and neglected 'system maintenance.' In the market-driven economy that has dominated the developed world through the 20th century, it is obvious that economic sustainability has been preeminent. There is a strong interdependence among economic, ecological and societal factors, however, and thus when ecological and/or social sustainability are compromised, economic sustainability is also in jeopardy. Narrow approaches to agronomy have resulted in short-term economic gains, but in order to sustain these, the approaches need to be broadened. For example, the new rice (*Oryza sativa* L.) cultivars introduced to India in the green revolution of the 1970s are now better integrated into the system, and at the same time, people have taken on more realistic expectations of the impact of a very specific agricultural innovation (Cantrell, 2002).

Inputs and technology have shifted agroecosystems to a state far removed from natural ecosystems. In current conventional approaches to agriculture, the crop is seen as simply an industrial product and improved production is achieved by refining a narrow and finite set of practices. Reductionist science, as developed by Francis Bacon and others, is well-equipped to pursue the "problems in agriculture" through appropriate testing of hypotheses. Cox, Picone, and Jackson (2004), however, draw our attention to the "problem of agriculture." Many agroecologists view the problem of agriculture as a whole (i.e., the overall approach taken to address issues in agriculture and its consequences) as one that requires a different framework from improving conventional agriculture, both scientifically and philosophically (Altieri, 1995; Gliessman, 1998; Gliessman, 2004). This need for a paradigm shift obviously presents many methodological difficulties, especially for a scientific establishment where small-plot research methods and Baconian science are firmly entrenched. The overall goal of the new philosophy is to integrate biological and cultural knowledge into agroecosystems to the greatest extent possible and to embrace a wider suite of criteria to use in evaluating success. This would involve moving away from a symptom focus to a vision for agroecosystem design.

Dimension 2: Systems Thinking

A systems perspective is intrinsic to an agroecological approach–it is not an option when one is attempting to develop a more holistic science. Through a systems approach, a broad range of biotic and human factors must be considered (Hill, 1998). Seeing an individual farm within regional, national, and international food systems can transform simplistic economic approaches to a more holistic view accounting for factors such as food security, social equity, and quality of life (Gliessman, 1998). Again, there are methodological difficulties. A systems approach requires a very extensive study, and is multidisciplinary in nature. Our research institutions are generally not designed for multidisciplinary approaches, but it is becoming clear that these institutions must be redesigned to encompass systems thinking (Francis, 2004; Hill and McRae, 1988).

How broadly should an agricultural system be defined? One example that we discussed previously (Shrestha and Clements, 2003) went beyond the traditional view of interactions at the field level, to also account for social, political, physical, biological, and economic factors. This expansion was based on a model presented by Cavigelli et al. (1998) to incorporate these external factors in the form of social, political, and regulatory environment as well as other factors such as farm size, labor, capital, debt, and land. Although agroecosystems formulated in these terms seem bewilderingly complex, the diverse challenges are largely a function of the complexity of modern society in general. In many developing countries, relatively simple systems exist analogous to those which pre-dated agricultural systems in developed countries and many of these are ecologically sustainable. In any case, it behooves us to understand the entire system, whether it is simple or complex.

Ecological theory leads us to explore complexity. For example, although simple models of ecological succession have been developed, they fail to incorporate all the elements that may be necessary to describe succession. When scientists attempt to apply the theory of ecological succession to processes in agriculture, it makes sense to apply comprehensive successional models like that of Pickett, Collins, and Armesto (1987) rather than more reductionist representations (Swanton, Clements, and Derksen, 1993). It comes down to the definition of an agroecosystem as a system. Ecologists like Eugene Odum have demonstrated time and time again that one cannot understand the system by simply studying its various parts, because the system takes on properties that would not be predicted based on studies of individual components. Many such emergent properties are revealed only if we study agroecosystems holistically, within their landscape and human contexts (Hill, 2002; Odum and Barrett, 2002).

Dimension 3. Local Adaptation

Although globalization is touted as a well-spring of economic growth, applying monolithic agricultural techniques or policies over broad geographic area can have severe and unintended consequences (Berry, 1977). From a purely technocratic point of view, it would make sense to develop efficient technologies that can effectively be transported to different regions within a country or around the world. Inevitably, this process has occurred through the history of agriculture, as various agricultural crops and products have been adapted successfully to new areas across bioregions and cultures. This process continued in the 20th century, accompanied by delivery of agronomic techniques that were often very specialized and input-intensive. It is not possible nor desirable to turn back the clock on this in each unique situation as we progress through the learning cycle. At the same time, it is important to recognize the uniqueness of place and to balance potential benefits of globalization of agriculture with the detrimental effects on local systems. Because agroecosystems are nested within larger ecosystems, local knowledge is often extremely valuable, and we ignore it at our peril. Rich reserves of specific knowledge can be incorporated readily into the design of agroecosystems suited for the local ecosystem (Gliessman, 2001; Altieri and Nicholls, 2004).

Dimension 4. The Non-Crop Biota

Despite enormous efforts, weeds, insects, fungal pathogens and other pests continue to require huge expenditures for control (Pimentel et al., 2000; Buhler, 2003). These pests are integral to the agroecosystem, and although their numbers are controlled to some extent, the term "eradication" is rarely applicable to the results of human efforts to control them. Some scientists have argued that eradication is not the ultimate aim, nor is it possible or desirable; in fact a more sustainable approach to agroecosystem management should be developed (Maxwell, 1999; Winston, 1997). For example, weeds play certain useful ecological roles within an agroecosystem, such that weed populations maintained below economic thresholds could be seen as providing a net benefit (Altieri, 1988; Clements, Weise, and Swanton, 1994). It is useful to recognize that the vast majority of non-crop species in the field are either beneficial or neutral, and many of these organisms are essential to balance in the agroecosystem. Many efforts to control non-desired pests are detrimental to this balance (Patriquin et al., 1986).

If the benefits from non-crop biota are to be realized, a more intimate biological knowledge of non-crop species must be obtained and applied. For example, knowledge of weed propagule production (Buhler, 2002; Dekker, 1999;), seedling emergence dynamics (Shrestha et al., 1999), phenological development (Swanton et al., 2000), thresholds of weed-crop competition (Jasieniuk et al., 2001; Swanton et al., 1999), and population biology (Maxwell, 1992; Maxwell and Ghersa, 1992; Maxwell and Luschei, 2004) would be useful in designing weed management strategies that consider the intentional presence of some weeds as desirable. To integrate means to embed something in the system. To be successful, integrated pest management (IPM) must involve a great variety of organisms and aspects built into the design of cropping systems (Landis and Menalled, 1998; Kennedy 1999a; Landis, Wratten, and Gurr, 2000). Goodell and Zalom (2000) have highlighted that a technical obstacle to adoption of IPM is the poor understanding and lack of appreciation of the basic biology of pests, beneficial organisms, and their interactions in agroecosystems. Adoption of truly integrated systems is complex, just as ecological interactions are complex, and thus extension of agroecological approaches to IPM requires innovative educational processes that encourage agricultural practitioners to buy into ecological pest control methods (Hill, 2004; Menalled, Landis, and Dyer, 2004). Just as pests are continually adapting to agroecosystems, practitioners would benefit from adopting an adaptive management paradigm (Menalled, Landis, and Dyer, 2004).

As well as weeds and other pests, non-crop biota includes countless beneficial organisms such as soil microbes (McClung, Dick, and Karns 1994; Kennedy 1999a; 1999b), mycorrhizae (Hamel et al., 1994, Hamel, 1996) or soil fauna, including earthworms (Stehouwer, Dick, and Traina 1994; Whalen and Parmelee, 1999; Whalen et al., 1999). Just as the pests need to be seen as part of the system, these beneficial organisms need to be accounted for and enhanced when possible. At the very least, we should not use chemical applications or other practices that reduce their populations and function. The benefits of biodiversity in agroecosystems can be seen in terms of genetic resources, edible plants and crops, livestock, soil organisms, naturally occurring biological control agents, and 'wild' species of natural habitats, all of which play valuable, but often overlooked roles (Thrupp, 2004).

Dimension 5. Crop Autecology

Through the development of an agricultural crop, via processes of selective breeding, genetic engineering and advances in agronomy, our current research strategies have lead agriculturalists to view the crop species the same way one would see an industrial product. Certainly the advances enabling us to manipulate crop physiology have provided opportunities to increase productivity substantially over the past century (Evans, 1998). However, there is a fallacy in the view of "plant as machine." No matter how refined growing techniques and plant genetics become, the plant growing in the natural environment (even within a greenhouse), is still an ecological entity. Given this fact, cropping systems can be developed for improving stability, resilience and other ecological properties. Under many circumstances, it is also useful to grow this ecological entity in polyculture rather than monoculture. Increasing crop diversity can produce and enhance diverse ecological interactions (Francis, 1986; Holmes and Barrett, 1997; Vandermeer, 1995; Vandermeer,

1998), resulting in a more robust ecological system. Thus, an agroecological approach embodies crop breeding, genetics and biotechnology within an ecological framework, rather than considering ecological factors after the fact to explain specific processes in the system. Within this broader paradigm, we are compelled to explore the implications of innovative possibilities for designing cropping systems, such as agroforestry, multicropping and other more diverse systems that are utilized in many areas in the tropics (Altieri and Nicholls, 2004; Thevathasan et al., 2004).

Dimension 6. Encompassing the Agricultural Landscape

As discussed previously, a more holistic view of agriculture involves conceptualizing agroecosystems within larger ecosystems. For example, the field margin may be considered as integral to the system (Maudsley et al., 1998; Thomas and Marshall, 1999; Marshall, 2002). Ideally, recognizing field margins as integral parts of the agricultural system will foster conscious designs and decisions that will promote ecological interactions among the biota in both realms. One net effect of a healthy environment between crop fields will be a more diverse landscape. Improvements in landscape diversity can reduce soil erosion, favor natural enemies of pests (Altierri, 1994; Barrett, 1999; Thomas and Marshall, 1999), or help protect wildlife (Thrupp, 2000). This is obviously a complex and poorly understood area of study, but it is clear that farm size, field size and shape, and nature of the resulting edges have implications for agroecosystems in terms of formation of buffer zones and corridors for protection of biodiversity (Ranney, Brunner, and Levenson, 1981; Gliessman, 1998). Furthermore, aesthetics are becoming an increasingly important consideration, in view of growing tensions in the rural-urban fringe in many parts of the world, as communities expand their boundaries into agricultural production areas.

Dimension 7. Closing the Materials Cycle: Crops, Livestock, and Local or Global Cycling

The current configuration of the global materials cycle, as determined by the type of agricultural systems that predominate, is a linear, flow-through process that obviates the potential for material recycling. Local systems provide a rational alternative. As well as integrating the non-crop biota, managers for a more holistic agriculture strive to include livestock as part of integrated systems. A clear benefit of this is a better closing of the circle of nutrient management, allowing manure produced in a local area to be utilized by crops in the same area. Other potential benefits include reduction of external inputs (e.g., nitrogen (N)), more efficient nutrient use of animal manure (e.g., increasing C:N ratio), incorporation of grasslands to reduce accumulation of soil N and improved utilization of mineralized N, broadening of crop rotation to improve pest control, optimal use of legumes for N fixation, and a more even economic distribution that can reduce financial risk (Lantinga, Oomen, and Shiere, 2004). Integration of livestock in agriculture is viewed as a mixed farming system.

A mixed farming approach is counter to mega-farm specialization. Given the difficulties with nutrient management (e.g., the spectre of groundwater pollution), new mixed farming approaches are being developed as alternative systems to those that separate livestock production from the crops that supply their feed (Clark, 2004; Clark, Buchanan-Smith, and Weise, 1993; Oomen et al. 1998; Lantinga, Deenen, and van Keulen, 1999). Although integrated or mixed farming approximates how livestock were produced historically, there is a dearth of knowledge regarding the potential impacts of present-day configurations of mixed systems. Long-term research is being conducted in The Netherlands and elsewhere to investigate the potential benefits of closing the materials cycles in this way (Lantinga, Oomen, and Shiere, 2004).

On a larger scale, agroecology is a field that is uniquely equipped to study the carbon (C) sequestering potential of sustainable agricultural practices (Hu et al., 1997b; Dick and Durkalski, 1998; Hu and Zhang, 2004). Complex interactions among increased atmospheric carbon dioxide, other nutrients and the soil biota will have implications for agricultural policy and how agroecosystems are viewed within global biogeochemical cycles (Hu et al. 1997a; Hu and Zhang, 2004).

Dimension 8. Technology and Ecology

Is agroecology necessarily a low-tech approach? Although agroecology generally fits best with low input, "natural" methods of agriculture (Vandermeer, 1995), cutting edge technology may work in concert with goals of ecological sustainability if used wisely. The potential credits and debits of biotechnology are seen in the challenge of managing insect resistance to transgenic plants (Anstad and Andow, 1995; 1996), or in preventing the development of resistance to glyphosate that may accompany extensive plantings of glyphosate-tolerant crops (White, Harmon, and Andow, 2004). In the latter case, there are clear ecological benefits to the genetically modified crops, but obviously it would be an oversimplification to see this one technique as a panacea for weed problems. Another example is the potential development of perennial grain crops; although the Land Institute in Kansas has not yet used biotechnological approaches, even the use of "traditional" breeding methods to develop perennial from annual crops requires considerable technological input (Cox et al., 2004). Similarly, adopting no-tillage agriculture is going back to a more natural method of growing crops without plowing, but requires specialized seeding equipment and most commonly involves application of the herbicide glyphosate (Stubbs et al., 2004). The main point here is that although agroecology has an inherent bias towards more natural or cultural means of agronomy, an agroecological approach does not preclude the technological innovations we have at our disposal in the 21st century. In fact, latest electronic tools such as geographic information systems and global positioning systems have been recommended to develop an understanding of interactions among complex and interrelated components of agroecosystems (Ellsbury et al., 2000). Agroecology is a holistic approach to agronomy that examines all agricultural practices and technologies within an ecological framework.

Dimension 9. Human Ecology

When agriculture is reduced to a defined suite of techniques the cultural part of agriculture gets left behind. Yet for thousands of years, human capital has been the essential driving force, as skilled farmers have tamed the intrinsic unpredictability of crop growth, the weather, and the market. If we are to incorporate more ecological knowledge into agriculture, it is difficult to reduce agriculture to mere techniques. For example, weed control with herbicides potentially could be accomplished by a robot. More ecological weed control techniques, however, may require detailed observations of weed phenology, adaptive use of cultural techniques, and detailed knowledge of weed biology. Thus, when technocratic methods such as herbicides are exchanged for or integrated with more cultural methods, re-integration of human capital must be considered in the process. In turn, this creates human benefits of sustainable agriculture and true integration of stakeholders into the process of producing food (Hill, 1999; Hill, 2004). Cultural factors are seldom considered, but can have major impacts. An Amish farmer in the northcentral US may have the same basic starting point as his neighbor (e.g., soils, climatic conditions), but may opt for a less technological approach that produces a very different ecological trajectory (Francis, 2004).

According to Hill (2001), increasing production regardless of cost neglects maintenance or caring, competition prevents collaboration, 'controlling' science prevents 'understanding' science, and transglobal managerialism inhibits regional self-reliance. These are just a few of many aspects to consider as we work to redesign a more human agriculture that may be more responsive to ecological factors in the face of irrepressible "progress" in the technology of production agriculture.

Dimension 10. The Natural Dimension

How does one measure the success of an agroecological approach? Because the basic idea behind agroecology is to employ ecological principles, it is logical to evaluate success based on how closely the behavior of the agroecosystem mimics natural systems (Ewel, 1986; Soule and Piper, 1992; Jackson and Jackson, 1999). Nature is thus the best measure of success (Cox, Picone and Jackson, 2004). There are various measures of successful performance in agriculture, ranging from economic profits to the potential of agricultural land to be converted back to the native habitat. To evaluate the social, economic and environmental sustainability of a given agricultural system, there is a need for appropriate indicators to evaluate agroecosystem health (Waltner-Toews, 1996; Clements and Swanton, 2001). If this "natural dimension" is the standard for comparison, appropriate indicators must be very similar to the approaches taken to evaluate the health of natural ecosystems. Factors such as biodiversity (Thrupp, 2004), soil structure and biology (Hamel et al., 2004; Wang and Dick, 2004; Whalen and Hamel, 2004), crop life history (Cox, Picone, and Jackson, 2004), nutrient cycling (Hu and Zhang, 2004; Lantinga, Oomen, and Shiere, 2004), social ecology (Hill, 2004) and landscape-level processes (Marshall, 2004; Odum and Barrett, 2004) all must be evaluated as intrinsic to the agroecosystem.

Agroecosystems are simplified versions of natural ecosystems, but nonetheless incorporate many elements and properties present in their natural counterparts. The current "industrial agriculture" can provide another useful concept for comparison, as newly designed, more sustainable agroecosystems are evaluated in the context of a broader agricultural landscape. Moving towards a "natural systems agriculture" (NSA) (Cox, Picone, and Jackson, 2004) will require us to incorporate more of these indicators in the assessment of agricultural health.

CONCLUSIONS

The combination of ten dimensions discussed here describe agroecology as a new way of thinking about agricultural issues, and indeed "the problem of agriculture" (Cox, Picone, and Jackson, 2004). Technical fixes from a 'reductionist' school of thought, such as specific new biotechnology applications (White, Harmon, and Andow, 2004), need to be subsumed within an agroecological framework to provide comprehensive scrutiny of ecological effects and potential unintended consequences. As Albert Einstein once said: "We can't solve problems by using the same kind of thinking we used when we created them." An agroecological framework explicitly incorporates biological and ecological factors as central to the issue, rather than as external contingencies that must be accommodated in order to maximize crop yield in an environmentally acceptable way. There is a growing network of institutions and scientists across North America and around the world following an agroecological approach. Change will not come easily, however, because it may involve fairly radical changes in the prevailing mindset and institutional structures (Hill, 2004). For those who are willing to think "outside the box" these new dimensions offer tremendous benefits in terms of total sustainability of our agricultural systems, including social, environmental and economic measures of sustainability. It is our hope that this volume will provide useful direction by reviewing the current status of agroecological research and education and pointing to future possibilities for research and application.

REFERENCES

Altieri, M.A. (1988). The impact, uses, and ecological role of weeds in agroecosystems. In *Weed Management in Agroecosystems*, ed. M.A. Altieri and M. Liebman, Boca Raton, FL: CRC Press, pp. 1-6.

Altieri, M.A. (1994). The influence of adjacent habitats on insect population in crop fields. In *Biodiversity and Pest Management in Agroeosystems*, ed. M.A. Altieri. New York, NY: Food Products Press. pp. 109-129.

Altieri, M.A. (1995). *Agroecology: The Science of Sustainable Agriculture*. 2nd edition. Boulder, CO: Westview Press.

Altieri, M.A. and C.I. Nicholls (2004). An agroecological basis for designing diversified cropping systems in the tropics. *Journal of Crop Improvement* 11(1/2):81-104.

Alstad, D.N. and D.A. Andow. (1995). Managing the evolution of insect resistance to transgenic plants. *Science* 268:1894-1896.

Alstad, D.N. and D.A. Andow. (1996). Implementing management of insect resistance to transgenic crops. *Agbiotech News and Information* 8:177N-181N.

Barrett, G.W. (2000). The impact of corridors on arthropod populations within simulated agrolandscapes. In *Interchanges of Insects Between Agricultural and Surrounding Landscapes*, eds. B. Ekbom, M.E. Irwin, and Y. Robert. Dordrecht, The Netherlands: Kluwer Academic Publishers, pp. 71-84.

Barrett, G.W. and E.P. Odum. (2000). The twenty-first century: The world at carrying capacity. *BioScience* 50:363-368.

Berry, W. (1977). *The Unsettling of America: Culture and Agriculture.* San Francisco, CA: Sierra Club Books.

Buhler, D.D. (2003). Weed biology, cropping systems, and weed management. *Journal of Crop Production* 8(1/2):245-270.

Cantrell, R. (2002). Rice: Why it's so essential for global security and sustainability. *Economic Perspectives* Vol 7 (May) [http://usinfo.state.gov/journals/ites/0502/ijee/rice.htm].

Cavigelli, M.A., S.R. Deming, L.K. Probyn, and R.R. Harwood, eds. (1998). Michigan Field Crop Ecology: Managing Biological Processes for Productivity and Environmental Quality. Michigan State University Extension Bulletin E-2646., East Lansing, MI: Michigan State University Extension.

Clark, E.A. (2004). Benefits of re-integrating livestock and forages in crop production systems. *Journal of Crop Improvement* 12(1/2):405-436.

Clark, E.A., J.G. Buchanan-Smith, and S.F. Weise (1993). Intensively managed pasture in the Great Lakes Basin: a future-oriented review. *Canadian Journal of Animal Science* 73:725-747.

Clements, D.R. and C.J. Swanton. (2001). Agriculture: Healthy, sick or left in the waiting room? In *Malthus and the Third Millennium*, eds. W. Chesworth, M.R. Moss, and V.G. Thomas. Guelph, ON: Faculty of Environmental Sciences, University of Guelph, pp. 125-140.

Clements, D.R., S.F. Weise, and C.J. Swanton. (1994). Integrated weed management and weed species diversity. *Phytoprotection* 75:1-18.

Conway, G.R. (1985). Agroecosystem analysis. *Agricultural Administration* 20:31-55.

Dekker, J. (1999). Soil weed seed banks and weed management. *Journal of Crop Production* 2:139-166.

Cox, T.A., Picone, J.C. and W. Jackson. (2004). Research priorities in natural systems agriculture. *Journal of Crop Improvement* 12(1/2):511-532.

Dick, W.A. and J.T. Durkalski. (1998). No tillage production agriculture and carbon sequestration in a typic fragiudalf soil of northeastern Ohio. In *Management of Carbon Sequestration in Soil*, ed. R. Lal, J.M. Kimble, R.F. Follet and B.A. Stewart, Boca Ration, FL: CRC Press. pp. 59-71.

Ellsbury, M.M., S.A. Clay, S.J. Fleischer, L.D. Chandler, and S.M. Schneider. (2000). Use of GIS/GPS systems in IPM: Progress and Reality. In *Emerging Technologies for Integrated Pest Management: Concepts, Research, and Implementation*, eds., G.G. Kennedy and T.B. Sutton, St. Paul, MN: APS Press. pp. 419-438.

Evans, L.T. (1998). *Feeding the Ten Billion: Plants and Population Growth.* Cambridge, UK: Cambridge University Press.

Ewel, J. (1986). Designing agricultural ecosystems for the humid tropics. *Annual Review of Ecology and Systematics* 17:245-271.

Francis, C.A., editor (1986). *Multiple Cropping Systems.* New York, NY: Macmillan Publ. Co.

Francis, C.A. (2003). Advances in the design of resource-efficient cropping systems. *Journal of Crop Production* 8(1/2):115-135.

Francis, C.A. (2004). Education in agroecology and integrated systems. *Journal of Crop Improvement* 11(1/2):21-43.

Francis, C., G. Lieblein, S. Gliessman, T.A. Breland, N. Creamer, R. Harwood, L. Salomonsson, J. Helenius, D. Rickerl, R. Salvador, M. Wiedenhoeft, S. Simmons, P. Allen, M. Altieri, C. Flora, and R. Poincelot. (2003). Agroecology: the ecology of food systems. *Journal of Sustainable Agriculture* 22(3):99-119.

Gliessman, S.R. (1998). *Agroecology: Ecological Processes in Sustainable Agriculture.* Chelsea, MI: Sleeping Bear Press.

Gliessman, S.R. (2001). *Agroecosystem Sustainability: Developing Practical Strategies.* Boca Raton, FL: CRC Press.

Gliessman, S.R. (2004). Integrating agroecological processes into cropping systems research. *Journal of Crop Improvement* 11(1/2):61-80.

Goldschmidt, W. (1998). Conclusion: The urbanization of rural America. In *Pigs, Profits, and Rural Communities*, eds., K.M. Thu and E.P. Durrenberger, Albany, NY: State University of New York Press, pp. 183-198.

Goodell, P.B. and F.G. Zalom. (2000). Delivering IPM: Progress and Challenges. In *Emerging Technologies for Integrated Pest Management: Concepts, Research, and Implementation*, eds., G.G. Kennedy and T.B. Sutton, St. Paul, MN: APS Press. pp. 483-496.

Hamel, C. (1996). Prospects and problems pertaining to the management of arbuscular mycorrhizae in agriculture. *Agriculture, Ecosystems and Environment* 60:197-210.

Hamel, C., F. Morin, J.A. Fortin, and D.L. Smith. (1994). Mycorrhizal colonization increase herbicide toxicity in apple. *Journal of the American Society for Horticultural Science* 119:1255-1260.

Hamel, C., C. Landry, A. Elmi, A. Liu, and T. Spedding. (2004). Nutrient dynamics: Utilizing biotic-abiotic interactions for improved management of agricultural soils. *Journal of Crop Improvement* 11(1/2):209-248.

Hill, S.B. (1998). Redesigning agroecosystems for environmental sustainability: A deep systems approach. *Systems Research* 15:391-402.

Hill, S.B. (1999). Landcare: A multi-stakeholder approach to agricultural sustainability in Australia. In *Sustainable Agriculture and Environment: Globalisation and the Impact of Trade Liberalisation*, eds. A.K. Dragun and C. Tisdell, Cheltenham, UK: Edward Elgar.

Hill, S.B. (2001). Sustainable agriculture by 2020? A social ecology approach. Eco-Farm and Garden Winter 2001: 28-31.

Hill, S.B. (2004). Redesigning pest management: A social ecology approach. *Journal of Crop Improvement* 12(1/2):491-510.

Hill, S.B. and R. MacRae. (1988). Developing sustainable agriculture education in Canada. *Agriculture and Human Values* 5(4): 92-95.

Holmes, D.M. and G.W. Barrett. (1997). Japanese beetle (*Popillia japonica*) dispersal behavior in intercropped vs. monoculture soybean agroecosystems. *American Midland Naturalist* 137:312-319.

Hu, S., D.C. Coleman, C.R. Carroll, P.F. Hendrix, and M.H. Beare. (1997a). Labile soil carbon pools in humid subtropical agricultural and forest ecosystems as influenced by management practices and vegetation types. *Agriculture, Ecosystems and Environment* 65:69-78.

Hu. S., N.J. Grünwald, A.H.C. van Bruggen, G. Gamble, L.E. Drinkwater, C. Shennan, and M.W. Demment. (1997b). Short-term effects of cover crop incorporation on soil C pools and N availability. *Soil Science Society of America Journal* 61:901-911.

Hu, S. and W. Zhang (2004). Impact of global change on biological processes in soil: Implications for agroecosystem management. *Journal of Crop Improvement* 12(1/2): 289-314.

Jackson, W. and L.L. Jackson. (1999). Developing high seed yielding perennial polycultures as a mimic of mid-grass prairie. In *Agriculture as a Mimic of Natural Systems*, eds. E. C. Lefroy, R. J. Hobbs, M. H. O'Connor, and J. S. Pate. Dordrecht, The Netherlands: Kluwer Academic Publishers, pp. 1-33.

Jackson, W. and J. Piper. (1989). The necessary marriage between ecology and agriculture. *Ecology* 70:1591-1593.

Jasieniuk, M., B.D. Maxwell, R.L. Anderson, J.O. Evans, D.J. Lyon, S.D. Miller, D.W. Morshita, A.G. Ogg, Jr., S.S. Seefeldt, P.W. Stahlman, F.E. Northam, P. Westra, Z. Kebede, and G.A. Wicks. (2001). Evaluation of models predicting winter wheat yield as a function of winter wheat and jointed goatgrass density. *Weed Science* 49:48-60.

Kennedy, A.C. (1999a). Soil microorganisms for weed management. *Journal of Crop Production* 2:123-138.

Kennedy, A.C. (1999b). Bacterial diversity in agroecosystems. *Agriculture, Ecosystems and Environment* 74:65-76.

Landis, D.A. and F.D. Menalled. (1998). Ecological considerations in the conservation of effective parasitoid communities in agricultural systems. In *Conservation Biological Control*, ed. P. Barbosa, San Diego, CA: Academic Press, pp. 101-121.

Landis, D.A., S.D. Wratten, and G.M. Gurr. (2000). Habitat management to conserve natural enemies of arthropod pests in agriculture. *Annual Review of Entomology* 45:175-201.

Lantinga, E.A., P.J.A.G. Deenen, and H. van Keulen. (1999). Herbage and animal production responses to fertilizer nitrogen in perennial ryegrass swards. II. rotational grazing and cutting. *Netherlands Journal of Agricultural Science* 47:243-261.

Lantinga, E.A., G.J.M. Oomen, and J.B. Schiere. (2004). Nitrogen efficiency in mixed farming systems. *Journal of Crop Improvement* 12(1/2):437-456.

Marshall, E.J.P. (2004). Agricultural landscapes: Field margin habitats and their interaction with crop production. *Journal of Crop Improvement* 12(1/2):365-404.

Maxwell, B.D. (1992). Weed thresholds: the space component and considerations for herbicide resistance. *Weed Technology* 6:205-212.

Maxwell, B.D. (1999). My view. *Weed Science* 47:129.

Maxwell, B.D. and C. Ghersa. (1992). The influence of weed seed dispersion versus the effect of competition on crop yield. *Weed Technology* 6:196-204.

Maxwell, B.D. and E. Luschei. (2004). The ecology of crop-weed interactions: Towards a more complete model of weed communities in agroecosystems. *Journal of Crop Improvement* 11(1/2):137-152.

Maudsley, M.J., T.M. West, H.P. Rowcliffe, and E.J.P. Marshall. (1998). Approaches to the restoration of degraded field boundary habitats in agricultural landscapes. In *Key Concepts in Landscape Ecology*, eds. J.W. Dover and R.G.H. Bunce, Aberdeen, Scotland, (UK): International Association of Landscape Ecology, pp. 387-392.

McClung, G., W.A. Dick, and J.S. Karns. (1994). EPTC degradation by isolated soil microorganisms. *Journal of Agricultural and Food Chemistry* 42:2926-2931.

Menalled, F.D., D.A. Landis, and L.E. Dyer. (2004). Research and extension supporting ecologically-based IPM systems. *Journal of Crop Improvement* 11(1/2):153-174.

Odum, E.P. (1984). Properties of agroecosystems. In *Agricultural Ecosystems*, eds. R. Lowrance, B.R. Stinner, and G.J. House, New York, NY: Wiley Interscience, pp. 5-12.

Odum, E.P. and G.W. Barrett. (2004). Redesigning industrial agroecosystems: Incorporating more ecological processes and reducing pollution. *Journal of Crop Improvement* 11(1/2):45-60.

Oomen, G.J.M., E.A. Lantinga, E.A. Goewie, and K.W. van der Hoek. (1998). Mixed farming systems as a way towards a more efficient use of nitrogen in European Union agriculture. *Environmental Pollution* 102:697-704.

Patriquin, D.G., N.M. Hill, D. Baines, M. Bishop, and G. Allen. (1986). Observations on a mixed farm during the transition to biological husbandry. *Biological Agriculture and Horticulture* 4:69-154.

Pearce, F. (1995). Poisoned waters. *New Scientist* 21(10):29-33.

Picket, S.T.A., S.L. Collins, and J.J. Armesto. (1987). A hierarchical consideration of causes and mechanisms of succession. *Vegetatio* 69:109-114.

Pimentel, D., C. Harvey, P. Resosudarmo, K. Sinclair, D. Kurz, M. McNair, S. Crist, L. Shpritz, L. Fitton, R. Saffouri, and R. Blair. (1995). Environmental and economic costs of soil erosion and conservation benefits. *Science* 267:1117-1123.

Pimentel, D., L. Lach, R. Zuniga, and D. Morrison. (2000). Environmental and economic costs of nonindigenous species in the United States. *BioScience* 50:53-65.

Ranney, J.W., M.C. Brunner, and J.B. Levenson. (1981). The importance of edge in the structure and dynamics of forest islands. In *Forest Island Dynamics in Man-Dominated Landscapes*, eds. R.L. Burgess and D.M. Sharp, New York, NY: Springer-Verlag. pp. 67-95.

Shrestha, A. and D.R. Clements. (2003). Emerging trends in cropping systems research. *Journal of Crop Production* 8(1/2):1-13.

Shrestha, A., E.S. Roman, A.G. Thomas, and C.J. Swanton. (1999). Modeling germination and shoot-radicle elongation of *Ambrosia artemisiifolia*. *Weed Science* 47:557-562.

Soule, J.D. and J.K. Piper. (1992). *Farming in Nature's Image*. Washington, DC: Island Press.

Stehouwer, R.C, W.A. Dick, and S.J. Traina. (1994). Sorption and retention of herbicides in vertically oriented earthworm and artificial burrows. *Journal of Environmental Quality* 23:286-292.

Stoskopf, N.C. (1981). *Understanding Crop Production.* Reston, Virginia: Reston Publishing.

Stubbs, T.L., A.C. Kennedy, and W.F. Schillinger. (2004). Soil ecosystem changes during the transition to no-till cropping. *Journal of Crop Improvement* 11(1/2): 105-136.

Swanton, C.J., D.R. Clements, and D.A. Derksen. (1993). Weed succession under conservation tillage: A hierarchical framework for research and management. *Weed Technology* 7:286-297.

Swanton, C.J., J.Z. Huang, A. Shrestha, M. Tollenaar, W. Deen, and H. Rahimian. (2000). Effects of temperature and photoperiod on the phenological development of barnyardgrass. *Agronomy Journal* 92:1125-1134.

Swanton, C.J., S. Weaver, P. Cowan, R. Van Acker, W. Deen, and A. Shrestha. (1999). Weed thresholds: Theory and applicability. *Journal of Crop Production* 2(1/2):9-30.

Thevathasan, N., A.M. Gordon, J.A. Simpson, and P.E. Reynolds. (2004). Biophysical and ecological interactions in a temperate tree-based intercropping system in southern Ontario, Canada. *Journal of Crop Improvement* 12(1/2):339-364.

Thomas, C.F.G. and E.J.P. Marshall. (1999). Arthropod abundance and diversity in differently vegetated margins of arable fields. *Agriculture, Ecosystems and Environment* 72:131-144.

Thrupp, L.A. (2000). Linking agricultural biodiversity and food security: the valuable role of agrobiodiversity for sustainable agriculture. *International Affairs* 76:265-281.

Thrupp, L.A. (2004). The importance of biodiversity in agroecosystems. *Journal of Crop Improvement* 12(1/2):315-338.

Vandermeer, J. (1995). The ecological basis of alternative agriculture. *Annual Review of Ecology and Systematics* 26:201-224.

Vandermeer, J. (1998). Maximizing crop yield in alley crops. *Agroforestry Systems* 40:199-206.

Wang, P. and W.A. Dick (2004). Microbial and genetic diversity in soil environments. *Journal of Crop Improvement* 12(1/2):249-288.

Waltner-Toews, D. (1996). Ecosystem health: A framework for implementing sustainability in agriculture. *BioScience* 46:686-689.

Whalen, J.K. and C. Hamel. (2004). Effects of key soil organisms on nutrient dynamics in temperate agroecosystems. *Journal of Crop Improvement* 11(1/2):175-208.

Whalen, J.K. and R.W. Parmelee. (1999). Quantification of nitrogen assimilation efficiencies and their use to estimate organic matter consumption by the earthworms *Aporrectodea tuberculata* (Eisen) and *Lumbricus terrestris* L. *Applied Soil Ecology* 13:199-208.

Whalen, J.K., R.W. Parmelee, D.M. McCartney, and J.L. VanArsdale. (1999). Movement of N from decomposing earthworm tissue to soil, microbial and plant N pools. *Soil Biology and Biochemistry* 31:487-492.

White, J.A., J.P. Harmon, and D.A. Andow (2004). Ecological context for examining the effects of transgenic crops in production systems. *Journal of Crop Improvement* 12(1/2):457-490.

Williams, A.E., L.J. Lund, J.A. Johnson, and Z.J. Kabala. (1998). Natural and anthropogenic nitrate contamination of groundwater in a rural community, California. *Environmental Science and Technology* 32:32-39.

Winston, M.L. (1997). *Nature Wars: People vs. Pests.* Cambridge, MA: Harvard University Press.

Zimdahl, R.L. (2002). Moral confidence in agriculture. *American Journal of Alternative Agriculture* 17:44-53.

Education in Agroecology and Integrated Systems

Charles A. Francis

SUMMARY. Courses and curricula in agriculture are traditionally organized within disciplinary boundaries, while study of integrated systems has been confined to aggregation of components and often superficial analysis. Agroecology is emerging as an integrative field that expands our focus to embrace the broad complexities of agricultural production and the entire food system. Education in agroecology must provide the skills and knowledge needed to design and evaluate new systems, as well as the capacity to vision into the future and anticipate the impacts of systems as well as new challenges that will face humanity. Agricultural universities need to be organized to prepare students to meet these increasingly complex challenges in the food and natural resource arena. Agroecology courses can build awareness and competence in using ecological principles to inform the design of future systems that are productive, economically sound, environmentally sustainable, and socially viable for the indefinite future. A model program in the Nordic region is presented as an example of one innovative curriculum. Study of the ecology of food systems can put natural resource, human population, and agricultural production into a balanced perspective that can guide our

Charles A. Francis is Professor, Department of Agronomy & Horticulture, University of Nebraska, Lincoln, NE 68583-0910 (E-mail: cfrancis@unlnotes.unl.edu).

[Haworth co-indexing entry note]: "Education in Agroecology and Integrated Systems." Francis, Charles A. Co-published simultaneously in *Journal of Crop Improvement* (Food Products Press, an imprint of The Haworth Press, Inc.) Vol. 11, No. 1/2 (#21/22), 2004, pp. 21-43; and: *New Dimensions in Agroecology* (ed: David Clements, and Anil Shrestha) Food Products Press, an imprint of The Haworth Press, Inc., 2004, pp. 21-43. Single or multiple copies of this article are available for a fee from The Haworth Document Delivery Service [1-800-HAWORTH, 9:00 a.m. - 5:00 p.m. (EST). E-mail address: docdelivery@haworthpress.com].

http://www.haworthpress.com/web/JCRIP

Digital Object Identifer: 10.1300/J411v11n01_02

research and development efforts toward a sustainable food system for the future.

KEYWORDS. Agroecosystems, sustainable agriculture, ecological agriculture, participatory learning, action education

INTRODUCTION: AGROECOLOGY AS THE ECOLOGY OF FOOD SYSTEMS

Education in agriculture has generally focused on the process of producing food, fiber, and raw materials for industrial and other manufacturing. Courses and curricula are organized along narrow disciplinary lines, in large part because that is the way research and departments were established many decades ago (Altieri and Francis, 1992). Although students have researched the components of agroecosystems, until recently there has been little attention to how the complexities of food and natural resource systems could be addressed through our classical offerings of courses and majors in current departments. The goal of this discussion is to expand the scope of our thinking about agroecology. Skills, knowledge, and vision can be achieved in educational programs that are explicitly focused on the complexity of food systems and how alternative approaches can be designed for the future.

Growing appreciation of the difficulties of solving global food challenges with current systems of crop and livestock production has merged with our awareness of resource scarcity in a world that is accelerating in materials transformation. Views differ on how long fossil fuels will be economically available for agriculture (Flavin and Lenssen, 1990), or if other substitutes will be developed (Dunn, 2001), and this makes the search for innovative educational systems and methods even more imperative. Other sectors add more value to fossil fuels and water or can afford to pay more for them than we can in agriculture, especially when food is perceived to be in surplus and prices are relatively low in most developed countries. Our comparative advantage in agriculture, as compared to other human industries, is the efficient capture of sunlight and water by crops and pasture plants across wide areas, and not in conversion of petroleum into food products. In fact the efficiency of energy use in producing many meat and specialty food products is quite low.

These are examples of issues that should be addressed as we study the food system, but such broad questions are not often raised in courses in agronomy focused on production or agricultural economics focused on short-term economic returns from single enterprises.

Agroecology is an integrative subject area with courses that deal with the ecological dimensions of farming, the description of agricultural systems in ecological terms, and the comparison of natural and managed ecosystems. Similar to other disciplines in the agricultural sciences, most courses deal very well with the production aspects and environmental impacts of agroecosystems. As stated in one of the most popular current texts, "Agroecology is defined as the application of ecological concepts and principles to the design and management of sustainable agroecosystems" (Gliessman, 1998). This falls short of addressing the total complexity of the food system, how energy and materials are invested, and how cycling of these materials could improve the long-term sustainability of human and natural ecosystems. The definition also does not emphasize the overwhelming role of humans in their role as managers of all cultivated and pasture lands as well as their impact on essentially all areas on the planet. Recently we have proposed a definition of *agroecology as the ecology of food systems* (Francis et al., 2003).

When agroecology is defined as an integrative area of study of the ecology of the food system, it is essential for us to examine the production process as well as the local and landscape ecological impacts, the broad economics of farms and communities, and other social dimensions of agriculture including involvement of families and organizations in the political process. In an increasingly urbanized society where a small proportion of the population is directly involved in production agriculture, for example only one percent in the U.S. today, the vast majority of citizens is becoming physically and psychologically separated from their sources of food. When a small proportion of income is spent for food and there are localized surpluses, it is easy to understand why people focus on fast food and quickly discount the importance of where and how that food is produced and the long-term impacts on our resources and environment (Schlosser, 2001). With growing consolidation of both the production process and the processing and distribution of food, and a growing disparity of incomes and access to resources on an international scale, it is obvious that the global food chain does not provide equal benefits to all consumers. Likewise, short-term economic analysis does not take into account the long-term impacts on resource use and the environment on which we depend. Agroecology defined as food systems ecology helps us focus on these broader issues, and puts into context the study of components of the system and how they can be improved.

ROLE OF AGRICULTURAL UNIVERSITIES

Current agricultural universities address these challenges through an organization or grouping of departments specialized in narrow aspects of the production process and its economic evaluation (Figure 1, from Lieblein, Francis, and King, 2000). On the upper story of this hypothetical university building are the basic sciences and humanities that our students take as general education requirements: chemistry, biology, sociology, mathematics, physics, economics, and others. These are taught from departments with those same names, and only infrequently are there team-taught courses across disciplines or explicit linkages between courses. The applied sciences such as crops, soils, animal pro-

FIGURE 1. Schematic diagram of current university structure with conventional departments, one-way flow of information, and disconnect among natural resource, farming, and urban environments (from Lieblein, Francis, and King, 2000; reprinted with permission of *Journal of Agricultural Education & Extension*, Wageningen, Netherlands).

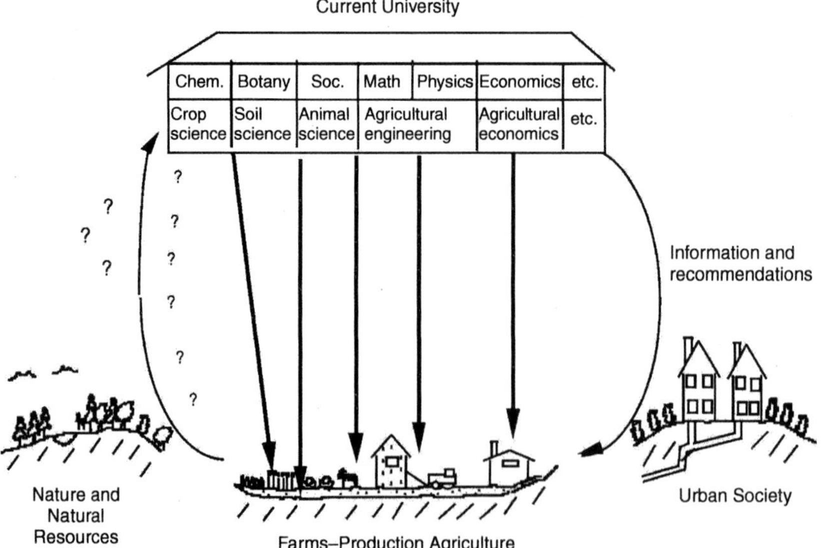

duction, agricultural engineering, agricultural economics, and others are likewise taught in discipline-specific departments with infrequent crossover or linkage among them. Exceptions are courses such as range and pasture management that may include instructors from crop and animal sciences, or environmental economics that taps expertise in both agricultural economics and environmental sciences. As shown in the figure, these courses are relatively disconnected from the real world except for examples brought in by instructors.

The principal educational connections between landgrant universities and their constituent communities are not in the classroom, but through Cooperative Extension. With bulletins and recommendations, summer field tours, and winter workshops our Extension specialists and educators reach out to both rural and urban audiences with information on special topics. Programs are designed to solve immediate problems in agricultural production, home crafts and skills, leadership, parenting, and a range of other topics. What characterizes these activities and products is a relatively strong one-way communication of information and recommendations. There is some feedback from audiences, especially through Extension educators. Our vision of the client community tends to separate the natural environment from the farming areas, and both of these from urban society, rather than making connections and studying interactions. One growing exception is the wealth of distance education offerings from some universities that attempts to reach a dispersed audience with courses for degrees or continuing education opportunities for professionals, yet these are also organized along discipline lines. Although somewhat overstated to make a point, Figure 1 clearly demonstrates the separation between departments on campus, the distance from campus to community, the one-way flow of information, and the relative isolation of students from the real world and their eventual job market. This model is in contrast to one described later that is designed to overcome some of the shortcomings of the current university structure.

It is these challenges of separation and specialization that university education in agroecology is designed to overcome. With focus on the larger food system, emphasis on components and interactions, and a systemic approach to study of resources, materials flows, cycles, and distribution of costs and benefits, agroecology broadly defined provides an educational platform for overcoming the problems of conventional agricultural education described above. Although this critique of educational offerings points out the narrowness of courses and curricula in most current university programs, there are some areas where systems approaches are found today. These are explored in the next section.

COURSES THAT ADDRESS ECOLOGICAL TOPICS

To suggest that all the topics addressed in the contributions to this volume are not considered at all in current courses and curricula would not be true. To suggest that they are covered well would be equally incorrect, and one of the goals of this volume is to explore the frontiers in science and education and build new linkages that are crucial for understanding agroecosystem structure and function, and how this understanding can help inform the design of more sustainable systems for the future.

There is an obvious division of interest between those departments and instructors interested in ecology and natural systems (Departments of Ecology and Systematics, Biological Sciences) and those focused on agricultural production (Departments of Agronomy, Animal Science, Entomology, Agricultural Economics, and others). According to Gliessman (1998), research and teaching in ecology and agronomy were closely linked in the early 1900s. Studies of crop ecology focused on where crops originated and where they were grown, and the climates and conditions that were best suited for the most popular crop species. Courses in ecological crop geography and systematics related to species in natural ecosystems were part of curricula, and course titles such as agricultural botany were common. Gliessman (1998) explained how scientists and educators diverged following World War II, when ecologists moved their focus to study of natural ecosystems and communities, while agronomists and other increasingly specialized agricultural scientists narrowed their focus to cultivated systems. Thus the concept of agroecology that was commonly accepted for some five decades went out of favor, only to coalesce again in the last 25 years as an integrative theme for education. A table of references, both books and technical articles, published over the past three decades is presented in Francis et al. (2003), including the most prominent recent books in agroecology used in the U.S., Latin America, and Europe (Altieri, 1983; Caporali, 1991; Carroll, Vandermeer, and Rosset, 1990; Gliessman, 2001; Montaldo, 1982).

There are courses in today's departments and curricula that address many of the themes presented in this volume (see Table 1). In fact, one strategy would be to incorporate the perspective of integrated systems, structure and function of agroecosystems, and the principles of ecology

TABLE 1. Current courses in temperate zone agricultural universities that deal with the topics covered in this volume, with examples in the area of agro-ecology.

Topic	Departments offering current courses
Natural versus agricultural systems	Environmental Science = natural systems Agronomy = agricultural systems
Ecology of cropping systems	Agronomy
Diverse tropical systems	Agronomy, Anthropology; frequently not offered
Ecology and tillage systems	Agronomy, Agricultural Engineering
Ecology + crop/weed interactions	Weed science, Agronomy
Integrated Pest Management	Entomology, Plant Pathology, Agronomy
Soil nutrient dynamics	Soils, in soil microbiology, Agronomy
Microbial diversity	Soils, in soil microbiology
Global climate change and ecology	Meteorology, Agronomy
Biodiversity in agroecosystems	Agronomy, Ecology
Biodiverse management strategies	Agronomy, Entomology
Agricultural landscapes	Rural Planning
Livestock and forages	Agronomy, Animal Science
Mixed farming systems	Agronomy, Agricultural Economics
Biotechnology	Genetics, Plant Breeding, Agronomy,
Social dimensions of IPM	Entomology; frequently not offered
Natural systems agriculture	Not covered, even in most agroecology courses

into a wide range of current courses (Francis and Altieri, 1992). Over the past decade, there is little evidence that such a change is happening in university agricultural education. There is more attention today to environmental impacts of farming practices, and some increase in interest in resource use efficiency. The integrated coverage of these topics is often most comprehensive in current agroecology courses.

Comparisons of natural and agricultural systems may occur in some courses in environmental science and in agronomy, but these are generally illustrative comments provided as examples: e.g., "Erosion from a natural grassland may be as low as 0.2 Mg of soil acre^{-1} year^{-1}, while that from highly erodible cropping fields in soybean (*Glycine max* L.)

the previous year could be as high as 10 Mg of soil $acre^{-1}$ from a single rain storm." Except for courses in agroecology, it is unusual to find whole system comparisons between natural ecosystems and agroecosystems. The ecology of cropping systems is covered in some advanced agronomy courses, often using the excellent reference by Loomis and Connor (1992), but this is rarely a major topic in such courses. Diversity in tropical cropping systems is an obvious topic of major interest in tropical agriculture courses, yet these are not taught in all landgrant universities due to limited demand by students.

Topics related to farming practices and ecology frequently find their way into current courses. Tillage has a direct impact on earthworm populations and strongly influences the rate of residue degradation, with both factors important to soil microbial and other biotic activity. Along with other aspects of soil biotic interactions, topics in soil microbiology classes deal with the dynamics of soil life and its contribution to nutrient availability and soil fertility. Likewise, weed seedbanks, germination, and success at reproduction are common topics in weed management and weed ecology courses. Along with Integrated Pest Management (IPM), these are probably the current courses offered in Agronomy and Entomology Departments that best incorporate ecological principles and language into study programs. Economic payoff for farmers from IPM philosophy and practices as well as an impressive reduction in negative environmental impacts of pest management due to reduction in pesticide applications have led to major acceptance over the past three decades of the IPM approach in mainstream research and teaching programs. Soil nutrient dynamics are so closely related to organic matter and interactions of soil biota with the inorganic materials found in soil that ecology is often brought into the discussions of nutrient budgeting, essential elements available for crop growth, and soil health or quality.

Broader topics such as global climate change, biodiversity on farms and in agroecosystems, and landscape level interactions are found in agronomy, meteorology, and ecology courses. Climate change is most frequently mentioned in relation to length of growing season or available rainfall, or related to changes in speciation in natural ecosystems. It is not often related to systems dynamics or to design of more diverse systems that would be more resilient to changes, unless this is found in an agroecology course. Current cropping systems are incredibly narrow in their biodiversity, since the intent is to plant and harvest large areas to single crops with the most efficient possible management and homogeneous treatment across the area, as well as elimination of all weeds that are perceived to reduce crop yields. Rotations, strip cropping patterns, and tropical intercropping are significant exceptions to this norm. Likewise, mixed farming systems are a common topic in tropical agriculture courses if these are part of the curriculum, but rarely from a temperate zone perspective.

Livestock and forage systems are often taught in cross-listed courses between agronomy or range science and animal science departments. Especially in the range science area, there is considerable attention to the performance of native grasslands and how their productivity can be modified through species improvement or intensive grazing systems, thus there is attention to the natural ecosystem in order to better understand the managed one.

Biotechnology is a current hot topic that stimulates much debate about the biological, ecological, and economic impacts of products that are new to the agricultural environment and may have some unintended long-term effects. Ecology is central to the study of gene escape, mating with wild relatives, and effects on the total ecosystem when new genetic combinations are introduced. There is concern from some scientists that the risk of transgenic crops arises from the fact that humans and other species currently in cultivated or natural ecosystems have no evolutionary experience to cope with the presence of entirely new genetic combinations.

One topic not often covered by current courses is the social dimensions of IPM or other agricultural strategies. The social impacts of different technologies, of the scale of agricultural operations and consolidation of farms into larger units, and of the near monopolies currently operating in livestock and commodity grains processing and marketing are considered as topics in some agricultural economics classes, and only infrequently in other production-oriented course offerings. These social dimensions are not on the radar screens of most people who teach agricultural courses, or at least they do not recognize the connections between social issues and specific technologies or systems.

Natural systems agriculture as promoted by The Land Institute in Kansas is a new approach that revives the characteristics of many indigenous systems, builds on principles of ecology, and uses nature as guide for design of agroecosystems (Jackson, 1980; Soule and Piper, 1992). This specific model is not considered outside of a few agroecology courses where the instructors are familiar with the concept and approach.

In conclusion, most of the topics included in contributions in this volume are not given comprehensive treatment in current discipline-specific courses nor in university curricula. They may be mentioned at times, providing examples of how agroecosystems differ from natural ecosystems, but there is little substantive attention given to the principles of ecology as they relate to design of agricultural systems. Agroecology courses provide one bright exception to this observation. At the very least, these courses deal with the complexity of interactions and the interconnectedness of all decisions on enterprises and design of the farm unit. Most incorporate economics and environmental impacts of the various systems used in agriculture, and ecology is often used as a guide to establish indicators for environmental health of a system. Social impacts are less often part of agroecology courses that are focused on the production process. For this reason, the new explicit definition of agroecology as the ecology of food systems will provide broader guidelines for design and implementation of future courses in this area (Francis et al., 2003).

DESIGN OF AGROECOLOGY COURSES

Agroecology courses are likely to bring together the important dimensions of agroecosystems that are discussed throughout this volume, topics that are currently ignored or underrepresented in the curriculum. In fact these courses have potential to become the umbrella or capstone experience that helps students relate a number of important but disparate details gleaned from other courses found in conventional, discipline-specific departments. Why is this possible with agroecology, and what makes these courses unique in the university educational environment?

Most of the principles established in the study of ecology and natural systems are relevant when applied to agroecosystems. For example, the hierarchy of spatial scale that is used to describe systems is useful as a focus for organization and interpretation of how systems function (see Figure 2). Farming systems could be characterized across a range of spatial scale from the field to the farm, and on to the landscape, the region, the continent and the world. Within the field, the individual plants and animals, the macro- and micro-organisms that inhabit the field, and the transient wild species that move through that area are all a part of the functioning agroecosystem. It is helpful to address research questions or to focus education at the proper level in the hierarchy.

In conventional agronomy or agricultural economics courses, for example, we address the individual production practices that will help

FIGURE 2. Spatial hierarchy of scale with identification of social systems decisions at each level, and the multiple dimensions at any level of scale (adapted from Olson, 1999).

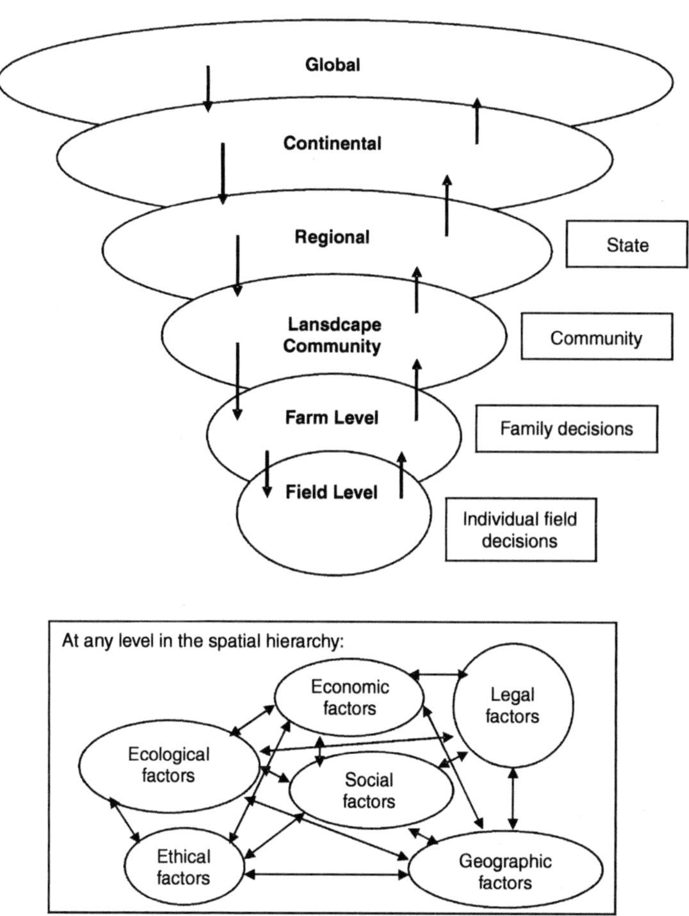

contribute to increased yields and profits. These are aggregated to provide measures of the individual farm's production and economic return. There is an implicit assumption that highest possible yields of crop commodities will lead to highest profits, and that most other factors such as environment or impacts of those commodities on nearby communities are externalities in the analysis. When the topic is regional or national production of a specific commodity, the interest is focused on annual total production, prices, production in competitor countries, global demand for the commodity, and predictions of futures markets and how to market strategically to maximize profits. There is little regard to the distribution of the benefits from the entire process, who gains and who loses in the home country or elsewhere, nor to the intangible impacts of production of that commodity on sustainability of farming systems nor of communities where the commodity is grown.

Agroecology courses take more factors into account in the evaluation of crop and livestock production. As shown along the right side of Figure 1, there are also social systems and decisions that are made at each level in the spatial hierarchy. Individuals and families make enterprise and technology decisions at the field and farm level. Community decisions such as well density, permitted nitrate levels in the water, or zoning for different land uses can directly affect farmers in that watershed or community. Regulations often come from state and national level, and policies on trade from national and international levels. Agroecology helps us focus on all these levels, and to make explicit where different decisions or practices will impact the individual farm or the landscape. An emerging appreciation of multifunctional rural landscapes and how they serve society helps inform our study at the landscape level (Francis et al., 2004).

Also important in agroecology is the concept that each level incorporates a number of factors beyond the ecology of production (Figure 3, lower right). At the farm level, there are economic realities that drive the use of technology, and ethical issues about the use of certain inputs or priorities on how to use resources. There are legal issues surrounding production of crops and livestock and how the specific practices or scale of an operation affects the neighbors. There are social issues such as available labor, distance to markets, intensity of involvement of different people in the family, and long-term aspirations that impact the choice of enterprises and design of systems. At other levels these same factors operate with greater or lesser importance. For example, details of the ecological functions of systems are most important at field, farm, and landscape levels. Ethical considerations occur at all levels, and extend from the farm family up through society to the national level. Most legal decisions are made at the federal or state levels, although there are some that are highly influenced by international rules and economics, and there are specific issues such as groundwater quality or concentrated livestock production that are decided and regulated at the landscape or community level.

FIGURE 3. Multiple characteristics of resources in the natural systems and social systems at the farm level (or other scale level), with primary interactions among these types of resources (from Francis et al., 2003; reprinted with permission of American Society of Agronomy, Madison, Wisconsin).

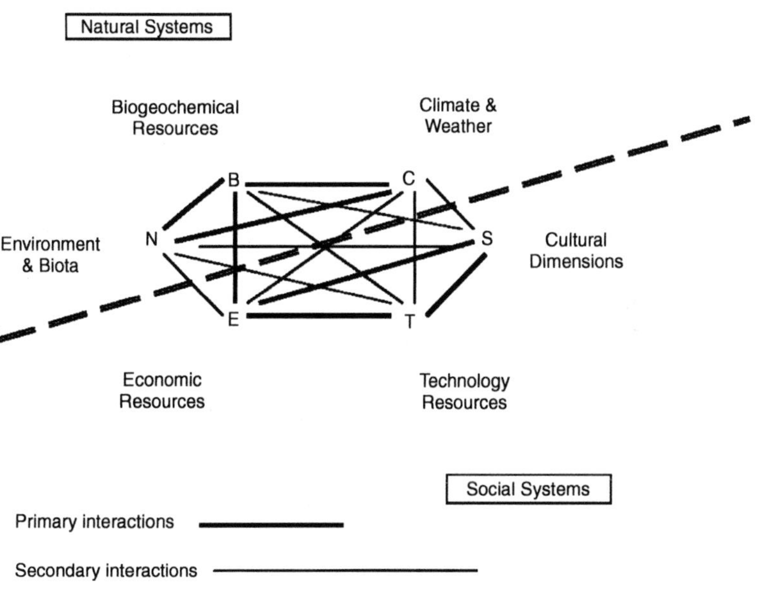

Hierarchy of scale is important in the design of research and teaching activities. To better understand the mechanisms of how a system operates at any one level, it is necessary to go to lower levels and study the components of the system and how they interact. Economics of the whole farm are determined by the returns to each of the enterprises on that farm. National production is an aggregate of the production on all the farms in the country. To better understand the context in which systems operate, or in other words the meaning to put on systems at any level, it is important to look to higher levels in the hierarchy. Field production of a given crop has little meaning unless this is considered in the context of a given year and the price for that commodity. To understand how the agroecosystem provides services to the larger landscape and society, it is necessary to look at the aggregate impacts of all the farms in the larger area. These are examples of how the principle of hierarchy of scale is useful to help us understand how systems function and how to improve them.

Another way to examine how agroecology helps us understand the complexity of systems is illustrated in Figure 3, at the farm level, where different resources in the natural system and the social system are listed. In the natural system, as well as the managed agricultural system, the soils, topography, long-term climate and short-term weather, and the plant and animal species present all interact to form the ecosystem and local environment. In the social system, there are economic and technological resources available to each farm family, and their use will be governed by many cultural and historical factors that shape the thinking, priorities, and plans for the family. What makes this analysis exciting are the complex interactions among factors, and major two-way interactions are shown in bold lines in the figure. The quality of soils and level of nutrients determine the amount of fertilizer to be applied, dependent on the economic capacity to buy that purchased input. Climate will strongly influence the natural species present on a farm, as well as the range of potential crops or animal species that can be grown there. Climate also has a strong influence on the evolution of soils in any given place. Cultural factors will determine to a large extent the decisions on use of technologies and how to allocate economic resources. For example, an Amish family could decide to use only human labor and animal manure rather than an herbicide for weed control and chemical fertilizer for nutrients, and put economic resources into a new barn instead; a non-Amish neighbor with similar soils and climate might decide to use the herbicide and the purchased fertilizer, even though many conditions on the farm are identical. Use of economic resources to access a certain technology depends on whether that technology is locally available and recommended for the specific conditions of the farm. Much more complex are the multiple interactions among these factors, often not made explicit nor understood by the decision maker. The framework used in agroecology is helpful to build an understanding of the farm in new ways.

Interactions of factors across levels in the hierarchy are also important to decisions and to farming success. Weather at the national and regional levels will influence crop productivity across a wide area and thus impact the price at the farm level. The economic situation for input companies will influence how much they invest in research, and this in turn result in fewer or more new technologies available to the individual farmer. In a democracy, the cultural factors that influence votes that are generated at the family, community, and state levels may lead to changes in farm legislation at the national level, and this in turn impacts the decisions on planting specific commodity crops at the farm level. It is the interactions across levels of the five dimensions shown in Figure 2 (lower left) that make the system complex, and it is the study of the interface between natural science and social science that makes agroecology (broadly defined) an especially valuable area of study in universities.

EXAMPLES OF AGROECOLOGY AND SYSTEMS COURSES

Agroecology courses have been part of the curriculum in agriculture in Mexico, and in environmental sciences at University of California at Santa Cruz (UCSC), for more than two decades. Programs have been initiated in the Postgraduate School in Chapingo and University of the Yucutan in Mexico, and in several universities in Chile and Spain. Courses are now found in many of the landgrant universities and other colleges in the U.S., and in the Nordic region. Examples drawn from several of these courses will illustrate the current emphases in teaching agroecology. In general, they are parallel to the topics covered in this volume, with major emphasis on biological and technological details of the production process and little mention of the social aspects of the system, except for the contribution on social dimensions of IPM that does address broad issues. The new Agroecology MSc degree program of the Nordic Forestry, Veterinary and Agricultural (NOVA) University in the Nordic Region is used to illustrate the broad approach used in education when agroecology is explicitly defined as the ecology of food systems.

Perhaps the most widely used textbook today for agroecology courses is Gliessman's (1998) *Agroecology: Ecological Processes in Sustainable Agriculture*. The goal of his course and the text is to prepare students to apply "ecological concepts and principles to the design and management of sustainable agroecosystems." The introduction outlines the current practices and problems inherent in conventional agriculture and describes a strategy to achieve greater sustainability in production. The parallels and differences between natural and agroecosystems are discussed in detail. The first major portion of the book deals with the natural resources, climate, and biotic components that impact agricultural systems: nutrients, light and water, temperature, wind, fire, and other organisms in the system. The second major section looks at components and their system level interactions: population processes, biodiversity, species interactions, stability, disturbance and succession, energy cycles, and the complexities of system dynamics. The concluding chapters address how to achieve greater sustainability, and how agriculture relates to the broader food system. Gliessman's course at UCSC and textbook are written from the point of view of an ecologist with extensive experience in multiple species systems in the tropics and temperate zones. His examples are taken primarily from the west coast of the U.S. and from Mexico. With only a few exceptions, his approach would appear to touch on most of the topics covered in this volume.

In contrast, a course we offer at University of Nebraska reflects our place on the Great Plains and the examples needed to make this educational experience meaningful to students in agronomy, horticulture, and natural resources. Using the Gliessman text, we expand the goals to help students access appropriate resources to define agroecology and its processes, evaluate key indicators of sustainability, compare conventional and alternative agricultural systems, and address real-world challenges through interdisciplinary team projects. Although topics and organization are similar to the course above, we focus on structure and function of the grasslands, dynamics of systems in response to stress, evaluation of conventional farming systems and design of alternatives, and grazing systems appropriate to the plains. The course includes economic evaluation of alternative systems and non-conventional valuation criteria, food processing and marketing, and global food production and food security. A practical farm design project is carried out parallel to the lectures and discussion periods. Again, most of the topics covered in this volume find their way into lectures or topical discussions through the course of the semester.

A rich literature has developed around agricultural systems research and education. The book by Checkland (1981) provides a theoretical framework and foundation for the systems thinking that underlies the practice. The conceptual article by Conway (1985) on analysis of agroecosystems and the practical application in education described by Bawden (1991) are among the many articles that apply the principles to agricultural systems. The journal *Agricultural Systems* published in U.K. brings current thinking together around this important topic. Perhaps the most comprehensive application of this concept in education has been the field-oriented, problem-solving program at the University of Western Sydney at Hawkesbury in New South Wales, Australia (Bawden et al., 1984). In this program, students work with individual farmers and families to analyze their complete operations, using this as a case study for problem solving and whole systems design. The students present their final farm plans to university faculty and to the farmers for evaluation. Sriskandarajah, Bawden, and Packham (1991) discuss the applications of these same ideas to the study of sustainability in Australian agriculture, and how this can be promoted through universities and extension.

The modular short course is another educational experience focused on agroecology and systems evaluation (Lieblein, Francis, and King, 1999). NOVA University in the Nordic Region has offered several such courses in Norway, Sweden, Estonia, and Lithuania in recent years. One model has students prepare by reading a number of journal articles and book chapters before participating in an intensive, one-week course that includes farm visits and interviews, small group projects, and in-depth discussion of issues. Following the oral presentation of a report on their analysis, student teams complete a group report to submit electronically for faculty evaluation. They also write an individual learner document that is a self-evaluation of their personal reaction to the learning environment in the course and their assessment of growth and learning in the overall activity. A similar course we have implemented in three states of the Midwest U.S. has been given excellent evaluations by students.

The most explicit course of study in agroecology and food systems is the new MSc program offered in NOVA University, starting with a full semester of two full-time study modules that are taken without other courses competing for student time. Objectives are to relate ecology to agriculture, to use systems thinking for complex problems and identification of emergent properties, to study how food is produced and moves from farm to consumer through case studies and comprehensive farm and food system plans, and to practice a range of communication skills. Farmers are involved as part of the teaching faculty, and much of the work is done on farms and in communities, integrating natural science and social science approaches. The two modules, *Agroecology and Farming Systems* and *Agroecology and Food Systems* are followed by two additional semesters of courses and a thesis directed toward a current issue in the food system. This degree program was initiated in 1999, and the fourth student group completed the modules from September to December, 2002.

The learning methods in these courses as well as others in agroecology include conventional lectures, class discussion, small group projects in class and outside, and presentations of written and oral reports. Emphasis on practical field observations, interviews, and synthesis of information is an application of what is called action learning (Schubert, 1995). What we have learned to date is that the conventional university departmental organization is often not well suited for this type of interdisciplinary course. Instructors are recruited and the reward system is designed to recognize contributions in specific and narrow areas of study. The generalists who are needed to teach a course such as agroecology and advise students across a complex array of subject matter areas, including both natural and social sciences, seem to have difficulty gaining acceptance in conventional departments. One alternative to the generalist faculty approach is to invite experts from several departments, each to present their discipline-specific topics and thus bring their expertise in each important component of the system. We have found it difficult to help both faculty and students make the vital connections to overall systems and the dynamic development unique to each course. To relate each topic to farming systems or food systems within the flow of the course is difficult to coordinate and make coherent in the continually evolving context of the student activities in and out of the classroom. We prefer to use these experts as resource people, and to depend on the continuity provided by generalist course instructors to carry the major load in catalyzing the learning process. The degree program is new, and we continue to adjust the program based on experience with each class.

DESIGNING UNIVERSITIES FOR THE FUTURE

One model for the structure and organization of learning in future agricultural universities is shown in Figure 4 (from Lieblein, Francis, and King, 2000). The future active learning university would feature courses in the building block sciences and humanities on the university campus. As shown, the courses would be integrated across subject matter areas, feature team teaching, and provide a seamless flow within this learning environment to make connections of the material to the real world through projects and case studies. For study of the applied technical and social sciences, students and faculty would move outside the campus environment into the natural resource, farming, and urban society context. Problem-solving activities, case studies, and internships would provide on-the-ground learning experiences where students and faculty in their learning communities confront and grapple with the complexity of current challenges found in the food system. The concept of faculty would be expanded to include people from farming, food processing and marketing, and government and non-profit organizations (Francis et al., 2001). Important to learning would be a focus on the group work process, since that is how most students will operate in the future job environment. All of these steps are being incorporated into the NOVA MSc program described above. There are obvious challenges in the con-

FIGURE 4. Idealistic structure of future universities designed for active learning and research, with movement of students, faculty, and information, and a close integration of natural resource, agroecosystem, and urban environments (from Lieblein, Francis, and King, 2000; reprinted with permission of *Journal of Agricultural Education & Extension*, Wageningen, Netherlands).

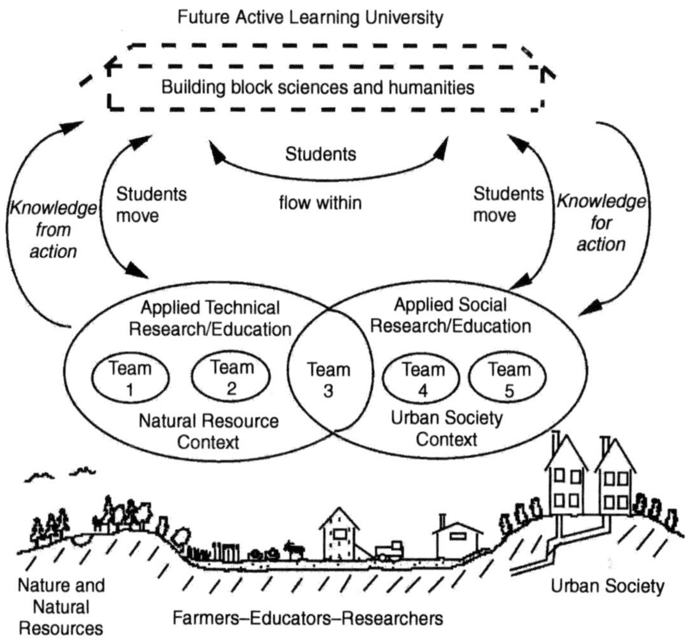

version of conventional university programs and structure to accommodate this type of learning, and these are being addressed by introducing innovative courses one at a time, each one pushing the university establishment to broaden the concept of education and learning.

CONCLUSIONS

When agroecology is defined as the ecology of food systems, the relevance, complexity, and impact of interconnections in the total system are obvious. We also can recognize the shortcomings of conventional classes in the curriculum to deal with this level of complexity and with larger issues that transcend our current disciplines. Such topics as energy efficiency of delivering food to the table, the economic plight of many small and medium sized farms, and the effects of current systems on the environment are central to long-term sustainability. Moreover, the impact of consolidation of input providers, processors, and global marketing corporations, and the influence of these macro changes on rural communities are among the issues that need to be studied as part of the food system. Agroecology and study of the entire food system can help education focus on a broad set of concerns.

Taking the topics presented in the contributions to this volume as a sample of "recent developments in agroecology," we observe that the majority deals with specific aspects or components of the production process, although most are framed in a systems context. There is a wide range of interpretations of the term "systems," from soil ecology to tillage to cropping pattern to landscape. Each topic in fact can describe a system at some level in the spatial scale. Many of the topics are included as minor components in current agronomy, entomology, or ecology courses. Yet crucial topics that are obvious by absence in this array of topics include (1) concern about the diversity of enterprises as they relate to economics of systems, (2) energy and other resource efficiencies of systems and the long-term perspectives on their availability, (3) any mention of processing, marketing, or consumer issues as part of the food system, and (4) the overwhelming importance of social issues and human decisions in determining functioning and success of agroecosystems. The only exception is mention of social factors in two contributions related to IPM. It is clear that agroecology thinking is still dominated by the natural sciences, and that major modifications in courses and curricula will be needed to achieve a focus on the entire food system. This is why our proposal to redefine agroecology takes on such importance.

Some of the key terms and concepts that should infuse our thinking about agroecology, especially planning for education in the processes of agroecosystems, are complexity, integration, holistic systems, interdisciplinarity, and long-term thinking. Methods that need to be included will come from both the natural sciences and the social sciences, and a rich blend of methodologies can result in equally robust understandings of how systems function and how they can be improved. It is especially important that each new generation of students be given the freedom to explore the alternatives, based on understanding of current systems and scientific principles, and not be constrained by courses that teach methods that are within the current human constructs that we call disciplines, majors, and departments. We need to think outside the box to explore viable options for the future.

There are some courses that meet these criteria. Based on pioneering educational efforts in a few universities in Mexico, Chile, Spain, Norway, and the U.S., courses in agroecology are emerging that will tackle the complexity of agricultural systems, including both biological production and social dimensions of systems and their impacts. The new MSc Program in Agroecology in NOVA University is one such program, and the intensive summer travel courses that emphasize farm visits and action learning add to the array of new options. There are graduate degrees in sustainable agriculture in several universities, and these increasingly take on the production, economic, environmental, and social dimensions that will be found in future agroecology courses.

Current traditional university organization into discipline-specific departments provides some disincentives for generalist approaches such as agroecology. A model for an active learning university with basic foundation courses on campus and applied problem solving courses conducted in the field and community is proposed for discussion. There is no doubt that the urgency of resource scarcity, burgeoning human population, and impacts of conventional agriculture on the environment compel us to seek new and innovative alternatives to our current educational offerings. Agroecology defined as the ecology of food systems provides one such model. We need a system of education that promotes the skills and knowledge plus the vision to identify and solve the most critical challenges in the future with regard to food systems. We owe our next generations nothing less than a system that expands rather than reduces their options.

REFERENCES

Altieri, M.A. (1983). Agroecology. Berkeley, CA: University of California Press.

Altieri, M.A. and C.A. Francis. (1992). Incorporating agroecology into the conventional agricultural curriculum. *American Journal of Alternative Agriculture* 7(1&2): 89-93.

Bawden, R.J. (1991). Systems thinking and practice in agriculture. *Journal of Dairy Science* 74:2362-2373.

Bawden, R.J., R.D. Macadam, R.J. Packham, and J. Valentine. (1984). Systems thinking and practice in the education of agriculturists. *Agricultural Systems* 13:205-225.

Caporali, F. (1991). Ecologia per L'Agricultura: Teoroia e Prática. UTET Libreria, Torino, Italy.

Carroll, C.R., J.H. Vandermeer, and P. Rosset, editors. (1990). *Agroecology*. New York, NY: McGraw-Hill Publications.

Checkland, P.B. (1981). *Systems Thinking, Systems Practice*. New York, NY: John Wiley.

Conway, G.R. (1985). Agroecosystems analysis. *Agricultural Administration* 20: 31-55.

Dunn, S. (2001). Hydrogen futures: Toward a sustainable energy system. WorldWatch Paper 157, Washington, DC: WorldWatch Institute. 90 p.

Flavin, C. and N. Lenssen. (1990). Beyond the petroleum age: designing a solar economy. Worldwatch Paper 100, Washington, DC: WorldWatch Institute. 65 p.

Francis, C.A. and M.A. Altieri. (1992). Agroecology and sustainable development: Innovative ideas for an effective university curriculum. *Journal of Sustainable Agriculture* 3:107-112.

Francis, C.A., G. Lieblein, J. Helenius, L. Salomonsson, H. Olsen, and J. Porter. (2001). Challenges in designing ecological agriculture education: A Nordic perspective on change. *American Journal of Alternative Agriculture* 16:89-95.

Francis, C., G. Lieblein, S. Gliessman, T.A. Breland, N. Creamer, R. Harwood, L. Salomonsson, J. Helenius, D. Rickerl, R. Salvador, M. Wiedenhoeft, S. Simmons, P. Allen, M. Altieri, C. Flora, and R. Poincelot. (2003). Agroecology: The ecology of food systems. *Journal of Sustainable Agriculture* 22(3):99-119.

Francis, C., L. Salomonsson, G. Lieblein, and J. Helenius. (2004). Serving multiple needs with rural landscapes and agricultural systems. In *Agroecosystems Analysis*, eds. D. Rickerl and C. Francis, Madison, WI: American Society of Agronomy. (in press).

Gliessman, S.R. (1998). *Agroecology: Ecological Processes in Sustainable Agriculture*. Chelsea, MI: Ann Arbor Press.

Gliessman, S.R., editor. (2001). *Agroecosystem Sustainability: Developing Practical Strategies*. Boca Raton, FL: CRC Press.

Jackson, W. (1980). *New Roots for Agriculture*. Lincoln, NE: University of Nebraska Press.

Lieblein, G., C. Francis, and J. King. (2000). Conceptual framework for structuring future agricultural colleges and universities in industrial countries. *Journal of Agricultural Education and Extension* 6:213-222.

Lieblein, G., C.A. Francis, L. Salomonsson, and N. Sriskandarajah. (1999). Ecological agriculture research: increasing competence through PhD courses. *Journal of Agricultural Education and Extension* 6:31-46.

Loomis, R.S. and D.J. Connor. (1992). *Crop Ecology: Productivity and Management in Agricultural Systems*. Cambridge, U.K.: Cambridge University Press.

Montaldo, P. (1982). *Agroecologia del Trópico Americano*. San Jose, Costa Rica: IICA.

Olson, R.K. (1999). Introduction. In *Under the Blade: The Conversion of Agricultural Landscapes*, eds. R.K. Olson and T.A. Lyson, Boulder, CO: Westview Press. pp. 1-13.

Schlosser, E. (2001). *Fast Food Nation: The Dark Side of the All-American Meal*. Boston, MA: Houghton Mifflin Company.

Schubert, W.H. (1995). Students as action researchers: Historical precedent and contradiction. *Curriculum and Teaching* 10(2):3-14.

Soule, J.D. and J.K. Piper. (1992). *Farming in Nature's Image: An Ecological Approach to Agriculture.* Washington, DC: Island Press.

Sriskandarajah, N., R.J. Bawden, and R.G. Packham. (1991). Systems agriculture: A paradigm for sustainability. *AFSRE Newsletter* 2(3):1-5.

Redesigning Industrial Agroecosystems: Incorporating More Ecological Processes and Reducing Pollution

Eugene P. Odum
Gary W. Barrett

SUMMARY. The development and mixed benefits of high-input industrial agriculture are reviewed, then illustrated with findings from a Georgia landscape study. As a basis for considering ways to reduce the serious negative aspects of industrial agriculture, we compared the basic ecological energetics and processes of natural ecosystems with pre-industrial, industrial, and reduced-input conservation tillage agroecosystems. We also reviewed our 20-year comparison of conventional tillage with no-till systems. The conclusion is that current trends to redesign industrial agriculture along the lines of reduced-input conservation tillage systems should be accelerated in order to maintain and restore soil quality, and to reduce pollution and soil erosion at the landscape scale. *[Article copies available for a fee from The Haworth Document Delivery Service: 1-800-HAWORTH. E-mail address: <docdelivery@haworthpress.com> Website: <http://www.HaworthPress.com> © 2004 by The Haworth Press, Inc. All rights reserved.]*

Eugene P. Odum was Director Emeritus, Institute of Ecology, University of Georgia, Athens, GA 30602 USA (deceased).

Gary W. Barrett is Odum Professor of Ecology, Institute of Ecology, University of Georgia, Athens, GA 30602 USA (E-mail: gbarrett@gbarrette.uga.edu).

The authors thank David Clements and Anil Shrestha for inviting them to contribute to this special volume on agroecology and colleagues Dave Coleman and Paul Hendrix for reviews of this manuscript. The authors thank Terry L. Barrett for preparation of this manuscript.

[Haworth co-indexing entry note]: "Redesigning Industrial Agroecosystems: Incorporating More Ecological Processes and Reducing Pollution." Odum, Eugene P., and Gary W. Barrett. Co-published simultaneously in *Journal of Crop Improvement* (Food Products Press, an imprint of The Haworth Press, Inc.) Vol. 11, No. 1/2 (#21/22), 2004, pp. 45-60; and: *New Dimensions in Agroecology* (ed: David Clements, and Anil Shrestha) Food Products Press, an imprint of The Haworth Press, Inc., 2004, pp. 45-60. Single or multiple copies of this article are available for a fee from The Haworth Document Delivery Service [1-800-HAWORTH, 9:00 a.m. - 5:00 p.m. (EST). E-mail address: docdelivery@haworthpress.com].

http://www.haworthpress.com/web/JCRIP

Digital Object Identifer: 10.1300/J411v11n01_03

KEYWORDS. Agroecosystem, agrolandscape, agricultural pollution, ecological energetics, industrial agroecosystem redesign, landscape ecology, soil building processes

INTRODUCTION

Natural ecosystems serve as an ideal model of how to manage agroecosystems based on ecological processes and theory. An understanding of natural ecosystems encourage us to question why conventional agriculture practices frequently result in environmental problems such as eutrophication at the landscape scale (Barrett and Skelton, 2002), reduction in soil quality (Coleman and Crossley, 1996), decreased biotic diversity (Collins and Qualset, 1999), and the need to integrate urban and agricultural landscapes (Barrett, Barrett, and Peles, 1999). Conventional or industrial agricultural traditionally has been based on the concept of increased crop yields (NCR, 1989). To achieve these yields, increased inputs of fertilizes, fossil fuels, and pesticides, frequently termed "subsidies," have been added in huge quantities as input to the agricultural enterprise (i.e., a change from a solar-powered, ecologic-based, food producing system to a more fuel-powered, economic-based, agricultural approach). Encompassing ecological processes in the food-producing system will help to alleviate these problems and restore a quality landscape.

Agroecosystems, in terms of the quantity and quality of energy sources, are intermediate between solar-powered natural ecosystems and fuel-powered human-organized techno-ecosystems such as cities (Odum, 1984). Solar energy is an abundant but dilute, low-powered source in terms of work potential per unit of quantity (e.g., joules, calories) as compared to concentrated, high-powered fossil fuels. It takes several thousand joules of solar energy to equal one joule of fossil fuel. Or to put it in other terms, the fossilization process has greatly concentrated ancient sunlight. In addition to differences in biological structure and material inputs (e.g., pesticides and fertilizers), it is this huge difference in energy power that makes current industrial agroecosystems so different from natural ecosystems, and so much more polluting.

Now we come to the main theme of this paper: to discuss the design of a reduced-input agroecosystem that incorporates some of the recycling and soil building properties of natural and pre-industrial systems. Or to put it in practical terms, designing sustainable crop and meat production systems that must replace current wasteful and unsustainable industrial agriculture if the world is to be nourished without major damage to the health and quality of our total environment (Odum, 1987).

AGRICULTURAL DEVELOPMENT

We present graphic models of energy flow and nutrient cycling of a natural ecosystem (Figure 1) and three kinds of agricultural systems, namely, pre-industrial, industrial, and reduced external-input sustainable agriculture or conservation tillage systems (Figure 2). The food web in natural ecosystems (Figure 1) consists of two basic chains: (1) the grazing energy flow path (i.e., plants, herbivores and their predators, and (2) the detritus microbial loop path. In most natural ecosystems more energy flows along the microbial loop than along the grazing chain, and most nutrient cycling is driven by this flow. A natural solar-powered system also includes internal pest control, an efficient recycling of nutrients, and the low rate of organic output (Figure 1).

Pre-industrial (also, termed "traditional") labor-intensive agriculture as still practiced in developing countries is very similar to natural ecosystems (Figure 2A). Farm animals replace wild animals in the grazing food chain and crop residues and manure maintain an active microbial

FIGURE 1. Diagram depicting a natural, solar-powered, ecosystem.

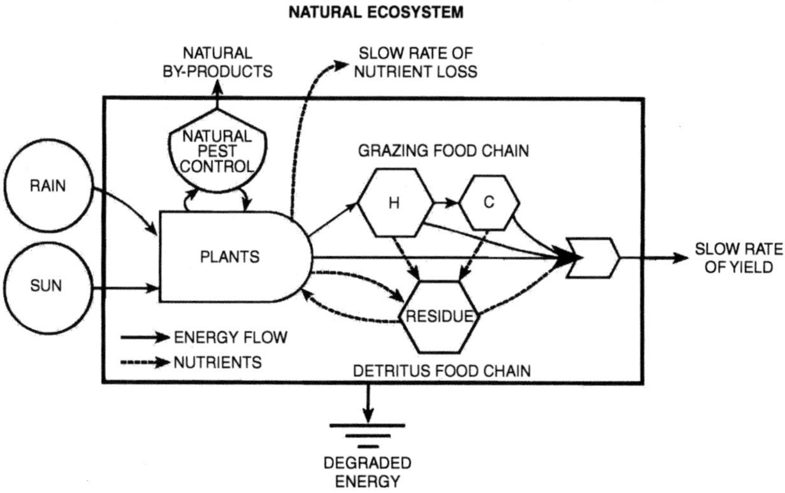

FIGURE 2. Diagram depicting pre-industrial, industrial, and low-input sustainable agriculture in the Western world based on solar, fossil fuel and material, subsidy inputs and outputs to the food production system.

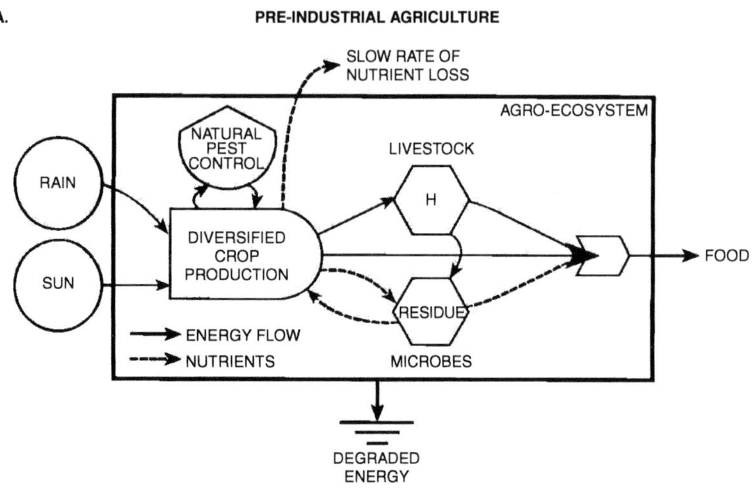

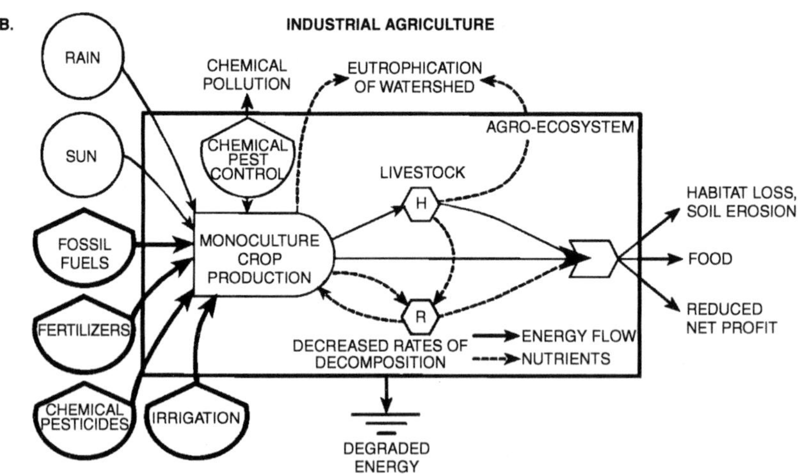

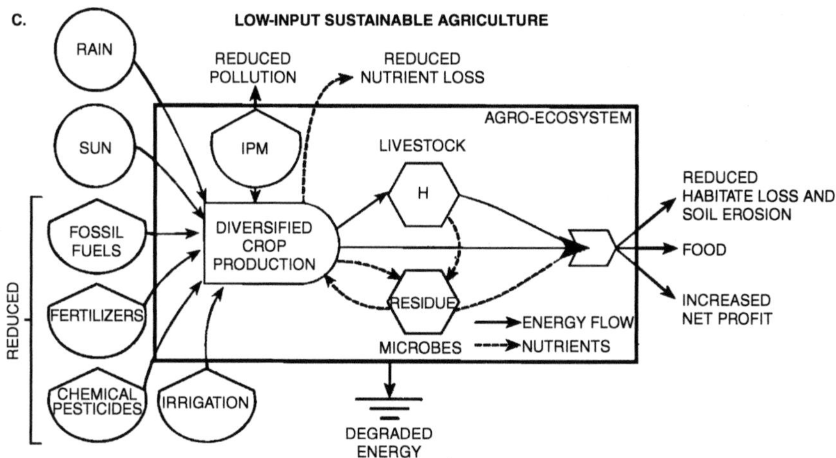

loop with little soil erosion, pollution, or loss of nutrients. Such systems are often very diverse with a variety of crops and farm animals, and even fish fed on plant residues. They are also relatively energy efficient in terms of energy input and output. Pre-industrial agriculture enriches the general landscape and provides adequate food at the village or local level, but not much to export to the big cities or to other countries.

In industrial agriculture (Figure 2B), fossil fuel driven machinery largely replaces human labor. Furthermore, very large subsidies of nutrients, fossil fuels, water and pesticides increase yields per unit of area, but also create extensive chemical and nutrient pollution which adversely affects both human health and the health of the natural life-supporting environment. Soil quality and erosion may also be adversely affected. There is also frequently extensive loss of habitat and biotic diversity. Since most of the crop residue is taken off with the harvest the microbial recycle is reduced and replaced with large inputs of fertilizers. In the United States of America today vast monocultures of field crops of grain and soybeans are not eaten locally, but are sold and exported for the money they will bring into the region.

Low-Input Sustainable Agriculture (LISA; Figure 2C) significantly reduces the amounts of human subsides, thus reducing pollution, stream eutrophication, soil erosion, and habitat loss. LISA must be accompanied by increased crop (and biotic) diversity, integrated pest management, and increased rates of nutrient recycling (i.e., a return to basing farming practices on natural ecosystem-based processes and diversified cropping systems).

Below we illustrate how one state, namely the State of Georgia (USA), encompassed industrial agriculture (Figure 2B) during the middle part of the 20th century. This section includes how research conducted at the University of Georgia HorseShoe Bend Experimental Site illustrates the ecological and economic significance of encompassing ecological processes in the food-producing system.

THE GEORGIA CASE STUDY

In the 1980s, the Georgia Institute of Ecology coordinated a comprehensive study of the fifty-year history of the Georgian environment including the development of industrial agriculture between the 1930s and 1980s (Odum and Turner, 1990). During this period, the area of farmland declined by half and yield of field crops increased four-fold, so state-wide presently twice the yield was obtained on half as much land (Figure 3). That is the good news. The downside is that during that 50-year period commercial fertilizer use increased 7-fold and nitrogen use 11-fold (Figure 3). At the same time, a good part of this huge subsidy runs off into streams creating severe nutrient pollution and increased stream eutrophication, indicating that the subsidies are excessive since the runoff does not help the crops! Even more serious is the fact that industrial agriculture is rapidly putting the small family farm out of business because the small farmer can not afford the expensive machinery, large tracts of land, and expensive subsidies. As a result, millions of farmers are being forced off the land into the already overcrowded cities worldwide. As further evidence for the excessive use of chemical subsides in the United States of America, Costanza and Greer (1998) report that pesticide and nitrogen use in the Chesapeake Bay watershed (which includes parts of eight states) has increased faster than the human population (4× for pesticides, 2× for nitrogen).

Most people are not aware that there are more domestic animals (in population equivalents) than people on Earth. Population equivalents as used here are based on food calorie consumption. An average beef cow consumes 3 times as many calories as an average person, thus the cow is equivalent to 3 people based on the amount of food calorie consumption. It takes about 73 short-lived broiler chickens to equal one person in food (energy) consumption. These animals produce a large amount of manure that can be either a resource or a pollutant. Domestic animals in human population equivalents increased faster than the human population in Georgia during the 50-year period resulting in an increase in the

FIGURE 3. Relationship of subsidy (fertilizer) input to crop yield in the State of Georgia (USA) between 1935 and 1984.

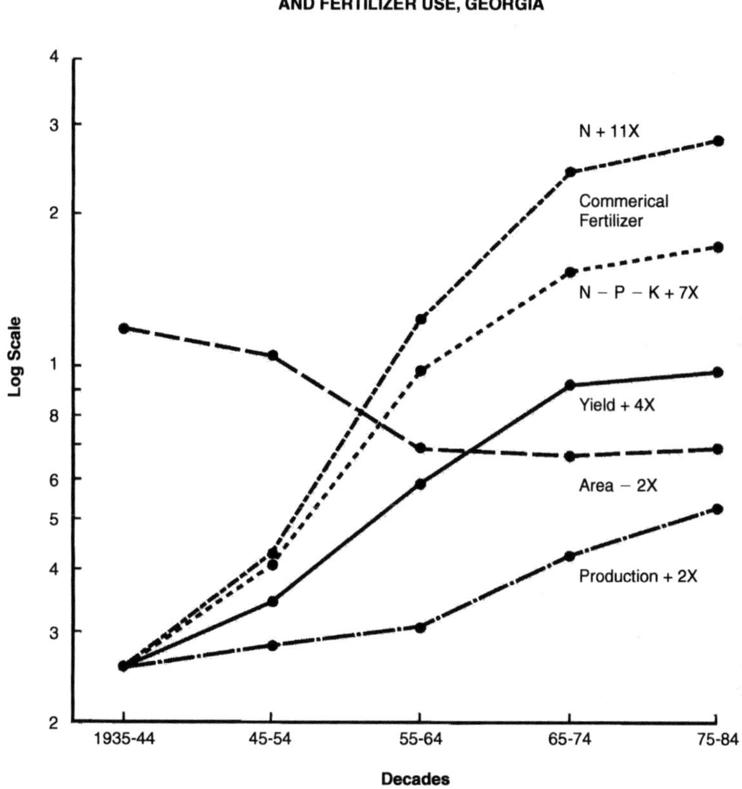

ratio of animals to people from 2.2 to 3.5 (Figure 4). When chickens are raised in "factories" and cows in feed lots, then manure becomes a pollutant. Free-range poultry, for example, represents an ecologically and economically healthy produce which addresses this potential problem (Thear, 1990).

Somewhat surprisingly Georgia produced one-fourth less vegetables in 1980 (in Calorie equivalents) than in 1930. This change represents a significant decrease in cropping (biotic) diversity and an increase cost of transportation necessary to move these crops to market. Today, in most states vegetables in the supermarket come from either California or Florida or, increasingly, from Mexico, Central and South America.

FIGURE 4. Relationship of human population growth to increase in domestic animal population increase in the State of Georgia (USA) between 1935 and 1984.

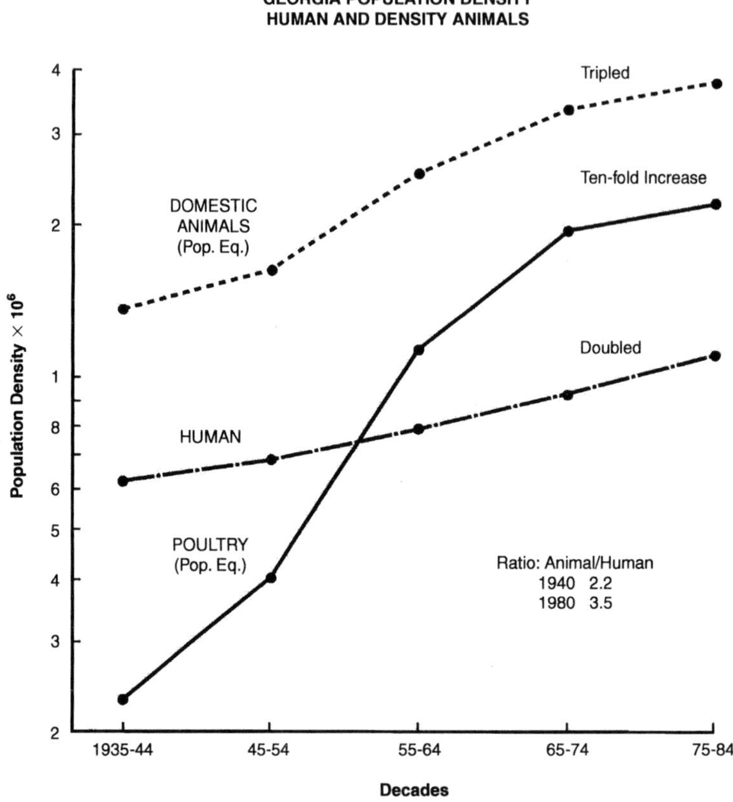

Redesign of an urban vegetable and fruit agriculture including specialty crops such as flowers may be one way to save the family farm (see Barrett and Peles, 1999 and Flora, 2001 regarding the redesign and integration of the agro-urban landscape). A recent study by Reganold, Glover, and Andrews (2001) revealed that organic apple production as a reduced-input system ranked higher than conventional systems regarding quality (sweeter and less tart apples), higher productivity, and greater energy efficiency.

The development of low-input sustainable agriculture includes (1) reduction in energy and material (chemical) subsidies, (2) increasing crop and landscape diversity, and (3) restoring the microbial recycle loop (Figure 2C). This stage of development also results in reduced soil erosion and habitat loss, less eutrophication and nutrient loss to steams, and an increase in food quality. Experience with these reduced-input systems suggests that we recommend to leaders in developing countries to skip industrial agriculture as now practiced in most Western countries and move directly to low-input sustainable agriculture. Genetically modified (GM) crops that are pest or herbicide resistant can play a part in this redesign if the potential problems with GM crops as reviewed by Altieri (2000) can be solved. Among crops targeted for genetically-engineered tolerance to herbicides or pesticides, as listed by Alteri (2000), include alfalfa (*Medicago sativa*), corn (*Zea mays*), oats (*Avena sativa*), soybeans (*Glycine max*), and wheat (*Triticum aestivum*). For meat production, there is a need to reduce or eliminate feed lot systems, to design sustainable pastures and forage production, and to increase species diversity of meat producers (e.g., bison, emus).

At our University of Georgia HorseShoe Bend Experimental Site (HSB), we have been researching reduced-input agroecosystems for more than 20 years with special emphasis on experimental comparison of no-tillage (NT) and conventional tillage (CT) systems. Plots exist plots that have not been plowed for 20 years (the NT plots) contiguous with plots that have been plowed twice each year (the CT plots). A study of comparing plots with and without mineral nitrate fertilizer revealed that the legume-fixed nitrogen was adequate for all crops is both systems (Groffman, Hendrix, and Crossley, 1987). Legume-fixed nitrogen is slower to become available to crops but is less prone to run-off. A minimum amount of other mineral fertilizer was applied, enough for plant needs but not enough for excess run-off. Only one herbicide treatment each year was applied just before planting of the summer crop in both systems and no pesticides have been needed over the years.

Faculty, post-doctorates, and graduate students from the Department of Soil Science (Agronomy) and the Institute of Ecology formed teams to compare crop yields, soil qualities, soil organisms, insects and weeds in the two systems. More than 100 journal papers have been published to date (see Hendrix et al., 2001 for a history of the agroecosystem program at HSB).

Over the years crop yields have not been statistically different in the two systems. A large part of organic matter in the top-soil of NT is in particulate form (POM) and in dissolved form (DOM) in CT. As a result, rather different detrital food chains develop-fungal based in NT and bacterial based in CT (Hendrix et al., 1986). However, in the deeper part of the total soil profile in NT the bacterial-based detritus chain persists, so both microbial recycle systems operate when plowing is reduced or eliminated (Beare et al., 1992). Microarthropods and earthworms, which improve the texture and porosity of soil, are much more abundant under NT cultivation (Blumberg and Crossley, 1983; House and Stinner, 1983; House and Parmelee, 1985; Crossley, Coleman, and Hendrix, 1989). An abundance of mulch (POM) the year around has been shown to reduce soil erosion even during hard rains and heavy sprinkler irrigation (Touchton et al., 1982).

Weeds are more abundant under NT, but have not been a serious problem, and they can even be an asset! Winter weeds are mostly small annuals that become part of clover-rye mulch. The spring herbicide treatment allows the summer crop to overtop weeds which as an understory harbor spiders, ground beetles and other predators which help keep insect herbivores under control (Crossley, Coleman, and Hendrix, 1989). A study comparing agricultural weeds with weedy growth in fallow vegetation (old-field succession) revealed that almost no species of noxious agricultural weeds persists in natural vegetation that follows abandonment of cropping (Odum, Park, and Hutcheson, 1994). Thus having fallow buffer strips or following a crop-fallow rotation does not contribute to the seedbank buildup of agricultural weeds.

In addition to these analytical studies, biocides to eliminate key components such as fungi or microarthropods were employed in order to clarify their role in these experimental agroecosystems (Parmelee et al., 1990). These experiments did not always work out as well as we had hoped because biocides often affect non-target organisms, or have unexpected side-effects (as do many medicines regarding human health). In summary, plowing interferes with many of the natural processes that increase soil quality.

Looking beyond these small-scale studies to the whole agricultural landscape, it is likely that increasing the scale of cropping to include a variety of soil types and climates will require more nitrogen, water, and pesticides (or biological control) than is required in small-scale research plots. However, the potential increases in subsidies must be beneficial to the landscape (i.e., based on the subsidy-stress gradient model; Odum, Finn, and Franz, 1979), rather than deleterious to the health of the system as a whole. In addition to *reducing excess* there is a need to seriously consider *increasing diversity* both within the crop field (monoculture to polyculture) and in the agricultural landscape. It has been shown that establishing buffer strips of fallow or other natural vegetation between crop fields greatly reduces the spread of insect pests (Kemp and Barrett, 1989; Holmes and Barrett, 1997; Thies and Tscharntke, 1999). As stated earlier, in the paper we need to design and establish an economically viable and sustainable vegetable and fruit urban agriculture to feed the rapidly growing cities. Finally, we probably can profit from reviewing the European experience (see, for example, Ryszkowski, French, and Kedziora, 1996) because cropping has been present in Europe for much longer than in the United States of America. With the use of inexpensive plastic greenhouses, vegetables are produced around the year in the Channel Islands in the English Channel and in the South of France. We could do the same in the southern half of the United States of America.

AN AGROLANDSCAPE OR REGIONAL APPROACH AND THE IMPORTANCE OF SCALE

During the past two decades, there has been increased emphasis on the importance of scale (Peterson and Parker, 1998; Schneider, 2001). Agricultural research and planning has been a significant component leading to the recognition of differences between observation on the large and small scale. For example, there has arisen greater emphasis on agrolandscape, rather than just on agroecosystem research and management (Barrett and Peles, 1994; Barrett, Barrett, and Peles, 1999), thus achieving a deeper understanding of the origins and role of agrodiversity both within agroecosystems at the regional and global scales (Smith, 1998; Collins and Qualset 1999; Brookfield, 2001). Such a top-down approach to sustainable agriculture and landscape ecology (Lowrance, Hendrix, and Odum, 1986; Sanderson and Harris, 2000) integrates human communities within the agrolandscape (Butler and Corkner 2001; Schoeneberger, Bentrup, and Francis, 2001).

We recognize that there exists processes that transcend all levels of organization, but also that each level exhibits processes, mechanisms, and principles specific to that level of organization (Barrett, Peles, and Odum, 1997). For example, concepts, principles, mechanisms, and challenges at the cellular, organismal, or population levels of organization (i.e., a "bottom-up" approach) include the genetics and biotechnology of transgenic crops, pesticide resistance in insect pests, and the invasion of exotic weed species. Analogous concepts, principles, mechanisms, and challenges at the community ecosystem, landscape, and global levels (i.e., a "top-down" approach) include the ecology of transgenic crops (Marvier, 2001), ecologically-based pest management (NRC,

2000), and the theory of scale-divergent dynamics (Schneider, 2001), respectively. Thus a "breakthrough" or solution of a problem at one level may depend on the situation at a larger scale. Pesticides, for example, may control certain pests at the field or crop levels, but create serious problems at greater trophic and geographical scales. Likewise, designing a sustainable agricultural field crop may come to naught if the farm system or landscape of which it is a part is unsustainable economically, or if the watershed or aquifer is so deteriorated that it cannot supply the water needed (Lowrance, Hendrix, and Odum, 1986).

In summary, long-term problems in ecology, especially in agroecosystem and agrolandscape ecology, often exist at the temporal/spatial scale of decades and over large areas. For example, alteration of the regional and global nitrogen cycle that results in global eutrophication (Vitousek et al., 1997). It is important to realize that patterns and processes that are measured at small scales do not necessarily hold at larger scales. For example, insect pest control at the field or single crop scale does not insure insect pest management at the landscape scale due to the interchanges of insects between agricultural and surrounding landscapes (Ekbom, Irwin, and Robert, 2000).

Biotic diversity, which is as important in agricultural as in natural systems, cannot be conserved by a single species approach (Salwasser, 1991). Neither can agricultural productivity be sustained by a single field (agroecosystem) approach (Barrett, 2000). Rather an agrolandscape approach is needed in which landscape elements (e.g., patches, corridors, and matrix) are patterned to optimize for a set of objectives related to integrative pest management, nutrient restoration and sequestration, optimum habitat fragmentation, increased biotic diversity, and patch/crop connectivity. Thus, the emerging field of agrolandscape ecology must continue to evolve if the marriage between ecology and agriculture is to become a reality (Jackson and Piper, 1989; Schoeneberger, Bentrup, and Francis, 2001). The emergence of this new integrative field of study will need to be based on an understanding of (a) natural ecosystems as models of design for sustainable agriculture, (b) the concept of scale (both temporal and spatial), and (c) the need for a new transdisciplinary integrative science to address problems and opportunities at these greater scales (Barrett, 2001). The concept of scale-divergent dynamics applies to a wide array of phenomena and processes, and has the potential to bring some theoretical unity to the marriage of basic and applied ecology, including the marriage of ecology and agriculture (Levin, 1992; Schneider, 2001).

CONCLUDING REMARKS

As the human population reaches 9-10 billion in the mid twenty-first century, which probably exceeds carrying capacity, and will be followed by a downsizing (Barrett and Odum, 2000; Lutz, Sanderson, and Scherbov, 2001), it is imperative that future generations understand Goodland's (1995) definition of sustainability as "maintaining natural capital" and "maintenance of resources." This understanding by the citizenry and by political leaders (i.e., ecological literacy) will insure that not only agricultural systems are based on an understanding of ecological systems, especially the natural processes that improve soil quality, but that the patterning of these systems at the landscape scale is understood and managed as well.

REFERENCES

Altieri, M. A. (2000). The ecological impact of transgenic crops on agroecosystem health. *Ecosystem Health* 6:13-23.

Barrett, G. W. (2000). The impact of corridors on arthropod populations within simulated landscapes. In *Interchanges of insects between agricultural and surrounding landscapes*, eds. B. Ekbom, M. E. Irwin, and Y. Robert, Dordrecht, Netherlands: Kluwer Academic Publishing, pp. 71-84.

Barrett, G. W. (2001). Closing the ecological cycle: The emergence of integrative science. *Ecosystem Health* 7:79-84.

Barrett, G. W., T. L. Barrett, and J. D. Peles. (1999). Managing agroecosystems as agrolandscapes: Reconnecting agricultural and urban landscapes. In *Biodiversity in agroecosystems*, eds. W. W. Collins and C. O. Qualset, Boca Raton, Florida, USA: CRC Press, pp. 197-213.

Barrett, G. W. and E. P. Odum. (2000). The twenty-first century: The world at carrying capacity. *BioScience* 50:363-368.

Barrett, G. W. and J. D. Peles. (1994). Optimizing habitat fragmentation: An agrolandscape perspective. *Landscape and Urban Planning* 28:99-105.

Barrett, G. W., J. D. Peles, and E. P. Odum. (1997). Transcending processes and the levels-of-organization concept. *BioScience* 47:531-535.

Barrett, G. W. and L. E. Skelton. (2002). Agrolandscape ecology in the 21st century. In *Landscape ecology in agroecosystems management*, ed. L. Ryszkowski, Boca Raton, Florida, USA: CRC Press, pp. 331-339.

Beare, M. H., R. W. Parmelee, P. F. Hendrix, W. Cheng, D. C. Coleman, and D. A. Crossley, Jr. (1992). Microbial and faunal interactions and effects on litter nitrogen and decomposition in agroecosystems. *Ecological Monographs* 62:569-591.

Blumberg, A. Y. and D. A. Crossley, Jr. (1983). Comparison of soil surface arthropod populations in conventional tillage, no-tillage and old field systems. *Agro-Ecosystems* 8:247-253.

Brookfield, H. (2001). *Exploring agrodiversity*, New York, New York, USA: Columbia University Press.

Butler, L. M. and R. Corkner. (2001). Bridges to sustainability: Links between agriculture, community and ecosystems. In *Interactions between agroecosystems and rural communities*, ed. C. B. Flora, Boca Raton, Florida, USA: CRC Press, pp. 157-173.

Costanza, R. and J. Greer. (1998). The Chesapeake Bay and its watershed: A model for sustainable ecosystem management. In *Ecosystem health*, eds. D. Rapport, R. Costanza, P. R. Epstein, C. Caubet, and R. Levins, London, UK: Blackwell Science, pp. 261-302.

Coleman, D. C. and D. A. Crossley, Jr. (1996). *Fundamentals of soil ecology*, San Diego, California, USA: Academic Press.

Collins, W. W. and C. O. Qualset. (1999). *Biodiversity in agroecosystems*, Boca Raton, Florida, USA: CRC Press.

Crossley, D. A., Jr., D. C. Coleman, and P. F. Hendrix. (1989). The importance of the fauna in agricultural soils: Research approaches and perspectives. *Agriculture, Ecosystems and Environment* 27:47-55.

Ekbom, B., M. E. Irwin, and Y. Robert. (2000). *Interchanges of insects between agricultural and surrounding landscapes*, Dordrecht, Netherlands: Kluwer Academic Publishers.

Flora, C. B. (2001). Shifting agroecosystems and communities. In *Interactions between agroecosystems and rural communities*, ed. C. Flora, Boca Raton, Florida, USA: CRC Press. pp. 5-13.

Goodland, R. (1995). The concept of environmental sustainability. *Annual Review of Ecology and Systematics* 26:1-24.

Groffman, P. M., P. F. Hendrix, and D. A. Crossley, Jr. (1987). Nitrogen dynamics in conventional and no-tillage agroecosystems with inorganic and legume nitrogen inputs. *Plants and Soil* 97:315-332.

Hendrix, P. F., E. P. Odum, D. A. Crossley, Jr., and D. C. Coleman. (2001). Horseshoe bend research: Old-field studies (1966-1975); agroecosystem studies (1976-2000). In *Holistic science: The evolution of the Georgia Institute of Ecology (1940-2000)*, eds. G. W. Barrett and T. L. Barrett, New York, New York, USA: Taylor & Francis. pp. 164-177.

Hendrix P. F., R. W. Parmelee, D. A. Crossley, Jr., D. C. Coleman, E. P. Odum, and P. M. Groffman. (1986). Detritus food webs in conventional and no-tillage agroecosystems. *BioScience* 36:374-380.

Holmes D. M. and G. W. Barrett. (1997). Japanese beetle (*Popillia japonica*) dispersal behavior in intercropped versus. monoculture soybean agroecosystems. *American Midland Naturalist* 137:312-319.

House, G. J. and B. R. Stinner. (1983). Arthropods in no-tillage soybean agroecosystems: Community composition and ecosystem interactions. *Environmental Management* 7:23-28.

House, G. J. and R. W. Parmelee. (1985). Comparison of soil arthropods and earthworms from conventional and no-tillage agroecosystems. *Soil & Tillage Research* 5:351-360.

Jackson, W. and J. Piper. (1989). The necessary marriage between ecology and agriculture. *Ecology* 70:1591-1593.

Kemp, J. C. and G. W. Barrett. (1989). Spatial patterning: Impact of uncultivated corridors on arthropod populations within soybean agroecosystems. *Ecology* 70:114-128.

Levin, S. A. (1992). The problem of pattern and scale in ecology. *Ecology* 73:1943-1967.

Lowrance, R., P. F. Hendrix, and E. P. Odum. (1986). A hierarchical approach to sustainable agriculture. *American Journal of Alternative Agriculture* 1:169-173.

Lutz, W., W. Sanderson, and S. Scherbov. (2001). The end of world population growth. *Nature* 412:543-545.

Marvier, M. (2001). Ecology of transgenic crops. *American Scientist* 89:160-167.

National Research Council (NRC). (1989). *Alternative agriculture*, Washington, DC, USA: National Academy Press.

National Research Council (NRC). (2000). *Professional societies and ecologically based pest management*, Washington, DC, USA: National Academy Press.

Odum, E. P. (1984). Properties of agroecosystems. In *Agricultural ecosystems unifying concepts*, eds. R. Lowrance, B. R. Stinner, and G. J. House, New York, New York, USA: John Wiley & Sons, pp. 5-11.

Odum, E. P. (1987). Reduced-input agriculture reduces nonpoint pollution. *Journal of Soil and Water Conservation* 42:412-414.

Odum, E. P. and M. G. Turner. (1990). The Georgia landscape: a changing resource. In *Changing landscapes: An ecological perspective*, eds. I. S. Zonneveld and R. T. T. Forman, New York, New York, USA: Springer-Verlag, pp. 138-164.

Odum, E. P., J. T. Finn, and E. H. Franz. (1979). Perturbation theory and the subsidy-stress gradient. *BioScience* 29:349-352.

Odum, E. P., T. Y. Park, and K. Hutcheson. (1994). Comparison of the weedy vegetation in old-field and crop fields on the same site reveals that fallowing crop fields does not result in seedbank buildup of agricultural weeds. *Agriculture, Ecosystems and Environment* 49:247-252.

Parmelee, R. W., M. H. Beare, W. Cheng, P. F. Hendrix, S. J. Rider, D. A. Crossley, Jr., and D. C. Coleman. (1990). Earthworm and enchytraeids in conventional and no-tillage agroecosystems: a biocide approach to assess their role in organic matter breakdown. *Biology of Fertilized Soils* 10:1-10.

Peterson, D. L. and V. T. Parker. (1998). *Ecological scale: Theory and applications*, New York, NY, USA: Columbia University Press.

Reganold, J. P., J. D. Glover, and P. K. Andrews. (2001). Sustainability of three apple production systems. *Nature* 410:926-930.

Ryszkowski, L., N. R. French, and A. Kedziora. (1996). *Dynamics of an agricultural landscape*, Poznan, Warsaw, Poland.

Salwasser, H. (1991). Roles and approaches of the USDA forest service. In *Landscape linkages and biodiversity*, ed. W. E. Hudson, Washington, DC, USA: Island Press, pp. 54-65.

Sanderson, J. and L. D. Harris. (2000). *Landscape ecology: A top-down approach*, Boca Raton, Florida, USA: Lewis Publishers.

Schneider, D. C. (2001). The rise of the concept of scale in ecology. *BioScience* 51: 545-553.

Schoeneberger, M. M., G. Bentrup, and C. A. Francis. (2001). Ecobelts: Reconnecting agriculture and communities. In *Interactions between agroecosystems and rural communities*, ed. C. B. Flora, Boca Raton, Florida, USA: CRC Press, pp. 239-260.

Smith, B. D. (1998). *The emergence of agriculture*, New York, New York, USA: W. H. Freeman and Company.

Thear, K. (1990). *Free-range poultry*, Ipswick, UK: Farming Press Books.

Thies, C. and T. Tscharntke. (1999). Landscape structure and biological control in agroecosystems. *Science* 285:893-895.

Touchton, J. T., W. A. Gardner, W. L. Hargrove, and R. R. Duncan. (1982). Reseeding crimson clover as a N source for no-tillage grain sorghum production. *Agronomy Journal* 74:283-287.

Vitousek, P. M., J. Aber, R. W. Howarth, G. E. Likens, P. A. Matson, D. W. Schindler, W. H. Schlesinger, and G. D. Tilman. (1997). Human alteration of the global nitrogen cycle: Causes and consequences. In *Issues in ecology, Number 1*, Washington, DC, USA: Ecological Society of America, p. 15.

Integrating Agroecological Processes into Cropping Systems Research

Stephen R. Gliessman

SUMMARY. Agroecology is defined as the application of ecological concepts and principles to the design and management of sustainable agroecosystems. By viewing cropping systems as agroecosystems, an understanding of the value of the emergent qualities of systems can become a guiding element in research design. A framework for applying this approach in cropping systems research in proposed. A protocol for researching the conversion to sustainable agriculture involves three levels of investigation. The first focuses on improving the efficiency of con-

---

Stephen R. Gliessman is Professor, Program in Community and Agroecology, Department of Environmental Studies, University of California, Santa Cruz, CA 95064 (E-mail: gliess@ucsc.edu).

The approach presented in this chapter is the result of a constant exchange of ideas and experiences over the years with a host of colleagues, students, and students who are now colleagues. The colleagues are an eclectic and challenging group of teachers, researchers, extensionists, development workers, farmers, family, and friends concerned about the future of our food and the farming communities from which this food comes. The author is extremely grateful to all of them, and especially recognizes the excellent editorial assistance of Dr. Joji Muramoto. The author is especially grateful to the support that has been provided by the Alfred Heller Endowed Chair in Agroecology. Thanks and appreciation also goes to the editors of this volume, for their foresight and patience.

Some of the material presented in this chapter is adapted from Gliessman 1998, published by CRC/Lewis Publishers, to whom the author is thankful for their support of agroecology.

[Haworth co-indexing entry note]: "Integrating Agroecological Processes into Cropping Systems Research." Gliessman, Stephen R. Co-published simultaneously in *Journal of Crop Improvement* (Food Products Press, an imprint of The Haworth Press, Inc.) Vol. 11, No. 1/2 (#21/22), 2004, pp. 61-80; and: *New Dimensions in Agroecology* (ed: David Clements, and Anil Shrestha) Food Products Press, an imprint of The Haworth Press, Inc., 2004, pp. 61-80. Single or multiple copies of this article are available for a fee from The Haworth Document Delivery Service [1-800-HAWORTH, 9:00 a.m. - 5:00 p.m. (EST). E-mail address: docdelivery@haworthpress.com].

http://www.haworthpress.com/web/JCRIP

Digital Object Identifer: 10.1300/J411v11n01_04

ventional farming inputs and practices in ways that reduce both their amounts and the environmental impacts of their use. The second focuses on substituting conventional inputs and practices with alternatives that meet broader environmental standards, such as certified organic. Since the problems addressed at these two levels continue to present themselves, a third level is proposed whereby the agroecosystem is redesigned so as to operate on the basis of a new set of ecological processes. At this third level, the emergent qualities of the system itself help prevent problems. Such an approach promotes the conversion to sustainability.

KEYWORDS. Agroecology, agroecosystem, conversion, emergent qualities, sustainability, sustainable agriculture

INTRODUCTION

The increases in agricultural yields over the past several decades have not come about without several costs (Altieri, 1995; Gliessman, 1998). With a primary focus on yield and profitability, agriculture has incurred a growing array of ecological, economic, and social problems (Kimbrell, 2002). Agroecology has stepped forward with an approach to food production systems that strive to balance the needs of ecological soundness, economic viability, and social equity. Defined as the application of ecological concepts and principles to the design and management of sustainable agroecosystems (Gliessman, 2001), agroecology is developing a perspective on agricultural cropping systems that establishes an ecological foundation for making the conversion to sustainable food systems (Francis et al., 2002).

Making the transition to sustainable cropping systems will not be a simple process. It is not just the adoption of a new practice or a new technology. There are no technological "silver bullets." The conversion process uses an ecological approach. In this process, cropping systems are perceived as parts of an ecosystem of interacting parts–an agroecosystem. The focus is on redesigning that system to promote the functioning of an entire range of different ecological processes (Gliessman et al., 1996; Gliessman, 1998). As the use of synthetic chemical inputs is reduced or eliminated, recycling is reemphasized, and ecological diversification once again plays a role, agroecosystem structure and function change as well. A range of processes and relationships begin to transform, beginning with aspects of basic soil structure, organic matter content, and diversity and activity of soil biota. Major changes begin to occur in the activity of and relationships among weed, insect, and pathogen populations, and in the functioning of natural control mechanisms. Ultimately, nutrient dynamics and cycling, energy use efficiency, and overall agroecosystem productivity are affected. Cropping systems are designed and managed with a view towards long-term sustainability.

When cropping system design is developed by agronomists, the focus is most often on ensuring maximum yields and return on investments. When cropping system design is developed by ecologists, the focus is most often on understanding the relationships between crops and environmental conditions, or the ecological impacts of cropping systems on the ecological landscape. But when a broader agroecological perspective is used to design and manage cropping systems, a focus is placed on developing complex ecological systems (agroecosystems) of complementary ecological processes and structures that integrate crop and non-crop organisms with their physical environment to promote the qualities of productivity, diversity, autonomy, and sustainability. The objective of this contribution is to review ways of using an agroecological perspective to bring about measurable changes.

AN AGROECOLOGICAL APPROACH

There has been a recent emergence in research activity on the ecology of cropping systems (Gliessman, 1990; Jackson, 1997; Tivy, 1990; Vandermeer, 1995). After a long history of separation and lack of interaction, ecologists and agronomists have begun to combine forces in order to study and help solve the problems confronting our food production systems. Out of this the field of agroecology has taken form.

Returning to the definition of agroecology given above, as the application of ecological concepts and principles to the design and management of sustainable agroecosystems, we can see the levels at which the science of ecology can be applied to the study of cropping systems (Figure 1). This definition has four important components that will be discussed below: (1) the ecological concepts and principles that serve as the foundation of agroecology; (2) designing and managing cropping systems with these concepts and principles in mind; (3) using a whole-systems, integrated approach that examines cropping systems as com-

FIGURE 1. A schematic diagram in which agroecology is represented as the multiple level analysis of cropping system design and management, with an ultimate goal of agroecosystem sustainability. The individual crop organism responds to the complex of environmental factors at the autecological level. At the higher synecological level, populations of crop species form the crop fields that together form the cropping community, or farm. As we consider abiotic and biotic ecosystem processes, we view the farm in the context of the larger agroecosystem. Ultimately, we can view the entire landscape in which farming communities are located as the interaction between agricultural and natural ecosystems. Our examination of the emergent qualities of systems allows us to determine long-term sustainability.

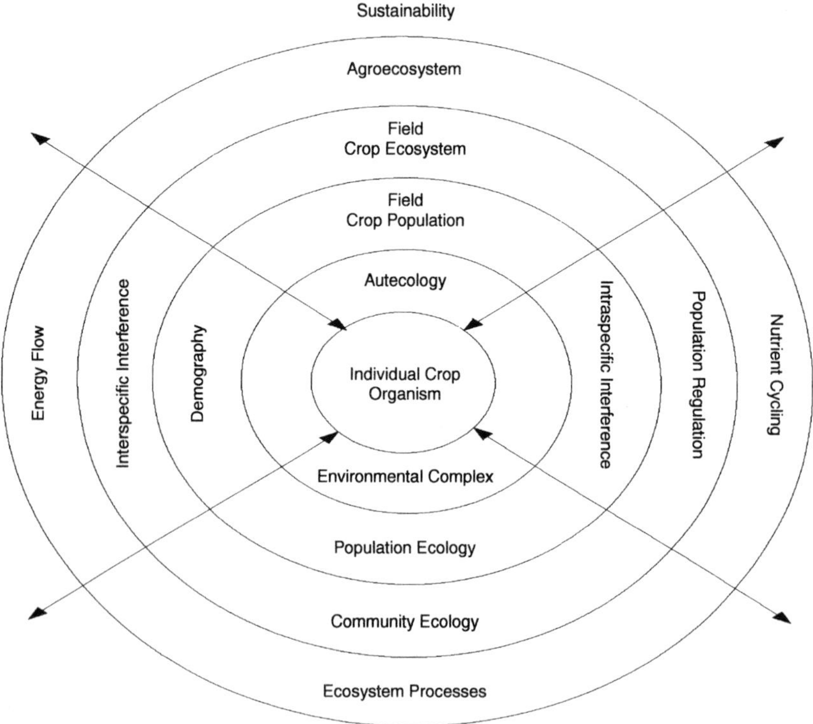

ponents of agroecosystems; and (4) the concept of sustainability where our focus is on the long term.

Ecological Concepts and Principles

Even in very simplified cropping systems, there are complex relationships among crop plants, noncrop plants, animals, and soil microorganisms, as well as between each of these kinds of organisms and their physical environment. In order to understand these relationships and interactions in their full complexity, it is useful to study cropping systems from a more specific perspective–that of the individual crop organism in relation to the factors in its immediate environment. Such a focus is known as autecology or physiological crop ecology.

Autecological study of cropping systems begins by breaking down the environment into individual factors and exploring how each factor affects the crop organism. It is important to understand how each factor, ranging from light, to temperature, to moisture, to soil, to fire, to the atmosphere and wind, and to other organisms, varies in time and space, as well as operates as ecological factors. Such an approach is essential for learning how to accommodate crops to each factor, as well as take advantage of the factor to improve the performance of the cropping system. It is also important for understanding crop distribution and where particular crops might grow and produce the best. It is important to understand the ranges of tolerance a crop organism has for each factor of the environment, as well as the optimum level or intensity that leads to the best crop response. Obviously, human selection for traits that help a crop organism better adapt to the crop environment can translate into yield increases and/or yield reliability.

Crop physiologists know, though, that although it is important to understand the impact that each factor has by itself, rarely does any one factor operate alone or in an unchanging manner on the crop organism. Moreover, all of the factors of the environment also interact with and affect each other. Therefore, the crop environment needs to be understood as a dynamic, constantly changing composite of all of the factors in interaction–otherwise known as the environmental complex. When the environment is examined as a complex of interacting factors, it is possible to observe characteristics that emerge only as a result of these interactions. Characteristics such as complexity, heterogeneity, and dynamic change of factors in the environment represent the final step in the autecological analysis of a cropping system, and sets the stage for the synecological study of the relationships of crop organisms and populations to biotic factors in the environmental complex.

Synecology begins by examining crop organisms as part of the complex of interactions going on between the living organisms in the environment. In crop ecology, the focus most often begins with the study of populations of individuals of the same species or variety. At this level, the primary concern in a cropping system is with rates of growth, development, and carrying capacity of the environment. Innumerable field tests have been conducted in agronomic research to develop the best formula for planting density and arrangement for single, monoculture crops where yield is the primary goal. The rest of the factors of the environment are "controlled" with a broad range of inputs and practices in order to increase the predictability of the cropping system output. Complex biological interactions between the crop species and other organisms are usually kept at a minimum.

But when the cropping system is viewed as an interactive community of populations of different crop as well as noncrop organisms, or from a community ecology perspective, the benefits of complexity begin to emerge. In ecological terms, a cropping system can be seen as a community formed by the diversity of interacting populations of crops, weeds, insects, and microorganisms. The interactions that go on between the populations of the crop system give the community characteristics, called emergent qualities, that exist only at the community level. To understand these qualities fully, an examination must be done beyond the level of the individual organism or populations of one species. Characteristics emerge such as system stability, productivity, and dynamic functioning. Interference between organisms takes the form of such interactions as competition, allelopathy, and synergistic mutualisms.

But the key difference of an agroecological approach goes beyond a primary focus on the negative types of interaction (i.e., those that "pest" organisms usually cause, leading to yield declines and the increased need for human-directed inputs and practices in order to bring about their "control"). Rather than wait for the problem interaction to present itself, a community ecology approach would try to learn how to create beneficial interactions and emergent qualities that not only reduce the need for external inputs, but also increase overall yields. A cropping system community, for example, that combines several crop populations, and as a result is more attractive to natural control agents of particular crop pests, permits the community level of interaction to effectively function. Ecologically-based cropping system management begins to invoke methods that increase spatial, temporal, and community diversity, and to require a diversity of approaches in order to minimize the impacts of negative interactions and maximize the impacts of the positive. The traditional corn (*Zea mays* L.)/bean (*Phaseolus vulgaris* L.)/ squash (*Cucurbita* spp.) intercrop of Mesoamerica is a well-known case (Amador and Gliessman, 1990). Many contributions to this volume also show the value of work at this ecological level of the cropping system.

Finally, with an awareness that the cropping system combines both the living organisms and the nonliving physical environment in which the organisms live, the cropping system must then be examined as an ecosystem. There is a complexity that characterizes the whole-system, and it becomes the foundation for designing cropping systems. Interactions between all parts, living and nonliving, become a function of the diversity of the system, and ultimately, of its sustainability. This diversity takes many forms, from the number of species present to the structural and functional diversity of the different parts of the system. Ecosystem diversity comes about as a result of the ways the different living and nonliving components of the system are organized and interact. Diversity can also take the form of the complex of biogeochemical cycles and the variety of living organisms present that make possible the organization and interactions going on within the system. Each organism fills a different ecological niche in the cropping system, occupying a specific habitat as well as carrying out specific ecological functions. Altogether, the connections begin to be made between ecological diversity, stability, and sustainability in terms of developing an agroecological framework for cropping system design and management.

Ecological Design and Management

A well-designed and managed cropping system is relatively stable, possesses many elements of autonomy, and is able to maintain productivity using mostly renewable energy inputs, especially from solar radiation. By examining a cropping system from an ecosystem perspective, we have a basis for looking beyond a primary focus on system outputs (yield or harvest). We can instead look at the complex set of biological, physical, and chemical interactions that determine the processes that permit us to achieve and sustain those yields. Hence, agroecology becomes much more process oriented. An understanding of the key ecological differences between natural ecosystems and agroecosystems can also be an important step in designing a more sustainable cropping system. Some of these differences are listed in Table 1. Our ability to incorporate ecosystem characteristics into the design and management of sustainable cropping systems can only come about when an ecological perspective is used in cropping system design and management.

Recently there have been some very useful examples of how to use ecological understanding in the design and management process. For example, intercropping and multiple cropping design have received considerable attention (Francis, 1987; Vandermeer, 1989). Populations of different species are combined to form crop communities where emergent qualities are expressed that are not manifest by each popula-

TABLE 1. Key ecological differences between natural ecosystems, conventional agroecosystems, and agroecosystems (adapted from Odum, 1984; Altieri, 1995; Gliessman, 1998).

	Natural Ecosystem	Agroecosystem	
		Sustainable	Conventional
Net productivity	Medium	Medium	High
Trophic interactions	Complex	Intermediate	Simple, linear
Species diversity	High	High	Low
Structural diversity	High	Moderate	Low
Genetic diversity	High	High	Low
Nutrient cycles	Closed	Semi-closed	Open
Stability (resilience)	High	High	Low
Human control	Independent	Interdependent	Dependent
Temporal permanence	Long	Long	Short
Habitat heterogeneity	Complex	Intermediate	Simple

tion alone. Returning to the example of the intercrop of corn, beans and squash (Amador and Gliessman, 1990; Gliessman, 1998), studies have shown that overyielding (higher combined yields in the mixtures as compared to equivalent areas planted to the monocrops) come about as a result of community level interactions (Table 2).

Altieri (1994) and Andow (1991) have reviewed the literature demonstrating the importance of system level interactions for developing alternative pest management strategies. Examples include intercropping, border plantings of attractive non-crop species, cover-cropping, and including important non-crop plant species in crop fields. Rosemeyer (2001) has reviewed the many mechanisms whereby plant-based organic mulches can improve the sustainability of cropping systems. A range of interactions, mechanisms of habitat modification, and management of system processes are impacted by the presence of these mulches. Soil erosion control, improved internal cycling of nutrients, improved efficiency of nutrient uptake, better moisture retention, promotion of root symbioses, weed and disease suppression, and enhanced biological control are some of the many benefits gained by the cropping system.

In the specific case of weeds, an extensive review of an ecologically-based approach to weed management demonstrates the potential for using the multiple factors that are known to influence weed popula-

TABLE 2. Community level interactions in a corn/bean/squash intercrop that contribute to overyielding in the crop mixture (adapted from Amador and Gliessman, 1990; Gliessman, 1998).

Interaction	Factors affected	Source
Mutualism	Higher biological nitrogen fixation	Boucher and Espinosa, 1982
Mutualism	Mycorrhizal bridge between crops allows for nutrient availability	Bethlenfalvay et al., 1991
Nutrient accumulation	Net gains in soil nitrogen in intercrop system	Gliessman, 1982
Competition	Shading by squash leaves restricts weed growth	Fujiyoshi, 1997
Allelopathy	Leaf leachates from squash inhibit weeds	Gliessman, 1983
Habitat modification	Herbivores reduced due to less concentration of resources in crop mixture	Risch, 1980
Biological control	More beneficial insects due to improved microclimate and better food resources	Letourneau, 1986

tion dynamics in cropping systems, rather than rely solely on the conventional use of herbicides (Liebman and Gallandt, 1997). This approach uses diverse types of information and a variety of control tactics that subject weeds to multiple, temporally variable stresses which in the long term reduce their presence and associated negative impacts on crops. There is particular value in understanding spatial and temporal patterns of weed abundance. How weeds actually reduce crop yields, weed life histories, weed niche characteristics, the mechanisms of possible resource competition, and possible emergent qualities of crop systems that are impacted by weed species is valuable information. Such information is essential for the successful design and management of crop/weed interactions that reduce the negative impacts while enhancing the positive impacts of the presence of weeds. These approaches range from the specific alteration of enhancement of ecological factors in the cropping system (e.g., soil moisture or nutrients), to crop rotations and cover cropping, to specific cultivation or tillage techniques and timing, to intercropping, to the introduction and enhancement of biological control mechanisms. But the most important message from this review is the need for the reliance on a combination of different management strategies, rather than a single control tactic. When these strategies incorporate the multiple levels of ecological knowledge discussed above, weed management shifts to one of many indirect controls and many possible interactions that can lead to successful management. This same approach applies to the management of other non-crop organisms that can be problems in cropping systems, including fungal and bacterial diseases, nematodes, insects, and even mammalian herbivores.

Cropping Systems as Agroecosystems

An agroecosystem is created when human manipulation and alteration of an ecosystem take place for the purpose of establishing agricultural production. This introduces several changes in the structure and function of the natural ecosystem, and as a result, changes in key system level qualities (Odum, 1984). These qualities are often referred to as the emergent qualities of systems–qualities that manifest themselves once all of the component parts of the system are organized. These same qualities can also serve as indicators of agroecosystem sustainability (Gliessman, 2001). Some of the key emergent qualities of ecosystems, and how they are altered in a cropping system context, are as follows.

Energy Flow: Energy flows through a natural ecosystem as a result of complex sets of trophic interactions, certain amounts being dissipated at different stages along the food chain, with the greatest amount of energy within the system ultimately moving along the detritus pathway (Odum, 1971). Annual production of the system can be calculated in terms of net primary productivity or biomass, each component with its corresponding energy content. Energy flow in agroecosystems is altered greatly by human interference (Pimentel and Pimentel, 1997). Although solar radiation is still a major source of energy, many inputs are derived from human sources and are often not self-sustaining. Cropping systems too often become through-flow systems, in which considerable energy is directed out of the system at the time of each harvest. Biomass is not allowed to otherwise accumulate within the system or contribute to driving important internal ecosystem processes (e.g., organic detritus returned to the soil serving as an energy source for microorganisms that are essential for efficient nutrient cycling). For sustainability to be attained, renewable sources of energy must be maximized, and energy supplied to fuel the essential internal trophic interactions needed to maintain other ecosystem functions.

Nutrient Cycling: Small amounts of nutrients continually enter an ecosystem through several hydrogeochemical processes. Through complex sets of interconnected cycles, these nutrients then circulate within the ecosystem, where they are most often bound in organic matter (Muramoto et al., 2001). Biological components of each system become very important in determining how efficiently nutrients move, ensuring that minimal amounts are lost from the system. In a mature ecosystem, these small losses are replaced by local inputs, maintaining a nutrient balance. Biomass productivity is linked very closely to the annual rates at which nutrients are able to be recycled. In a cropping system, recycling of nutrients is minimal and considerable quantities are lost from the system with the harvest or as a result of leaching or erosion due to a great reduction in permanent biomass levels held within the system (Tivy, 1990). The frequent exposure of bare soil between crop plants during the season, or from open fields between cropping seasons, creates "leaks" of nutrients from the system. Modern agriculture has come to rely heavily upon nutrient inputs derived from, or obtained with, petroleum-based sources to replace these losses. Sustainability requires that these "leaks" be reduced to a minimum and recycling mechanisms be reintroduced and strengthened. Ultimately, human societies need to find ways to return nutrients consumed in agricultural products back to where they came from–the cropping systems that produced them in the first place.

Population Regulating Mechanisms: Through a complex combination of biotic interactions and limits set by the availability of physical resources, population levels of the various organisms are controlled, and thus link eventually to the productivity of the ecosystem. Selection through time tends towards the establishment of the most complex structure biologically possible within the limits set by the environment, permitting the establishment of diverse trophic interactions and niche diversification. Due to human-directed genetic selection and domestication, as well as the overall simplification of cropping systems (i.e., the loss of niche diversity and a reduction in trophic interactions), populations of crop plants or animals are rarely self-reproducing or self-regulating. Human inputs in the form of seed or control agents, often dependent on large energy subsidies, determine population sizes. Biological diversity is reduced, natural pest control systems are disrupted, and many niches or microhabitats are left unoccupied. The danger of catastrophic pest or disease outbreak is high, often even despite the availability of intensive human interference and inputs. A focus on sustainability requires the reintroduction of the diverse structures and species relationships that permit the functioning of natural control and regulation mechanisms (Altieri, 1994). Mass selection and local adaptation processes must continue to provide the foundation for agrobiodiversity, rather than the accelerated erosion of such diversity that comes from centralized crop breeding programs or the genetic modification of crops through the insertion of single traits focused on specific responses. We must learn to work with diversity, rather than focus on cropping system simplification

Dynamic Equilibrium: The species richness or diversity of mature ecosystems permits a degree of resistance to all but very damaging perturbations. In many cases, periodic disturbances ensure the highest diversity, and even highest productivity (Connell, 1978). System stability is not a steady state, but rather a dynamic and highly fluctuating one that permits ecosystem recovery following disturbance. This permits the establishment of an ecological equilibrium that functions on the basis of sustained resource use which the ecosystem can maintain indefinitely or shift if the environment changes. At the same time, rarely do we witness what might be considered large-scale disease outbreaks in healthy, balanced ecosystems. But due to the reduction of natural structural and functional diversity, much of the resilience of the system is lost, and constant human-derived external inputs must be maintained. The overemphasis on maximizing harvest outputs that has come about as individual farms get bigger and the number of farmers gets smaller, upsets the former ecological equilibrium. This imbalance can only be maintained if human interference continues to intensify. To reintegrate sustainability, the emergent qualities of system resistance and resiliency must once again play a determining role in cropping system design and management. Agroecosystem sustainability requires that economic efficiency be weighed against the benefits of ecological efficiency and the accompanying quality of life for the human communities living and working in the agricultural landscape.

Moving Towards Sustainability

The final component in the ecological approach to cropping system design and management is sustainability. By this we accept the requirement of examining cropping systems for both their ability to produce acceptable yields as well as to maintain the ability to produce these yields for the long term in ways that are ecologically sound, economically viable, and socially equitable (Gliessman, 1998). Ultimately, to achieve the goal of sustainability, we need to be able to analyze both the immediate and future impacts of cropping system design and management so that we can identify the key points in each system on which to focus the search for alternatives or solutions to problems. We must learn to be more competent in our agroecological analysis in order to avoid problems or negative changes before they occur, rather than struggle to reverse the problems after they have been created. Once an ecosystem is disturbed for the purpose of converting it to agricultural production, the original equilibrium and resilience is altered and replaced by something that reflects a combination of ecological and socio-economic constraints and factors. The challenge for agroecology is to find a research approach that maximizes the reliance on natural ecosystem processes and minimizes the dependence on external interference with these processes. Ecological sustainability is built on this premise, and the concepts of ecosystem resilience and stability are measurable with the tools and methods developed for the study of ecosystem structure and function. The agroecological approach provides us with this alternative (Gliessman, 2001).

MAKING THE CONVERSION

Farmers have always had a reputation for being innovators and experimenters, willingly adopting new practices when they perceive that some benefit will be gained. This has been especially true in alternative and organic agriculture, where over the past 20 years creative farmers have made bold moves into a manner of farming that challenges conventional wisdom on how agriculture should be done (National Research Council, 1989). The kind of agricultural products consumers are willing to buy have also undergone transformation. The remarkable increases in area devoted to organic agriculture during the past decade are good examples (USDA, 2000).

As this transition occurs, we are constantly faced with the question of how sustainable these new cropping systems really are. When we examine farming systems as ecological systems, and use the science of agroecology for their design and management, we begin to realize that farmers and researchers must work together very closely to ensure that these new cropping systems are not just trading one set of problems for others. Agroecology offers a set of guiding principles for making sure that sustainability is part of our framework while we make the conversion to alternative production. We should not be satisfied with an approach that merely substitutes conventional inputs and practices with more environmentally benign alternatives. We should not be satisfied with an approach that is determined primarily by market demands and does not include the economic and social health of the agricultural communities in which food is produced. And we should not be satisfied with an approach that does not ensure food security for all consumers in all parts of the world. A much broader set of tools must be developed to evaluate the conversion process. Agroecology also provides the ecological foundations for such an evaluation.

Principles Guiding the Conversion Process

The conversion process can be complex, requiring changes in field practices, day to-day management of the farming operation, planning, marketing, and even philosophy. The principles listed in Table 3 can serve as general guidelines for navigating the overall transformation. The integration of these principles creates a synergism of interactions and relationships in cropping systems that eventually leads to the development of the properties of sustainable agroecosystems. Emphasis on particular principles will vary, but all of them can contribute greatly to the conversion process.

For many farmers, rapid conversion to alternative farming is neither possible nor practical. The re-establishment of many ecological processes and relationships may take many years. As a result, many conversion efforts proceed in slower steps toward the ultimate goal of sustainability, and meanwhile make the minimal changes necessary. Studies on the conversion process are still very limited (for examples see Swezey et al., 1994, 1999; Hendricks, 1995; Gliessman et al., 1996). They tell us that there is a lot of research that still needs to be done to improve yields and pest management, as well as improve the indicators of sustainability. Current research efforts point out three distinct levels of conversion. These levels help us describe the steps that farmers actually take in converting from conventional cropping systems, and they can serve as a map outlining a step-wise, evolutionary conversion process alternative systems should take in order to achieve sustainability. They are also helpful for categorizing agricultural research as it relates to conversion (Gliessman, 1998).

Level 1: Increase the Efficiency of Conventional Practices in Order to Reduce the Use and Consumption of Costly, Scarce, or Environmentally Damaging Inputs

This approach is what we might call the "pre-alternative." Its goal is to use conventional inputs more efficiently so that fewer inputs will be

TABLE 3. General principles guiding the process for the conversion of cropping systems to ecologically based design and management (Adatped from Gliessman, 1998).

1. Shift from throughflow nutrient management to recycling of nutrients, with increased dependence on natural processes such as biological nitrogen fixation and mycorrhizal relationships.

2. Use renewable sources of energy instead of non-renewable sources.

3. Eliminate the use of non-renewable off-farm human inputs that have the potential to harm the environment or the health of farmers, farm workers, or consumers.

4. When materials must be added to the system, use naturally-occurring materials instead of synthetic, manufactured inputs.

5. Manage pests, diseases, and weeds instead of "controlling" them.

6. Reestablish the biological relationships that can occur naturally in cropping systems instead of reducing and simplifying them.

7. Make more appropriate matches between cropping patterns and the productive potential and physical limitations of the farm landscape.

8. Use a strategy of adapting the biological and genetic potential of agricultural plant and animal species to the ecological conditions of the farm rather than modifying the farm to meet the needs of the crops and animals.

9. Value most highly the overall health of the agroecosystem rather than the outcome of a particular crop system or season.

10. Emphasize conservation of soil, water, energy, and biological resources.

11. Incorporate the idea of long-term sustainability into overall cropping system design and management.

needed and the negative impacts of their use will be reduced as well. This approach has been the primary emphasis of much conventional agricultural research, through which numerous agricultural technologies and practices have been developed. Examples include optimal crop spacing and density, improved machinery, pest monitoring for improved pesticide application, improved timing of operations, and precision farming for optimal fertilizer and water placement. Although these kinds of efforts reduce the negative impacts of conventional agriculture, they do not help break its dependence on external human inputs, and do not qualify for the environmental certification required for many alternative crop production strategies, such as certified organic or certified environmentally friendly.

Level 2: Substitute Conventional Inputs and Practices with Alternative Practices

We might call this approach the "commercial alternative." The goal at this level of conversion is to replace resource-intensive and environment-degrading products and practices with those that are more environmentally benign. Most organic farming research, for example, has emphasized such an approach. Examples of alternative practices include the use of nitrogen-fixing cover crops and rotations to replace synthetic nitrogen fertilizers, the use of biological control agents rather than pesticides, and the shift to reduced or minimal tillage. At this level, the basic cropping system structure is not greatly altered, hence many of the same problems that occur in conventional systems also occur in those with input substitution.

Level 3: Redesign the Cropping System So That It Functions on the Basis of a New Set of Ecological Processes

We might call this level the "sustainable alternative." At this level, overall system design eliminates the root causes of many of the problems that still exist at Levels 1 and 2. Thus rather than finding sounder ways of solving problems, the problems are prevented from arising in the first place. Whole-system conversion studies allow for an understanding of yield-limiting factors in the context of agroecosystem structure and function. Problems are recognized, and thereby prevented, by internal site- and time-specific design and management approaches, instead of by the application of external inputs. An example is the diversification of farm structure and management through the use of rotations, multiple cropping, and agroforestry.

In terms of research, agronomists and other agricultural researchers have done a good job of transitioning from Level 1 to Level 2, but the transition to Level 3 has really only just begun. Agroecology provides the basis for this type of research. And eventually it will help us find answers to larger, more abstract questions, such as what sustainability is and how we will know we have achieved it.

On Farm Conversions

As farmers undertake to convert their cropping systems to alternative management, it becomes important to develop systems for evaluating and documenting the success of these efforts and the changes they engender in the functioning of the farm system. Such evaluation systems will help convince a larger segment of the agricultural community that conversion to sustainable practices is possible and economically feasible.

The study of the process of conversion begins with identifying a study site. This should be a functioning, on-farm, commercial crop production unit whose owner-operator wishes to convert to alternative management and wants to participate in the design and management of the farm system during the conversion process (Swezey et al. 1994; Gliessman et al., 1996). Such a "farmer-first" approach is considered essential in the search for viable farming practices that eventually have the best chance of being adopted by other farmers.

The amount of time needed to complete the conversion process depends greatly on the type of crop or crops being farmed, the local ecological conditions where the farm is located, and the prior history of management and input use. For short-term annual crops, the time frame might be as short as three years, and for perennial crops and animal systems, the time period is probably at least five years or longer.

Study of the conversion process involves several levels of data collection and analysis:

- Examine the changes in ecological factors and processes over time through monitoring and sampling.
- Observe how yields change with changing practices, inputs, designs, and management.
- Understand the changes in energy use, labor, and profitability that accompany the above changes.
- Based on accumulated observations, identify key indicators of sustainability and continue to monitor them well into the future.
- Identify indicators that are "farmer-friendly" and can be adapted to on-farm, farmer-based monitoring programs, but that are linked to our understanding of ecological sustainability.

Each season, research results, site-specific ecological factors, farmer skill and knowledge, and new techniques and practices can all be examined to determine if any modifications in management practices need to be made to overcome any identified yield-limiting factors. Ecological components of the sustainability of the system become identifiable at this time, and eventually can be combined with an analysis of economic and social sustainability as well. The key to developing sustainability is to build a strong ecological foundation under the cropping system, using the ecosystem knowledge of agroecology described above. This foundation then serves as the framework for producing the sustainable harvests needed by humans. In order to maintain sustainable harvests, though, human management is a requirement. Agroecosystems are not self-sustaining, but rely on natural processes for maintenance of their productivity. Cropping system use of natural ecosystem knowledge allows the system to sustain over the long-term human removal of biomass without large subsidies of non-renewable energy and without detrimental effects on the surrounding environment.

FUTURE PERSPECTIVES

The problems we are facing today in agriculture are creating the pressures for the changes that are needed convert to alternative, more sustainable cropping systems. It is one thing to express the need for these changes, and yet another to actually make the changes that are required. Designing and managing sustainable cropping systems, as an approach, is in its formative stages. It builds initially upon the fields of ecology and agricultural science, and emerges as the science of agroecology. This combination can play an important role in taking the steps necessary for the transition to sustainable alternatives.

Conventional cropping systems can no longer be viewed as strictly production systems driven primarily by economic pressures. We need to reestablish an awareness of the strong ecological foundation upon which cropping systems originally developed and ultimately depend. In the broader context of sustainability, we must study the environmental background of the entire agroecosystem in which the cropping systems operate, as well as the complex of processes involved in the maintenance of long-term productivity. We must first establish the ecological basis of sustainability in terms of resource use and conservation, including soil, water, genetic resources, and air quality. Then we must examine the interactions between the many organisms of the agroecosystem, beginning with interactions at the individual species level, and culminating at the ecosystem level as our understanding of the dynamics of the entire system is revealed.

Our understanding of ecosystem level processes should then integrate the even more complex aspects of the social, economic and political systems within which cropping systems function. Such an integration of ecosystem and social system knowledge about agricultural processes will not just lead to a reduction in synthetic inputs used for maintaining productivity. It will also permit the evaluation of such qualities of agroecosystems as the long-term effects of different input/output strategies, the importance of the environmental services provided by agricultural landscapes, and the relationship between economic and ecological components of sustainable agroecosystem management. By properly selecting and understanding the design and management of cropping systems, we can be ensured that will promote a sustainable future.

REFERENCES

Altieri, M.A. (1994). *Biodiversity and Pest Management in Agroecosystems.* New York, NY: Haworth Press.

Altieri, M.A. (1995). *Agroecology: The Scientific Basis of Alternative Agriculture.* 2nd Edition. Boulder, CO: Westview Press.

Amador, M.F. and S.R. Gliessman. (1990). An ecological approach to reducing external inputs through the use of intercropping. In *Agroecology: Researching the Ecological Basis for Sustainable Agriculture*, ed. S.R. Gliessman, New York, NY: Springer-Verlag. pp. 146-159.

Andow, D.A. (1991). Vegetational diversity and arthropod population responses. *Annual Review of Entomology* 36:561-586.

Bethlenfalvay, G.J., M.G. Reyes-Solis, S.B. Camel, and R. Ferrera-Cerrato. (1991). Nutrient transfer between the root zones of soybean and maize plants connected by a common mycorrhizal inoculum. *Physiologia Plantarum* 82:423-432.

Boucher, D.H. and J. Espinosa. (1982). Cropping systems and growth and nodulation responses of beans to nitrogen in Tabasco, Mexico. *Tropical Agriculture* 59:279-282.

Connell, J.H. (1978). Diversity in tropical rain forests and coral reefs. *Science* 199: 1302-1310.

Francis, C.A. (ed.). (1987). *Multiple Cropping.* New York, NY: MacMillan.

Francis, C., G. Lieblein, S. Gliessman, T.A. Breland, N. Creamer, R. Harwood, L. Salomonsson, J. Helenius, D. Rickerl, R. Salvador, M. Wiedenhoeft, S. Simmons, P. Allen, M. Altieri, C. Flora, and R. Poincelot. (2003). Agroecology: The ecology of food systems. *Journal of Sustainable Agriculture* 22(3):99-119.

Fujiyoshi, P. (1997). Ecological aspects of interference by squash in a corn/squash intercropping agroecosystem. PhD thesis in Biology, University of California, Santa Cruz.

Gliessman, S.R. (1982). Nitrogen cycling in several traditional agroecosystems in the humid tropical lowlands of southeastern Mexico. *Plant and Soil* 67:105-117.

Gliessman, S.R. (1983). Allelopathic interactions in crop-weed mixtures: Applications for weed management. *Journal of Chemical Ecology* 9:991-999.

Gliessman, S.R. (ed.) 1990. *Agroecology: Researching the Ecological Basis for Sustainable Agriculture.* New York, NY: Springer-Verlag.

Gliessman, S.R. (1998). *Agroecology: Ecological Processes in Sustainable Agriculture.* Boca Raton, FL: CRC/Lewis Publishers.

Gliessman, S.R. (ed.) (2001). *Agroecosystem Sustainability: Towards Practical Strategies.* Series in Advances in Agroecology. Boca Raton, FL: CRC Press.

Gliessman, S.R., M.R. Werner, S. Swezey. E. Caswell, J. Cochran, and F. Rosado-May. (1996). Conversion to organic strawberry management changes ecological processes. *California Agriculture* 50:24-31.

Hendricks, L.C. (1995). Almond growers reduce pesticide use in Merced County field trials. *California Agriculture* 49:5-10.

Jackson, L.E. (ed). (1997). *Ecology in Agriculture*. San Diego, CA: Academic Press.

Kimbrell, A. (ed.) (2002). *Fatal Harvest: The Tragedy of Industrial Agriculture*. Washington, DC: Island Press.

Letourneau, D.K. (1986). Associational resistance in squash monoculture and polycultures in tropical Mexico. *Environmental Entomology* 15:285-292.

Liebman, M. and E.R. Gallandt. (1997). Many little hammers: ecological approaches for management of crop-weed interactions. In *Ecology in Agriculture*, ed. L.E. Jackson, San Diego, CA: Academic Press. pp. 291-343.

Muramoto, J., E.C. Ellis, Li, Zhengfang, R.M. Machado, and S.R. Gliessman (2001). Field-scale nutrient cycling and sustainability: comparing natural and agricultural ecosystems. In *Agroecosystem Sustainability: Towards Practical Strategies*, ed. S.R. Gliessman, Series in Advances in Agroecology, Boca Raton, FL: CRC Press. pp. 121-134.

National Research Council. (1989). *Alternative Agriculture*. Washington, DC: National Academy Press.

Odum, E.P. (1971). *Fundamentals of Ecology*. Philadelphia, PA: W.B. Saunders.

Odum, E.P. (1984). Properties of agroecosystems. In *Agricultural Ecosystems: Unifying Concepts*, eds. R. Lowrance, B.R. Stinner, and G.J. House, New York, NY: John Wiley & Sons, pp. 5-12.

Pimentel, D. and M. Pimentel. (eds.) (1997). *Food, Energy, and Society*. 2nd Edition, Niwot, CO: University Press of Colorado.

Risch, S. (1980). The population dynamics of several herbivorous beetles in a tropical agroecosystem: the effect of intercropping corn, beans, and squash in Costa Rica. *Journal of Applied Ecology* 17:593-612.

Rosemeyer, M.E. (2001). Improving agroecosystem sustainability using organic (plant-based) mulch. In *Agroecosystem Sustainability: Towards Practical Strategies*, ed. S.R. Gliessman, Series in Advances in Agroecology, Boca Raton, FL: CRC Press. pp. 67-91.

Swezey, S.L., J. Rider, M. Werner, M. Buchanan, J. Allison, and S.R. Gliessman. (1994). Granny smith conversions to organic show early success. *California Agriculture* 48:36-44.

Swezey, S.L., P. Goldman, R. Jergens, and R. Vargas. 1999. Preliminary studies show yield and quality potential of organic cotton. *California Agriculture* 53:9-16.

Tivy, J. (1990). *Agricultural Ecology*. London, UK: Longman Scientific and Technical.

USDA. (2000). U.S. Organic Agriculture. Economic Research Service Issues Center. Washington, DC. www.econ.ag.gov/whatsnew/issues/organic/.

Vandermeer, J. (1989). *The Ecology of Intercropping*. Cambridge, MA: Cambridge University Press.

Vandermeer, J. (1995). The ecological basis of alternative agriculture. *Annual Review of Ecology and Systematics* 26:201-224.

An Agroecological Basis for Designing Diversified Cropping Systems in the Tropics

Miguel A. Altieri
Clara I. Nicholls

SUMMARY. Small scale diversified systems which rely mostly on local resources and complex crop arrangements, are reasonably productive and stable, exhibiting a high return per unit of labor and energy. In many ways complex polycultures and agroforestry systems used by small tropical farmers mimic the structure and function of natural communities therefore acquiring many features typical of such communities, such as tight nutrient cycling, resistance to pest invasion, vertical structure, and high levels of biodiversity.

An agroecological approach to improve tropical small farming systems must ensure that promoted systems and technologies are suited to the specific environmental and socio-economic conditions of small farmers, without increasing risk or dependence on external inputs. Rather, agroecological development projects should incorporate elements of traditional agricultural knowledge and modern agricultural science, featuring resource-conserving yet highly productive systems such as polycultures, agroforestry, and the integration of crops and livestock.

It is ecologically futile to promote mechanized monocultures in areas

---

Miguel A. Altieri is Associate Professor and Clara I. Nicholls is Research Fellow, Division of Insect Biology, University of California, 201 Wellman Hall, Berkeley, CA 94720 (E-mail: agroeco3@nature.berkeley.edu).

[Haworth co-indexing entry note]: "An Agroecological Basis for Designing Diversified Cropping Systems in the Tropics." Altieri, Miguel A., and Clara I. Nicholls. Co-published simultaneously in *Journal of Crop Improvement* (Food Products Press, an imprint of The Haworth Press, Inc.) Vol. 11, No. 1/2 (#21/22), 2004, pp. 81-103; and: *New Dimensions in Agroecology* (ed: David Clements, and Anil Shrestha) Food Products Press, an imprint of The Haworth Press, Inc., 2004, pp. 81-103. Single or multiple copies of this article are available for a fee from The Haworth Document Delivery Service [1-800-HAWORTH, 9:00 a.m. - 5:00 p.m. (EST). E-mail address: docdelivery@haworthpress.com].

http://www.haworthpress.com/web/JCRIP

Digital Object Identifer: 10.1300/J411v11n01_05

of overwhelming biotic intricacy where pests flourish year-round and nutrient leaching is a major constraint. Here, it pays to imitate natural cycles rather than struggle to impose simplistic ecosystems that are not inherently complex. For this reason, many researchers think that successional ecosystems can be particularly appropriate templates for the design of sustainable tropical agroecosystems.

KEYWORDS. Tropical agroecosystems, biodiversity, pest management, ecosystem mimicry

INTRODUCTION

Of all the regions where agriculture is practiced, the tropics are where novel production approaches are most urgently needed. This region has not benefited significantly from modern technologies that led to high agricultural productivity in the temperate regions. Abundant rainfall and high temperatures promote competition from weeds, pest outbreaks and nutrient leaching; constraints that constantly plague the large-scale plantations and annual crop monocultures that cover large areas in the tropics (Beets, 1990).

In many tropical areas, agriculture is highly mechanized and has implied the simplification of the structure of the environment over vast areas, replacing nature's diversity with a small number of cultivated plants and domesticated animals. Genetically, monocultures are shockingly dependent on a handful of crop varieties. Researchers have repeatedly warned about the extreme vulnerability associated with this genetic uniformity, claiming that ecological homogeneity in agriculture is closely linked to pest invasions (Adams, Ellingbae, and Rossineau, 1971; Robinson, 1996). Many scientists argue that the drastic narrowing of cultivated plant diversity has put tropical food production in greater peril. In vain, farmers have tried to overcome these biotic constraints typical of the less seasonal tropics by applying large amounts of chemical fertilizers and pesticides, but this approach has been limited by scarce and expensive fossil fuels, but mostly by ecological backlash in the form of significant environmental and health externalities (Conway, 1997).

On the other hand, small farmers, especially those living in more marginal environments, and who were bypassed by agricultural modernization have not relied on agrochemicals to sustain production. Although estimates vary considerably, about 1.9-3.3 billion rural people in the developing world remain directly untouched by modern agricultural technology. The great majority of these people are peasants, indigenous people and small family farmers, who mostly still farm the valleys and slopes of rural landscapes with traditional and/or subsistence methods. About 370 million of these, are extremely poor people whose livelihoods depend on the vast, diverse, and risk prone marginal environments in the south (Conway, 1997). Most of these people cultivate in small scale diversified systems which rely on local resources and complex crop arrangements. Research has shown that such systems are reasonably productive and stable, exhibiting a high return per unit of labor and energy (Netting, 1993). For example, in Latin America, peasant production units reached about 16 million in the late 1980s occupying close to 60.5 million ha, or 34.5% of the total cultivated land which reaches about 175 million ha (De Grandi, 1996). The peasant population includes 75 million people representing almost two thirds of the Latin America's total rural population (Ortega, 1986). Average farm size of these units is about 1.8 ha, although the contribution of peasant agriculture to the general food supply in the region is significant. In the 1980s it reached approximately 41% of the agricultural output for domestic consumption, and is responsible for producing at the regional level 51% of the maize (*Zea mays*), 77% of the beans (*Phaseolus* spp.), and 61% of the potatoes (*Solanum tuberosum*).

Searching for ways to develop more sustainable agroecosystems, several researchers have posited that tropical agroecosystems should mimic the structure and function of natural communities (a practice followed by thousands of indigenous farmers for centuries), as these systems exhibit tight nutrient cycling, resistance to pest invasion, vertical structure and preserve biodiversity (Ewel, 1986; Soule and Piper, 1992).

If such ecological approach is used, it is important to ensure that promoted systems and technologies are suited to the specific environmental and socio-economic conditions of small farmers, without increasing risk or dependence on external inputs. Rather, agroecological development projects should feature resource-conserving yet highly productive systems such as polycultures, agroforestry and the integration of crops and livestock (Altieri, 1995).

The ecological futility of promoting mechanized monocultures in tropical areas of overwhelming biotic intricacy where pests flourish year-round and nutrient leaching is a major constraint has been amply demonstrated (Browder, 1989). A more reasonable approach is to imitate natural cycles rather than struggle to impose horticultural simplicity in ecosystems that are inherently complex. Ewel (1986) argues that successional ecosystems can be particularly appropriate templates for the design of sustainable tropical agroecosystems. Building on this idea and the contributions of modern agroecology, we provide principles for agroecosystem design emphasizing the development of cropping systems that enhance nutrient capture, and confer associational resistance to pests, thus reducing agroecosystem vulnerability while providing biological stability and productivity.

COMPARING NATURAL AND AGRICULTURAL ECOSYSTEMS

Many agroecologists have argued that by understanding the structural and functional differences between natural systems and agroecosystems we can learn much about the underlying processes that make crop systems more vulnerable to insect pests, dependent on external inputs, and inefficient in the use of local resources (Carrol, Vandermeer, and Rosset, 1990).

The dominant components of an agroecosystem are plants (and animals) selected, propagated, tended, and harvested by humans for the purpose of food and fiber production. In comparison to unmanaged systems, the composition and structure of agroecosystems is simple. The plant biomass is composed of stands usually dominated by one major crop plant within well-defined field boundaries.

The net result is an artificial monoculture system that requires constant human intervention. Commercial seed-bed preparation and mechanized planting replace natural methods of seed dispersal; chemical pesticides replace natural controls on populations of weeds, insects, and pathogens; and genetic manipulation replaces natural processes of plant evolution and selection. Even decomposition is altered since plant growth is harvested and soil fertility maintained, not through nutrient recycling, but with fertilizers (Cox and Atkins, 1974).

Human manipulation and alteration of ecosystems for the purpose of establishing agricultural production makes agroecosystems structurally and functionally very different from natural ecosystems (Table 1). Agroecosystems are artificial ecosystems that are solar powered, as are natural ecosystems, but differ in that (1) the auxiliary energy sources that enhance productivity are processed fuels (along with animal and human labor) rather than natural energies; (2) diversity is greatly reduced by

TABLE 1. Structural and functional differences between natural ecosystems and agroecosystems (modified after Gliessman, 1998).

Characteristics	Agroecosystem	Natural ecosystem
Net productivity	High	Medium
Trophic chains	Simple, linear	Complex
Species diversity	Low	High
Genetic diversity	Low	High
Mineral cycles	Open	Closed
Stability (resilience)	Low	High
Entropy	High	Low
Human control	Definite	Not needed
Temporal permanence	Short	Long
Habitat heterogeneity	Simple	Complex
Phenology	Synchronized	Seasonal
Maturity	Immature, early successional	Mature, climax

human management in order to maximize yield of specific food and other products; (3) the dominant plants and animals are under artificial rather than natural selection; and (4) control is external and goal oriented rather than internal via subsystem feedback as in natural ecosystems (Gliessman, 1998).

Perhaps of major significance in relation to the instability of tropical monocultures are the following processes that alter ecosystem structure and function.

Landscape and Field Simplification

With agriculture the original flora and fauna are completely replaced over a vast area decreasing habitat heterogeneity. Where patches of natural vegetation persist they often occur on sites unsuitable for agriculture and contribute only minimally to the ecological stability of the area. As biological diversity is reduced, trophic structures tend to become simplified, and many niches are left unoccupied (Thies and Tscharntke,

1999). The danger of increased invasions and catastrophic pest or disease outbreaks is high, despite the intensive human input in the form of agrochemicals.

Monocultures are unfavorable environments for natural enemies of pests, due to high levels of disturbance and lack of ecological infrastructure the capacity. The capacity of predators and parasites to control invaders is lower than in more diverse agroecosystems (Landis, Wratten, and Gurr, 2000). Most agroecologists agree that due to their reduced structural and functional diversity in relation to natural ecosystems, agroecosystems have lower resiliency than natural ecosystems.

Disruption of Succession

Intensive agriculture prevents normal succession from taking place. Every newly planted crop represents the first stage of succession that is neither persistent nor steady state. The objective of growing a crop is to obtain the greatest possible harvest. Constant disturbance keeps the agroecosystem at the early stages of succession, where a greater proportion of gross productivity is available as net productivity or harvestable biomass. To maintain a system of this type, it is necessary for humans to assume responsibility for the costs of maintenance and regulation normally taken care of by the natural processes that lead to the establishment of a climax ecosystem.

Lowering of Plant Defenses

In natural ecosystems, the assemblage of organisms is the result of natural selection and coevolution. Agroecosystems consist of unnatural assemblages of human selected domesticated species and an assortment of native or imported opportunistic species that manage to invade the site. These two groups have not been integrated into a steady-state system by the process of coevolution, and many opportunistic species frequently become weed, insect, and disease pests that must be dealt with by the farmer.

Throughout the crop domestication process, humans tended to select plants with fewer morphological and chemical defenses. Such intense human selection for rapid growth and high reproductive output resulted in a general lowering of the plants' allocation to defense. Of course, significant amounts of toxic secondary compounds remain in many edible crops, but the general trend has been the gradual reduction of those chemicals and morphological features that protected plants from arthropod herbivores. This often left the plants more vulnerable than their wild relatives, and it largely explains the widespread belief that there are more outbreaks of insects in agroecosystems than natural ecosystems (Altieri, 1994). In addition, excessive use of chemical fertilizers can create nutrient imbalances in crop plants, which in turn, reduce resistance to insect pests (Luna, 1998).

Inefficient Nutrient Cycling

Recycling of nutrients is minimal in most agroecosystems and considerable quantities are lost from the system with the harvest or as a result of leaching or erosion due to a great reduction in permanent biomass levels held within the system. Lower levels of soil organic matter (SOM) accumulation and reduced biological activity in monoculture is also a key factor explaining low soil fertility in deeply weathered and leached tropical soils. The frequent exposure of bare soil between cropping seasons also creates "leaks" of nutrients from the system. Instead of using locally recycled nutrients, farmers have come to rely heavily on petroleum-based nutrient inputs to replace these losses (Magdoff and van Es 2000).

BUILDING ON TRADITIONAL AGRICULTURE

Many agricultural scientists have argued that the starting point in the development of new pro-poor agricultural development approaches are the very systems that traditional farmers have developed and/or inherited throughout centuries. Such complex farming systems, adapted to the local conditions, have helped small farmers to manage harsh environments sustainably and to meet their subsistence needs, without depending on mechanization, chemical fertilizers, pesticides or other technologies of modern agricultural science (Denevan, 1995). The persistence of millions of hectares under traditional agriculture in the form of raised fields, terraces, polycultures, agroforestry systems, etc., document a successful indigenous agricultural strategy and comprises a tribute to the "creativity" of small farmers throughout the developing world (Wilken, 1997). These microcosms of traditional agriculture offer promising models for other areas as they promote biodiversity, thrive without agrochemicals, and sustain year-round yields. It is estimated that about 50 million individuals belonging to about 700 different ethnic indigenous groups live and utilize the humid tropical regions of the world.

About two million of these live in the Amazon and southern Mexico (Toledo, 2000). In Mexico, half of the humid tropics is utilized by indigenous communities and "ejidos" featuring integrated agriculture-forestry systems aimed at subsistence and local regional markets.

Traditional farming systems commonly support a high degree of plant diversity in the form of polycultures and/or agroforestry patterns (Gliessman, 1998). This strategy of minimizing risks by planting several species of plants and varieties of crops stabilizes yields over the long term, promotes diet diversity and maximizes returns even under low levels of technology and limited resources (Harwood, 1979).

Most peasant systems are productive despite their low use of chemical inputs. Generally, agricultural labor has a high return per unit of input. The energy return to labor expended in a typical peasant farm is high enough to ensure continuation of the present system. Also in these systems, favorable rates of return between inputs and outputs in energy terms are realized. For example, on Mexican hillsides, maize yields in hand-labor dependent swidden systems are about 1940 kg ha^{-1}, exhibiting an output/input ratio of 11:1. In Guatemala, similar systems yield about 1066 kg ha^{-1} of maize, with an energy efficiency ratio of 4.84. When animal traction is utilized, yields do not necessarily increase but the energy efficiency drops to values ranging from 3.11-4.34. When fertilizers and other agrochemicals are utilized, yields can increase to levels of 5-7 Mg ha^{-1}, but energy rations start exhibiting inefficient values (less than 2.0) (Netting, 1993).

In most multiple cropping systems developed by smallholders, productivity in terms of harvestable products per unit area is higher than under sole cropping with the same level of management (Francis, 1986). Yield advantages can range from 20 to 60% and accrue due to reduction of pest incidence and more efficient use of nutrients, water and solar radiation.

Undoubtedly, the ensemble of traditional crop management practices used by many resource-poor farmers represent a rich resource for modern workers seeking to create novel agroecosystems well adapted to the local agroecological and socioeconomic circumstances of peasants. Peasants use a diversity of techniques, many of which fit well to local conditions. The techniques tend to be knowledge-intensive rather than input-intensive, but clearly not all are effective or applicable, therefore modifications and adaptations may be necessary. The challenge is to maintain the foundations of such modifications grounded on peasants' rationale and knowledge.

An example of this is efforts to develop alternatives to "slash and burn." Slash and burn is perhaps one of the best examples of an ecological strategy to manage agriculture in the tropics. By maintaining a mosaic of plots under cropping and some in fallow, farmers capture the essence of natural processes of soil regeneration typical of any ecological succession. These systems, however, are reaching their limits for a variety of reasons. By understanding the rationale of the slash and burn, a contemporary discovery, the use of "green manures," has provided an ecological pathway to the intensification of the "milpa," in areas where long fallows are not possible anymore due to population growth, land scarcity or conversion of forest to pasture.

Experiences in Central America show that "mucuna" [*Mucuna pruriens* (L.) DC.] based maize systems are fairly stable allowing respectable yield levels (usually 2-4 Mg ha^{-1}) every year. In particular, the system appears to greatly diminish drought stress because the mulch layer helps conserve water in the soil profile. With enough water around, nutrients are made readily available, in good synchronization with major crop uptake. In addition, the mucuna suppresses weeds, either because this plant physically prevents them from germinating and emerging or from surviving very long during the mucuna cycle, or because a shallow rooting of weeds in the litter layer-soil interface makes them easier to control. Data shows that this system grounded in farmers knowledge, involving the continuous annual rotation of mucuna and maize, can be sustained for at least 15 years at a reasonably high level of productivity, without any apparent decline in the natural resource base (Buckles, Triomphe, and Sain, 1998).

As illustrated with the "mucuna" system, an increased understanding of the agroecology and ethnoecology of traditional farming systems is necessary to continue developing contemporary systems. This can only occur from integrative studies that determine the myriad of factors that condition how farmers perceive their environment and subsequently how they modify it to later translate such information to modern scientific terms.

DESIGNING SUCCESSION ANALOG AGROECOSYSTEMS

As traditional farmers have done, natural successional communities can be used as models for agroecosystem design because they offer several traits of potential value to agriculture: (i) high resistance to pest invasion and attack, (ii) high retention of soil nutrients, (iii) enhanced agrobiodiversity, and (iv) reasonable productivity (Ewel, 1999).

As stated by Gliessman (1998) a major challenge in the tropics is to design agroecosystems that, on the one hand, take advantage of some of the beneficial attributes of the early stages of succession yet, on the other hand, incorporate some of the advantages gained by allowing the system to reach the later stages of succession. As shown in Table 2, only one desirable ecological characteristic of agroecosystems–high net primary productivity–occurs in the early stages of successional development; all the others do not become manifest until the later stages of development, an important reason to create more permanent agroecosystems through the inclusion of perennials.

Ecological Principles for Design

- *Increasing species diversity* as this promotes fuller use of resources (nutrients, radiation, water, etc.), protection from pests and compensatory growth. Many researchers have highlighted the importance of various spatial and temporal plant combinations to facilitate complementary resource use or to provide intercrop advantage such as in the case of legumes facilitating the growth of cereals by supplying it with extra nitrogen (N). Compensatory growth is another desirable trait as if one species succumbs to pests, weather or harvest, another species fills the void maintaining full use of available resources. Crop mixtures also minimize risks especially by creating the sort of vegetative texture that controls specialist pests.
- *Enhance longevity* through the addition of perennials that contain a thick canopy thus providing continual cover that protects the soil. Constant leaf fall builds OM and allows uninterrupted nutrient circulation. The establishment of dense, deep root systems of long-lived woody plants represents an effective mechanism for nutrient capture offsetting the negative losses through leaching.
- *Impose a fallow* to restore soil fertility through biomass accumulation and biological activation, and to reduce agricultural pest populations as life cycles are interrupted with a rotation of fallow vegetation and crops.
- *Enhance additions of organic matter* by including legumes, biomass producing plants and incorporating animals. Accumulation of both "active" and "slow fraction" SOM is key for activating soil biology, improving soil structure and macroporosity and elevating the nutrient status of soils.

- *Increase landscape diversity* by having in place a mosaic of agroecosystems representative of various stages of succession. Risk of complete failure is spread among, as well as within, the various cropping systems. Improved pest control is also linked to spatial heterogeneity at the landscape level.

MANAGEMENT OPTIONS FOR NATURAL SUCCESSION MIMICRY

Under a scheme of managed succession, natural successional stages are mimicked by intentionally introducing agricultural plants, animals, practices, and inputs that promote the development of interactions and connections between component parts of the agroecosystem. Plant species (both crop and noncrop) are planted that capture and retain nutrients in the system and promote good soil development. These plants include legumes, with their N-fixing bacteria, and plants with phospho-

TABLE 2. Desirable ecological characteristics of agroecosystems in relation to successional development (adapted from Gliessman 1998).

	Successional stage of greatest development			
Characteristic	**Early**	**Middle**	**Late**	**Benefit to agroecosystem**
High species diversity				Reduced risk of catastrophic crop loss
High total biomass				Larger source of soil organic matter
High net primary productivity				Greater potential for production of harvestable biomass
Complexity of species interactions				Greater potential for biological control
Efficient nutrient cycling				Diminished need for external nutrient inputs
Mutualistic interference				Greater stability; diminished need for external inputs

rus-trapping mycorrhizae. As the system develops, increasing diversity, food web complexity, and level of mutualistic interactions all lead to more effective feedback mechanisms for pest and disease management. The emphasis during the development process is on building a complex and integrated agroecosystem with less dependence on external inputs.

There are many ways that a farmer, beginning with a recently cultivated field of bare soil, can allow successional development to proceed beyond the early stages. One general model is to begin with an annual monoculture and progressing to a perennial tree crop system, as follows (Gliessman, 1998):

1-2 year: The farmer begins by planting a single annual crop that grows rapidly, captures soil nutrients, gives an early yield, and acts as a pioneer species in the developmental process.

3 year: As a next step (or instead of the previous one), the farmer can plant a polyculture of annuals that represent different components of the pioneer stage. The species would differ in their nutrient needs, attract different insects, have different rooting depths, and return a different proportion of their biomass to the soil. One might be a N-fixing legume. All of these early species would contribute to the initiation of the recovery process, and they would modify the environment so that non-crop plants and animals–especially the macro- and microorganisms necessary for developing the soil ecosystem–can also begin to colonize.

4 year: Following the initial stage of development, short-lived perennial crops can be introduced. Taking advantage of the soil cover created by the pioneer crops, these species can diversify the agroecosystem in important ecological aspects. Deeper root systems, more SOM stored in standing biomass, and greater habitat and microclimate diversity all combine to advance the successional development of the agroecosystem.

5 year: Once soil conditions improve sufficiently, the ground is prepared for planting longer-lived perennials, especially orchard or tree crops, with annual and short-lived perennial crops maintained in the areas between them. While the trees are in their early growth, they have limited impact on the environment around them. At the same time, they ben-

efit from having annual crops around them, because in the early stages of growth they are often more susceptible to interference from aggressive weedy species that would otherwise occupy the area.

6 year: As the tree crops develop, the space in between them can continue to be managed with annuals and short-lived perennials.

7 year and beyond: Eventually, once the trees reach full development, the end point in the developmental process is achieved. This last stage is dominated by woody plants which are key to the site-restoring powers of fallow vegetation because of their deep, permanent root systems.

Once a successionally developed agroecosystem has been created, the problem becomes one of how to manage it. The farmer has three basic options:

- Return the entire system to the initial stages of succession by introducing a major disturbance, such as clear-cutting the trees in the perennial system. Many of the ecological advantages that have been achieved will be lost and the process must begin anew.
- Maintain the system as a perennial or tree crop agroecosystem.
- Reintroduce disturbance into the agroecosystem in a controlled and localized manner, taking advantage of the dynamics that such patchiness introduces into an ecosystem. Small areas in the system can be cleared, returning those areas to earlier stages in succession, and allowing a return to the planting of annual or short-lived crops. If care is taken in the disturbance process, the belowground ecosystem can be kept at a later stage of development, whereas the aboveground system can be made up of highly productive species that are available for harvest removal.

One example of a crop successional design comes from Costa Rica where researchers conducted spatial and temporal replacements of wild species by botanically/structurally/ecologically similar cultivars. Successional members of the natural system such as *Heliconia* spp., cucurbitaceous vines, *Ipomoea* sp., legume vines, shrubs, grasses and small trees were replaced by plantain (*Musa* sp.), squash (*Cucurbita*) varieties, and yams (*Dioscorea*). By years two and three, fast growing tree crops Brazil nuts (*Bertholletia excelsa* Humb. & Bonpl.), peach [*Prunus persica* (L.) Batsch.], palm (*Palmae*), and rosewood (*Dalbergia nigra* Allem. Ex. Benth.) formed an additional stratum, thus maintaining a continuous crop cover, avoiding site degradation and nutrient leaching, and providing nutrients throughout the year (Ewel, 1999).

DIVERSIFICATION STRATEGIES

In the process of emulating nature's diversity, various strategies to restore agricultural diversity in time and space can be utilized (Altieri, 1994; Gliessman, 1998; Finch and Sharp, 1976; Francis, 1986; Nair, 1982; Pearson and Ison, 1987):

- *Crop rotations:* Temporal diversity in the form of leguminous green manures are incorporated into cropping systems, providing crop nutrients and breaking the life cycles of several insect pests, diseases, and weed species.
- *Variety mixtures:* Increasing plant genetic diversity at the field level through the use of variety mixtures and/or multi-lines increases genetic heterogeneity, reducing the vulnerability of monoculture crops to diseases.
- *Polycultures:* Complex cropping systems in which two or more crop species are planted within sufficient spatial proximity to result in competition or complementation, thus enhancing yields and minimizing risks.
- *Agroforestry systems:* A system where trees are grown together with annual crops and/or animals, providing the benefits of perennials and resulting in enhanced, complementary relations between components while increasing multiple use of the agroecosystem.
- *Cover crops:* The use of pure or mixed stands of legumes or other annual plant species under fruit trees for the purpose of providing soil cover, improving soil fertility, enhancing biological control of pests, and modifying the orchard microclimate.
- *Animal integration through crop-livestock mixtures*, which aids in achieving high biomass output and optimal recycling.

All of the above forms of agroecosystem diversification strategies share in common the following features (Altieri, 1995):

- Maintain high vegetative cover as an effective soil and water conserving measure, achieved through the use of no-till practices, mulch farming, and use of cover crops and other appropriate methods.
- Provide a regular supply of SOM through the addition of plant biomass, manure, compost, which serves as a source of nutrients and fuel to microbial populations.
- Enhance nutrient recycling mechanisms through the use of systems based on legumes, trees, and incorporation of livestock.
- Promote pest regulation through enhanced activity of biological control agents achieved by conserving natural enemies and antagonists through establishment of an ecological infrastructure associated to diversified cropping patterns.

The mechanisms that result in higher productivity in diverse agroecosystems are embedded in the process of facilitation. Facilitation occurs when one crop modifies the environment in a way that benefits a second crop, for example, by lowering the population of a critical herbivore, or by releasing nutrients that can be taken up by the second crop (Vandermeer, 1989). Facilitation may result in overyielding even where direct competition between crops is substantial. The combined effects or synergies of complex agroecosystems can best be understood when one examines the findings of research on the effects of plant diversity and soil fertility on insect pest populations.

Vegetational Diversity and Pest Outbreaks

Experiments testing the theory that decreased plant diversity in agroecosystems leads to enhanced herbivorous insect abundance, have shown that mixing certain plant species with the primary host of a specialized herbivore gives a fairly consistent result: specialized insect pest species usually exhibit higher abundance in monoculture than in diversified crop systems (Altieri and Letourneau, 1982; Andow, 1991).

Several reviews have been published documenting the effects of within-habitat diversity on insects (Altieri, 1994). Two main ecological hypotheses (natural enemy hypothesis and the resource concentration hypothesis) have been offered to explain why insect communities in agroecosystems can be stabilized by constructing vegetational architectures that support natural enemies and/or directly inhibit pest attack. The literature is full of examples of experiments documenting that diversification of cropping systems often leads to reduced pest populations. Andow (1991) reviewed 150 published studies documenting the effects of agroecosystem diversification on insect pest abundance, examining 198 herbivore species in total. Fifty-three percent of these species were found to be less abundant in the more diversified system, 18% were more abundant in the diversified system, 9% showed no difference, and 20% showed a variable response.

Many of these studies have transcended the research phase and have found applicability to control specific pests such as the lepidopteran stemborers in Africa. Scientists at ICIPE developed a habitat management system that uses two kinds of crops that are planted together with maize: a plant that repels these borers (the push) and another that attracts (pulls) its natural enemies (Khan et al., 1998). The push-pull system has been tested on over 450 farms in two districts of Kenya and has now been made available to the national extension systems in East Africa. Participating farmers in the breadbasket of Trans Nzoia are reporting a 15-20% increase in maize yield. In the semi-arid Suba district–plagued by both stemborers and the parasitic weed *Striga* (Scrophulariaceae)–a substantial increase in milk yield has occurred in the last four years, with farmers now being able to support grade cows on the fodder produced. When farmers plant maize (*Zea mays* L.), napier (*Pennisetum purpureum* Schumach.) and desmodium (Leguminosae) together, a return of US$ 2.30 for every dollar invested is made, as compared to only $1.40 obtained by planting maize as a monocrop. Two of the most useful trap crops that pull the borers' natural enemies are napier grass (*P. purpureum*) and Sudan grass [*Sorghum sudanense* (Piper) Stapf.], both important fodder plants; these are planted in a border around the maize. Two excellent borer-repelling crops which are planted between the rows of maize are molasses grass (*Melinis minitifolia* Beauv.), which also repels ticks, and the leguminous silverleaf, *Desmodium*, which can suppress the parasitic weed, *Striga*, by a factor of 40 compared to maize monocrop; its N-fixing ability increases soil fertility; and it is an excellent forage. As an added bonus, sale of *Desmodium* seed is proving to be a new income-generating opportunity for women in the project areas.

It is clear that both empirical data and theoretical arguments suggest that differences in pest abundance between diverse and simple annual cropping systems can be explained by differences in the movement, colonization and reproductive behavior of herbivores and by the activities of natural enemies. The studies further suggest that the more diverse the agroecosystems and the longer this diversity remains undisturbed, the more internal links develop to promote greater insect stability (Altieri and Nicholls, 1999). Research along these lines is crucial to a vast majority of small farmers who rely on the rich complex of predators and parasites associated with their mixed cropping systems for insect pest control. Any changes on the levels of plant diversity within such systems can lead to disruptions that may enhance dependence on pesticides. Regardless, more studies are needed to determine the underlying elements of plant mixtures that disrupt pest invasion and that favor natural enemies.

Integrating Effects of Soil Management: Healthy Soils–Healthy Plants

For resource-poor farmers, crop diversification strategies must be complemented by regular applications of organic amendments (crop residues, animal manures, and composts) to maintain or improve soil quality and productivity. Despite the fact that this is a common practice of tropical smallholders, little is known about the multifunctional effects of organic amendments on other agroecosystem components, beyond the documented effects on improved soil structure and nutrient content. Well-aged manures and composts can serve as sources of growth-stimulating substances, such as indole-3-acetic acid and humic and fulvic acids (Magdoff and van Es, 2000). Beneficial effects of humic acid substances on plant growth are mediated by a series of mechanisms, many similar to those resulting from the direct application of plant growth regulators.

The ability of a crop plant to resist or tolerate pests is tied to optimal physical, chemical and biological properties of soils. Adequate moisture, good soil tilth, moderate pH, right amounts of SOM and nutrients, and a diverse and active community of soil organisms all contribute to plant health. Organic rich soils generally exhibit good soil fertility as well as complex food webs and beneficial organisms that prevent infection by disease-causing organisms such as *Pythium* and *Rhizoctonia*. On the other hand, farming practices such as high applications of N fertilizer can create nutrition imbalances, and render crops susceptible to diseases such as *Phytophtora* and *Fusarium* and stimulate outbreaks of Homopteran insects such as aphids and leafhoppers (Campbell, 1989). In fact, there is increasing evidence that crops grown in organic rich and biologically active soils are less susceptible to pest attack. Many studies suggest that the physiological susceptibility of crops to insect pests and pathogens may be affected by the form of fertilizer used (organic vs. chemical fertilizer).

The literature is abundant on the benefits of organic amendment additions that encourage resident antagonists thus enhancing biological control of plant diseases. Several bacteria species of the genus *Bacillus* and *Pseudomonas*, as well as the fungus *Trichoderma* are key antagonists that suppress pathogens through competition, lysis, antibiosis, and hyperparasitism (Palti, 1981).

Studies documenting lower abundance of several insect herbivores in low-input systems, have partly attributed such reduction to a low N content in organically farmed crops (Luna, 1988). In Japan, density of immigrants of the planthopper *Sogatella furcifera* Horváth was significantly lower while settling rates of female adults and survival rate of immature stages of ensuing generations were lower in organic rice (*Oryza sativa* L.) fields. Consequently, the density of planthopper nymphs and adults in the ensuing generations decreased in organically farmed fields (Kajimura, 1995). In England, conventional winter wheat (*Triticum aestivum* L.) fields developed a larger infestation of the aphid *Metopolophium dirhodum* (Walker) than its organic counterpart. This crop also had higher levels of free protein amino acids in its leaves during June, which were believed to have resulted from a N top dressing of the crop early in April. However, the difference in the aphid infestations between crops was attributed to the aphid's response to relative proportions of certain non-protein to protein amino acids in the leaves at the time of aphid settling on crops (Kowalski and Visser, 1979). In greenhouse experiments, when given a choice of maize grown on organic versus chemically fertilized soils, European corn borer (*Ostrinia nubilalis* Hübner) females preferred to significantly lay more eggs in chemically fertilized plants (Phelan, Maan, and Stinner, 1995).

Such findings are of key importance to tropical resource-poor farmers such as Cakchiquel farmers in Patzum, Guatemala who have experienced increased pest populations (aphids and corn earworms) in maize since they abandoned organic fertilization and adopted synthetic fertilizers (Morales, Perfecto, and Ferguson, 2001). Many farmers undergoing modernization may be facing similar impacts due to higher fertilizer use, which in turn may create subtle imbalances in the agroecology of specific farming systems.

CONCLUSIONS

Technological innovation in the tropics has been characterized by the transfer of agricultural systems from temperate regions without due consideration to their ecological fit. Monoculture agriculture (i.e., extensive grain and plantation crops) is basically a cultural baggage of early colonial times, which still make short-term economic sense, but in the long-term constitutes a total ecological mismatch. It is time to use ecological principles as part of the design criterion of agroecosystems, thus replacing what has become a strictly economic decision-making process with one that includes ecological ideas, and especially local farmers' perspectives (Vandermeer, 1995).

An important challenge is to apply such ideas to design new agroecosystems using nature as a model. Such mimics, like their models, can be productive, pest resistant and conservative of nutrients and other resources, and consequently more cost-effective and less risky to farmers, especially poor peasants. As discussed above, a key strategy for sustainable tropical agriculture is to reincorporate diversity into the agricultural landscape and manage it more effectively. Emergent ecological properties develop in diversified agroecosystems that allow the system to function in ways that maintain soil fertility, encourage pest regulation and sustain productivity.

Surely there are no simple links between species diversity and ecosystem stability. Apparently functional characteristics of component species are as important as the total number of species. Recent studies with grassland plots conclude that functionally different roles represented by many plants are at least as important as the total number of species in determining processes and services in ecosystems (Tilman, Wedin, and Knops, 1996). It is far easier to mimic specific ecosystem processes then to try to duplicate all the complexity of nature. All that is needed is to select the right kind of diversity (adding one or two plant species), to achieve herbivore resistance, enhanced productivity and nutrient supply.

The main limitation of promoting species-rich agroecosystems is that they are difficult to manage. The biggest challenge in managing a successionally developed system is to learn how to introduce disturbance in ways that stimulate system productivity on the one hand and, on the other, provide resistance to change and variation within the ecosystem. This can be done in many different ways depending on local environmental conditions, the structure of mature natural ecosystems normally present, and the feasibility of maintaining modifications of those conditions over the long term.

Some authors contend that there is a trade-off between high-diversity and low yield, and that farmers will always have to choose between systems that confer low risk, low productivity, and high productivity but high risk. According to Ewel (1986) the very attributes that make diverse agroecosystems attractive, seem to have biological costs that are incompatible with high yield.

The literature is somewhat divided on this issue, although a significant number of scientists highlight the overyielding advantages of polycultures and the multifunctionality of small and diversified farms (Francis, 1986; Vandermeer, 1989). The very practice of millions of small tropical farmers that favor polycultures, agroforestry and diversified patters give credibility to a more agroecological approach. Regardless, the task at hand for tropical agroecologists will be to design complex agroecosystems that simultaneously sustain harvestable products and ecological functions.

Given a range of economic and environmental circumstances, it is possible that for capitalized farmers with access to inputs, a relatively simple rotation or intercrop may be all that is needed. For resource-poor farmers, where crop failure cannot be tolerated, diverse cropping systems should be the agroecosystem of choice. Whatever the preferred system, diversity will be of value in a large or small-scale agroecosystem for a variety of reasons (Altieri, 1994; Gliessman, 1998).

- As diversity increases, so do opportunities for coexistence and beneficial interactions between species that can enhance agroecosystem sustainability.
- Greater diversity often allows better resource-use efficiency in an agroecosystem. There is better system-level adaptation to habitat heterogeneity, leading to complementarity in crop species needs, diversification of niches, overlap of species niches, and partitioning of resources.
- Ecosystems, in which plant species are intermingled, possess an associated resistance to herbivores. As in diverse systems, there is a greater abundance and diversity of natural enemies of pest insects, keeping in check the populations of individual herbivore species.
- A diverse crop assemblage can create a diversity of microclimates within the cropping system that can be occupied by a range of noncrop organisms–including beneficial predators, parasites, pollinators, soil fauna and antagonists–that are of importance for the entire system.
- Diversity in the agricultural landscape can contribute to the conservation of biodiversity in surrounding natural ecosystems.

- Diversity in the soil performs a variety of ecological services such as nutrient recycling and detoxification of noxious chemicals and regulation of plant growth.
- Diversity reduces risk for farmers, especially in marginal areas with more unpredictable environmental conditions. If one crop does not do well, yield and income from others can compensate.

REFERENCES

Adams, M.W., A.H. Ellingbae, and E.C. Rossineau. (1971). Biological uniformity and disease epidemics. *BioScience* 21:1067-1070.

Altieri, M.A. and D.K. Letourneau. (1982). Vegetation management and biological control in agroecosystems. *Crop Protection* 1: 405-430.

Altieri, M.A. (1991). Ecology of tropical herbivores in polycultural agroecosystems. In *Plant-Animal Interactions: Evolutionary Ecology in Tropical and Temperate Regions*, eds. P.W. Price, T.M. Lewinshon, and W.W. Benson, New York, NY: John Wiley and Sons, Inc., pp. 607-616.

Altieri, M.A. (1995). *Agroecology: The Science of Sustainable Agriculture*. Boulder, CO: Westview Press.

Altieri, M.A. (1994). *Biodiversity and Pest Management in Agroecosystems*. New York, NY: The Haworth Press.

Altieri, M.A. (1999). Applying agroecology to enhance the productivity of peasant farming systems in Latin America. *Environment, Development and Sustainability* 1:197-217.

Altieri, M.A. and C.I. Nicholls. (1999). Biodiversity, ecosystem function and insect pest management in agricultural systems. In *Biodiversity in Agroecosystems*, eds. W.W. Collins and C.O. Qualset, Boca Raton, FL: CRC Press.

Altieri, M.A., D.K. Letourneaou, and J.R. Davis. (1983). Developing sustainable agroecosystems. *BioScience* 33: 45-49.

Andow, D.A. (1991). Vegetational diversity and arthropod population response. *Annual Review of Entomology* 36:561-586.

Barbosa, P. (1998). *Conservation Biological Control*, San Diego, CA: Academic Press.

Beets, W.C. (1990). *Raising and Sustaining Productivity of Smallholder Farming Systems in the Tropics*, Alkmaar, Holland: AgBé Publishing.

Browder, J.O. (1989). *Fragile Lands in Latin America: Strategies for Sustainable Development*. Boulder, CO: Westview Press.

Buckles, D., B. Triomphe, and Gustavo Sain. (1998). *Cover Crops in Hillside Agriculture: Farmer Innovation with Mucuna*, Ottawa, Canada: International Development Research Center.

Campbell, R. (1989). *Biological Control of Microbial Plant Pathogens*. Cambridge, UK: Cambridge University Press.

Carrol, C. R., J.H. Vandermeer, and P.M. Rosset. (1990). *Agroecology*. New York, NY: McGraw Hill Publishing Company.

Conway, G.R. (1997). *The Doubly Green Revolution.* London, UK: Penguin Books.

Cox, G.W. and N.D. Atkins. (1974). *Agricultural Ecology: An Analysis of World Food Production Systems.* San Francisco, CA: W.H. Freeman and Company.

De Grandi, J.C. (1996). El desarrollo de los sistemas de agricultura campesina en America Latina. Serie FAO-Gestion de Sistemas de Explotacion Agricola #12. Rome, Italy:FAO. p. 83. (In Spanish).

Denevan, W.M. (1995). Prehistoric agricultural methods as models for sustainability. *Advances in Plant Pathology* 11:21-43.

Ewel, J.J. (1986). Designing agricultural ecosystems for the humid tropics. *Annual Review of Ecological Systems* 17: 245-71.

Ewel, J.J. (1999). Natural systems as models for the design of sustainable systems of land use. *Agroforestry Systems* 45:1-21.

Finch, C.V. and C.W. Sharp. (1976). *Cover Crops in California Orchards and Vineyards.* Washington, DC: USDA Soil Conservation Service.

Francis, C.A. (1986). *Multiple Cropping Systems.* New York, NY: MacMillan Press.

Gliessman, S.R. (1998). *Agroecology: Ecological Processes in Sustainable Agriculture.* Ann Arbor, MI: Ann Arbor Press.

Harwood, R.R. (1979). *Small Farm Development–Understanding and Improving Farm Systems in the Humid Tropics.* Boulder, CO: Westview Press.

Hendrix, P.H., D.A.Jr. Crossley, and D.C. Coleman. (1990). Soil biota as components of sustainable agroecosystems. In *Sustainable Agricultural Systems*, eds. C.A. Edwards, R. Lal, P. Madden, R. Miller, and G. House, Ankeny, IA: Soil and Water Conservation Society.

Kajimura, T. (1995). Effect of organic rice farming on planthoppers. *Population Ecology* (Japan) 37:219-224.

Khan, Z.R., K. Ampong-Nyarko, A. Hassanali, and S. Kimani. (1998). Intercropping increases parasitism of pests. *Nature* 388:631-632.

Kowalski, R. and P.E. Visser. (1979). Nitrogen in a crop-pest interaction: Cereal aphids. In *Nitrogen as an Ecological Parameter*, ed. J.A. Lee, Oxford, UK: Blackwell Scientific Publications. pp. 67-74.

Landis, D.A., S.D. Wratten, and G.A. Gurr. (2000). Habitat management to conserve natural enemies of arthropod pests in agriculture. *Annual Review of Entomology* 45:175-201.

Liebman, M. and T. Ohno. (1998). Crop rotation and legume residue effects on weed emergence and growth: Implications for weed management. In *Integrated Weed and Soil Management*, eds. J.L. Hatfield and B.A. Stwerrt, Ann Arbor, MI: Ann Arbor Press. pp. 181-221.

Luna, J. (1998). Influence of soil fertility practices on agricultural practices on agricultural pests. In *Global Perspectives in Agroecology and Sustainable Agricultural Systems*, eds. P. Allen and D.Van Dusen, Proceedings of the VI Scientific Conference, Santa Cruz, CA: International Organic Agriculture Movements (IFOAM). pp. 589-600.

Magdoff, F. and H. van Es. (2000). *Building Soils for Better Crops.* Beltsville, MD: Sustainable agriculture networks.

Morales, H., I. Perfecto, and B. Ferguson. (2001). Traditional soil fertilization and its impact on insect populations in corn. *Agriculture, Ecosystems and Environment* 84:145-155.

Nair, P.K.R. (1982). *Soil Productivity Aspects of Agroforestry.* ICRAF, Nairobi.

Netting, R. McC. (1993). *Smallholders, Householders: Farm Families and the Ecology of Intensive, Sustainable Agriculture*, Stanford, CA: Stanford University Press.

Norman, M.J.T. (1979). Annual cropping systems in the tropics. Gainesville, FL: University Presses of Florida.

Ortega, E. (1986). *Peasant Agriculture in Latin America.* Santiago, Chile: Joint ECLAC-FAO Agriculture Division.

Palti, J. (1981). *Cultural Practices and Infectious Plant Diseases.* New York, NY: Springer.

Pearson, C.J. and R.L. Ison. (1987). *Agronomy of Grassland Systems.* Cambridge, UK: Cambridge University Press.

Phelan, P.L., J.F. Maan, and B.R. Stinner. (1995). Soil fertility management and host preference by European corn borer on maize: a comparison of organic and conventional farming. *Agriculture, Ecosystems and Environment* 56:1-8.

Pretty, J.N. (1994). *Regenerating Agriculture.* London, UK: Earthscan Publications Ltd.

Reijntjes, C., B. Haverkort and A. Waters-Bayer. (1992). *Farming for the Future: An Introduction to Low-External-Input and Sustainable Agriculture*, London, UK: MacMillan Press Ltd.

Richards, P. (1985). *Indigenous Agricultural Revolution.* Boulder, CO: Westview Press.

Robinson, R.A. (1996). *Return to Resistance: Breeding Crops to Reduce Pesticide Resistance.* Davis, CA: AgAccess.

Soule, J.D. and J.K. Piper. (1992). *Farming in Nature: A Image.* Washington DC: Island Press.

Sumner, D.R. (1982). Crop rotation and plant productivity. In *Handbook of Agricultural Productivity, Vol. I*, ed. M. Rechcigl, Boca Raton, FL: CRC Press.

Thies, C. and T. Tscharntke. (1999). Landscape structure, and biological control in agroecosystems. *Science* 285: 893-895.

Tilman, D., D. Wedin, and J. Knops. (1996). Productivity and sustainability influenced by biodiversity in grassland ecosystems. *Nature* 379:718-720.

Toledo, V.M. (2000). La Paz in chiapas: Ecologia, luchas indigenas y modernidad alternativa. Ediciones Quinto Sol. Mexico, D.F.

Vandermeer, J. (1989). *The Ecology of Intercropping.* Cambridge, UK: Cambridge University Press.

Vandermeer, J. (1995). The ecological basis of alternative agriculture. *Annual Review of Ecological Systems* 26: 201-224.

Wilken, G.C. (1987). *Good Farmers: Traditional Agricultural Resource Management in Mexico and Guatemala.* Berkeley, CA: University of California Press.

Soil Ecosystem Changes During the Transition to No-Till Cropping

Tami L. Stubbs
Ann C. Kennedy
William F. Schillinger

SUMMARY. Growers in the United States and worldwide are adopting no-tillage (no-till) cropping to reduce soil erosion, improve soil quality, increase water infiltration, and reduce number of passes with farm equipment over their fields. Soil erosion from dry farmed (i.e., non-irrigated) cropland in most regions of the United States exceeds the tolerable rate. An understanding of the changes in the soil ecosystem with changing tillage practices is needed to minimize the impact of agriculture on the environment and foster the use of sustainable agricultural practices. The soil biota is critical to the functioning of any agro-ecosystem, but studying the soil biota is difficult due to the diversity and the challenges associated with isolating and identifying these organisms. Soil disturbance or lack of disturbance can have a profound effect on biotic populations, processes and community structure. This contribution examines changes that occur in soil during the transition to no-till crop-

Tami L. Stubbs is Associate in Research, Department of Crop and Soil Sciences, Washington State University, Pullman, WA 99164-6420.

Ann C. Kennedy is Soil Scientist, USDA-Agricultural Research Service, Pullman, WA 99164-6421 (E-mail: akennedy@wsu.edu).

William F. Schillinger is Associate Scientist, Department of Crop and Soil Sciences, Washington State University, Lind, WA 99341.

[Haworth co-indexing entry note]: "Soil Ecosystem Changes During the Transition to No-Till Cropping." Stubbs, Tami L., Ann C. Kennedy, and William F. Schillinger. Co-published simultaneously in *Journal of Crop Improvement* (Food Products Press, an imprint of The Haworth Press, Inc.) Vol. 11, No. 1/2 (#21/22), 2004, pp. 105-135; and: *New Dimensions in Agroecology* (ed: David Clements, and Anil Shrestha) Food Products Press, an imprint of The Haworth Press, Inc., 2004, pp. 105-135. Single or multiple copies of this article are available for a fee from The Haworth Document Delivery Service [1-800-HAWORTH, 9:00 a.m. - 5:00 p.m. (EST). E-mail address: docdelivery@haworthpress.com].

http://www.haworthpress.com/web/JCRIP

Digital Object Identifer: 10.1300/J411v11n01_06

ping, interrelations among organisms in the soil food web, and the relationships between organisms and their environment. As interest grows in sustainable cropping systems that mimic processes and soil organic matter turnover of native, undisturbed systems, it is imperative to understand how the transition to no-till affects an organism's niche, or functional role within the soil environment. Ecosystem investigations will enhance the understanding of changes that occur with the adoption of reduced tillage and no-till cropping systems so that these systems become increasingly viable.

KEYWORDS. No-till, conventional tillage, conservation tillage, microorganisms, fauna, residue management

INTRODUCTION

Degradation of agricultural soils caused by excessive tillage has spurred interest in minimum tillage and no-till cropping practices. No-till cropping creates the physical conditions of surface-managed residues and undisturbed soil that leave soil less susceptible to wind and water erosion (Baker, Saxton, and Ritchie, 1996). Maintenance of surface residues often increases microbial populations and diversity. Soil organic matter (OM) levels increase with no-till and soil may sequester C that would otherwise be released to the atmosphere as CO_2. In spite of the advantages of no-till farming, making the transition to no-till from tillage-based cropping is not without challenges. Research to determine soil biotic and agronomic changes, and economic feasibility during the transition from conventional tillage to no-till is ongoing. The study of soil biota is difficult due to the vast number and diversity of organisms, and the problems associated with isolating and identifying the soil's biotic community (Hawksworth and Mound, 1991).

With less soil disturbance, changes in soil nutrient status and the plant/microorganism interaction within the soil environment occur that may change the type and number of organisms present. As interest grows in developing sustainable cropping systems that mimic the processes and soil OM turnover of native, undisturbed systems, it is imperative to understand how the shift to no-till practices affects an organism's niche, or functional role within the soil environment.

TILLAGE PRACTICES

We define no-till cropping as planting directly into the residue of the previous crop without any tillage operations that mix or stir the soil prior to planting. No-till, also known as direct seeding, opens the door in all production zones for energy savings, improved soil quality, and excellent control of wind and water erosion. Adoption of minimum tillage and a continued movement toward no-till is needed to reduce soil erosion to tolerable levels. Acceptance of no-till is often slowed by concerns about transition costs, lack of experience and expert knowledge of no-till practices, grower resistance to change, uncertainties with crop yields, and risks of crop loss resulting from unpredictable agronomic factors. However, resource conservation and environmental benefits are favored by no-till and provide incentives for a gradual, continuing shift to this technology that is currently viewed as the farming practice of the future (Papendick and Parr, 1997). Surface residue from the previous crop is preferably left in the field, but may be swept into windrows, or removed by baling or burning in high residue situations. With conventional tillage there are many different management techniques and implements that produce differing degrees of disturbance. Moldboard plowing involves the greatest soil disturbance where the soil is inverted and about 90% of the residue buried. Minimum or reduced tillage involves some type of tillage implement (disk, chisel, cultivator, power harrow), but maintains 30% or more residue cover. In ridge tillage, ridges are made from tilled soil, and crops are planted into the ridges. With no-till there are varying levels of disturbance depending upon the type of no-till drill used. In other words, there are many gradations within no-till systems, but all cause considerably less soil disturbance compared to conventional tillage systems.

As growers reduce tillage, they may experience reduced crop yield due to interference from residue, which may reduce stand establishment due to lack of suitable no-till drills (Baker, Saxton, and Ritchie, 1996; Rasmussen, Rickman, and Klepper, 1997), increased incidence of disease (Cook and Haglund, 1991), nutrient immobilization (Elliott and Papendick, 1986), and more weeds (Kettler et al., 2000). The economics of no-till compared to tillage-based systems are often mixed, even within the same geographic region. For example, in eastern Washington State, the relative economics of no-till in the high (more than 450 mm annually) precipitation zone are generally positive (Camara, Young, and Hinman, 1999), whereas no-till cropping systems in the low (less than 300 mm annually) precipitation zone are not yet as profitable as conventional practices (Young, Hinman, and Schillinger, 2001).

Crop species diversity, cropping intensity, and crop rotation may change with tillage system, and in turn affect microbial diversity (di Castri and Younes, 1990; Hawksworth and Mound, 1991; Thomas and Kevan, 1993). Tillage changes soil physical characteristics and residue decomposition as well as soil chemical and structural characteristics. These factors both directly or indirectly impact the soil biology and the soil biotic community (Figure 1). Soil microbial populations are not static or homogeneous (Mikola and Setala, 1998), but rather highly sensitive to disturbance (Elliott and Lynch, 1994), and the soil biota may be the first group of organisms to shift with changes in tillage management. Soil organisms may be used as early indicators of changes in soil quality due to tillage and other perturbations (Kennedy and Papendick, 1995). Changes in soil parameters, especially microbial factors, during the transition from conventional tillage, to minimum tillage, and ultimately to no-till vary with soil type, implement used, cropping system, precipitation, or other variables (Kennedy and Smith, 1995; Lupwayi, Rice, and Clayton, 1998).

FIGURE 1. The relationships between tillage and soil ecology.

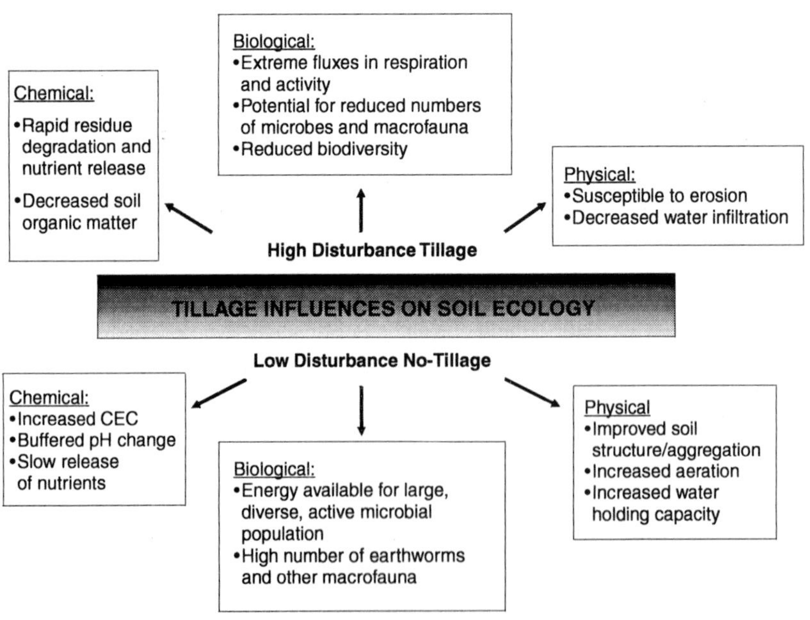

THE TRANSITION PERIOD

The study of soil microbial ecology is critical for our overall understanding of tillage management impacts. We refer to the changes occurring in soil physical, chemical and biological characteristics with increasing time under no-till, before reaching a new equilibrium, as the "transition period." Changes during the transition from conventional tillage to no-till may occur almost immediately or may require many years before differences can be detected (Figure 2). Effective management during the transition period is critical for the initial success of no-till systems and often dictates whether a grower continues to use no-till for the long term. Understanding the changing ecological relationships in the soil during the transition period is key to effective management. Change in soil quality with no-till compared to conventional tillage occurs most rapidly (during 1-4 years) in the surface soil (0-5 cm), and includes increased mineralized carbon (C), active microbial biomass (Alvarez and Alvarez, 2000), soil OM, aggregate stability, exchangeable Ca and extractable P, Mn, and Zn, and less extractable K, Fe, and Cu (Rhoton,

FIGURE 2. Changes in soil quality characteristics with time during the transition from tillage to no-till.

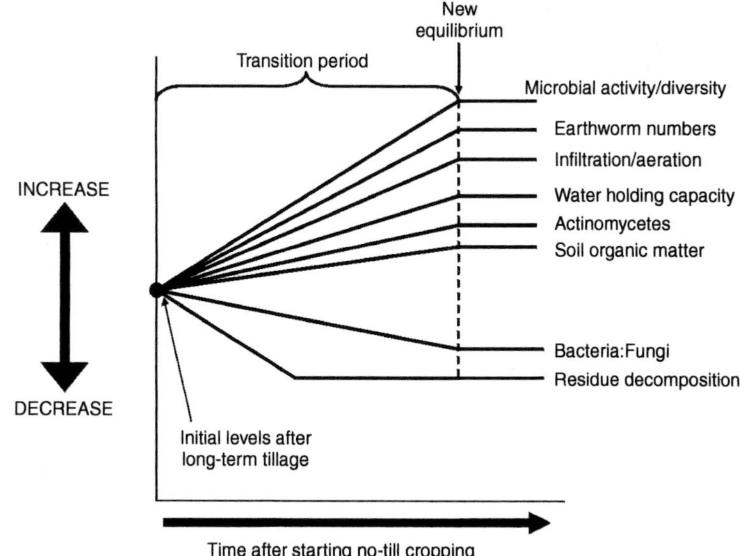

2000). After three years of no-till farming on land that was previously conventionally tilled in a cereal-fallow system in Washington State, increased electrical conductivity and readily mineralized C levels, and decreased pH were observed at two sites. Shifts in the fatty acid methyl ester (FAME) profile were associated with crop type and tillage (Stubbs et al., unpublished data). Long-term changes with no-till vs. conventional tillage include more total microbial biomass, higher contribution of microbial biomass C to organic C, and greater fungal hyphal length (Ananyeva et al., 1999). Soil quality changes are evident during the transition to no-till, but their significance in predicting the duration of this transition period or the positive or negative impacts of no-till is not yet known. The flux in soil parameters evident during this transition period will vary with soil type, climate and cropping system, thus long-term cropping system studies in numerous agro-environments are needed.

CROP SELECTION, ROTATIONS, AND RESIDUE

Crop species, rotation sequence, and degree of cropping intensity can directly or indirectly affect the soil biotic community. Plants and their exudates influence the soil microorganisms and soil microbial community that are found near roots (Duineveld et al., 1998; Ibekwe and Kennedy, 1998; Ohtonen et al., 1999). Soil microorganisms may affect plant growth and influence plant competition among species (Westover, Kennedy, and Kelley, 1997). In turn, plants may be a selective force for rhizosphere microbial populations through their influences on exudation patterns (Meharg and Killham, 1995) and soil nutrients (Droge, Puhler, and Selbitschka, 1999; Jensen and Nybroe, 1999; Pennanen et al., 1999). The composition of the plant community may drive the composition of the soil microbial community (Achouak et al., 2000; Fulthorpe, Rhoades, and Tiedje, 1998; Minamisawa, Nakatsuka, and Isawa, 1999); microbial diversity is more responsive to differences in soil characteristics rather than plant type (Degens et al., 2000; Gross, Pregitzer, and Burton, 1995; Wardle et al., 1999).

The transition to no-till often necessitates changes in crop rotation for residue and disease management, or to manage economic risk. Increased cropping intensity and planting specific crop species, combined with reduced tillage, have been shown to improve soil quality characteristics. Drijber et al. (2000) found that Nebraska soils planted to winter wheat (*Triticum aestivum* L.) had greater microbial biomass than fallow soil, regardless of tillage intensity and that during the fallow portion of a wheat-fallow rotation, no-till fields had greater microbial biomass than fields that had been moldboard plowed. The microbial diversity of soils under wheat preceded by a legume crop [red clover (*Trifolium pratense* L.) or peas (*Pisum sativum* L.)] was higher than in wheat preceded by summer fallow or continuous wheat (Lupwayi, Rice, and Clayton, 1998). In a study in Brazil where fields had not received inoculant of the nitrogen (N)-fixing bacterium *Bradyrhizobium* for more than 15 years, no-till combined with crop rotations containing soybeans (*Glycine max* L.) resulted in higher populations and greater diversity and activity of *Bradyrhizobium* than cropping systems without soybeans (Ferreira et al., 2000).

No-till and conventional tillage systems have vastly different quantities and types of surface residue distribution. Surface residue retention improves soil quality characteristics (Dalal, Henderson, and Glasby, 1991; Doran, Sarrantonio, and Liebig, 1996; Karlen et al., 1994; Ladd et al., 1994) by increasing OM accumulation (Nyakatawa, Reddy, and Sistani, 2001). Other benefits of surface residues include greater fungal biomass (Holland and Coleman, 1987; Karlen et al., 1994), soil C levels (Holland and Coleman, 1987; Karlen et al., 1994), earthworm populations, and microbial activity (Karlen et al., 1994). Soil aggregates are more water-stable when residue levels are doubled than when the residue is maintained or removed in continuously cropped corn in Wisconsin (Karlen et al., 1994). In a double-cropped wheat and soybean system, non-burned, no-till soil had greater numbers of algae, actinomycetes and fungi initially (Harris et al., 1995). Although increases in microbial numbers and activities may be beneficial, this may not always be the case (Kennedy and Smith, 1995). It may be more constructive to study the changes, either plus or minus, in populations or activities as indicators of flux within the ecosystem.

PHYSICAL AND CHEMICAL SOIL CHARACTERISTICS

Tillage will fundamentally change the physical and chemical characteristics of the soil and profoundly alter the matrix supporting the growth, functioning and survival of the soil biota (Table 1). Physical parameters of soil altered by tillage have a great influence on the soil ecology. Disturbance of the soil system, intact versus homogenized soil, infiltration, and residue placement are some of the changes that will affect soil organisms. Compaction of soils under long-term no-till management is a physical change that concerns growers, and greater surface

TABLE 1. Physical and chemical attributes of changing tillage systems in select studies.

SOIL ATTRIBUTE[§]	*LENGTH OF STUDY/ YEARS OF SAMPLING*	*SYSTEMS COMPARED**	*RESULTS*	*LOCATION*	*REFERENCE*
AS, SOM, pH	0, 4, 8 yr/0, 4, 8 yr	NT, CT, cotton, sorghum-corn, and soybean-wheat rotations.	After 4 yrs NT 0-2.5 cm ↑ AS, SOM, pH unchanged.	Mississippi	Rhoton, 2000
AS, SOM	8 yr/1 yr	MP (20-25 cm) and cultivation to 12 cm.	Shallow tillage ↑ AS and SOM.	Sweden	Stenberg, Stenberg and Rydberg, 2000
SOM, pH	3 yr/3 yr	NT, CT cotton with poultry litter and winter rye cover crop.	Reduced till or NT ↑ surface SOM; winter rye cover crop and poultry litter, pH unchanged.	Alabama	Nyakatawa, Reddy and Sistani 2001
SOM	7 yr/1 yr	NT and stubble mulch tillage, winter wheat-sorghum-fallow.	NT ↑ SOM.	Texas	Unger, 1991
Distribution of C with aggregate size classes	26 (NE), 33 (OH), 9 (MI), 16 (KY)/1 yr	Native vegetation, NT, CT.	Macro- and micro-aggregate C: Native vegetation > NT > CT.	Nebraska, Ohio, Michigan, Kentucky	Six et al., 2000
pH, POM C, TOC	8 yr/2 yr	NT, chisel plow, MP	NT 0-5 cm ↑ pH; NT had greatest amount of TOC as POM, NT had greatest amount of TOC.	Illinois	Hussain, Olson and Ebelhar, 1999
TOC, pH, EC	NT 20+ yr; tillage treatments 5 yr/1 yr	NT plowed 5 yr previously, NT undisturbed, and CT	NT plow ↓ TOC 20%; NT plow ↑ pH; EC NT plow = NT undisturbed.	Nebraska	Kettler et al., 2000
TOC, microbial biomass N, pH	20 yr/1 yr	CT and NT with residue or burned	NT surface 0-2.5 cm ↑ TOC and microbial biomass N; NT with residue and urea ↓ pH.	Australia	Dalal, Henderson and Glasby, 1991
BD	3 yr/0, 3yr	NT and disk, soybean-corn rotation.	No significant changes in BD for NT soils, disking ↑ BD in 67% of cases.	Iowa	Logsdon and Cambardella, 2000
BD, water infiltration	NT 20+ yr; Tillage 5 yr/1 yr	NT plowed 5 yr previously, NT undisturbed, and CT.	NT ↑ BD; water infiltration for the second 2.5 cm of water applied was greater for NT undisturbed than NT plow or CT.	Nebraska	Kettler et al., 2000
Gravimetric moisture, pH, total C	Cultivated land cropped since 1950, forest cleared and mowed annually since 1960, sampled for 1 yr	Cultivated, annually cropped, forest that was cleared, not plowed or cropped	Cultivated ↓ C by half, ↑ moisture and pH.	Michigan	Robertson, Crum and Ellis, 1993
Denitrification and nitrification potential, inorganic N, pH	100+ yr/1 yr	Undisturbed prairie, CT wheat-barley-pea rotation.	Prairie soils ↑ denitrification and nitrification potential; cultivated ↑ inorganic N; = pH.	Washington	Kennedy and Smith, 1995

[§] AS = aggregate stability, SOM = soil organic matter, BD = bulk density, POM = particulate organic matter, TOC = total organic carbon, EC = electrical conductivity.
* NT = no-till, CT = conventional tillage, MP = moldboard plow.

bulk density measurements have been found in many no-till soils (Pierce, Fortin, and Staton, 1994). However, these fears may be unfounded for some soil types. The loss of OM associated with intensively tilled, eroded soils may lead to increased bulk density (Edwards 1991; Karlen et al., 1990). Logsdon and Cambardella (2000) showed that in several fine-loam soils, soil bulk density remained constant for the first three years of no-till management. In three of four rice producing study sites in Bangladesh, use of no-till with mulch, compost and green manure resulted in lower bulk density than systems using conventional tillage and inorganic fertilizers (White et al., 2001). Changes in bulk density with during the transition to no-till will most likely vary by location, and are dependent on such variables as soil type, OM content, cropping system and climate (Kettler et al., 2000). Detrimental increases in soil bulk density can often be avoided in no-till regimes by operating equipment under proper soil moisture conditions.

Beneficial soil quality changes are often observed between long-term deep cultivation, and shallow tillage practices. Stenberg, Stenberg, and Rydberg (2000) showed that in fields cultivated to a depth of 12 cm, soil OM and microbial biomass increased and led to improved aggregate stability compared to conventional moldboard plowing at 20-25 cm depth. Greater aggregate stability is most likely the result of increased OM supporting greater microbial biomass that exuded extracellular metabolites such as polysaccharides and proteins to cement soil particles into larger, more stable aggregates (Stenberg, Stenberg, and Rydberg, 2000).

Worm and root channels may enhance water infiltration and reduce runoff (Edwards et al., 1992). In the steeply sloping Palouse region of southeastern Washington and northern Idaho, soil loss by water erosion during a single winter season may exceed 50 Mg ha^{-1}. The worst erosion typically results when snowmelt and/or rain occurs on thawed soil overlying a subsurface frozen layer (McCool, 1990). No-till growers in this region credit worm channels, that extend from the soil surface to depth, for improved soil water infiltration and reduced erosion in their fields (John Aeschliman, personal communication). Intact channels from reduced soil disturbance and lack of soil mixing each season result in less mixture of substrates for microbial growth. Tillage may lead to soils with a greater proportion of microaggregates, which are C-depleted, compared to macroaggregates, which contain more C than microaggregates (Six et al., 2000).

Despite all of the advantages of reduced tillage, physical soil disturbance is beneficial in some agricultural systems. Abawi and Widmer

(2000) cite numerous examples where yield of beans due to disease was reduced when the soil was subjected to reduced tillage or no-till compared to intensive tillage. The increase in yield with tillage was attributed to reduced compaction, improved drainage, and higher soil temperature that led to improved bean root competition against pathogens. In several wheat production regions of the world, Rhizoctonia root rot caused by *Rhizoctonia solani* (Kühn) AG8, is an important disease of crops planted into cereal stubble with no-till (Weller et al., 1986). Rhizoctonia is a minor disease of cereals grown under conventional tillage, but can be devastating for these crops in no-till cropping systems. The most effective practices that limit the severity of this disease in no-till cropping systems are (i) elimination of volunteer and other grass weeds that serve as hosts for the pathogen during the "green bridge" period of 2-3 weeks and preferably 2-3 months prior to planting wheat (Smiley et al., 1992), and (ii) soil disturbance in the seed row 5-6 cm below the seed at the time of planting (Roget et al., 1996). There are presently no wheat cultivars that are resistant to *Rhizoctonia*.

Some soil chemical changes that vary with tillage exert a subtle effect on soil organisms. Long-term depletion and homogenization of nutrients may affect the soil community, but it is difficult to measure these changes when nutrient content is altered by the distribution of the substrate and OM. Organic matter and N concentrations are more uniform throughout the soil zone mixed by tillage than under no-till (Unger, 1991). Organic matter content is greater in surface soils of no-till systems compared to tilled systems; however, the increase in OM with reduced tillage is not always evident at lower soil depths (Unger, 1991). It is possible that the differences between cultivated and undisturbed lands are due to C and N depletion (Cavigelli and Robertson, 2000; Paul et al., 1999). With long-term cultivation, C and N pools are homogenized and C pools have been depleted by 89% and 75%, respectively (Knops and Tilman, 2000). In intensively tilled cropping systems, the exhausted C and N pools are not readily re-established even with conservation practices (Drinkwater, Wagoner, and Sarrantonio, 1998; Knops and Tilman, 2000). As would be expected, the homogenization of nutrients is greatest with increased tillage intensity (Robertson, Crum, and Ellis, 1993). Bacterial populations tend to decline under low pH, whereas low pH has less effect on fungal communities. Soils with similar C and N levels may have similar microbial community composition (Wardle et al., 1999; Zak et al., 2000).

Stubbs (2000) showed that residue quality, and decomposition of wheat straw in soil varied widely among cultivars and that decomposition is correlated with acid detergent fiber, lignin, total N and C:N ratio. Knowledge of differences in wheat cultivar decomposition rates is important for growers in both high and low precipitation regions where excessive straw production may interfere with planting operations or, conversely, where straw may not be in sufficient quantity to control erosion.

Response of soil enzymes to tillage is site-specific, and more dependent on soil characteristics and sampling protocol than management (Bergstrom, Monreal, and King, 1998). Two sites in Ontario were assayed for the soil enzymes dehydrogenase, urease, glutaminase, phosphatase, arylsulfatase, and β-glucosidase. At the site with no-till and prior history of forages including alfalfa (*Medicago sativa* L.) in the crop rotation, there were increases in all of the soil enzymes assayed. At a separate site, differences in enzyme concentration were not consistent with management, and depended more upon soil texture and sampling depth than differences in tillage (Bergstrom, Monreal, and King, 1998).

SPECIFIC GROUPS OF ORGANISMS

The organisms of the biotic community in soil can be viewed as an interconnected web, where each portion of the web interacts with other members of the soil community (Figure 3). A soil food web contains different trophic levels ranging from bacteria and fungi to carnivorous animals. The members of the soil food web have a profound effect on crop production through processes such as residue decomposition and nutrient cycling. Plants, with their spreading root systems, are able to access a large volume of soil and supply soil organisms with nutrients. This is especially true in the vicinity of the rhizosphere, where microbial populations may be 10- to 100-fold higher than in soil without growing plants (Bottomley, 1998). High populations of microorganisms in the rhizosphere lead to larger numbers of nematodes and protozoa that feed on bacteria and fungi, and in turn, higher populations of microarthropods that prey on nematodes and protozoa.

A soil food web's composition and complexity varies with management (Moore, 1994). The root-microorganism and root-macrofaunal interactions after minimum tillage are vastly different from that after moldboard plowing (Table 2). In no-till cropping systems, microbial activities differ immensely with depth, with the greatest microbial activity

FIGURE 3. The soil food web. Adapted from Hunt et al., 1987.

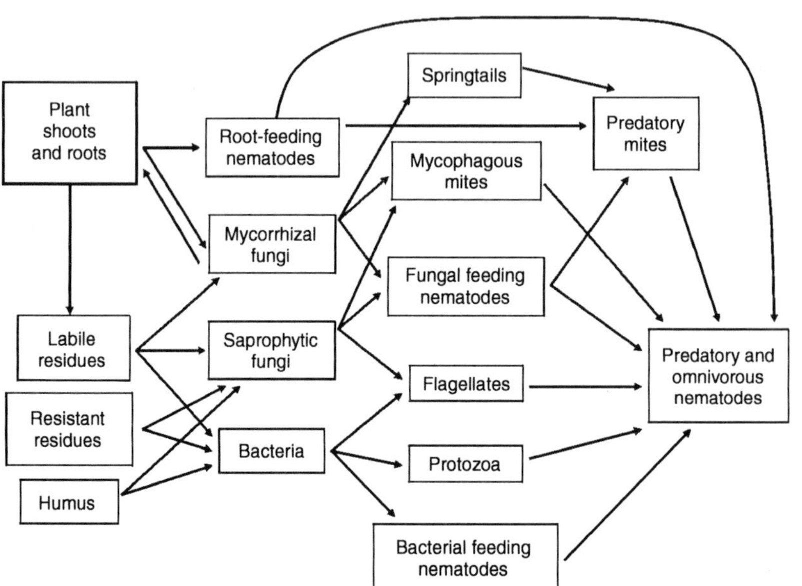

occurring near the surface, whereas in the tilled system activities are more evenly distributed throughout the tillage layer (Doran, 1980). The type and functional characteristics of the species present in agroecosystems may be as important as the number of species (Grime, 1997; Hopper and Vitousek, 1997). Bacteria dominate a tillage-based system, whereas fungi and earthworms play a major role in no-till systems (Verhoef and Brussaard, 1990). Microbial community analyses are useful in distinguishing among specific organism populations associated with conventional, minimum and no-tillage cropping systems (Drijber et al., 2000). Invertebrates influence each level of the soil food web, and play a key role in soil OM retention and turnover (Wolters, 2000). Residue quality affects the populations of micro- and mesofauna that colonize residue (vanVliet et al., 2000), and thus rate of residue decomposition and N release. The above ground, herbivorous members of a food web, for example grasshoppers, may alter nutrient cycling within an ecosystem by changing plant abundance and the rate of residue decomposition (Belovsky and Slade, 2000). Soil food dynamics will vary with degree of disturbance and fluctuate with changes in management.

TABLE 2. Biological attributes of changing tillage systems in select studies.

SOIL ATTRIBUTE§	LENGTH OF STUDY/ YEARS OF SAMPLING	SYSTEMS COMPARED*	RESULTS	LOCATION	REFERENCE
Microbial biomass	8 yr/1 yr	MP (20-25 cm) and cultivation to 12 cm	Shallow tillage ↑ microbial biomass.	Sweden	Stenberg, Stenberg and Rydberg, 2000
Microbial biomass, fungal hyphal length	22 yr (mollisol soil) and 39 yr (spodosol soil)/3 yr	NT and CT	NT ↑ microbial biomass, greater contribution of C_{mic} to C_{org}, and greater fungal hyphal length.	Russia	Ananyeva et al., 1999
Microbial biomass, community structure	25 yr/1 yr	NT, sub-till, MP wheat fallow and native sod	Cropping ↑ microbial biomass, no tillage differences in surface 0-15 cm, in fallow NT ↑ microbial biomass. NT ↑ FAME biomarker for AM fungi, PL- and EL-FAMEs differed among tillage treatment in wheat-fallow.	Nebraska	Drijber et al., 2000
Total and active microbial biomass, mineralized C	1 yr/1 yr	NT and CT	NT ↑ active microbial biomass and mineralized C in the top 5 cm, CT and NT = for total microbial biomass.	Argentina	Alvarez and Alvarez, 2000
Microbial biomass C	9 yr/5 yr	Continuous wheat, wheat-pasture, wheat-beans, wheat-fallow with residues burned, incorporated or retained	↑ microbial biomass C when wheat-pasture for first 3 yr and when residues retained or incorporated.	Australia	Ladd et al., 1994
Biomass C, phosphatase, dehydrogenase, microbial diversity	100+ yr/1 yr	Undisturbed prairie, CT wheat-barley-pea rotation	Higher biomass C, phosphatase and dehydrogenase in prairie soil, ↑ microbial diversity in cultivated soil	Washington	Kennedy and Smith, 1995
Soil enzymes	7 to 13 yr/2 yr	NT and CT, varying crop rotations at two sites	Site 1: Higher concentrations of all enzymes with NT and previous history of alfalfa; Site 2: No consistent response of enzymes to tillage practice.	Ontario	Bergstrom, Monreal and King, 1998
Glomalin, arbuscular mycorrhizal fungi	1 to 4 yr/1 yr.	NT and plow tillage	NT ↑ glomalin, leading to greater AS.	Maryland	Wright, Starr and Paltineau, 1999
Fungal communities	10 yr/1 yr	NT and CT, soil and sorghum residue	Residue: CT ↑ density of fungal hyphae, but fewer CFU^{Ψ} than NT Soil: NT ↑ density of fungal hyphae, no differences in CFU^{Ψ}.	Georgia	Beare et al., 1993

TABLE 2 (continued)

SOIL ATTRIBUTE[§]	*LENGTH OF STUDY/ YEARS OF SAMPLING*	*SYSTEMS COMPARED*[*]	*RESULTS*	*LOCATION*	*REFERENCE*
Fungal biomass, residue decomposition, N immobilization	2yr/2 yr	CT with surface residue and incorporated residue	Surface residue ↑ fungal biomass, ↑ immobilization, ↓ residue decomposition.	Colorado	Holland and Coleman, 1987
Glucosamine:muramic acid, fungal biomass, POM C,TOC, water stable aggregates	12 (ND), 11 (CO), 15 (TX), 26 (NE), 22 (KS), and 26 (KY) yr/1 yr	CT and NT, except CT and tall grass prairie in KS, and CT and bluegrass sod in KY	All values higher for NT.	North Dakota, Nebraska, Colorado, Texas, Kansas, Kentucky	Guggenberger et al., 1999
Soil-borne fungal pathogens of beans	Review article		Tillage ↓ several root diseases.		Abawi and Widmer, 2000
Actinomycetes, algae, bacteria, fungi, nitrifiers	3 yr/3 yr	Burning, NT, CT with soybean-wheat	Bacteria and nitrifiers no change; NT ↑ actinomycetes, algae and fungi early in growing season.	Georgia	Harris et al., 1995
Bradyrhizobia populations and diversity	17 yr/1 yr	NT and CT soybean, wheat-maize, soybean-wheat, maize rotations	NT and soybean rotations > CT and cont. cropping.	Brazil	Ferreira et al., 2000
Microbial diversity	3 yr/2 yr	NT and CT, continuous wheat, red clover-wheat, pea-wheat, fallow-wheat rotations	↑ diversity under NT, and red clover-wheat and pea-wheat rotations than with continuous wheat or fallow-wheat.	Alberta	Lupwayi, Rice and Clayton, 1998
CO_2 flux	13 yr/2 yr	NT and CT continuous wheat and wheat-fallow	Continuous wheat: NT < CT. Wheat-fallow: NT < CT in wheat.	Saskatchewan	Curtin et al., 2000
Collembola, Acarina	11 yr/2 yr	NT, chisel plow, ridge till corn	NT > CT, populations in CT recovered with time.	Ontario, Canada	Neave and Fox, 1998
Mites	1 yr/1 yr	MP, chisel plowing and light cultivation	Tillage ↓ populations.	Germany	Hülsmann and Wolters, 1998
Orbatid mites	Review article	NT and CT	NT > CT		Behan-Pelletier, 1999

Arthropods	1 yr /10 months	NT and CT wheat	Predators, biomass > in NT; phytophagous CT = NT.	Argentina	Marasas, Sarandon and Cicchino 2001
Arthropods	3 yr/1 yr	NT and CT sorghum with winter clover or winter rye cover crops	Biomass was affected more by type of N source than tillage. Winter rye > winter clover.	Georgia	Blumberg, Hendrix and Crossley, 1997
Ant diversity	Established fields/ 2 weeks	Reduced tillage and CT	Reduced tillage > CT.	North Carolina	Peck, McQuaid and Campbell, 1998
Invertebrates (herbivores, detritovores and predators)	5 yr/5 yr	NT, reduced tillage, CT grain production	NT ↑ detritovores and predators, no differences in herbivores.	NE Australia	Robertson, Kettle, Simpson 1994
Earthworms	5 yr/1 yr	NT and CT, corn-soybean rotation	NT > CT 57% of sites; NT = CT 29% of sites; CT > NT 14% of sites.	Indiana and Illinois	Kladivko, Akhouri and Weesies, 1997
Earthworms	1 yr/1 yr	Alfalfa, NT corn, CT corn	Alfalfa = NT corn > CT corn.	Wisconsin	Gallagher and Wollenhaupt, 1997
Earthworms	5 yr/5 yr	Low/normal tractor traffic	↑ earthworms with low tractor traffic.	Norway	Hansen and Engelstad, 1999
Earthworms	NT 3 yr, chisel disk 3 wk/1 yr	Chisel disk and NT, corn-soybean, cont. corn, cont. soybean rotations	NT > chisel-disk, no difference with rotation or N fertilizer.	Missouri	Jordan et al., 1997
Earthworms	1 yr/1 yr	CT/NT, soybean-corn and wheat-corn rotations	NT soybean-corn > NT wheat-corn > CT	Missouri	Hubbard, Jordan and Stecker, 1999
Earthworms	24 yr/2 yr	NT and CT wheat-fallow	NT > CT	Alberta	Clapperton et al., 1997

§ FAME = fatty acid methyl esters; POM = particulate organic matter; TOC = total organic carbon
* NT = no-till, CT = conventional tillage, MP = moldboard plow
Ψ CFU = colony-forming units

Microorganisms

The presence of a large and diverse soil microbial community is crucial to the productivity of any agroecosystem, regardless of management. The composition of the microbial community influences the rate of residue decomposition and nutrient cycling in both no-till and tillage-based systems (Beare et al., 1993).

One of the primary roles of fungi is decomposition of plant residues, especially lignin, which is more resistant to decomposition than other plant residue components. This is especially important in a successful no-till system that requires specialized fungi to degrade surface residue, such as cellulose (*Chaetomium*, *Fusarium*, *Trichoderma*) and lignin (Basidiomycetes) decomposers. No-till systems provide for a more diverse population of residue decomposers, and a slower release of nutrients (Curtin et al., 2000; Hussain, Olson, and Ebelhar, 1999). Soil fungi may be responsible for the net N immobilization that often occurs in the surface residue of no-till fields (Frey et al., 2000). Fungal hyphae contribute to higher soil OM content and aggregate stability in surface soils under no-till, and reduce the potential for soil erosion (Guggenberger et al., 1999).

Specialized fungi that form symbiotic relationships with plant roots are mycorrhizae. Mycorrhizal fungi are responsible for translocation of nutrients, especially soil P, and are known to help alleviate water stress during drought conditions (Morton, 1998). Glomalin, a glycoprotein produced by arbuscular mycorrhizae, is well correlated with increased aggregate stability. As the transition period with no-till lengthens, so does aggregate stability and glomalin production (Wright, Starr, and Paltineanu, 1999). Moldboard plowing in a wheat-fallow crop rotation in Nebraska has been shown to markedly reduce the fatty acid methyl ester (FAME) biomarker for mycorrhizal fungi compared to that from no-till (Drijber et al., 2000).

Tillage may negatively affect fungi to a greater extent than other microorganisms, such as bacteria, because of the physical severing of the hyphal mat or hyphal strands that can form after long periods with little disturbance. Because of the physical disruption of fungal hyphae with cultivation, movement of fungi through cultivated soils is reduced. In no-till systems a higher proportion of fungal decomposers are found, while cultivated systems favor higher populations of bacterial decomposers (Hendrix et al., 1990). At six sites (North Dakota, Nebraska, Colorado, Texas, Kansas and Kentucky) of a study comparing no-till and conventional tillage, fungal biomass was greater than bacterial biomass in the no-till fields (Guggenberger et al., 1999). Fungi are responsible for causing many soil-borne plant diseases (Cook and Haglund, 1991). Diseases such as Rhizoctonia may be more devastating in no-till systems where there is no severing of hyphal strands (Weller et al., 1986).

Bacteria are also decomposers in soil, especially in tillage-based systems (Beare et al., 1993), and are crucial for mineralization of nutrients, making them available to plants and other organisms. Soil bacteria also aid in weathering soil minerals, contribute to soil formation, and secrete polysaccharides to hold soil particles together and promote aggregate stability. Some bacteria produce antibiotics that affect other organisms. Root-colonizing bacteria affect plant growth by producing root growth-promoting hormones, and some have the potential for use as biological control agents (Alexander, 1998). Actinomycetes are a specialized group of soil bacteria. They are able to degrade plant materials such as cellulose, mineralize nutrients, and some produce antibiotics. Actinomycetes may tolerate low soil water potential better than other bacteria, but they are not tolerant of low soil pH (Alexander, 1998). Early in the growing season, undisturbed fields in Georgia with soybean surface residue had the highest numbers of actinomycetes and fungi compared to burned or tilled treatments (Harris et al., 1995).

The dynamics of bacterial functions in soil can change with tillage. In a study of the diversity of prairie and cultivated soils in Washington, microbial diversity indices were greater in cultivated systems compared to undisturbed grassland (Kennedy and Smith, 1995). With tillage, greater residue surface area was in contact with the soil and more substrate was available for colonization and more microbial activity occurred. The increase in microbial diversity with disturbance indicates a change in the microbial community to one that exhibits a greater range of substrate utilization and resistance to the stress of cultivation. In contrast to the Kennedy and Smith (1995) results, Lupwayi, Rice, and Clayton (1998) showed that crop rotation and reduced tillage support greater microbial diversity than conventional tillage as shown by a reduction in substrate richness and evenness. Their study promoted no-till as a component of sustainable agriculture systems. This is one example of the many contradictory results that may be due to differences in soils, climates and vegetation as well as degree of soil disturbance.

Algae are photoautotrophic organisms that are found in soil at populations of 10^3-10^4 per gram of soil, far fewer than bacteria and fungi. Numbers of algae are positively related to soil moisture and light (Shimmel and Darley, 1985), and negatively related to soil depth. Algae assist in N fixation and soil aggregate stability. They are susceptible to soil disturbance and may be good indicators of soil quality. Algae are positively affected by reduced tillage and maintaining surface residue which allows for a higher moisture regime for longer periods of time, and fosters algal growth on the soil surface (Harris et al., 1995).

Micro- and Mesofauna

In any soil community the presence of the various trophic groups is critical to the functioning of that soil. The intermediate-size group of micro- and mesofauna function as both predators of microorganisms and a food source for macrofauna. Protozoa, nematodes, and mites are important members of the soil food web because of their contributions to residue decomposition and nutrient cycling. Protozoa are considered to be microfauna because of their size (< 200 μm long), while nematodes and mites are categorized as mesofauna (200-1000 μm long).

Protozoa are crucial to the functioning of the soil food web and other ecosystems because of their roles as consumers of bacteria (Wood, 1989) and for nutrient cycling and aiding in providing energy for microorganisms, plants, and animals (Foissner, 1999). The fluctuations in microbial populations with tillage will also affect protozoan populations since protozoa feed on these organisms. Protozoa may be useful indicators of changes in soil quality because they react rapidly to changes in the environment; however, there is a need for improved methodology to accurately count and identifying protozoa and other soil organisms (Foissner, 1999).

Nematodes are multicellular, round worms that may occur in numbers as large as several million meter^{-2} of surface soil (Ingham, 1998). They can be sub-divided into groups based on their functional roles: bacterial feeding, fungal feeding, root feeding, predatory, and omnivorous. They are key in the turnover of microbial communities through predation and nutrient release. Nematodes and protozoa may graze discriminately upon bacteria and fungi, thus altering the microbial community in soil, and altering residue decomposition rates (Ingham, 1998). Some are plant parasites; however, most rely on other organisms as a food source (Wood, 1989). Tillage negatively affects populations of larger nematodes (Jones, Larbey, and Parrott, 1969; Oostenbrink, 1964).

Mites (Acari) are very sensitive to the presence of litter or residue. Mechanical disturbance of surface residue has a negative effect on mite populations and species diversity. No-till cropping had a positive effect on abundance and richness of Collembola and Acarina, with the residue regulating extremes of high temperature and loss of moisture compared to tillage (Neave and Fox, 1998). Hulsmann and Wolters (1998) also found the greatest negative effects of tillage on soil mites soon after tillage had occurred. While no-till may enhance the residue levels in surface soils, thus enhancing mite populations, use of the herbicide Atrazine in a no-till situation had a negative effect on mite populations (Moore et al., 1984).

Macrofauna

Arthropods such as millipedes (Diplopoda) and centipedes (Chilopoda) play a critical role in soil quality maintenance. Arthropods are important in OM formation, soil structure and soil pore formation due to their mechanical activity (Zunino, 1991). Tillage systems affect arthropod fauna by altering the proportion of functional arthropod groups found in soil (Marasas, Sarandon, and Cicchino, 2001), with the abundance and diversity of animal species decreasing with increasing tillage intensity (Alderweireldt, Konjev, and Polett, 1991; Pfiffner and Niggli, 1996). There is an increase in biodiversity of edaphic taxa of arthropods under no-till, with predators comprising the highest numbers (Robertson, Kettle, and Simpson, 1994; Stinner and House, 1990). Moldboard plowing decreases predator numbers, while no-till favors more predatory activity due to reduced mechanical disturbance and greater residue cover.

Tillage impacts the numbers of arthropods, with greater numbers found under no-till than under minimum or conventional tillage (Marasas, Sarandon, and Cicchino, 2001; Roberston, Kettle, and Simpson, 1994). In both of those studies, functional groups (predators, detritovores, phytophagous) were affected to different degrees as tillage level was reduced. Cropping system differences may confound clear conclusions about the effect of tillage on arthropods. In some cases, arthropod populations may respond more to cropping practices than to tillage. In a 7-year study in New Zealand, intensive tillage did not adversely affect arthropod populations compared to hand-hoeing, but that effect was most likely due to higher quality litter present with high weed infestations prior to cultivation (Wardle et al., 1999). Presence of greater weed biomass was more favorable to arthropods than either a perennial crop [asparagus (*Asparagus officinalis* L.)] or an annual crop [corn (*Zea mays* L.)]. Tillage treatment (no-till or conventional) played a smaller role in determining arthropod biomass than did type of cover crop, with arthropod numbers and biomass greater on sorghum (*Sorghum bicolor* L.) that followed a winter rye (*Secale cereale* L.) cover crop in Georgia, compared to a winter clover cover crop (Blumberg, Hendrix, and

Crossley, 1997). Neave and Fox (1988) found that cryptozoic invertebrates became more abundant under no-till especially in summer, but their numbers in conventional tillage, chisel or ridge tillage treatments increased later in the season.

Ants (Hymenoptera: Formicidae) may be useful as bioindicators of the effects of management practices in agroecosystems (Peck, McQuaid, and Campbell, 1998). In North Carolina where minimum and conventional tillage were compared, most species of ants preferred the undisturbed field margin to the cultivated areas. Of the ant species inhabiting cropland, 76% were found in fields prepared with minimum tillage (Peck, McQuaid, and Campbell, 1998).

Earthworms may be early and visual indicators of reduced tillage in some soils. No-till systems have the potential to increase populations of earthworms in agricultural fields (Clapperton et al., 1997; Jordan et al., 1997). The populations of shallow-dwelling earthworms (*Apporectodea tuberculata* and *A. trapezoides*) were greater in no till soils at 8 of 14 sites in Indiana and Illinois when compared with paired conventional tillage soils (Kladivko, Akhouri, and Weesies, 1997). Greater quantity of residue, and lower C:N ratio of residue in a no-till corn-soybean rotation in Missouri led to higher populations of earthworms (*A. trapezoides*) than a no-till wheat-corn rotation, and both of these treatments had more earthworms than the conventionally tilled treatments (Hubbard, Jordan, and Stecker, 1999). Cropping systems in Norway that reduced compaction through low tractor traffic compared to normal tractor traffic had higher populations of earthworms; however, species composition did not change regardless of traffic (Hansen and Engelstad, 1999). Earthworm populations are higher in soils after chisel-type implements that slice the soil without the mixing and inversion of soil by disc plows (Paoletti, Schweigel, and Favretto, 1995).

High populations (40-60 earthworms m^{-2}) of earthworms (*Lubricus terrestris* L.) impact residue management strategies for no-till fields in Wisconsin due to their ability to bury large quantities of surface residue (122 g biomass m^{-2} from October to May) which may expose the soil surface to the potential of rain drop and splash erosion (Gallagher and Wollenhaupt, 1997). Earthworm middens at the soil surface of a no-till field in Ohio were found to enhance soil quality and productivity because they contain greater soil water, coarse organic litter (>2 mm), total C and N, C:N ratio, microbial activity, NH_4-N, and dissolved organic N than bulk soil from outside the cornfield (Subler and Kirsch, 1998). These middens contribute to the spatial heterogeneity in soils. While earthworms burial of residue may impact residue management, many of the positive attributes associated with earthworm activity affect agroecosystem processes of water infiltration (Edwards et al., 1992), residue decomposition (Subler and Kirsch, 1998), and N-cycling (Parkin and Berry, 1999).

METHODS FOR MEASURING CHANGES IN SOIL ECOSYSTEMS WITH TILLAGE

Many methods are currently used by researchers to help quantify biological diversity and to compare the effects of management practices on the soil ecosystem. Along with plate counts and emergence traps, other methods are now available, enhancing our ability to measure changes in the soil biota. Microbial plate counts do not provide accurate community structure information of soils due to the issue of culturability; however, they still provide a measure of a portion of the population (Dodds et al., 1996; Kennedy and Smith, 1995). Substrate utilization patterns have been used to study community structure and indicate functional diversity and metabolic potential (Bossio and Scow, 1995; Degens, 1999; Garland, 1996; Haack et al., 1995; Zak et al., 1994). To qualitatively characterize soil microbial communities, whole soil fatty acid methyl ester (FAME) analysis is used in microbial community investigations (Cavigelli, Robertson, and Klug, 1995; Kennedy and Busacca, 1995; Zelles et al., 1994).

Molecular genetic techniques are the most recent tool for characterizing soil microbial communities. Nucleic acid analyses have shown that several thousand independent genomes can be present in 1 g of soil (Torsvik, Goksoyr, and Daae, 1990). Methods such as denaturing gradient gel electrophoresis (DGGE) (Muyzer and Uitterlinden, 1993) and single strand conformation polymorphism (SSCP) (Lee, Zo, and Kim, 1996) allow comparisons of microbial similarities among whole soil community samples. DNA reassociation (diversity), percent G+C composition of community DNA (species composition), and DNA:DNA hybridization (similarity and relative diversity) (Griffiths, Ritz, and Glover, 1996) can also be used in community analyses. DNA microarray technology has been used to rapidly analyze microbial communities based on phylogenetic groupings (Guschin et al., 1997).

There are several methods of analysis for larger soil biota including those for nematodes (Yeates and Bongers, 1999), protozoa (Foissner, 1999), and mites (Behan-Pelletier, 1999; Koehler, 1999). Traps of various designs are used routinely for collection of soil fauna that are soil quality indicators (Coleman et al., 1999; Paoletti, 1999). These methods require an inventory of the material collected with standard methods of collection and taxonomic identification. Protozoa are enumerated by two methods, either direct counts or by a culture technique (Coleman and Crossley, 1996). In the culture technique, population density is determined by the activity after a bacterial inoculum is added to soil suspensions as a food source for protozoa. Several procedures are acceptable for extracting nematodes from soil, including the Baermann funnel method, filtration, decanting and sieving, elutriation, or flotation/centrifugation (Coleman and Crossley, 1996; Nieminen and Setala, 1998; Yeates and Bongers, 1999). Soil collembolans and mites can be extracted from soil cores or litterbags using a Tullgren apparatus (Nieminen and Setala, 1998), Berlese funnel (Coleman and Crossley, 1996), or some modification of the two (Behan-Pelletier, 1999). Mites may be extracted from soil using funnel extraction with warming, or by mechanical extraction with heptane flotation (Koehler, 1999).

Cryptozoic fauna may be sampled by randomly placing cryptozoa boards (Neave and Fox, 1998) on the soil surface. Peck et al. (1998) used pitfall traps partially filled with glycerin, ethanol and water to collect ants for counts and identification. Earthworms can be extracted from soil following a surface application of dilute formalin solution (Subler and Kirsch, 1998), but the most widely used method involves hand sorting from soil cores (Gallagher and Wollenhaupt, 1997; Hansen and Engelstad, 1999; Hubbard, Jordan, and Stecker, 1999; Paoletti, 1999).

The methods that can be used in ecological studies are as varied as the organisms found in the soil, yet these ecological investigations are still hindered by the available techniques. As the technology advances in the area of molecular ecological investigations, so too will advance our understanding of the dynamics and complexity of soil ecology.

CONCLUSION

With sustainable agricultural practices becoming a priority for growers and the general public alike, an in-depth understanding of the soil ecosystem is needed. Studies of agroecosystem changes during the transition to no-till are needed to minimize the impact of agriculture on the environment. The soil biota, its interactions and distribution are critical to the functioning of any agricultural system. Long-term cultivation has a profound effect on biotic populations, processes and community structure. There are varying levels of disturbance between conventional tillage and no-till, and each level exerts a different effect on the soil environment. The "transition period" of no-till is characterized by soil physical, chemical and biological changes that may affect in soil productivity. In order to better manage this time of transition, growers may increase cropping intensity, experiment with different crop rotations, and introduce new crop species–all of which may further alter the soil environment and lead to changes in the soil biota. Thorough examination of soil environmental changes during the transition period is warranted so that growers may successfully reduce tillage, while reducing economic risk.

Almost all groups of soil organisms are negatively impacted by soil disturbance. Only a small portion of the microbial or faunal groups can currently be collected and studied, which provides only a partial picture of soil communities. However, several new methods are available that will enhance our knowledge of what is occurring in soil. What is presently known about the living soil is minuscule, but as laboratory techniques become more sophisticated and more research is conducted, we will gain a better understanding. The degree of soil disturbance, and the quantity, type and quality of residue in a cropping system all add to the complexity of the system and result in varying results and conclusions found in the literature. Population and process level studies, as well as investigations at the ecosystem and functional level, are needed to develop management systems that include soil biota for successful sustainable cropping systems. Ecological investigations will enhance the understanding of changes that occur with the adoption of minimum tillage and no-till cropping systems so that these systems become increasingly viable.

REFERENCES

Abawi, G. S. and T. L. Widmer. (2000). Impact of soil health management practices on soilborne pathogens, nematodes, and root diseases of vegetable crops. *Applied Soil Ecology* 15:37-47.

Achouak, W., J. M. Thiery, P. Roubaud, and T. Heulin. (2000). Impact of crop management on intraspecific diversity of *Pseudomonas corrugata* in bulk soil. *FEMS Microbial Ecology* 31:11-19.

Alderweireldt, M., K. Desender, and M. Polett. (1991). Abundance and dynamics of adult and larval Coleoptera in different agro-ecosystems. In *Advances in Coleopterology*, eds. M. Zunino, X. Bellés and M. Blas, Barcelona, Spain: European Association of Coleopterology, pp. 223-232.

Alexander, D. B. (1998). Bacteria and archaea. In *Principles and Applications of Soil Microbiology*, eds. D. M. Sylvia, J. J. Fuhrmann, P. G. Hartel, and D. A. Zuberer, Upper Saddle River, New Jersey: Prentice-Hall, Inc., pp. 44-71.

Alvarez, C. R. and R. Alvarez. (2000). Short-term effects of tillage systems on active soil microbial biomass. *Biology and Fertility of Soils* 31:157-161.

Ananyeva, N. D., T. S. Demkina, W. J. Jones, M. L. Cabrera, and W. C. Steen. (1999). Microbial biomass in soils of Russia under long-term management practices. *Biology and Fertility of Soils* 29:291-299.

Baker, C. J., K. E. Saxton, and W. R. Ritchie. (1996). *No-Tillage Seeding: Science and Practice.* CAB International Wallingford. UK 258 p.

Barns, S. M., R. E. Fundyga, M. W. Jeffries, and N. R. Pace. (1994). Remarkable archaeal diversity detected in a Yellowstone National Park hot spring environment. *Proceedings of the National Academy of Sciences* 91:1609-1613.

Beare, M. H., B. R. Pohlad, D. H. Wright, and D. C. Coleman. (1993). Residue placement and fungicide effects on fungal communities in conventional and no-tillage soils. *Soil Science Society America Journal* 57:392-399.

Behan-Pelletier, V. M. (1999). Oribatid mite biodiversity in agroecosystems: Role for bioindication. *Agriculture Ecosystems and Environment* 74:411-423.

Belovsky, G. E. and J. B. Slade. (2000). Insect herbivory accelerates nutrient cycling and increases plant production. *Proceedings of the National Academy of Science* 97:14412-14417.

Bergstrom, D. W., C. M. Monreal, and D. J. King. (1998). Sensitivity of soil enzyme activities to conservation practices. *Soil Science Society of America Journal* 62: 1286-1295.

Blumberg, A. J. Y., P. F. Hendrix, and D. A. Crossley, Jr. (1997). Effects of nitrogen source on arthropod biomass in no-tillage and conventional tillage grain sorghum agroecosystems. *Environmental Entomology* 26:31-37.

Bossio, D. A. and K. M. Scow. (1995). Impact of carbon and flooding on the metabolic diversity of microbial communities in soils. *Applied and Environmental Microbiology* 61: 4043-4050.

Bottomley, P. J. (1998). Microbial ecology. In *Principles and Applications of Soil Microbiology*, eds. D. M. Sylvia, J. J. Fuhrmann, P. G. Hartel, and D. A. Zuberer, Upper Saddle River, New Jersey: Prentice-Hall, Inc., pp. 149-167.

Camara, O. M., D. L. Young, and H. R. Hinman. (1999). Economic case studies of eastern Washington and northern Idaho no-till farmers growing wheat, barley, lentils, and peas in the 19- to 22-inch precipitation zone. Washington State University Extension Bulletin 1886, Pullman, WA. http://farm.mngt.wsu.edu/PDFDocuments/EB1886.pdf

Cavigelli, M. A. and G. P. Robertson. (2000). The functional significance of denitrifier community composition in a terrestrial ecosystem. *Ecology* 81:1402-1414.

Cavigelli, M. A., G. P. Robertson, and M. J. Klug. (1995). Fatty acid methyl ester (FAME) profiles as measures of soil microbial community structure. In *The Significance and Regulation of Soil Biodiversity*, eds. H. P. Collins, G. P. Robertson, and M. J. Klug, Netherlands: Kluwer Academic Publishers, pp. 99-113.

Clapperton, M. J., J. J. Miller, F. J. Larney, and C. W. Lindwall. (1997). Earthworm populations as affected by long-term tillage practices in southern Alberta, Canada. *Soil Biology and Biochemistry* 29:631-633.

Coleman, D. C., J. M. Blair, E. T. Elliott, and D. H. Wall. (1999). Soil Invertebrates. In *Standard Soil Methods for Long-Term Ecological Research, LTER Network Series*, eds. G. P. Robertson, D. C. Coleman, C. S. Bledsoe, and P. Sollins, New York, NY: Oxford University Press, pp. 349-377.

Coleman, D. C. and D. A. Crossley, Jr. (1996). Secondary production: Activities of heterotrophic organisms–the soil fauna. In *The Fundamentals of Soil Ecology*, eds. D. C. Coleman, and D. A. Crossley, Jr., San Diego, CA: Academic Press, pp. 51-72.

Cook, R. J. and W. A. Haglund. (1991). Wheat yield depression associated with conservation tillage caused by root pathogens in the soil not phytotoxins from the straw. *Soil Biology and Biochemistry* 23:1125-1132.

Curtin, D., H. Wang, F. Selles, B. G. McConkey, and C. A. Campbell. (2000). Tillage effects on carbon fluxes in continuous wheat and fallow-wheat rotations. *Soil Science Society of America Journal* 64:2080-2086.

Dalal, R. C., P. A. Henderson, and J. M. Glasby. (1991). Organic matter and microbial biomass in a Vertisol after 20 years. of zero-tillage. *Soil Biology and Biochemistry* 23:435-441.

Degens, B. P. (1999) Catabolic response profiles differ between microorganisms grown in soils. *Soil Biology and Biochemistry* 31:475-477.

Degens, B. P., L. A. Schipper, G. P. Sparling, and M. Vojvodic-Vukovic. (2000). Decreases in organic C reserves in soils can reduce the catabolic diversity of soil microbial communities. *Soil Biology and Biochemistry* 32:189-196.

di Castri, F. and T. Younes. (1990). Ecosystem function of biological diversity. *Biology International Special Issue* 22:1-20.

Dodds, W., M. K. Banks, C. S. Clenan, C. W. Rice, D. Sotomayor, E. A. Strauss, and W. Yu. (1996). Biological properties of soil and subsurface sediments under abandoned pasture and cropland. *Soil Biology and Biochemistry* 28:837-846.

Doran, J. W., M. Sarrantonio, and M. A. Liebig. (1996). Soil health and sustainability. *Advances in Agronomy* 56:1-54.

Doran, J. W. (1980.) Soil microbial and biochemical changes associated with reduced tillage. *Soil Science Society of America Journal* 44:765-771.

Drijber, R. A., J. W. Doran, A. M. Parkhurst, and D. J. Lyon. (2000). Changes in soil microbial community structure with tillage under long-term wheat-fallow management. *Soil Biology and Biochemistry* 32:1419-1430.

Drinkwater, L. E., P. Wagoner, and M. Sarrantonio. (1998). Legume-based cropping systems have reduced carbon and nitrogen losses. *Nature* 396:262-265.

Droge, M., A. Puhler, and W. Selbitschka. (1999). Horizontal gene transfer among bacteria in terrestrial and aquatic habitats as assessed by microcosm and field studies. *Biology and Fertility of Soils* 29:221-245.

Duineveld, B. M., A. S. Rosado, J. D. VanElsas, and J. A. vanVeen. (1998). Analysis of the dynamics of bacterial communities in the rhizosphere of the chrysanthemum via denaturing gradient gel electrophoresis and substrate utilization patterns. *Applied and Environmental Microbiology* 64:4950-4957.

Edwards, W. M. (1991). Soil structure: Processes and management. In *Soil Management for Sustainability*, eds. R. Lal and F. J. Pierce, Ankeny, IA: Soil and Water Conservation Society, pp. 7-14.

Edwards, W. M., M. J. Shipitalo, S. J. Traina, C. A. Edwards, and L. B. Owens. (1992). Role of *Lumbricus terrestris* (L.) burrows on quality of infiltrating water. *Soil Biology and Biochemistry* 24:1555-1561.

Elliott, L. F. and J. M. Lynch. (1994). Biodiversity and soil resilience. In *Soil Resilience and Sustainable Land Use*, eds. D. J. Greenland and I. Szabolcs, Wallingford, UK: CAB International, pp. 353-364.

Elliott, L. F. and R. I. Papendick. (1986). Crop residue management for improved soil productivity. *Biological Agriculture and Horticulture* 3:131-142.

Ferreira, M. C., D. de S. Andrade, K. M. deO. Chueire, S. M. Takemura, and M. Hungria. (2000). Tillage method and crop rotation effects on the population sizes and diversity of Bradyrhizobia nodulating soybean. *Soil Biology and Biochemistry* 32:627-637.

Foissner, W. (1999). Soil protozoa as bioindicators: Pros and cons, methods, diversity, representative examples. *Agriculture, Ecosystems and Environment* 74:95-112.

Frey, S. D. and E. T. Elliott, K. Paustian, and G. A. Peterson. (2000). Fungal translocation as a mechanism for soil nitrogen inputs to surface residue decomposition in a no-tillage agroecosystem. *Soil Biology and Biochemistry* 32:689-698.

Fulthorpe, R. R., A. N. Rhoades, and J. M. Tiedje. (1998). High levels of endemicity of 3-chlorobenzoate-degrading soil bacteria. *Applied and Environmental Microbiology* 64:1620-1627.

Gallagher, A. V. and N. C. Wollenhaupt. (1997). Surface alfalfa residue removal by earthworms *Lumbricus terrestris* L. in a no-till agroecosystem. *Soil Biology and Biochemistry* 29:477-479.

Garland, J. L. (1996). Patterns of potential C source utilization by rhizosphere communities. *Soil Biology and Biochemistry* 28: 223-230.

Griffiths, B. S., K. Ritz, and L. A. Glover. (1996). Broad-scale approaches to the determination of soil microbial community structure: Application of the community DNA hybridization technique. *Microbial Ecology* 31:269-280.

Grime, J. P. (1997). Biodiversity and ecosystem function: The debate deepens. *Science* 277:1260-1261.

Gross, K. L., K. S. Pregitzer, and A. J. Burton. (1995). Spatial variation in nitrogen availability in three successional plant communities. *Journal of Ecology* 83:357-367.

Guggenberger, G. S., D. Frey, J. Six, K. Paustian, and E. T. Elliott. (1999). Bacterial and fungal cell-wall residues in conventional and no-tillage agroecosystems. *Soil Science Society of America Journal* 63:1188-1198.

Guschin, D. Y., B. K. Mobarry, D. Proudnikov, D. A. Stahl, B. Rittman, and A. D. Mirzabekov. (1997). Oligonucleotide microchips as genosensors for determinative and environmental studies in microbiology. *Applied and Environmental Microbiology* 63:2397-2402.

Haack, S. K., H. Garchow, M. J. Klug, and L. J. Forney. (1995). Analysis of factors affecting the accuracy, reproducibility, and interpretation of microbial community carbon source utilization patterns. *Applied and Environmental Microbiology* 61: 1458-1468.

Hansen, S. and F. Engelstad. (1999). Earthworm populations in a cool and wet district as affected by tractor traffic and fertilisation. *Applied Soil Ecology* 13:237-250.

Harris, P. A., H. H. Schomberg, P. A. Banks, and J. Giddens. (1995). Burning, tillage and herbicide effects on the soil microflora in a wheat-soybean double-crop system. *Soil Biology Biochemistry* 27:153-156.

Hawksworth, D. L. and L. A. Mound. (1991). Biodiversity databases: The crucial significance of collections. In *The Biodiversity of Microorganisms and Invertebrates: Its Role in Sustainable Agriculture*, ed. D. L. Hawksworth, Melksham, UK: CAB International, Redwood Press Ltd., pp.17-29.

Hendrix, P. F., D. A. Crossley, J. M. Blair, and D. C. Coleman. (1990). Soil biota as components of sustainable agroecosystems. In *Sustainable Agricultural Systems*, eds. C. A. Edwards, R. Lal, P. Madden, R. H. Miller, and G. House, Ankeny, IA: Soil and Water Conservation Society, pp. 637-654.

Holland, E. A. and D. C. Coleman. (1987). Litter placement effect on microbial and organic matter dynamics in an agroecosystem. *Ecology* 68:425-433.

Hooper, D. U. and P. M. Vitousek. (1997). The effects of plant composition and diversity on ecosystem processes. *Science* 277:1302-1305.

Hubbard, V. C., D. Jordan, and J. A. Stecker. (1999). Earthworm response to rotation and tillage in a Missouri claypan soil. *Biology and Fertility of Soils* 29:343-347.

Hulsmann, A. and V. Wolters. (1998). The effects of different tillage practices on soil mites, with particular reference to Oribatida. *Applied Soil Ecology* 9:327-332.

Hunt, H. W., D. C. Coleman, E. R. Ingham, R. E. Ingham, E. T. Elliott, J. C. Moore, S. L. Rose, C. P. P. Reid, and C. R. Morley. (1987). The detrital food web in a shortgrass prairie. *Biology and Fertility of Soils* 3:57-68.

Hussain, I., K. R. Olson, and S. A. Ebelhar. (1999). Long-term tillage effects on soil chemical properties and organic matter fractions. *Soil Science Society of America Journal* 63:1335-1341.

Ibekwe, A. M. and A. C. Kennedy. (1998). Phospholipid fatty acid profiles and carbon utilization patterns for analysis of microbial community structure under field and greenhouse conditions. *FEMS Microbiology Ecology* 26:151-163.

Ingham, E. R. (1998). Protozoa and nematodes. In *Principles and Applications of Soil Microbiology*, eds. D.M. Sylvia, J.J. Fuhrmann, P.G. Hartel, and D.A. Zuberer, Upper Saddle River, New Jersey: Prentice-Hall, Inc., pp. 114-131.

Jensen, L. E. and O. Nybroe. (1999). Nitrogen availability to *Pseudomonas fluorescens* DF57 is limited during decomposition of barley straw in bulk soil and in the barley rhizosphere. *Applied and Environmental Microbiology* 65:4320-4328.

Jones, F. G. W., D. W. Larbey, and D. M. Parrott. (1969). The influence of soil structure and moisture on nematodes, especially *Xiphinema, Longgidorus, Trichodorus* and *Heterodera* spp. *Soil Biology and Biochemistry* 1:153-165.

Jordan, D., J. A. Stecker, V. N. Cacnio-Hubbard, F. Li, C. J. Gantzer, and J. R. Brown. (1997). Earthworm activity in no-tillage and conventional tillage systems in Missouri soils: A preliminary study. *Soil Biology and Biochemistry* 29:489-491.

Karlen, D. L., N. C. Wollenhaupt, D. C. Erbach, E. C. Berry, J. B. Swan, N. S. Eash, and J. L. Jordahl. (1994). Crop residue effects on soil quality following 10 years of no-till corn. *Soil and Tillage Research* 31:149-167.

Karlen, D. L., D. C. Erbach, T. C. Kaspar, T. S. Colvin, E. C. Berry, and D. R. Timmons. (1990). Soil tilth: A review of past perceptions and future needs. *Soil Science Society of America Journal* 54:153-161.

Kennedy, A. C. and A. J. Busacca. (1995). Microbial analysis to identify source of PM-10 material. In *Particulate Matter: Health and Regulatory Issues*, eds. Roy J. Weiskircher, Jr. and James R. Zwikl, Pittsburgh, PA: Air and Waste Management Association, pp. 670-675.

Kennedy, A. C. and K. L. Smith. (1995). Soil microbial diversity and the sustainability of agricultural soils. *Plant and Soil* 170:75-86.

Kennedy, A. C. and R. I. Papendick. (1995). Microbial characteristics of soil quality. *Journal of Soil and Water Conservation* 50:243-248.

Kettler, T. A., D. J. Lyon, J. W. Doran, W. L. Powers, and W. W. Stroup. (2000). Soil quality assessment after weed-control tillage in a no-till wheat-fallow cropping system. *Soil Science Society of America Journal* 64:339-346.

Kladivko, E. J., N. M. Akhouri, and G. Weesies. (1997). Earthworm populations and species distributions under no-till and conventional tillage in Indiana and Illinois. *Soil Biology and Biochemistry* 29:613-615.

Knops, J. M. H. and D. Tilman. (2000). Dynamics of soil nitrogen and carbon accumulation for 61 years after agricultural abandonment. *Ecology* 81:88-98.

Koehler, H. H. (1999). Predatory mites (Gamasina, Mesostigmata). *Agriculture, Ecosystems and Environment* 74: 395-410.

Ladd, J. N., M. Amato, Z. Li-Kai, and J. E. Schultz. (1994). Differential effects of rotation, plant residue and nitrogen fertilizer on microbial biomass and organic matter in an Australian Alfisol. *Soil Biology and Biochemistry* 26:821-831.

Lee, D. H., Y. G. Zo, and S. J. Kim. (1996). Nonradioactive method to study genetic profiles of natural bacterial communities by PCR-single-strand-conformation polymorphisms. *Applied and Environmental Microbiology* 62:3112- 3120.

Logsdon, S. D. and C. A. Cambardella. (2000). Temporal changes in small depth-incremental soil bulk density. *Soil Science Society of America Journal* 64:710-714.

Lupwayi, N. Z., W. A. Rice, and G. W. Clayton. (1998). Soil microbial diversity and community structure under wheat as influenced by tillage and crop rotation. *Soil Biology and Biochemistry* 30:1733-1741.

Marasas, M. E., S. J. Sarandon, and A. C. Cicchino. (2001). Changes in soil arthropod functional group in a wheat crop under conventional and no tillage systems in Argentina. *Applied Soil Ecology* 18:61-68.

McCool, D. K. (1990). Crop management effects on runoff and soil loss from thawing soil. In *Proc. Intl. Symp. Frozen Soil Impacts on Agricultural, Range and Forest Lands*, In ed. K. R. Cooley, CRREL Special Rpt. 90-1. Hanover, NH:US Army Corps of Engineers, Cold Regions Res. And Eng. Lab., pp. 171-176.

Meharg, A. A. and K. Killham. (1995). Loss of exudates from the roots of perennial ryegrass inoculated with a range of microorganisms. *Plant and Soil* 170:345-349.

Mikola, J. and H. Setala. (1998). No evidence of trophic cascades in an experimental microbial-based soil food web. *Ecology* 79:153-164.

Minamisawa, K., Y. Nakatsuka, and T. Isawa. (1999). Diversity and field site variation of indigenous populations of soybean Bradyrhizobia in Japan by fingerprints with repeated sequences RS alpha and RS beta. *FEMS Microbiology Ecology* 29:171-178.

Moore, J. C. (1994). Impact of agricultural practices on soil food web structure: Theory and application. *Agriculture, Ecosystems and Environment* 51:239-247.

Moore, J. C., R. J. Snider, and L. S. Robertson. (1984). Effects of different management practices on Collembola and Acarina in corn production systems. 1. The effects of no-tillage and Atrazine. *Pedobiologia* 26:143-152.

Morton, J. B. (1998). Fungi. In *Principles and Applications of Soil Microbiology*, eds. D. M. Sylvia, J. J. Fuhrmann, P. G. Hartel, and D. A. Zuberer, Upper Saddle River, New Jersey: Prentice-Hall, Inc., pp. 72-93.

Muyzer, G., E. C. deWaal and A. G. Uitterlinden. (1993). Profiling of complex microbial populations by denaturing gradient gel electrophoresis analysis of polymerase chain reaction-amplified genes coding for 16S rRNA. *Applied and Environmental Microbiology* 59:695-700.

Neave, P. and C. A. Fox. (1998). Response of soil invertebrates to reduced tillage systems established on a clay loam soil. *Applied Soil Ecology* 9:423-428.

Nieminen, J. K. and H. Setala. (1998). Enclosing decomposer food web: Implications for community structure and function. *Biology and Fertility of Soils* 26:50-57.

Nyakatawa, E. Z., K. C. Reddy, and K. R. Sistani. (2001). Tillage, cover cropping, and poultry litter effects on selected soil chemical properties. *Soil and Tillage Research* 58:69-79.

Ohtonen, R., H. Fritze, T. Pennanen, A. Jumpponen, and J. Trappe. (1999). Ecosystem properties and microbial community changes in primary succession on a glacier forefront. *Oecologia* 119:239-246.

Oostenbrink, M. (1964). Harmonious control of nematode infestation. *Nematologica* 10:49-56.

Paoletti, M. G. (1999). The role of earthworms for assessment of sustainability and as bioindicators. *Agriculture, Ecosystems and Environment* 74:137-155.

Paoletti, M. G., U. Schweigel, and M. R. Favretto. (1995). Soil microinvertebrates, heavy metals and organochlorines in low and high input apple orchards and coppiced woodland. *Pedobiologia* 39:20-33.

Papendick, R. I. and J. F. Parr. (1997). No-till farming: The way of the future for a sustainable dryland agriculture. *Annals of Arid Zone* 36:193-208.

Parkin, T. B. and E. C. Berry. (1999). Microbial nitrogen transformations in earthworm burrows. *Soil Biology and Biochemistry* 31:1765-1771.

Paul, E. A., D. Harris, H. P. Collins, U. Schulthess, and G. P. Robertson. (1999). Evolution of CO_2 and soil carbon dynamics in biologically managed, row-crop agroecosystems. *Applied Soil Ecology* 11:53-65.

Peck, S. L., B. McQuaid, and C. L. Campbell. (1998). Using ant species (Hymenoptera: Formicidae) as a biological indicator of agroecosystem condition. *Environmental Entomology* 27:1102-1110.

Pennanen, T., J. Liski, E. Baath, V. Kitunen. J. Uotila, C. J. Westman, and H. Fritze. (1999). Structure of the microbial communities in coniferous forest soils in relation to the site fertility and stand development stage. *Microbial Ecology* 38:168-179.

Pfiffner, L. and U. Niggli. (1996). Effects of bio-dynamic, organic and conventional farming on ground beetles (Coleoptera:Carabidae) and other epigaeic arthropods in winter wheat. *Biological Agriculture and Horticulture* 12:353-364.

Pierce, F. J., M. C. Fortin, and M. J. Staton. (1994). Periodic plowing effects on soil properties in a no-till farming system. *Soil Science Society of America Journal* 58:1782-1787.

Rasmussen, P. E., R. W. Rickman, and B. L. Klepper. (1997). Residue and fertility effects on yield of no-till wheat. *Agronomy Journal* 89:563-567.

Rhoton, F. E. (2000). Influence of time on soil response to no-till practices. *Soil Science Society of America Journal* 64:700-709.

Robertson, G. P., J. R. Crum, and B. G. Ellis. (1993). The spatial variability of soil resources following long-term disturbance. *Oecologia* 96:451-456.

Robertson, L. N., B. A. Kettle, and G. B. Simpson. (1994). The influence of tillage practices on soil macrofauna in a semi-arid agro-ecosystem in northeastern Australia. *Agriculture, Ecosystems and Environment* 48:149-156.

Roget, D. K., S. M. Neate, and A. D. Rovira. (1996). Effect of sowing point design and tillage practice on the incidence of rhizoctonia root rot, take-all, and cereal cyst nematode in wheat and barley. *Australian Journal of Experimental Agriculture* 36:683-693.

Saiki, R. K., S. J. Scharf, F. Flaoona, K. B. Mullis, G. T. Horn, H. A. Erlich, and N. Arnheim. (1985). Enzymatic amplification of beta-globin genomic sequences and restriction site analysis for diagnosis of sickle cell anemia. *Science* 230:1350-1354.

Shimmel, S. M. and W. M. Darley. (1985). Productivity and density of soil algae in an agricultural system. *Ecology* 66:1439-1447.

Six, J., K. Paustian, E. T. Elliott, and C. Combrink. (2000). Soil structure and organic matter: I. Distribution of aggregate-size classes and aggregate-associated carbon. *Soil Science Society of America Journal* 64:681-689.

Smiley, R. W., A. G. Ogg, Jr., and R. J. Cook. (1992). Influence of glyphosate on severity of rhizoctonia root rot and growth and yield of barley. *Plant Disease* 76: 937-942.

Stenberg, M., B. Stenberg, and T. Rydberg. (2000). Effects of reduced tillage and liming on microbial activity and soil properties in a weakly structured soil. *Applied Soil Ecology* 14:135-145.

Stinner, B. R. and G. J. House. (1990). Arthropods and other invertebrates in conservation-tillage agriculture. *Annual Review of Entomology* 35: 318.

Stubbs, T. L. (2000). Soil quality and residue decomposition potential in conservation farming systems. MS Thesis, Pullman, WA: Washington State University.

Subler, S. and A. S. Kirsch. (1998). Spring dynamics of soil carbon, nitrogen, and microbial activity in earthworm middens in a no-till cornfield. *Biology and Fertility of Soils* 26:243-249.

Thomas, V. G. and P. G. Kevan. (1993). Basic principles of agroecology and sustainable agriculture. *Journal of Agricultural and Environmental Ethics* 5:1-19.

Torsvik, V., J. Goksoyr, and F. L. Daae. (1990). High diversity in DNA of soil bacteria. *Applied and Environmental Microbiology* 56:782-787.

Unger, P. W. (1991). Organic matter, nutrient, and pH distribution in no- and conventional-tillage semiarid soils. *Agronomy Journal* 83:186-189.

van Vliet, P. C. J., V. V. S. R. Gupta, and L. K. Abbott. (2000). Soil biota and crop residue decomposition during summer and autumn in south western Australia. *Applied Soil Ecology* 14:111-124.

Veres, G., R. A. Gibbs, S. E. Scherer, and C. T. Caskey. (1987). The molecular basis of the sparse fur mouse mutation. *Science* 237: 415-417.

Verhoef, H. A. and L. Brussaard. (1990). Decomposition and nitrogen mineralization in natural and agroecosystems: The contribution of soil animals. *Biogeochemistry* 11:175-211.

Wardle, D. A., G. W. Yeates, K. S. Nicholson, K. I. Bonner, and R. N. Watson. (1999). Response of soil microbial biomass dynamics, activity and plant litter decomposition to agricultural intensification over a seven-year period. *Soil Biology and Biochemistry* 33:1707-1720.

Weller, D. M., R. J. Cook, G. MacNish, E. N. Bassett, R. L. Powelson, and R. R. Petersen. (1986). Rhizoctonia root rot of small grain favored by reduced tillage in the Pacific Northwest. *Plant Disease* 70:70-73.

Westover, K. M., A. C. Kennedy, and S. E. Kelley. (1997). Patterns of rhizosphere microbial community structure associated with co-occurring plant species. *Journal of Ecology* 85:863-873.

White, S. K., M. F. Hossain, N. Sultana, S. F. Elahi, M. H. Choudhury, S. Sarker, Q. K. Alam, J. A. Rother, and J. L. Gaunt. (2001). Low-input ecological rice farming in Bangladesh. In *Sustainable Management of Soil Organic Matter*, eds. R. M. Rees, B. C. Ball, C. D. Campbell, and C. A. Watson, New York, NY, CAB International, pp. 201-206.

Wolters, V. (2000). Invertebrate control of soil organic matter stability. *Biology and Fertility of Soils* 31:1-19.

Wood, M. (1989). *Soil Biology*. Chapman & Hall, New York. 154 pp.

Wright, S. F., J. L. Starr, and I. C. Paltineanu. (1999). Changes in aggregate stability and concentration of glomalin during tillage management transition. *Soil Science Society of America Journal* 63:1825-1829.

Yeates, G. W. and T. Bongers. (1999). Nematode diversity in agroecosystems. *Agriculture, Ecosystems and Environment* 74:113-136.

Young, D. L, H. R. Hinman, and W. F. Schillinger. (2001). Economics of winter wheat-summer fallow vs. continuous no-till spring wheat in the Horse Heaven Hills, Washington. Washington State University Extension Bulletin. 1907, Pullman, WA. 35 p. http://farm.mngt.wsu.edu/PDFDocuments/EB1907.pdf

Zak, D. R., K. S. Pregitzer, P. S. Curtis, and W. E. Holmes. (2000). Atmospheric CO_2 and the composition and function of soil microbial communities. *Ecological Applications* 10:47-59.

Zak, J. C., M. R. Willig, D. L. Morrehead, and H. G. Wildman. (1994). Functional diversity of microbial communities: a quantitative approach. *Soil Biology and Biochemistry* 26:1101- 1108.

Zelles, L., Q. Y. Gai, R. X. Ma, R. Rackwitz, K. Winter, and F. Beese. (1994). Microbial biomass, metabolic activity and nutritional status determined from fatty acid patterns and poly-hydrozybutyrate in agriculturally managed soils. *Soil Biology and Biochemistry* 26:439-446.

Zunino, M. (1991). Food relocation behaviour: a multivalent strategy of Coleoptera. In *Advances in Coleopterology*, eds. M. Zunino, X. Bellés, and M. Blas, Barcelona, Spain: European Association of Coleopterology, pp: 297-314.

The Ecology of Crop-Weed Interactions: Towards a More Complete Model of Weed Communities in Agroecosystems

Bruce D. Maxwell
Edward Luschei

SUMMARY. Understanding the ecology of crop-weed interactions includes a wide array of population and community ecology theory. The first thorough application of theory in Weed Ecology was predicting weed impacts on crops. The generality of crop yield response to weeds has been well supported with a wealth of empirical studies, however, inclusion of these models into weed management decision support systems (DSS) has been less fruitful. Parameterization of the empirical models has indicated extreme variability over space and time, even in experimental plots. Thus, further development and implementation of ecologically based weed management must depend on understanding sources of crop response variation, understanding complexity of interactions among the many sources of variation, and the realization that forecasting crop response will always have some level of uncertainty. Ecological theory may help us understand how to best construct concep-

Bruce D. Maxwell is Professor, Department of Land Resources and Environmental Science, Montana State University, Bozeman, MT 59717 (E-mail: bmax@montana.edu).

Edward Luschei is Professor, Department of Agronomy, University of Wisconsin, Madison, WI 59706.

[Haworth co-indexing entry note]: "The Ecology of Crop-Weed Interactions: Towards a More Complete Model of Weed Communities in Agroecosystems." Maxwell, Bruce D., and Edward Luschei. Co-published simultaneously in *Journal of Crop Improvement* (Food Products Press, an imprint of The Haworth Press, Inc.) Vol. 11, No. 1/2 (#21/22), 2004, pp. 137-151; and: *New Dimensions in Agroecology* (ed: David Clements, and Anil Shrestha) Food Products Press, an imprint of The Haworth Press, Inc., 2004, pp. 137-151. Single or multiple copies of this article are available for a fee from The Haworth Document Delivery Service [1-800-HAWORTH, 9:00 a.m. - 5:00 p.m. (EST). E-mail address: docdelivery@haworthpress.com].

http://www.haworthpress.com/web/JCRIP

Digital Object Identifer: 10.1300/J411v11n01_07

tual and predictive models of crop-weed communities to accomplish our applied goals.

KEYWORDS. Competitive interactions, weed ecology, niche assembly, neutral theory

INTRODUCTION

The interaction between crops and weeds has received extensive study. The primary purpose of these studies has been to quantify the impact of weeds on crops so that management of the weeds could be justified (Cousens, 1987; Zimdahl, 1980). In this contribution, we examine the ecological theories on interactions that may be important in structuring plant communities in agricultural systems. Our focus is applied in that we are interested in interactions that have tangible results on our ability to achieve management objectives over short- and long-term time horizons. We argue that while the historic focus on understanding of competitive relationships between weeds and crops has lead to a more accurate understanding of the short-term cost of weeds, the importance of competitive interactions in structuring weed communities in agroecosystems remains unclear. We will briefly outline the logic behind the ecological models of plant interactions and dynamics from the scale of the individual through the community, highlighting areas we consider to be particularly relevant to the design of ecologically sound management practices.

THEORIES OF PLANT COMMUNITY STRUCTURE

Relatively few studies have identified positive interactions between crops and weeds (Andow, 1988). In crop-weed competition studies, if crop yields were found to increase in the presence of weeds, there has been a tendency to deem the data anomalous and provide an explanation for why the data did not comply with accepted beliefs (i.e., weeds reduce crop yield and/or quality). In our own work, the abilities to continuously measure yields with field monitoring devices on harvesting equipment allowed us to identify many regions, particularly where resources are limiting, that crop yields may be positively correlated with weed presence (Luschei et al., 2001). Perhaps these observations reflect a spatial variability in resources (water, fertility, etc.) to which both crops and weeds were similarly responding. Weeds may have a net positive influence on crops by providing habitat for crop pest predators or pathogens or for crop pollinators (Andow, 1988). Thus, interactions between crops and weeds may have many different results and can be categorized (Table 1).

The acceptance of multiple crop-weed interaction outcomes has allowed for a more ecological approach to understanding the crop-weed mix as a plant community. Two general views have emerged that offer explanation for how plant communities are structured. The most prevalent view is that plant communities are groups of interacting species whose presence, absence, or relative abundance can be deduced from "assembly rules" that are based on the functional roles (inherently defined ecological niches) of each species in the community (Booth and Swanton, 2002; Diamond, 1975; Levin, 1970; MacArthur, 1970; Weiher and Keddy, 1999). According to this theory, species coexist in interactive equilibrium with the other species in the community. The stability of the community and its resistance to invasion is derived from the adaptive equilibrium of member species, each of which has evolved to be the best competitor in its own ecological niche (Pontin, 1982). Even though this theory would place crop-weed interactions in a disequilibrium state, because of the high frequency of disturbance in most annual agroecosystems, it would still maintain that the species that can co-occur within a community are mostly determined by interspecific competition for limited resources and other biotic interactions and abiotic factors (Booth and Swanton, 2002). This theory is particularly appeal-

TABLE 1. Types of interaction that can occur between two species that share the same environment (adapted from Burkholder, 1952).

	Result of Interaction Between 2 Species (A & B)	
Types of Interference	**A**	**B**
competition	–	–
mutualism	+	+
commensalism	+	0
amensalism	–	0
parasitism	+	–

ing to farmers and weed scientists, because it confirms the notion that weeds by their very coexistence with the crop, compete for resources and thereby reduce crop yields. It is also appealing to agronomists that promote the use of inter-crops to fill the open niches left by a monoculture crop that otherwise would get filled by weeds. The niche assembly theory would also suggest that disturbance that removes plants could open niches and thus be a major mechanism allowing invasion of weeds. Thus, disturbed areas or areas accidentally not planted to the crop would be expected to be good predictors of weed occurrence.

The other general theory that has emerged to explain community structure is based on the assumption that communities are open, nonequilibrium assemblages of species largely thrown together by chance, variability in demographic processes, and random dispersal (Hubbell, 2001). Thus, this theory assumes that species come and go, and that their presence, absence or relative abundance is dictated primarily by random dispersal and stochastic local extinction. The intrinsic and extrinsic variability in ecological processes that would define an ecological niche preclude natural selection. Hubbell (2001) called this the *Unified Neutral Theory* and contrasted it with the more prevalent view described above, which he called the *Niche-Assembly Theory*. The *Unified Neutral Theory* is appealing as a null hypothesis for explaining agroecosystem plant community structure because most cropping systems are nonequilibrium as a result of high frequency disturbance from tillage and planting. Thus, selection for traits that increase colonization potential (e.g., high fecundity and dispersal) over competitive ability would be expected, although the intensity of selection for any particular traits will be low in more diverse cropping systems (Jordan and Jannink, 1997). Still it is difficult to build a solid argument for either theory as dominant in explaining agroecosystem plant communities. For example, the *Niche-Assembly Theory* would seem to better account for the consistent observation that weed species tend to aggregate, indicating that specific processes associated with habitat requirements, dispersal, response to management, and/or relative competitive ability, may interact to determine distribution and community structure.

Empirical evidence for any single factor determining plant community structure is rare and often the variation in response is more indicative of a large random component in the processes, which would reject all but a null hypothesis that may be stated as the *Unified Neutral Theory*. Thus, we may conclude that ecological processes consistent with *Niche Assembly Theory* are acting on crop-weed communities, but a large random component in these processes will often obscure them and/or the interaction among the processes may cloud the view of the mechanisms. It is important to recognize the large random component and adopt a philosophical approach to understanding the ecology of crop-weed interactions as probabilistic outcomes that have potentially complex interactions as well as spatial and temporal variation. Ecological theories that describe plant community dynamics are useful in providing a language within which we can frame hypotheses (Kareiva, 1989). Even in the presence of large amounts of intrinsic stochasticity, there exists tremendous potential to manipulate agricultural systems to optimize interactions at different time scales (Jordan and Jannink, 1997; Mohler, 2001). However, deriving agronomic recommendations from ecological theory requires careful consideration and explicit representation of risk.

Diversity-Stability Hypothesis

Structural simplicity in agricultural production systems decreases the number of possible interactions between species. One consequence of designing agricultural systems with low species diversity is that fewer interactions are possible. Having fewer connections between elements of an agricultural system is not intrinsically harmful or beneficial. However, it has been speculated that the productivity and stability of agricultural systems may be adversely impacted by low diversity. The diversity-stability hypothesis put forward by Elton (1958) stimulated much discussion before being criticized (Peters, 1995) and systematically dismantled (Goodman, 1975). Then Lehman and Tilman (2000) came full-circle to argue that both sides of the debate may be correct; diversity can stabilize communities while simultaneously destabilizing component populations. The design of improved agriculture systems often mirrors these issues (Swift and Anderson, 1993).

The natural connotation of stability being a desirable system-level property has led to its incorporation into the lexical foundations of IPM approaches to weed management (Norris, 1982; Clements et al., 1994). Maxwell (1999) observed that from the point of view of stability, conventional management trades the natural population regulating mechanisms that potentially exists in diverse communities for regulation based on high-levels of chemical inputs. In addition, it was predicted that the transition to lower input systems would probably be met with increased complexity in overall system behavior, ruling out prescriptive management.

Ecological theory supports the intuitive notion that species diversity and temporal stability in total biomass production are linked. Lehman and Tilman (2000) observed that three very different classes of models predict a positive influence of plant species diversity on the stability of total production. They also note that with respect to individual species, stability tends to decrease with increasing diversity. There are three effects that combine to produce the relationship between diversity and stability. First is the "over-yielding" effect, which results from the fact that species mixtures are capable of more completely utilizing a resource spectrum than single species. Just as a disjoint resource use can enhance stability, resource similarity or the "covariance effect" can also increase productivity by allowing one species to utilize resources foregone by a second. Lastly, to the extent that individual species environmental responses are uncorrelated, the "portfolio effect" can hedge the total productivity against any particular environment and thereby enhance stability.

None of the above mechanisms aid the cash grain farmer interested in the performance of a single crop species. The models examined by Lehman and Tilman (2000) would predict a purely detrimental effect of additional species from the perspective of the farmer. We argue that the theoretical support for this effect results from a fundamental inability of deterministic formalisms to capture a process which must be regarded as intrinsically stochastic at the farm-scale. Weather can greatly influence the life-cycles of weeds, either by directly altering resource pools or germination timing with respect to the crop. We have witnessed heavy wild oat (*Avena fatua* L.) infestations that can essentially die of drought where sparse populations survive. The inability of deterministic models to capture complex interactions does not make them intractable, it simply places a constraint on their use as a predictive tool for prescriptive management. They can still be very useful for developing an understanding of the first principles of system behavior. In recent discussions about a shift from prediction to ecological forecasting, Clark et al. (2001) present a strong case for the need to use models with "fully specified" uncertainties to address system behavior with large amounts of intrinsic uncertainty. They further note that landscape processes are often unpredictable from fine-grained (small plot experiment) studies (Carpenter et al., 1996; Clark et al., 2001). Interpolation of experimental results from small plots must be weighed with appropriate skepticism. For example, small plot experiments that are conducted to quantify the affect of weed density on crop yield will almost always deliver a negative hyperbolic response. One must be careful to interpret this response as a first principle with appropriate levels of variation if the quantitative function is scaled up and thus logically will incur more interactions with other processes that can influence crop yield (Figure 1).

Theoretical Models of Ecological Interactions

Most theoretical models of agroecosystem interactions are simply applications of general ecology theory that was more focused on natural

FIGURE 1. Spring wheat yield in response to wild oat density in small plot experiment (a) in Bozeman, MT and in a 150 ha production field (b) approximately 300 km away near Sun River, MT in 1999.

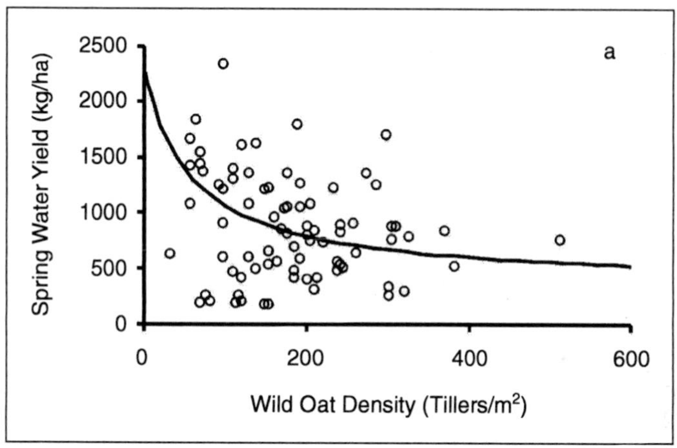

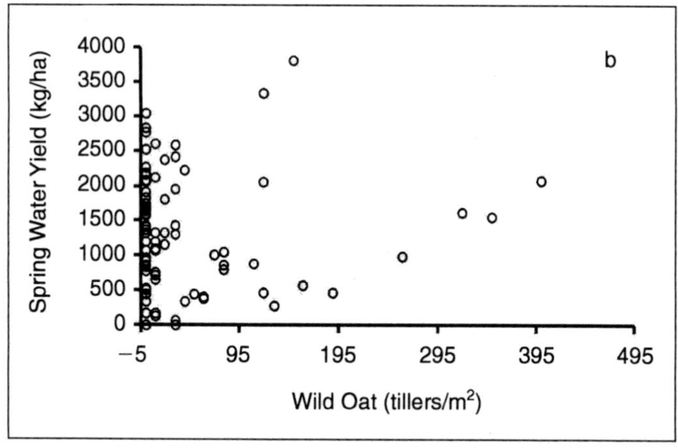

ecosystems. Agroecosystems are typically less complex than natural systems and thus one may conclude that the simplifications in the models may not be as constrained when applied in agroecosystems. Generally, however, the models are still gross simplifications. Simplifications are most often introduced to enhance generality or make models analytically tractable. Simplified models seldom improve predictive power except at larger scales than the predictions were intended (i.e., Ockham's razor). Thus simplified models further confirm first principles, but may have reduced utility for understanding the variability in responses and subsequent probability of outcomes. Parsimony mandates that the burden of proof, for the need to adopt an increase in complexity, rests with the advocates for constructing more complex models. It must be recognized that evidence provided by fine-grained studies is necessarily weak due to the large intrinsic uncertainties in agricultural systems. This is not to say that the "higher order" interactions are not important, merely that the mathematical formalism so successful at describing physical and chemical dynamics is limited in its suitability for use in natural or agroecosystems. We propose that the simplifications miss an important set of affects arising from ecosystem services and interactions provided by more diverse communities. For example, weeds may provide habitat for crop pest predators or pathogens or for crop pollinators (Andow, 1983). These kinds of effects are neglected, not because they are not important, but because they are difficult to study because they depend so strongly on the particulars of a given situation. Any successful study of small size and short duration will avoid phenomenon requiring extensive replication in space or time. Indirect interactions that are extremely variable on the spatial or temporal scale at which we can study them will be interpreted as experimental noise.

The lack of an ability to observe or quantify interactions, which may well stem from cultural, political and practical constraints on scientists, leads to conceptual inertia that predisposes us to discount complex interactions in favor of simple, direct and consistent phenomena. The immediate search for explanations involving direct mechanisms can lead us to not give due weight to alternative explanations. Sing (2002) found wild oat could serve as a population sink for the wheat-stem sawfly (*Cephus cinctus*), a significant pest of wheat (*Triticum aestivum* L.) production in Montana. Van Wychen (2002) speculated that the role of population demographic processes and seed dispersal may be more important in the formation of weed spatial pattern than direct competitive effects of wild oat in spring wheat crops.

Investigating interactions in weed-crop systems necessitates choosing a spatial and temporal scale to study as well as specifying the observations that will serve to define the state of a system. In the least complex models, individuals are represented as ensembles in a state-space and are described by a statistical property of the ensemble. More complicated individual-based models (McGlade, 1999) can be useful tools in discovering how the entire ensemble behaves rather than simply tracking the state of the population through a statistical property like the mean individual response. The behavior of individuals, populations and communities is necessarily linked by their nested membership. Lotka (1925) described the dynamics of these aggregates as an extension of a field of statistical mechanics to "aggregates of living organisms."

> It so happens that many of the components that play an important role in nature, both organic and inorganic, are built up of large numbers of individuals, themselves very small as compared with the aggregations they form. Accordingly, the study of systems of this kind can be taken up in two separate aspects, namely, first with attention centered upon the phenomena displayed by the component aggregates in bulk . . . and secondly conducted with the attention centered primarily upon the phenomena displayed by the individuals of which the aggregates are composed . . . It is evident that between these two branches or aspects of the general discipline there is an inherent relation, arising from the fact that the bulk effects observed are the nature of a statistical manifestation or the resultant of the detail working of the micro-individuals. A.J. Lotka (1925)

Describing the behavior of a system state defined on one level by levels directly below is referred to as hierarchical modeling (Allen and Star, 1982). Maurer (1999) goes so far as to refer to the modeling of community dynamics as "modeling kinetics in hierarchical systems." If we are to carry Lotka's comparison between the dynamics of ensembles of organisms and inert matter further, then the dynamics are overwhelmingly driven by two-species interactions (the so-called "pair approximation" in analogy with the relative rarity of three particle collisions in a gas). By making assumptions about the system being near "dynamical equilibrium," one can derive (Maurer, 1999) the form of the well-known set of differential equations called the Lotka-Volterra equations that describe the statistical properties of the ensemble of organisms. There has been some discussion in the ecological literature on the significance of "higher order" interactions (Kareiva, 1994), but models to capture any generality about these interactions are rare.

Lotka-Volterra Model

While the Lotka-Volterra model worked remarkably well for describing animal populations, it was less successful for plants for a number of reasons (Harper, 1977). These reasons include: stationarity, lack of "mixing" for plants, ambiguity about what constitutes an individual (genets versus ramets) and census difficulties, primarily because of the seed phase in the life history of plants. Driven by local interaction and a general lack of mixing, some noted that plant interactions were inherently spatial. From a modeling point of view local interconnectedness can be represented explicitly using agent based models (Berger et al. 2002), cellular automata (Wissel, 2000; Wolfram, 2002) or "neighborhood" models (Pacala and Silander, 1990; Stoll and Weiner, 2000).

Tilman (1982) took a different tact and extended the differential-equations of Lotka and Volterra to explicitly describe the changing resource levels. He demonstrated that the dynamics of (n) species utilizing (m) resources could be described by (m + n) differential equations and that the interaction was solely through shared resource pools (so that the species could be studied independently with respect to their resource utilization). With a host of assumptions including the existence of "equilibrium" dynamics, he described the conditions necessary for coexistence or competitive exclusion.

Competitive Exclusion Principle

The domination of ecological theory and experimentation centered on competitive interactions began with Gause (1934) and the development of his *Competitive Exclusion Principle*. The origins of the *Competitive Exclusion Principle* are said to be partly based upon the mathematical model created by Lotka and Volterra in the 1920s (Begon, Harper, and Townsend, 1996). The theoretical underpinnings of this principle are pervasive in plant and weed ecology. Gause (1934), a Russian microbiologist, grew different yeast species together in petri dishes on uniform media and measured the population growth rate of each species. Initially, growth rates (*dN/dt*) were depressed for both species compared to rates when grown in isolation, indicating competition for limited resources. However, over time one species became dominant in the mixture and the other species became extinct. Thus, the *Competitive Exclusion*

Principle suggests that two species cannot coexist when they have identical needs of a limited resource and species that coexist in nature must evolve ecological differences in order to share an environment. Stable coexistence of two species was only possible where intraspecific competition was greater than interspecific competition for the species in a community (Harper, 1977; May, 1981; Ricklefs, 1979).

Other concepts that grew from the *Competitive Exclusion Principle* include ideas about how to define the multidimensional space (habitat) needs of a species based on resource needs and environmental tolerances including response to predation, herbivory and pathogens. *Niche Assembly Theory* holds that natural selection drives species within a community to partition the environment (use different parts of the environment), with the result that competition is minimized, thereby defining a species' niche.

The *Competitive Exclusion Principle* has not met with total acceptance, and it is perhaps the blurred relationship with competition and the other biotic and abiotic influences that prevents its approval (Crawley, 1997; Begon, Harper, and Townsend, 1996). Nevertheless, in many respects the theory and models used to characterize the interactions among plants developed following Gause (1934) are simply variants on the *Competitive Exclusion Principle*. They have sought to include the effects of variable environments and species adaptations to the environment to account for species coexistence (Chesson, 1986; Grime, 1984; Huston, 1979; Hutchinson, 1959; Levin, 1979; Ricklefs and Schluter, 1993). Clearly, one should not be surprised by the prevalence of the assumption of competition for resources as the driving first principle mechanism determining the structure and function of weed communities in crops. The various strengths and weaknesses of the variants on the *Competitive Exclusion Principle* can be argued, but all represent some potential for explaining weed-crop community structure. They can be generally grouped into equilibrium versus non-equilibrium assumptions about the plant community. Depending on the time scale, equilibrium (long-time scale) or non-equilibrium (short time scale) theories may be most suitable for understanding interactions among weeds and crops.

CONCLUSIONS

We have reviewed the hypotheses developed to explain the structure of plant communities and by association weed communities in annual crops. It has generally become accepted that no single putative mechanism gives rise to communities but we can treat hypotheses as *Community Assembly Rules* that we can draw upon to both explain and manage plant community structure (Clements et al. 1994; Booth and Swanton, 2002). In general the above hypotheses can be summarized as follows: If communities are equilibrium assemblages then the maintenance of species richness depends on rare species having some advantage over common ones by the common ones suffering density-dependent reductions in survival or fecundity, the rare species obtaining some frequency-dependent advantage in recruitment, or frequency independent spatial stochasticity providing an ephemeral, but reliable refuge from superior competitors. Thereby, we encourage the use of *Assembly Rule Theory* as first principles, but maintain that one must be acutely aware of complex interactions and stochastic variability that can shroud mechanism and lead to abandonment of predictive models based on the first principles. Therefore, we suggest continuing to employ first principle models to understand and even forecast outcomes for management, but placing greater focus on study of variation across time and space so that management alternatives can be placed in the context of probabilities of improvement over accepted or conventional approaches to management. First principle models purposely ignore or assume constant many interacting factors in order to expose the first principles. Thus, a good way to incorporate first principles into models (DSS) that will be applied in management without explicitly knowing how the interactions influence responses, is to empirically characterize the variation in the first principle responses under typical management conditions and report outcomes as probabilities. Eventually, more complex models that capture specific interactions can be added to improve forecasting of management outcomes.

We have attempted to describe the importance of both theory and simplicity to the conceptualization and implementation of agroecological principles. Redesigning agricultural systems may necessitate encouraging interactions that are inherently more complicated, yet the framework we employ to describe and understand dynamics should not be forced to match this complexity.

Thus, agroecology is the science of integrating the biological, agronomic and socioeconomic disciplines to understand and predict the behavior of agroecosystems (Gliesman, 1998). The ecological approach to understanding agronomic systems begins with embracing the complexity of interactions across time and space scales and hierarchical systems of organization (Clements and Shrestha, 2004). This does not mean that we cannot forecast outcomes until we fully understand the complex interactions, it only means that we must provide management related forecasts of system response in the form of probabilities.

REFERENCES

Allen, T.F.H. and B. Starr. (1982). *Hierarchy: Perspectives for Ecological Complexity*. Chicago, IL: University of Chicago Press.

Andow, D.A. (1983). Effects of agricultural diversity on insect populations. In *Environmentally Sound Agriculture*, ed. W. Lockeretz, New York, NY: Praeger Publishers.

Andow, D.A. (1988). Management of weeds for insect manipulation in agroecosystems. In *Weed Management in Agroecosystems: Ecological Approaches*, eds. M.A. Altieri and M. Liebman, Boca Raton, FL: CRC Press. pp. 265-301.

Begon, M., J.L. Harper, and C.R. Townsend. (1996). *Ecology*, 3rd ed. Oxford, UK: Blackwell Science.

Berger, U., H. Hildenbrandt, and V. Grimm. (2002). Towards a standard for the individual-based modeling of plant populations: Self-thinning and the field-of-neighborhood approach. Nat. Res. Modeling 15:39-54.

Booth, B.D. and C.J. Swanton. (2002). Assembly theory applied to weed communities. *Weed Science* 50:2-13.

Burkholder, P.R. (1952). Cooperation and conflict among primitive organisms. *American Science* 40:601-631.

Carpenter, S.R., J.F. Kitchell, K.L. Cottingham, D.E. Schindler, D.L. Christensen, D.M. Post, and N. Voichick. (1996). Chlorophyll variability, nutrient input and grazing: Evidence from whole-lake experiments. *Ecology* 77:725-735.

Chesson, P.L. (1986). Environmental variation and the coexistence of species. In *Community Ecology*, eds. J. Diamond and T.J. Case, New York, NY: Harper and Row. pp. 240-256.

Clark, J.S., M. Lewis, and L. Horvath. (2001). Invasion by extremes: Variation in dispersal and reproduction retards population spread. *American Naturalist* 157: 537-554.

Clark, J.S., S. R. Carpenter, M. Barber, S. Collins, A. Dobson, J. Foley, D. Lodge, M. Pascual, R. Pielke, Jr, W. Pizer, C. Pringle, W. V. Reid, K. A. Rose, O. Sala, W. H. Schlesinger, D. Wall, and D. Wear. (2001). Ecological forecasts: An emerging imperative. *Science* 293:657-660.

Clements, D.R., S.F. Weise, and C.J. Swanton. (1994). Integrated weed management and wee species diversity. *Phytoprotection* 75:1-18.

Clements, D.R. and A. Shrestha. (2004). New dimensions in agroecology for developing a biological approach to crop production. *Journal of Crop Improvement* 11(1/2): 1-20.

Cousens, R. (1987). Theory and reality of weed control thresholds. *Plant Protection Quarterly* 2:13-20.

Crawley, M.J. (1997). The structure of plant communities. In *Plant Ecology*, 9th ed., ed. M.J. Crawley, Oxford, UK: Blackwell Science. pp. 475-531.

Diamond, J.M. (1975). Assembly of species communities. In *Ecology and Evolution of Communities*, eds. M.L. Cody and J.M. Diamond, Cambridge, MA: Harvard University Press. pp. 342-444.

Elton, C.S. (1958). *The Ecology of Invasions by Animals and Plants*. London, UK: Methuen. (please check publisher).

Gause, G.F. (1934). *The Struggle for Existence*. Baltimore, MD: Williams and Wilkins Co.

Goodman, D.G. (1975). The theory of diversity-stability relationships in ecology. *Quarterly Review of Biology* 50:237-266.

Gliessman, S.R. (1998). *Agroecology: Ecological Processes in Sustainable Agriculture*. Chelsea, MI: Ann Arbor Press.

Grime, J.P. (1973). Competitive exclusion in herbaceous vegetation. *Nature* 242: 344-347.

Harper, J.L. (1977). *Population Biology of Plants*. New York, NY: Academic Press.

Hubbell, S.P. (2001). The unified Neutral Theory of Biodiversity and Biogeography. Monographs in Population Biology, No. 32. Princeton, NJ: Princeton University Press.

Hutchinson, G.E. (1959). Homage to Santa Rosalia or why are there so many kinds of animals? *The American Naturalist* 93:145-159.

Huston, M.A. (1994). *Biological Diversity: The Coexistence of Species on Changing Landscapes*. Cambridge, UK: Cambridge University Press.

Jordan, N. R. and J.L. Jannink. (1997). Assessing the practical importance of weed evolution: a research agenda. *Weed Research* 37:237-246.

Kareiva, P. (1989). Renewing the dialogue between ecology theory and experiments in ecology. In *Perspectives in Ecological Theory*, eds. J. Roughgarden, R.M. May, and S.A. Levin, Princeton, NJ: Princeton University Press. pp. 68-88.

Kareiva, P. (1994). Higher-order interactions. *Ecology* 75:1527-1528.

Lehman, C.L. and D. Tilman. (2000). Biodiversity, stability, and productivity in competitive communities. *The American Naturalist* 156:534-552.

Levin, R. (1970). Extinction. In *Some Mathematical Problems in Biology*, ed. M. Gerstenhaber, Providence, RI: American Mathematical Society. pp. 75-107.

Levin, R. (1979). Coexistence in a variable environment. *The American Naturalist* 114:765-783.

Lotka, A.J. (1925). *Elements of Mathematical Biology*, New York, NY: Williams & Wilkins.

Luschei, E.C., L.R. Van Wychen, B.D. Maxwell, A.J. Bussan, D. Buschena, and D. Goodman. (2001). Implementing and conducting on-farm weed research with the use of GPS. *Weed Science* 49: 536-542.

MacArthur, R.H. (1970). Species packing and competitive equilibrium for many species. *Theoretical Population Biology* 1:1-11.

Maurer, B.A. (1999). *Untangling Ecological Complexity*. Chicago, IL: University of Chicago Press.

Maxwell, B.D. (1999). My view: A perspective on ecologically based pest management. *Weed Science* 47:129.

May, R.M., ed. (1981). *Theoretical Ecology: Principles and Applications*. Oxford, UK: Sinauer Associates.

McGlade, J. (1999). Individual-based models in ecology. In *Advanced Ecological Theory: Principles and Applications*, ed. J. McGlade, Oxford, UK: Blackwell Science, Inc.

Mohler, C.L. (2001). Weed evolution and community structure. In *Ecological Management of Agricultural Weeds*, eds. M. Liebman, C.L. Mohler, and C.P. Staver, New York, NY: Cambridge University Press. pp. 444-493.

Norris, R.F. (1982). Interactions between weeds and other pests in the agroecosystem. In *Biometeorology in Integrated Pest Management*, eds. J.L. Hatfield and I.J. Thomason, New York, NY: Academic Press. pp. 343-406.

Pacala, S.W. and J.A. Silander, Jr. (1990). Field tests of neighborhood population dynamics models of two annual weed species. *Ecological Monographs* 60:113-134.

Peters, R.H. (1995). *A Critique for Ecology*. Cambridge, UK: Cambridge University Press.

Pontin, A.J. (1982). *Competition and Coexistence of Species*. London, UK: Pitman.

Ricklefs, R.E. and D. Schluter (eds.). (1993). *Species Diversity in Ecological Communities: Historical and Geographical Perspectives*. Chicago, IL: University of Chicago Press.

Ricklefs, R.E. (1979). *Ecology*. New York, NY: Chiron Press.

Sing, S. (2002). Spatial and biotic interactions of the wheat stem sawfly with wild oat and Montana dryland spring wheat. PhD Thesis, Bozeman, MT: Montana State University.

Stoll, P. and J. Weiner. (2000). A neighborhood view of interactions among individual plants. In *The Geometry of Ecological Interactions*, eds. U. Dieckmann, R. Law and J.A.J. Metz, Cambridge, UK: Cambridge University Press. pp. 94-114.

Swift, M.J. and J.M. Anderson. (1993). Biodiversity and ecosystem function in agricultural systems. In *Biodiversity and Ecosystem Function*, eds. E.D. Schulze and H.A. Mooney, Berlin, Germany: Springer-Verlag. pp. 15-38.

Tilman, D. (1982). *Resource Competition and Community Structure*. Princeton, NJ: Princeton University Press.

Van Wychen, L. (2002). Field-scale spatial distribution, water use and habitat of wild oat in the semiarid Northern Great Plains. PhD Thesis, Bozeman, MT: Montana State University.

Weiher, E. and P. Keddy. (1999). *Ecological Assembly Rules: Perspectives, Advances, Retreats*. Cambridge, UK: Cambridge University Press.

Wissel, C. (2000). Grid-based models as tools for ecological research. In *The Geometry of Ecological Interactions*, eds. U. Dieckmann, R. Law and J.A.J. Metz, Cambridge, UK: Cambridge University Press. pp. 94-114.

Wolfram, S. (2002). *A New Kind of Science*. Champaign, IL: Wolfram Research.

Zimdahl, R.L. (1980). *Weed Crop Competition: A Review*. Corvallis, OR: International Plant Protection Center.

Research and Extension Supporting Ecologically Based IPM Systems

Fabián D. Menalled
Douglas A. Landis
Larry E. Dyer

SUMMARY. Integrated Pest Management (IPM) has a long history of developing pest management strategies based on ecological principles.

Fabián D. Menalled was affiliated with the Department of Entomology and Center for Integrated Plant Systems, 204 CIPS, Michigan State University, East Lansing, MI 48824. Current affiliation: 308 National Soil Tilth Laboratory, 2150 Pammel Drive, Ames, IA 50011-4420.

Douglas A. Landis is affiliated with the Department of Entomology and Center for Integrated Plant Systems, 204 CIPS, Michigan State University, East Lansing, MI 48824 (E-mail: landisd@pilot.msu.edu).

Larry E. Dyer is affiliated with Kellogg Biological Station, Extension Land & Water Program, 3700 East Gull Lake Drive, Hickory Corners, MI 49060.

The authors wish to thank Jana Lee and Dora Carmona for their work on the studies described here. The authors also acknowledge the contributions of Dale Mutch who has helped to shape their approach to IPM education, and Paul Marino who provided numerous insights into the work described here. Comments from Matt Liebman, Karen Renner, Richard Harwood, David Clements and two anonymous reviewers have helped to improve this manuscript.

This research was made possible through funding provided by USDA SARE grant LNC 95-85, USDA NRI Grant 99-35316-7911 and by the Michigan Agricultural Research Station.

[Haworth co-indexing entry note]: "Research and Extension Supporting Ecologically Based IPM Systems." Menalled, Fabián D., Douglas A. Landis, and Larry E. Dyer. Co-published simultaneously in *Journal of Crop Improvement* (Food Products Press, an imprint of The Haworth Press, Inc.) Vol. 11, No. 1/2 (#21/22), 2004, pp. 153-174; and: *New Dimensions in Agroecology* (ed: David Clements, and Anil Shrestha) Food Products Press, an imprint of The Haworth Press, Inc., 2004, pp. 153-174. Single or multiple copies of this article are available for a fee from The Haworth Document Delivery Service [1-800-HAWORTH, 9:00 a.m. - 5:00 p.m. (EST). E-mail address: docdelivery@haworthpress.com].

http://www.haworthpress.com/web/JCRIP

Digital Object Identifer: 10.1300/J411v11n01_08

While IPM systems differ in their reliance on chemical controls, an ecological approach to IPM offers opportunities both to test ecological theory and to develop novel pest management techniques. We review ecological concepts relevant to habitat management as a tool for managing pest and natural enemy populations in annual cropping systems, and we describe a case study of the impacts of habitat management on predatory ground beetles (Coleoptera: Carabidae) in annual cropping systems. Experimental manipulation of carabid population density revealed a positive relationship between the abundance of carabids in cropland and prey removal. Provision of stable refuge habitats in annual crops increased carabid species richness and abundance. Further studies demonstrated that the presence of refuge habitats mitigated the impacts of insecticide disturbance on carabid abundance and community structure in adjacent cropland. These results were used as part of a program to educate extension agents and producers in the principles of agroecology and ecologically based pest management. We encourage extension agents and producers to test novel practices using adaptive management. In this approach, predictions are formulated regarding specific management practices and evaluated against the results. Based on these observations, management is then adapted to yield the desired outcome. Such an approach recognizes the inherent uncertainty of multifactor ecological manipulation while providing producers with methods to manage this uncertainty.

KEYWORDS. Agroecology, ecologically based pest management, biological control, adaptive management

INTRODUCTION

In a review of the history of Integrated Pest Management (IPM), Kogan (1998), states that IPM has emerged as one of the most recognizable and robust concepts in modern agriculture. In spite of, or perhaps due to its wide acceptance, there are many definitions of IPM. Most modern definitions of IPM include an emphasis on; (1) use of appropriate pest control measures either singly or in concert, (2) economic benefits to producers and society, (3) environmental benefits, and (4) the need to consider multiple pests (Kogan, 1998).

Prior to the advent of synthetic organic pesticides, pest control was achieved primarily through knowledge of pest biology integrated with multiple control practices such as variety selection, crop rotation, cultural and biological controls (National Research Council, 1996). A sound understanding of the interactions between a pest and its crop environment was fundamental for the success of such systems. However, with the advent of highly effective chemical pest controls in the late 1940s, emphasis began to shift towards their use in a single tactic approach (Newsom, 1980). The concepts of integrated control and eventually IPM grew out of the recognition that chemical controls needed to be integrated into a broader concept of pest management (Stern et al., 1959).

In spite of this recognition, it has been argued that most IPM systems are still dominated by an over-reliance on chemical pest controls (National Research Council, 1996). In addition, because it is possible to implement such approaches with little consideration of the characteristics, processes, and dynamics of agroecosystems (Levins and Wilson, 1980), producers have lost some of their former appreciation for the ecology of their production systems. This situation poses a challenge to agroecologists. How do we conduct IPM research in a way that enhances our fundamental understanding of agroecosystems and, at the same time, allows us to effectively educate growers regarding their management options?

In this contribution we specifically examine the potential for incorporating an ecological approach to aid in the development and application of IPM systems for sustainable annual crop agriculture. We utilize a case study from our work on carabid beetles (Coleoptera: Carabidae) dynamics as an example of the potential importance of incorporating an ecological perspective in the design of IPM programs. We do so by first, introducing the ecological concepts and research background related to this particular case study. We then describe a series of experiments aimed at understanding the ecology of carabid beetles as generalist predators of weeds and insects in annual row crop agriculture. Finally, we discuss how adaptive management, a concept developed by ecosystem managers, may be useful in the implementation of such ecologically based IPM systems.

We acknowledge that due to the case-specific nature of many ecological processes, there are undoubtedly exceptions to any ecological principles presented here (Letourneau, 1998). However, despite this limitation, ecological properties emerging from natural systems have been successfully used to model perennial grain cropping systems (Jackson and

Piper, 1989) and tropical agricultural systems (Ewel, 1986). We also acknowledge an entomological bias on our part and an emphasis on annual row-crop agriculture that we hope will not distract from the main points of the manuscript. Our goal is that the concepts presented here might stimulate critical thinking on how ecological principles may be further used to define and implement ecologically-based IPM systems for a variety of pests and cropping systems.

ECOLOGICAL CONCEPTS AND INTEGRATED PEST MANAGEMENT

Modern industrial agriculture can be described as a human-dominated enterprise in which the pressure towards increasing crop yield generates ecosystems characterized by biological simplification, output specialization, subsidized off-farm chemical and energetic inputs, and decreasing environmental heterogeneity (Altieri and Nicholls, 1999; Letourneau and Altieri, 1999; Swift and Anderson, 1993; Tscharntke and Kruess, 1999). There is an increasing body of ecological principles and empirical studies showing that such farming systems may pose long-term risks to sustainability and environmental health (Matson et al., 1997). In this section, we analyze three ecological concepts we believe are relevant to the successful implementation of IPM strategies in annual crop production systems: (1) resource predictability, (2) habitat simplification and fragmentation, and (3) disturbance regimes. A brief summary of the ecological concepts discussed in the next sections is presented in Table 1.

Resource Predictability and Arthropod Abundance

Crop populations in large-scale annual monocultures represent a stable and predictable resource for herbivores. However, because of density-independent oscillations in their populations, herbivores in annual crops represent a less predictable resource for their predators and parasites. This differential predictability in resources for herbivores versus natural enemies contributes to the increased extinction rates of organisms occupying higher trophic levels of annual crops (Tscharntke and Kruess, 1999). From an applied point of view, this low resource predictability may partially explain the lower rates of success of imported biological control agents in annual crop agriculture (Hall and Ehler, 1979; Hall, Ehler, and Bisabri-Ershadi, 1980).

TABLE 1. Summary description of several ecological concepts related to the development and implementation of ecologically based IPM.

Ecological Concept	Reference	Applicability to IPM
Resource predictability	Tscharntke and Kruess, 1999	While crops represent a stable and predictable resource for herbivores, herbivores present lower predictability for higher trophic levels resulting in increased local extinction of predators and parasitoids.
Enemy impact hypothesis	Pimentel, 1961	Increased plant community diversity correlates with higher food and host resources for natural enemies resulting in higher probability of maintaining pest populations at reduced levels.
Resource concentration hypothesis	Root, 1973	Pest outbreaks are commonly observed in large monocultures because herbivores can easily find food in systems with low plant species diversity.
Cyclic colonization	Wissinger, 1997	Stable habitats such as woodlots provide resources that promote natural enemy survivorship. Early in the growing season, natural enemies colonize crop fields from these stable habitats.
Island biogeography	MacArthur and Wilson, 1967	Maintaining large stable habitats in agricultural landscapes increases natural enemy species immigration rates and decreases extinction favoring natural enemy conservation.
Metapopulations	Levins, 1969	A network of stable habitats scattered across agricultural landscapes maintains a series of genetically interconnected local populations of natural enemies.
Disturbance	Pickett and White, 1985	Agricultural management practices such as tillage, pesticide application, nutrient management, cultivation, and harvest represent ecological disturbances that kill natural enemies.
Habitat management	Landis, Wratten, and Gurr, 2000 Lee, Menalled, and Landis, 2001	Stable habitats in agroecosystems can increase natural enemy survivorship and community recovery following disturbance.

Two complementary hypotheses have been proposed to explain the relationship between resource accessibility and pest outbreak in low diversity systems: the enemy impact hypothesis (Pimentel, 1961), and the resource concentration hypothesis (Root, 1973). The enemy impact hypothesis postulates three mechanisms responsible for the relationship between pest regulation and system diversity. First, an increase in plant community diversity is correlated with higher diversity of herbivore species as well as pollen and nectar that provide alternative food or host supplies to predators or parasitoids. Second, an associated increase in the stability of parasitoid or predator resources secures higher diversity at the primary carnivore trophic level and a greater probability that herbivore species will be maintained below threshold levels. Finally, diverse arthropod communities typically contain many generalist predators with potential to suppress a variety of pest species. The resource concentration hypothesis also proposes three mechanisms to explain the observed high rate of pest outbreaks occurring in large monocultures. First, herbivores can easily find food resources in these conspicuous and low diversity systems. Second, due to the existence of large amount of resources, herbivores often remain in such habitats. Finally, the resource availability in low diverse systems translates into increased herbivore reproduction and greater potential for population (i.e., pest) outbreaks.

Habitat Simplification and Fragmentation

As noted in the previous section, simplified annual cropping systems provide stable and predictable resources for herbivores. For predators and parasitoids, however, the simplification of the agricultural landscape can be viewed as habitat fragmentation because the crop habitat does not by itself provide sufficient resources for their survival. In an agricultural landscape, non-crop habitats can serve as refuges for natural enemies that colonize agricultural fields, kill pests, and return to refuges following a 'cyclic colonization pattern' (Wissinger, 1997). In this context, fencerows, woodlots, and herbaceous strips can be regarded as isolated islands where beneficial organisms can thrive within otherwise inhospitable environments. In a similar way, the theory of island biogeography (MacArthur and Wilson 1967) has been applied to help understand the occurrence of organisms in agricultural landscapes (Kruess and Tscharntke, 1994; Price and Waldbauer, 1994; Simberloff, 1986; Thies and Tscharntke; 1999; Zabel and Tscharntke, 1998). However, in highly simplified agroecosystems, a large species donor area [i.e., the continent as defined by MacArthur and Wilson (1967)] may not exist and the metapopulation model proposed by Levins (1969) can be used as an appropriate theoretical framework for pest control.

In agroecosystems, a metapopulation can be regarded as a series of genetically interconnected local populations living in hedgerows, fencerows, woodlots, and riparian strips. It has been shown that in a metapopulation, dispersal between patches allows local nonequilibrium populations to persist at the regional scale (Murdoch, Chesson, and Chesson, 1985; Murdoch and Briggs, 1996). Although the metapopulation approach has been widely adopted by conservation biologists (McCullough, 1996), agroecologists have yet to incorporate these ideas strongly in the development of IPM systems (Landis and Menalled, 1998; Letourneau, 1998). However, several empirical studies have demonstrated that habitat fragmentation can disrupt ecosystem functions such as host-parasitoid interactions (Kruess and Tscharntke, 1994; Roland, 1993; Tscharntke, 2000), and alter overall natural enemy abundance (Marino and Landis, 1996; Menalled et al., 1999; Nicholls, Parrella, and Altieri, 2001; Thies and Tscharntke, 1999), community richness and diversity (Gut et al., 1982; Colunga-Garcia, Gage, and Landis, 1997), and fecundity (Bommarco, 1998).

Disturbance in Annual Crop Systems and Generalist Predators

In the ecological literature, disturbance is defined as a "relatively discrete event in time that disrupts ecosystems, community or population structure, and changes resources, substrate availability, or the physical environment" (Pickett and White, 1985). In this context, common agricultural practices such as primary and secondary tillage, herbicide and pesticide applications, nutrient management, cultivation, and harvest can each be characterized as disturbance events. Ecologists have long recognized the importance of magnitude, frequency, and predictability of disturbances in determining the species composition and other characteristics of biological communities (Sousa, 1984). At a larger scale, the area and distribution of disturbances influence the spatial arrangement of plant communities in landscapes. Due to bottom-up influences of vegetation on insect communities, plant species composition has been hypothesized to determine the structural and chemical environment for beneficial insects (Askew, 1980; Hawkins and Lawton, 1987; Price, 1991). Therefore, the disturbance regime associated with a given agricultural management system is of vital importance in determining the possibility of implementing an IPM program that incorporates natural enemies as a primary mechanism of pest control.

Landis and Menalled (1998) discuss the importance that agricultural disturbances can have on the abundance, diversity, and effectiveness of beneficial organisms. They argue that management practices can have both direct and indirect effects, and can act at various spatial scales:

within crop fields, at the farm-level, and/or at the landscape-level. The impact that a disturbance regime has on the organisms living in a certain area and on the physical environment can be observed at different time intervals. The short-term impact of agricultural disturbances is to disrupt some portion of the microbial, plant, and animal communities present in the crop field (Booij and Noorlander, 1988; Croft, 1990; Los and Allen, 1983). This is followed by a gradual recolonization of the disturbed area by individuals remaining in the systems or from surrounding areas. Several studies have shown that agricultural practices such as insecticide applications can also have long-term impacts on natural enemy community structure and function (Basedow, 1990; Burn, 1989).

HABITAT MANAGEMENT IN ANNUAL CROP IPM SYSTEMS

From the previous discussion it is clear that a common outcome of modern annual crop agriculture is a reduction in natural enemy effectiveness in association with high abundance of resources that favor herbivore colonization and population growth. This situation ultimately leads to poor pest control and the need for additional disturbances (e.g., pesticide applications) to restrict pest damage (Brust, Stinner, and McCartney, 1985; 1986; Edwards, Sunderland, and George, 1979). To use predators and parasitoids successfully for pest management, it is necessary to modify agroecosystems to conserve natural enemies and enhance their impacts. While use of "soft pesticides" and practices such as application of food sprays can mitigate the immediate impact of disturbance, a comprehensive solution should address the ultimate cause, i.e., the disturbance regime and resulting habitat simplification that reduces habitat suitability for beneficial organisms (Landis and Menalled, 1998). Habitat management, defined as a series of practices meant to "alter habitats to improve availability of resources required by natural enemies for optimal performance," represents a comprehensive approach to reduce the negative impact of agricultural management practices and favor natural enemies in annual crop agriculture (Landis, Wratten, and Gurr, 2000).

Habitat management practices can be implemented at the within-field level, farm level, or landscape level and include practices such as conservation tillage, cover crops, intercropping, and creation of sown plant strips (Altieri and Letourneau, 1982; Landis, Wratten, and Gurr, 2000; Pickett and Bugg, 1998). A common feature of these practices is an increase in the planned biodiversity (organisms purposely included in the agroecosystem) and the associated biodiversity (soil flora and fauna, herbivores, carnivores, etc.) that colonize the crop fields (Vandermeer and Perfecto, 1995). However, increasing diversity *per se* is not the final objective of habitat management. A key component is to identify the elements of diversity that enhance ecological services and then determine the management practices that will allow establishment, survivorship, and reproduction of these organisms (Altieri and Nicholls, 1999; Nentwig, Frank, and Lethmayer, 1998).

Can Habitat Management Enhance Carabid Communities? A Case Study

Carabid beetles (Colepotera: Carabidae) are polyphagous predators commonly found in agricultural systems that have the potential of consuming many pest species (Sopp et al., 1992; Sunderland, 1975; Sunderland et al., 1987; Vickerman and Sunderland, 1975). Several studies have demonstrated that common agricultural practices such as tillage, cultivation and pesticide application frequently reduce carabid abundance (Basedow, 1990; Brust, 1994; House and Parmelee, 1985; Stinner and House, 1990; Terry, Potter, and Spicer, 1993), and alter community characteristics (Cárcamo, Niemelä, and Spence, 1995; Clark, Gage, and Spence, 1997; Kromp, 1989).

We utilized ground beetles as a model system to evaluate the response of beneficial organisms to different types of agricultural management practices. We first evaluated if altering carabid beetle abundance *per se* would influence rates of prey removal in the field. To do so, we manipulated carabid numbers with minimal habitat alteration by creating plots surrounded by different boundaries that selectively affected carabid dispersal (Menalled, Lee, and Landis, 1999). Three treatments were established: (1) naturally occurring communities, (2) augmented communities using ingress boundaries, and (3) reduced communities using egress boundaries (Figure 1). In comparison to the no boundary treatment, carabids increased 54% and decreased 83% in plots surrounded by ingress and egress boundaries respectively. Predation of sentinel onion fly pupae (*Delia antiqua* (Meigen)) was positively correlated with carabid abundance ($r^2 = 0.70$, $P < 0.0001$) confirming the importance of developing a habitat management strategy aimed at increasing the abundance of carabid beetles in crop fields.

Next, we tested the concept that providing less disturbed habitats in close spatial association with crop fields represents a viable approach to

FIGURE 1. Experimental manipulation of ground-dwelling arthropod density. Top: natural occurring density with no boundary. Center: reduced density, where the number of invertebrates is decreased using egress boundaries. Bottom: augmented density by means of ingress boundaries, i.e., devices that allow invertebrates to move into but not out of a plot.

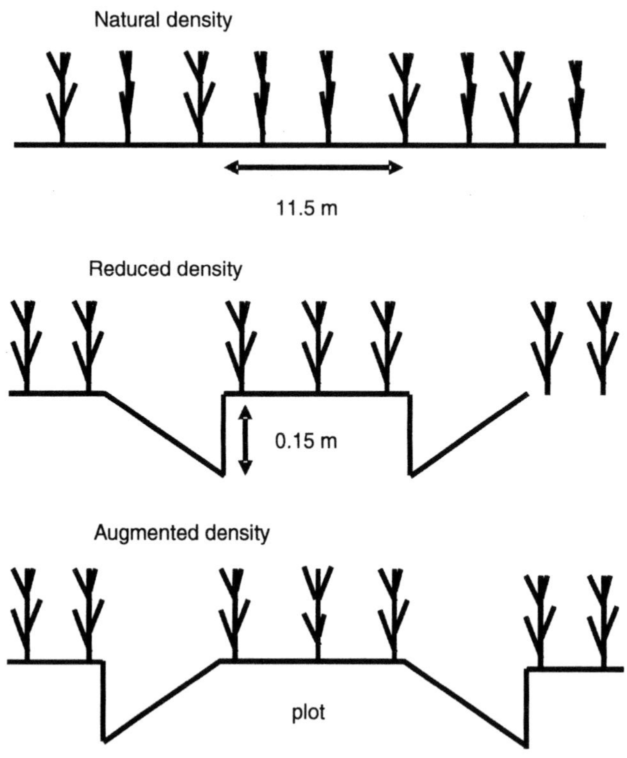

conserving carabid communities. Undisturbed vegetative refuge habitats such as grasslands, hedgerows, field margins, or grassy strips within a field can enhance carabid abundance, fecundity, and species diversity by supplying overwintering sites, food, and shelter (Lys and Nentwig, 1992; Zangger, Lys, and Nentwig, 1994). We evaluated the importance of herbaceous refuge strips in carabid beetle conservation in a 1.4 ha site arranged in a split-plot design with four 32 m × 66 m blocks. Each block contained two main plots, one with a 3 m wide refuge strip in the center and the other with a control strip planted with corn (*Zea mays* L.). The refuge strips were established in 1995 using orchard grass (*Dactylis glomerata* L.), white clover (*Trifolium repens* L.), and sweet clover (*Melilotus officinalis* L.) surrounding a central strip of perennial flowering plants to provide supplementary food for predators and parasitoids (Carmona and Landis, 1999). Soybean (*Glycine max* L. Merr.) was planted in the crop areas in 1995 followed by oats (*Avena sativa* L.) in 1996 and then three successive years of corn. Between 1996 and 1999, we used a grid of pitfall traps to sample activity-density of adult carabid beetles. Over this period, we detected a steady increase in the number of carabid species (Figure 2) with consistently more individuals trapped in refuge strips than in control strips comprised of corn only (Figure 3) (Carmona and Landis, 1999; Lee, Menalled, and Landis, 2001).

The next step was to assess if refuge habitats could act as sources of natural enemies colonizing adjacent fields, thereby mitigating the consequences of insecticide disturbance. We hypothesized that insecticide application would reduce carabid activity-density, species richness and alter community composition, but that the presence of refuges would allow carabid assemblages to recover in treated crop areas. We utilized the field plots described above to test these hypotheses (Lee, Menalled, and Landis, 2001). Between 1998 and 1999, main plots were divided into two subplots, one of which was treated with terbufos (S-[[(1,1-dimethylethyl) thio] methyl]; Counter™ 20 CR American Cyanamid), a soil insecticide commonly used in non-rotated corn to control corn

FIGURE 2. Species richness (number of species) of carabid beetles captured in pitfall traps over time in refuge habitat study, E. Lansing, Michigan. Number per year represents total species capture in refuges and crop areas combined. Crop areas planted to oats in 1996 and corn 1997-1999.

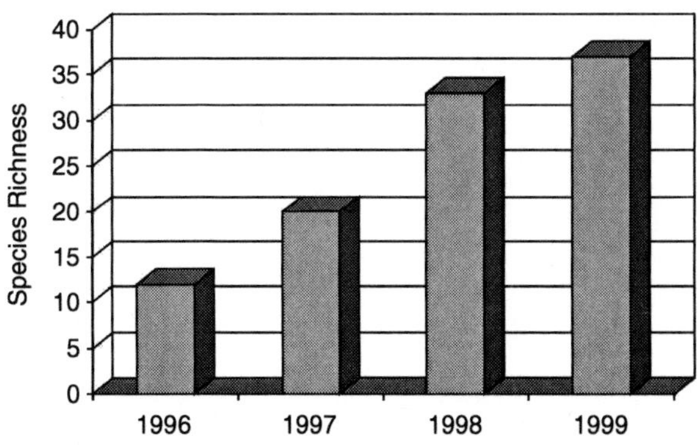

FIGURE 3. Activity density of carabids in refuge strips and corresponding control (crop) strips in refuge habitat study, E. Lansing, Michigan, May-October 1995-1999. Crop areas planted to oats in 1996 and corn 1997-1999. An * indicates a significant difference (p > 0.05) in season long abundance between refuge and control plots within a year as indicated by a t-test.

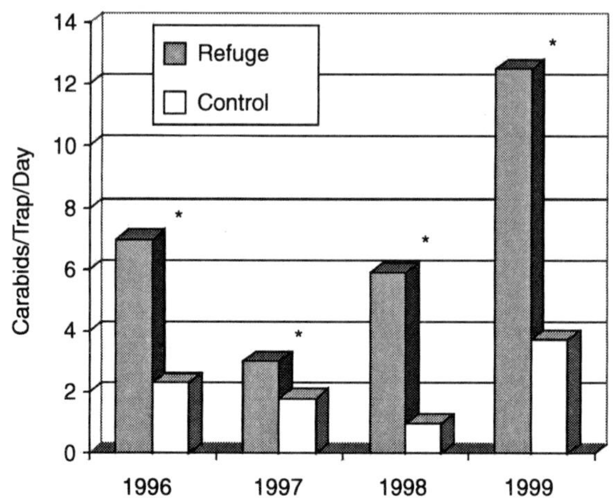

rootworm larvae (*Diabrotica* spp.). Four treatments were established and are referred to as: (1) refuge-untreated crop, (2) refuge-treated crop, (3) control-untreated crop, and (4) control-treated crop.

Before planting and insecticide application in 1998 and 1999, carabid activity-density in the crop area was similar across all treatments. In both years, insecticide application significantly reduced carabid activity-density, species richness and altered community composition in the crop area. Contrary to our predictions, refuge strips did not consistently augment carabid numbers in areas where insecticide was applied. As insecticide toxicity declined, carabid captures in insecticide treated crop areas adjacent to refuges surpassed those where insecticide was not applied, suggesting an interaction between refuges and disturbances that determine the net movement of beneficial organisms in agricultural landscapes (Lee, Menalled, and Landis, 2001).

A Refined View

Results of our studies support the concept of habitat management maintaining higher species diversity and abundance of generalist predators in annual crops. Specifically, our work reveals that refuges can increase carabid movement into disturbed crop habitats, moderating the impact of disturbance on communities of these natural enemies. However, contrary to our predictions, the interactions between refuges and adjacent crop areas were not straightforward. Two different mechanisms may be responsible for the observed results. If insecticide disturbance reduced the long-term availability of prey, high rates of carabid captures in the crop areas could be due to an increase in activity of hungry carabids (Chiverton, 1984; Frampton et al., 1995; Mauremooto et al., 1995; Wheater, 1991). Alternatively, insecticide disturbance may have actually increased prey availability in crop areas in the long term. When insecticide inputs eliminate early-season predators, certain prey species may thrive and become abundant once toxicity levels have declined. Under this scenario, the abundance of prey and chemical cues emanating from prey may attract or arrest more beetles into the crop area (Kielty et al., 1996; Wheater, 1991). Despite the ultimate cause, while refuges harbor many carabids, more carabids might eventually be drawn into treated crop areas than untreated crop areas. The mechanisms responsible for the observed results represent both testable hypotheses for future studies on habitat management and challenges to refinement of our ecological understanding of these systems. Although lack of correspondence between ecological concepts and practical application can lead to underutilization of ecological knowledge as a source of useful IPM guidelines (Dyer and Gentry, 1999), we believe that collaboration among ecologists, agronomists, and pest management specialists can address these challenges in a productive fashion.

EXTENSION OF ECOLOGICALLY BASED IPM SYSTEMS

In our extension programs, we have attempted to provide producers and extension agents with an understanding of ecological processes occurring in annual cropping systems. Two publications have been particularly directed towards this effort; "Michigan Field Crop Ecology" (Cavigelli et al., 1998) and "Michigan Field Crop Pest Ecology and Management" (Cavigelli et al., 2000). Both provide a general framework of ecological principles as a basis for crop production and pest management. Participants in our extension programs have appreciated this conceptual framework, but they have also asked for specific management practices. However, the recommendations we can provide often do not resemble the "recipes" producers have come to expect in the conventional IPM approach. The ecological management tools we suggest are expected to affect many species and influence processes at different levels of organization. Although implementation of these tools is based on generally applicable ecological principles, the outcomes will be influenced by site and case specific variables. How can we as researchers and educators help producers to manage this complexity? Developments in the concept of ecosystem management over the last decade (Christensen et al., 1996) may provide a guide.

Ecosystem management has emerged as a process for managing public lands with multiple stakeholders, such as rangelands, forested areas and watersheds. It has been adopted as a land management policy by several federal agencies including U.S. Fisheries and Wildlife (Beattie, 1996), Bureau of Land Management (Dombeck, 1996) and U.S. Forest Service (Thomas, 1996). It is also becoming policy for many state government agencies and non-governmental organizations (Yaffee et al., 1996). Although there is no single definition or process for ecosystem management, some basic characteristics are emerging (Christensen et al., 1996). The system must be managed to achieve measurable economic, environmental and social goals while maintaining ecosystem integrity. Sustainability of ecosystem structure and function is a central tenet of ecosystem management, so practices that deplete the system are unacceptable. Ecosystems are dynamic and our knowledge of them is incomplete and constantly changing. Managers must make decisions and take action in this environment of uncertainty. Adaptive management is the key element that transforms ecosystem management from a philosophical or policy framework into a management process.

An adaptive management process acknowledges uncertainty, makes decisions based on the best available knowledge, formulates testable hypotheses about management outcomes with specific, measurable predictions, and proceeds with monitoring to test those management hypotheses (Christensen et al., 1996; Walters and Holling, 1990). Walters and Holling (1990) describe three approaches to adaptive management. First, in an evolutionary approach management is gradually improved as better management options emerge through trial and error. Second, a passive adaptive approach strives for one best model of ecosystem function, and modifies the model as experiments and observation of management outcomes indicate inadequacies. Third, an active adaptive approach acknowledges uncertainty, admits several alternative models and generates hypotheses that would distinguish between those models. They argue that learning will be most rapid with an active adaptive approach.

The need for an ecosystem management approach to management of agricultural pests has been widely recognized (Lewis et al., 1997; Speight, 1983; Sustainable Agriculture Network, 2000). As with other land managers, farmers must make decisions and act in the face of uncertainty when incorporating ecological tools for pest management into their farming operations. Adapting a process developed largely for public lands to agroecosystems under private ownership presents challenges, especially for those changes that are more meaningful if made over significant portions of landscapes (Roberts and Parker, 1998; Jackson and Jackson 2002). Nevertheless, adaptive management may provide a useful process for making management decisions and evaluating outcomes.

Staver (2001) proposed a model for farmer, extensionist and scientist collaboration to implement adaptive management of agricultural weeds. In this approach a group meets at critical points in the crop cycle to identify problems and make management plans, do experiments and monitoring, evaluate the observed results of management, and revise management plans. A great strength of this approach is combining the knowledge and ways of learning of scientists and farmers. Periodic interactions enables the group to revise their understanding of ecosystem functioning, critically examine hypotheses for management, and propose outcomes from diverse perspectives. Farmers, extensionists, and scientists together decide what experiments and observations would best test those hypotheses.

An adaptive management approach thus represents a suitable framework by which ecological knowledge can be applied to IPM decisions. Farmers' intuitive understanding of pest ecology has formed part of crop production technology from the beginning of agriculture (Staver, 2001). Through a process of on-farm observation, logic, experimentation, extrapolation, risk calculation, and communication farmers have accumulated a vast understanding of how an agricultural system works. This "evolutionary" adaptation can be enhanced by an "active" adaptive management approach (Walters and Holling, 1990). The adaptive management hypothesis-testing approach is a decision-making process that may accelerate the application of ecological knowledge to farming practices. This process requires effective interaction between scientists, extension educators, and farmers with all participants agreeing on research priorities, temporal and spatial scales of trials, and decision-making processes (Staver, 2001).

CONCLUSIONS

The development of IPM reveals a long history of using ecological principles as the basis for understanding agricultural systems and the manipulations necessary to manage pests. However, even within an IPM context, the level of necessary ecological knowledge varies from system to system. In IPM programs where scouting, thresholds and use of chemical pesticides dominate, the need to understand all of the ecological interactions of the system may be reduced. However, advocates of sustainable agriculture generally favor a more ecologically based approach to pest management.

Research on ecologically based IPM opens opportunities to test current ecological concepts in managed systems. In our work, this has yielded confirmation of hypotheses in some cases and lack of correspondence in others. This is to be expected given the complexity of interactions at work in these systems. By formulating applied studies that test ecological predictions we hope to contribute to both refinement of ecological knowledge and to generate specific management recommendations for producers to test.

One of the greatest continuing challenges for sustainable agriculture lies in how to educate producers and extension educators in the principles of agroecology. Teaching ecological concepts in extension education represents a major paradigm shift and requires new approaches. Many involved in extension education (ourselves included) have been used to "having all the answers" regarding a particular pest situation. Unfortunately, ecologically based IPM approaches seldom lend themselves to such easy recipes. Recognizing this, we have adopted the approach of educating clientele in the principles of agroecology and sharing the results of our ongoing studies, even if they have not yet resulted in a series of clearly stated management recommendations. Often this approach is effective at stimulating producers to evaluate similar practices on their own farms. We encourage and support these efforts by teaching producers how to interpret their findings using an adaptive management paradigm.

REFERENCES

Altieri, M. A. and D. Letourneau. (1982). Vegetation management and biological control in agroecosystems. *Crop Protection* 1:405-430.

Altieri, M. A. and C. I. Nicholls. (1999). Biodiversity, ecosystem function, and insect pest management in agricultural systems. In *Biodiversity in Agroecosystems*, eds., W. W. Collins and C. O. Qualset, Boca Raton, FL: CRC Press, pp. 69-84.

Askew, R. R. (1980). The diversity of insect communities in leaf-mines and plant galls. *Journal of Animal Ecology* 49:817-829.

Basedow, Th. (1990). Effects of insecticides on Carabidae and the significance of these effects for agriculture and species number. In *The Role of Ground Beetles in Ecological and Environmental Studies*, ed. N. E. Stork, Andover, UK: Intercept, pp. 115-125.

Beattie, M. (1996). An ecosystem approach to fish and wildlife conservation. *Ecological Applications* 6: 696-699.

Bommarco, R. (1998). Reproduction and energy reserves of a predatory carabid beetle relative to agroecosystem complexity. *Ecological Applications* 8: 846-853.

Booij, C. J. H. and J. Noorlander. (1988). Effects of pesticide use and farm management on carabids in arable crops: Environmental effects of pesticides. *Agriculture, Ecosystems and Environment* 40:199-125.

Brust, G. E. (1994). Natural enemies in straw mulch reduce Colorado potato beetle populations and damage in potato. *Biological Control* 4:163-169.

Brust, G. E., B. R. Stinner, and D. A. McCartney. (1985). Tillage and soil insecticides effects on predator-black cutworm (Lepidoptera: Noctuidae) interactions in corn agroecosystems. *Journal of Economic Entomology* 78:1389-1392.

Brust, G. E., B. R. Stinner, and D. A. McCartney. (1986). Predator activity and predation in corn agroecosystems. *Environmental Entomology* 15:1017-1021.

Burn, A. J. (1989). Long-term effects of pesticides on natural enemies of cereal crop pests. In *Pesticides and Non-Target Invertebrates*, ed., P. C. Jepson, Andover, UK: Intercept, pp. 177-193.

Cárcamo, H. A., J. K. Niemelä, and J. R. Spence. (1995). Farming and ground beetles: Effects of agronomic practice on populations and community structure. *The Canadian Entomologist* 127:123-140.

Carmona, D. A. and D. A. Landis. (1999). Influence of refuge habitats and cover crops on seasonal activity-density of ground beetles (Coleoptera: Carabidae) in field crops. *Environmental Entomology* 28:1145-1153.

Cavigelli, M. A., S. R. Deming, L. K. Probyn, and D. R. Mutch (eds.). (2000). *Michigan Field Crop Pest Ecology and Management.* East Lansing, MI: Michigan State University Extension Bulletin E-2704, 108 pp.

Cavigelli, M. A., S. R. Deming, L. K. Probyn, and R. R. Harwood (eds.). (1998). *Michigan Field Crop Ecology: Managing Biological Processes for Productivity and Environmental Quality.* East Lansing, MI: Michigan State University Extension Bulletin E-2646, 92 pp.

Colunga-Garcia, M., S. H. Gage, and D. A. Landis. (1997). Response of an assemblage of Coccinellidae (Coleoptera) to a diverse agricultural landscape. *Environmental Entomology* 26:797-804.

Chiverton, P. A. (1984). Pitfall-trap catches of the carabid beetle *Pterostichus melanarius*, in relation to gut contents and prey densities, in insecticide treated and untreated spring barley. *Entomologia Experimentalis et Applicata* 36:23-30.

Christensen, N. L., A. M. Bartuska, J. H. Brown, S. Carpenter, C. D'Antonio, R. Francis, J. F. Franklin, J. A. MacMahon, R. F. Noss, D. J. Parsons, C. H. Peterson, M. G. Turner, and R. G. Woodmansee. (1996). The report of the Ecological Society of

America Committee on the scientific basis for ecosystem management. *Ecological Applications* 6:665-691.

Clark, M. S., S. H. Gage, and J. R. Spence. (1997). Habitat and management associated with common ground beetles (Coleoptera: Carabidae) in a Michigan agricultural landscape. *Environmental Entomology* 26:519-527.

Croft, B. A. (1990). *Arthropod Biological Control Agents and Pesticides*. New York, NY: Wiley Interscience.

Dombeck, M. P. (1996). Thinking like a mountain: BLM's approach to ecosystem management. *Ecological Applications* 6:699-702.

Dyer, L. A. and G. Gentry. (1999). Predicting natural-enemy responses to herbivores in natural and managed systems. *Ecological Applications* 9:402-408.

Edwards, C. A., K. D. Sunderland, and K. S. George. (1979). Studies on polyphagous predators of cereal aphids. *Journal of Applied Ecology* 16: 811-823.

Ewel, J. J. (1986). Designing agricultural ecosystems for the humid tropics. *Annual Review of Ecology and Systematics* 17:245-271.

Frampton, G. K., T. Çilgi, G. L. A. Fry, and S. D. Wratten. (1995). Effects of grassy banks on the dispersal of some carabid beetles (Coleoptera: Carabidae) on farmland. *Biological Conservation* 71:347-355.

Gut, L. J., C. E. Jochums, P. H. Westigard, and W. J. Liss. (1982). Variations in pear psylla (*Psylla pyricola* Foerster) densities in southern Oregon orchards and its implications. *Acta Horticulterea* 124:101-111.

Hall, R. W. and L. E. Ehler. (1979). Rate of establishment of natural enemies in classical biological control. *Bulletin of the Entomological Society of America* 25:280-82.

Hall, R. W., L. E. Ehler, and B. Bisabri-Ershadi. (1980). Rate of success in classical biological control of arthropods. *Bulletin of the Entomological Society of America* 26:111-114.

Hawkins, B. A., and J. H. Lawton. (1987). Species richness for parasitoids of British phytophagous insects. *Nature* 326:788-790.

House, G. J. and R. W. Parmelee. (1985). Comparison of soil arthropods and earthworms from conventional and no-tillage agroecosystems. *Soil Tillage Research* 5:351-360.

Jackson, D. L. and L. L. Jackson. (2002). *The Farm as Natural Habitat*. Washington, DC: Island Press.

Jackson, W. and J. Piper (1989). The necessary marriage between ecology and agriculture. *Ecology* 70:1591-1593.

Kielty, J. P., L. J. Allen-Williams, N. Underwood, and E. A. Eastwood. (1996). Behavioral responses of three species of ground beetle (Coleoptera: Carabidae) to olfactory cues associated with prey and habitat. *Journal of Insect Behavior* 9:237-250.

Kogan, M. (1998). Integrated pest management: Historical perspectives and contemporary developments. *Annual Review of Entomology* 43:243-270.

Kromp, B. (1989). Carabid beetle communities (Carabidae: Coleoptera) in biologically and conventionally farmed ecosystems. *Agriculture, Ecosystems and Environment* 27:241-251.

Kruess, A. and T. Tscharntke. (1994). Habitat fragmentation, species loss, and biological control. *Science* 264:1581-1584.

Landis, D. A. and F. D. Menalled. (1998). Ecological considerations in conservation of parasitoids in agricultural landscapes. In *Conservation Biological Control*, ed., P. Barbosa, San Diego, CA: Academic Press. pp. 101-121.

Landis, D. A., S. D. Wratten, and G. M. Gurr. (2000). Habitat management to conserve natural enemies of arthropod pests in agriculture. *Annual Review of Entomology* 45:175-201.

Lee, J. C., F. Menalled, and D. Landis. (2001). Refuge habitats modify impact of insecticide disturbance on carabid beetle communities. *Journal of Applied Ecology* 38:472-483.

Letourneau, D. K. (1998). Conservation biology: Lessons for conserving natural enemies. In *Conservation Biological Control*, ed., P. Barbosa, San Diego, CA: Academic Press. pp. 9-38.

Letourneau, D. K. and M. A. Altieri. (1999). Environmental management to enhance biological control in agroecosystem. In *Handbook of Biological Control*, eds., S. Bellows and T. W. Fisher, San Diego, CA: Academic Press. pp. 319-354.

Levins, R. (1969). Some demographic and genetic consequences of environmental heterogeneity for biological control. *Bulletin of the Entomology Society of America* 15:237-240.

Levins, R. and M. Wilson. (1980). Ecological theory and pest management. *Annual Review of Entomology* 25:287-308.

Lewis, W. J., J. C. van Lenteren, S. C. Phatak, and J. H. Tumlinson. (1997). A total system approach to sustainable pest management. *Proceedings of the National Academy of Science USA* 94:12243-12248.

Los, L. M. and W. A. Allen. (1983). Abundance and diversity of adult Carabidae in insecticide-treated and untreated alfalfa fields. *Environmental Entomology* 12:1068-1072.

Lys, J. A. and W. Nentwig. (1992). Augmentation of beneficial arthropods by strip-management. 4. Surface activity, movements and activity density of abundant carabid beetles in cereal fields. *Oecologia* 92:373-382.

Marino, P. C. and D. A. Landis. (1996). Effect of landscape structure on parasitoid diversity and parasitism in agroecosystems. *Ecological Applications* 6:276-284.

Matson, P. A., W. J. Parton, A. G. Power, and M. J. Swift. (1997). Agricultural intensification and ecosystem properties. *Science* 277:504-509.

Mauremooto, J. R., S. D. Wratten, S. P. Worner, and G. L. A. Fry. (1995). Permeability of hedgerows to predatory carabid beetles. *Agriculture, Ecosystems and Environment* 52:141-148.

MacArthur, R. H. and E. O. Wilson. (1967). *The Theory of Island Biogeography*. Princeton, NJ: Princeton University Press.

McCullough, D. R. (1996). *Metapopulations and Wildlife Conservation*. Washington, DC: Island Press.

Menalled, F. D., P. C. Marino, S. H. Gage, and D. A. Landis. (1999). Does agricultural landscape structure affect parasitism and parasitoid diversity? *Ecological Applications* 9:634-641.

Menalled, F., J. Lee, and D. Landis. (1999). Manipulating carabid beetle abundance alters prey removal rates in corn fields. *Biocontrol* 43:441-456.

Murdoch, W. W., J. Chesson, and P. L. Chesson. (1985). Biological control in theory and practice. *American Naturalist* 125:344-366.

Murdoch, W. W. and C. J. Briggs. (1996). Theory for biological control: Recent developments. *Ecology* 77:2001-2013.

National Research Council. (1996). *Ecologically Based Pest Management: New Solutions for a New Century*. Washington, DC: National Academy Press.

Nentwig, W., T. Frank, and C. Lethmayer. (1998). Sown weed strips: artificial ecological compensation areas as an important tool in conservation biological control. In *Conservation Biological Control*, ed. P. Barbosa, San Diego, CA: Academic Press. pp. 133-153.

Newsom, L. D. (1980). The next rung up the integrated pest management ladder. *Bulletin of the Entomological Society of America* 26:369-374.

Nicholls, C. I., M. Parrella, and M. A. Altieri. (2001). The effects of a vegetational corridor on the abundance and dispersal of insect biodiversity within a northern California organic vineyard. *Landscape Ecology* 16: 133-146.

Pickett, C. H. and R. L. Bugg. (1998). *Enhancing Biological Control: Habitat Management to Promote Natural Enemies of Agricultural Pest*. Berkeley, CA: University of California Press.

Pickett, S. T. A. and P. S. White. (1985). *The Ecology of Natural Disturbance and Patch Dynamic*. San Diego, CA: Academic Press.

Pimentel, D. (1961). Species diversity and insect population outbreaks. *Annals of the Entomological Society of America* 54: 76-86.

Price, P. W. (1991). Evolutionary theory of host and parasitoid interactions. *Biological Control* 1:83-93.

Price, P. W. and G. P. Waldbauer. (1994). Ecological aspects of pest management. In *Introduction to Pest Management*, eds., R. L. Metcalf and W. H. Luckmann, New York, NY: Wiley-Interscience, pp. 33-65.

Roberts, S. D. and G. R. Parker. (1998). Ecosystem management: Opportunities for private landowners in the central hardwood region. *Northern Journal of Applied Forestry* 15:43-48.

Roland, J. (1993). Large-scale forest fragmentation increases the duration of tent caterpillar outbreak. *Oecologia* 93:25-30.

Root, R. B. (1973). Organization of a plant-arthropod association in simple and diverse habitats: The fauna of collards (*Brassicae oleraceae*). *Ecological Monographs* 43:95-124.

Simberloff, D. (1986). Island biogeography and integrated pest management. In *Ecological Theory and Integrated Pest Management Practices*, ed., M. Kogan, New York, NY: John Wiley and Sons, pp. 19-35.

Sopp, P. I., K. D. Sunderland, J. S. Fenlon, and S. D. Wratten. (1992). An improved quantitative method for estimating invertebrate predation in the field using and enzyme-linked immunosorbent assay (ELISA). *Journal of Applied Ecology* 29: 295-302.

Sousa, W. O. (1984). The role of disturbance in natural communities. *Annual Review of Ecology and Systematics* 15:353-391.

Speight, M. R. (1983). The potential of ecosystem management for pest control. *Agriculture, Ecosystems and Environment* 10:183-199.

Staver, C. P. (2001). Knowledge, science, and practice in ecological weed management: farmer-extensionist-scientist interactions. In *Ecological Management of Agricultural Weeds*, eds., M. Liebman, C. Mohler, and C. Staver, Cambridge, UK: Cambridge University Press. pp. 99-138.

Stern, V. M., R. F. Smith, R. van den Bosch, and K. S. Hagen. (1959). The integrated control concept. *Hilgardia* 29:81-101.

Stinner, B. R. and G. J. House. (1990). Arthropods and other invertebrates in conservation-tillage agriculture. *Annual Review of Entomology* 35:299-318.

Sunderland, K. D. (1975). The diet of some predatory arthropods in cereal crops. *Journal of Applied Ecology* 12:507-515.

Sunderland, K. D., N. E. Crook, D. L. Stacey, and B. J. Fuller. (1987). A study of feeding by polyphagous predators on cereal aphids using ELISA and gut dissection. *Journal of Applied Ecology* 24:907-933.

Sustainable Agriculture Network. (2000). *Naturalize Your Farming System: A Whole-farm Approach to Managing Pests*. Burlington, VT: University of Vermont, Sustainable Agriculture Network. 20 pp.

Swift, M. J. and J. M. Anderson. (1993). Biodiversity and ecosystem function in agroecosystems. In *Biodiversity and Ecosystem Function*, eds., E. Schultz and H. A. Mooney, New York, NY: Springer-Verlag, pp. 57-83.

Terry, L. A., D. A. Potter, and P. C. Spicer. (1993). Insecticides affect predatory arthropods and predation on japanese beetle (Coleoptera: Scarabaeidae) eggs and fall armyworm (Lepidoptera: Noctuidae) pupae in turfgrass. *Horticultural Entomology* 86:871-878.

Thies, C. and T. Tscharntke. (1999). Landscape structure and biological control in agroecosystems. *Science* 285:893-895.

Thomas, J. W. (1996). Forest Service perspective on ecosystem management. *Ecological Applications* 6:703-705.

Tscharntke, T. (2000). Parasitoid populations in the agricultural landscape. In *Parasitoid Population Biology*, eds., M. E. Hochberg and A. R. Ives, Princeton, NJ: Princeton University Press, pp. 235-253.

Tscharntke, T. and A. Kruess. (1999). Habitat fragmentation and biological control. In *Theoretical Approaches to Biological Control*, eds., B. A. Hawkins and H. W. Cornell, Cambridge, UK: University Press, pp. 190-205.

Vandermeer, J. and I. Perfecto. (1995). *Breakfast of Biodiversity: The Truth About Rainforest Destruction*. Oakland, CA: Food First Book.

Vickerman G. P. and K. D. Sunderland. (1975). Arthropods in cereal crops: Nocturnal activity, vertical distribution and aphid predation. *Journal of Applied Ecology* 12:755-766.

Walters, C. J. and C. S. Holling. (1990). Large-scale management experiments and learning by doing. *Ecology* 71:2060-2068.

Wheater, C. P. (1991). Effects of starvation on locomotor activity in some predacious Coleoptera (Carabidae, Staphylinidae). *Coleopterists Bulletin* 45:371-378.

Wissinger, S. A. (1997). Cyclic colonization in predictably ephemeral habitats: A template for biological control in annual crop systems. *Biological Control* 10:4-15.

Yaffee, S. L., A. F. Phillips, I. C. Frentz, P. W. Hardy, S. M. Maleki, and B. E. Thorpe. (1996). *Ecosystem Management in The United States: An Assessment of Current Experience*, Washington, DC: Island Press.

Zabel, J. and T. Tscharntke. (1998). Does fragmentation of *Urtica* habitats affect phytophagous and predatory insects differentially? *Oecologia* 116:419-425.

Zangger, A., J. A. Lys, and W. Nentwig. (1994). Increasing the availability of food and the reproduction of *Poecilus cupreus* in a cereal field by strip-management. *Entomologia Experimentalis et Applicata* 71:11-120.

Effects of Key Soil Organisms on Nutrient Dynamics in Temperate Agroecosystems

Joann K. Whalen
Chantal Hamel

SUMMARY. Soil organisms are a diverse group that can influence the nutrient dynamics in temperate agroecosystems profoundly. Many organisms interact in a symbiotic or mutualistic way with plants, and these relationships have co-evolved, permitting plants and soil organisms to flourish in the soil environment. Numerous controlled lab or small plot-scale studies have demonstrated that soil organisms can mobilize or transfer substantial quantities of nutrients to crops, in relationship to crop requirements. However, the simple scaling up of such results to explain conditions on a large field scale is very much constrained by a lack of information on the spatio-temporal distribution of soil organisms in temperate agroecosystems. The numbers, diversity and activity of soil organisms in temperate agroecosystems are affected by agricultural management practices such as tillage operations, but our knowledge of the key organisms or groups of organisms that contribute to nutrient cy-

---

Joann K. Whalen is affiliated with the Department of Natural Resource Sciences and McGill School of Environment, McGill University, Ste Anne de Bellevue, QC, Canada, H9X 3V9 (E-mail: whalenj@nrs.mcgill.ca).

Chantal Hamel is affiliated with Agriculture and Agri-Food Canada, Swift Current, SK, Canada, S9H 3X2.

[Haworth co-indexing entry note]: "Effects of Key Soil Organisms on Nutrient Dynamics in Temperate Agroecosystems." Whalen, Joann K., and Chantal Hamel. Co-published simultaneously in *Journal of Crop Improvement* (Food Products Press, an imprint of The Haworth Press, Inc.) Vol. 11, No. 1/2 (#21/22), 2004, pp. 175-207; and: *New Dimensions in Agroecology* (ed: David Clements, and Anil Shrestha) Food Products Press, an imprint of The Haworth Press, Inc., 2004, pp. 175-207. Single or multiple copies of this article are available for a fee from The Haworth Document Delivery Service [1-800-HAWORTH, 9:00 a.m. - 5:00 p.m. (EST). E-mail address: docdelivery@haworthpress.com].

http://www.haworthpress.com/web/JCRIP

Digital Object Identifer: 10.1300/J411v11n01_09

cling and crop production under different sets of management practices is limited. Better management of nutrients in temperate agroecosystems requires better knowledge of soil biota, their effects on nutrient cycling and their contribution to crop production.

KEYWORDS. Soil fauna, soil biological activity, symbiotic, free-living, nutrient uptake, crop production

INTRODUCTION

Soil organisms are a diverse group that can influence the nutrient dynamics in temperate agroecosystems profoundly. The cycling of nutrients through their biomass is a key process that, along with photosynthesis, maintains life on earth. Even the smallest and least mobile of the soil organisms has evolved mechanisms to derive energy and nutrients from soil organic matter (OM) and minerals. Many also interact in a symbiotic or mutualistic way with plants, and these relationships have co-evolved, permitting plants and soil organisms to flourish in the soil environment. The numbers and diversity of plants and belowground organisms in a particular soil are strongly influenced by soil physical and chemical properties, but these organisms can also modify the physical and chemical characteristics of soil and change the trajectory of soil development.

Although it is well known that soil organisms can mobilize or transfer substantial quantities of nutrients to crops, in relationship to crop requirements, it is difficult to simply scale up these results to the field scale. In temperate agroecosystems, the contribution of soil organisms to nutrient cycling and crop production is largely ignored, and many producers rely on synthetic fertilizers to provide the nutrients required by their crops. Environmental concerns for soil, water and air quality are prompting producers to re-evaluate their agricultural practices and fertilizer use. For agriculture to be sustainable, we must devise better management practices to maintain or increase crop yields without degrading soils and the environment. Nutrient management is central to the development of more sustainable agricultural systems, but better management of nutrients requires better knowledge of soil biota and their role in nutrient cycling. This article presents an overview of the organisms that inhabit soils and how they transform nutrients in OM and soil minerals into forms that can be used by plants. We will discuss briefly how agricultural practices affect the activity of soil organisms, with an emphasis on tillage practices.

THE BIOLOGY AND ECOLOGY OF SOIL ORGANISMS

Soil Microorganisms

There are five distinct groups of microorganisms found in agricultural soils (in order of increasing size and cellular complexity): archea, bacteria, fungi, eukaryotic algae and cyanobacteria. Archea are unique organisms adapted to extreme conditions, such as elevated temperature or saline environments, and the only group found in agricultural soils are methanogens that use CO_2 as a source of carbon (C) and as an electron acceptor. The diversity and activity of these organisms are not well known, but their impact on nutrient cycling in temperate agricultural soils is likely quite small (Alexander, 1998). Bacteria are among the most numerous soil microorganisms (10^8 to 10^9 bacteria can be found in 1 g of soil), but they generally account for less than half of the soil microbial biomass (Wollum, 1998). Most bacteria are single cell organisms but the actinomycetes and several species of cyanobacteria are filamentous. These sedentary organisms often live in colonies on soil surfaces and within soil aggregates, attached by the exopolysaccharides secreted on their cell walls. The soil bacteria have diverse metabolic capabilities and mediate many of the biochemical transformations that convert nutrients from organic to inorganic forms during OM decomposition.

The soil fungi have a filamentous growth habit. They are a major component of the soil microbial biomass and their biomass can exceed that of crop roots (Olsson et al., 1999). These organisms "move" through soil as their hyphae grow toward soil patches rich in nutrients or organic debris, but long distance displacement of the fungi is generally achieved through spore dispersal by wind, water and animal vectors. Some fungi also produce motile spores that can swim towards a food source (Morton, 1998). The majority of soil fungi are aerobic heterotrophs capable of decomposing the most complex and recalcitrant organic compounds found in soils (Paul and Clark, 1996). Another important group of soil fungi are the arbuscular mycorrhizal (AM) fungi, obligate biotrophs that derive their energy from living plants rather than from decomposing OM. Both heterotrophic and biotrophic soil fungi make an important contribution to crop nutrition in temperate agroecosystems.

Eukaryotic algae and cyanobacteria possess chlorophyll and derive their energy from photosynthesis; some cyanobacteria can also fix nitrogen (N_2) from the atmosphere. These organisms are best known as pioneer species that rapidly colonize bare soils, such as those disturbed by glaciation or volcanic activity. They contribute to soil development and nutrient cycling by forming crusts that stabilize the soil structure, particularly in desert and arid soils. However, algae and cyanobacteria probably have a negligible effect on the nutrient dynamics of temperate agroecosystems (Paul and Clark, 1996).

The numbers and activity of soil microorganisms are influenced by environmental factors (e.g., temperature, moisture, aeration, soil texture, pH, salinity, concentration of dissolved nutrients) and biological factors (e.g., predation, competition, symbiosis, mutualistic interactions) (Grayston et al., 1998). Perhaps the most important factor governing the size of soil microbial populations is the amount and quality of the organic substrates available to support their growth and reproduction. The availability of organic substrates varies spatially and temporally in all terrestrial ecosystems, leading to fluctuations in microbial activity, and ultimately, the quantities of nutrient available for plant uptake. At the same time, plants exude C through their roots into the soil, which can alter microbial activity and increase nutrient availability in their root zone (rhizosphere).

Soil Microfauna and Mesofauna

The microfauna and mesofauna of soils are a diverse group of organisms that range in size from 0.02 to 2 mm. Soil microfauna include protozoa and nematodes, while the mesofauna may include tardigrada, collembola, acari, protura, diplurans, symphylids, and pauropoda. The focus of this discussion is the role of protozoa, nematodes, collembola and acari on nutrient dynamics in agricultural soil, as the effect on nutrient cycling of most other soil micro- and meso-fauna is not well known. Interested readers are advised to consult Dindal's (1990) 'Soil Biology Guide' for more information on the less-well studied soil mesofauna.

Protozoa are unicellular eukaryotic organisms ranging in size from 2 to 1000 μm long that inhabit the water films around soil particles, OM and roots. The free-living soil protozoa include flagellates, ciliates, naked amoebae and testate amoebae, and generally number more than 10^6 individuals m^{-2}, with the greatest numbers and biomass found in the top 5 cm of soils (Ekelund, Rønn, and Christensen, 2001). Small flagellates and naked amoebae inhabit the smallest water-filled pore spaces in soil; ciliates, testate amoebae, rotifers and nematodes share larger pore spaces. The distribution of protozoa varies spatially and temporally, and more protozoa tend to be found within aggregates and in earthworm casts, in the rhizosphere, near decomposing plant and animal residues, and in the drilosphere (lining of earthworm burrows) than in other soil habitats (Bamforth, 1997; Griffiths, 1994; Tiunov et al., 2001). These habitats, which are often "hot spots" of protozoan activity, constitute less than 10% of the soil volume but support more than 90% of the biological activity in most soils (Coleman and Crossley, 1996).

Much of the study on nematodes in agricultural soils has focused on those that are parasites of agricultural plants, and numerous texts have been devoted to their study and control (e.g., Evans, Trudgill, and Webster, 1993). However, both parasitic and non-plant parasitic soil nematodes have a role in nutrient cycling in temperate agroecosystems; many plant parasitic nematodes alter plant root morphology and function, which changes nutrient transformations in the rhizosphere, whereas non-plant parasitic nematodes influence nutrient cycling through their interactions with other organisms in soil food webs. Soil nematodes are classified as bacterivorous, fungivorous, predatory, omnivorous or plant parasite nematodes, depending on their feeding habits. They have been used as indicators of ecosystem health due to the number and diversity of nematodes (more than 10^6 individuals m^{-2} from more than 30 taxa) commonly found in soils (Freckman and Ettema, 1993).

Nematodes inhabit water-filled pores and water films, and tend to be most numerous in the rhizosphere, near decomposing residues and in the drilosphere than in other soil habitats (Griffiths and Caul, 1993; Robertson and Freckman, 1995; Tiunov et al., 2001). Ingham et al. (1985) found that up to 70% of the bacterial and fungal feeding nematodes inhabited soil located 1 to 2 mm from root surfaces. Environmental conditions that affect moisture content in pore spaces, soil temperature, and the availability of food sources cause seasonal fluctuations in nematode populations. Nematode numbers in water-filled pore space increase as soil matric potential decreases (e.g., soil becomes drier), which can stimulate microbivorous grazing and perhaps lead to competition in nematode populations (Savin et al., 2001). Nematode taxa have different temperature optima, but most nematodes grow and survive best when soil temperatures are below 25°C (Ferris, Lau, and Venette, 1995). OM composition affects nematode dynamics; for example, Sohlenius and Boström (1984) found that populations of fungal feeding nematodes increased during the later stages of barley (*Hordeum vulgare* L.) straw decomposition, which coincided with an increase in fungal biomass and proportion of lignin in the residue.

Two major groups of mesofauna that have an important role in OM decomposition and nutrient cycling in agroecosystems are the springtails (Collembola) and soil mites (Acari). They range from 0.1 to 6 mm in length and include herbivores, bacterivores, fungivores, predators and omnivores, although individuals may change their feeding habits when their preferred food source is limited (Beare et al., 1992; Coleman and Crossley, 1996). Springtails are primitive insects whose populations can reach 10^5 or more individuals m^{-2}. The preferred diet of springtails appears to be decaying plant and animal residues and the microorganisms, particularly the fungi, associated with these residues. Soil mites are chelicerate arthropods related to spiders, and between 10^3 and 10^5 m^{-2} can be found in agroecosystems (Coleman and Crossley, 1996). The suborders of mites found in agricultural soils include the Oribata, Prostigmata, Mesostigmata and Astigmata. Soil mites have diverse feeding habits and consume decaying plant and animal residues (Oritbatida and Astigmata), small arthropods and nematodes (Mesostigmata) and microorganisms, particularly fungi (Prostigmata, Astigmata).

Springtails and mites can be classified based on their body size and habitat in the soil: the largest individuals that live on the soil surface are epedaphic or epigeic, whereas medium-sized individuals that live in the upper 5 to 10 cm of soil are hemiedaphic, and those living in the 15 to 20 cm soil layer are euedaphic (Larink, 1997). Many springtails and mites prefer to live in the rhizosphere rather than away from plant roots in row-cropped agroecosystems (Garrett et al., 2001). Springtail and mite numbers and activities are generally highest during the spring and autumn in temperate agroecosystems when temperature and moisture conditions are most favorable and organic residues are abundant (Larink, 1997).

Soil Macrofauna

In contrast to microfauna and mesofauna, whose range is confined to water films and existing air-filled pore spaces, macrofauna have the capability to create their own niches in soil through their burrowing activities, which can alter soil structure and soil nutrient cycling significantly. Numerous insects and burrowing animals spend part of their life, or specific stages of their life cycle, in the soil, but it is beyond the scope of this contribution to discuss how such organisms affect nutrient dynamics in agroeosystems. Dindal's (1990) 'Soil Biology Guide' provides an excellent description of the insects that reside in soils on a part- and full-time basis.

The major groups of soil macrofauna that may influence nutrient dynamics in agricultural soils include isopods, millipedes, centipedes, harvestmen (Opiliones), enchytraeids and earthworms. The distribution and ecology of most of these organisms in agroecosystems is not well known, and the reader should consult reviews by Didden, Fründ, and Graefe (1997); Wolters and Ekschmitt (1997); and Halaj, Cady, and Uetz (2000) for more information on isopods, millipedes, centipedes, harvestmen, and enchytraeids. The discussion here is limited to the effects of earthworms on nutrient dynamics in temperate agroecosystems.

Earthworms are probably the most important soil-inhabiting invertebrates due to their roles in OM decomposition and soil formation. Earthworms influence nutrient cycling in terrestrial ecosystems directly, through the release of nutrient from their tissues via excretion and mortality, and indirectly, by modifying soil physical, chemical, and biological properties (Blair, Parmelee, and Lavelle, 1995; Edwards and Bohlen, 1996). The earthworm family Lumbricidae is the most common taxa found in temperate agroecosystems. Most of the Lumbricidae found in North American agroecosystems were introduced from Europe. They are the dominant earthworms either because of a lack of endemic (native) species or because they out-compete endemic species for available resources (Kalisz and Wood, 1995). Earthworm populations may contain up to 200 individuals m^{-2} (Edwards, 1983). The LOMBRI-ASSESS database of earthworm populations from approximately 350 study sites found cultivated fields contained, on average, about 90 individuals m^{-2} with a biomass of 7.2 g dry weight m^{-2} (Paoletti, 1999). Generally, there are not more than four to six different species in the earthworm communities in temperate row-cropped agroecosystems (Edwards, 1983).

The main factors that affect the size and distribution of earthworms are environmental stresses on their habitat (climate, soil properties, vegetation, and food resources) and biotic interactions within soil faunal communities (competition, predation, parasitism, and disease) (Edwards and Bohlen, 1996; Curry, 1998). Seasonal fluctuations in earthworm populations related to soil temperature and moisture have been well documented (Edwards and Bohlen, 1996). Earthworms are thought to inhabit discrete patches in soil, but the size of patches occupied by earthworms varies among species and may be related to soil properties such as OM, texture and vegetation (Hendrix et al., 1992; Poier and Richter, 1992).

NUTRIENT TRANSFORMATIONS BY SOIL ORGANISMS

Soil organisms are responsible for transforming nutrients into forms that can be used by plants and can be considered to belong to two groups, plant-symbiotic organisms and free-living organisms. The plant-symbiotic organisms are obligate biotrophs that require a plant host for growth and reproduction and directly transfer nutrients from soils to plants. The free-living organisms are heterotrophs that contribute to plant nutrition indirectly by releasing nutrients from soil OM and mineral surfaces that can be absorbed by plant roots. Although free-living microorganisms and microfauna are not intimately associated with plant roots, their numbers and activity tend to be much greater in the rhizosphere than in bulk soil due to the abundance of C released from plant roots. An estimated 10 to 30% of photosynthates are released into soil through passive and active root excretions, the sloughing off of root cap cells and epidermal cell senescence (Bowen and Rovira, 1999). When rhizodeposition stimulates the activity of free-living soil organisms in a way that causes net mineralization or mobilization of nutrients, it may enhance crop production in temperate agroecosystems.

Plant-Symbiotic Soil Organisms: Mycorrhizal Fungi

The majority of crop plants found in temperate agroecosystems are associated with AM fungi (Olsson et al., 1999). The symbiotic relationship between AM fungi and plant roots involves connections between the fungal cell wall and plant cytoplasmic membranes that permit energy transfer from the plant to the fungi and nutrient and water transfer from the fungi to the plant (Smith and Read, 1997). Nutrient and water absorption by the host plant is improved because the fungal symbiont develops an extensive network of microscopic hyphae attached to plant roots (mycorrhizosphere) that permits the exploitation of a larger volume of soil than possible by the plant roots alone (Marschner, 1995; Smith and Read, 1997).

The AM fungi-plant symbiosis forms early in plant development, and up to 80% of the fungal hyphae isolated from soils under corn (*Zea mays* L.) and barley production may be from AM fungi within 5 weeks after seeding the field (Kabir et al., 1997). Since the concentrations of available nutrients in agricultural soils are highest in surface soils and decline with increasing soil depth, most mycorrhizal hyphae are found in the top 25 cm of the soil profile (Kabir et al., 1998a). Development of the AM fungi-plant symbiosis is affected by the level of plant-available phosphorus (P) in soils, and degree of mycorrhizal infection of plant roots typically declines when soil P concentrations are high enough to support crop production (Koide and Li, 1990). The degree of mycorrhizal colonization of plant roots and hyphal development also varies with crop species and genotypes (Estaun, Calvet, and Hayman, 1987; Liu et al. 2000a; Peterson and Bradbury, 1995).

As much as 40 to 50% of the C derived from photosynthesis is channelled directly to AM fungi (Harris and Paul, 1987). The quantity of C transferred from the host to the fungal symbiont changes during the growing season as plant nutritional requirements change. Kabir et al. (1997) found the number of mycorrhizal hyphae associated with corn increased to a maximum in mid-summer and then declined. A portion of plant photosynthates accumulates in mycorrhizal biomass, but significant quantities may be exuded from fungal hyphae into the soil. In a laboratory study, the growth and reproduction of the plant pathogenic fungi *Fusarium oxysporum* increased significantly in the presence of active mycorrhizal hyphae, probably due to its use of C substrates that were exuded from mycorrhizal hyphae (St-Arnaud et al., 1995). The transfer of C from plants to soils via AM fungi may stimulate the growth of free-living bacteria, fungi and other soil organisms in the mycorrhizosphere (Finlay and Söderström, 1992). There is growing evidence that the free-living microbial communities in the mycorrhizosphere are specific to the type of plant and arbuscular mycorrhizal fungi present (Andrade, Linderman, and Bethlenfalvay, 1998; Westover, Kennedy, and Kelly, 1997). Further work is needed to determine how these free-living microorganisms, in association with mycorrhizal fungi, contribute to plant growth and nutrition.

Mycorrhizal fungi are best known for their ability to increase plant uptake of nutrients that are relatively immobile, such as P, Cu and Zn, and water (Bolan, 1991; Liu et al. 2000b). They increase the volume of soil exploited by plants through their hyphal networks, and may produce phosphatases that convert organically-bound P into inorganic P that can be absorbed by roots and perhaps produce organic acids that solubilize P bound to soil minerals (Marschner, 1995). Bethlenfalvay and Franson (1989) showed that mycorrhizal fungi inhibit Mn uptake in a mycorrhizal soybean (*Glycine max* L. Merr.), and they also inhibit Fe and Mn uptake by corn, which may help to maintain an optimal nutrient balance in crop tissues (Liu et al., 2000b). Mycorrhizal fungi can also enhance plant uptake of mobile nutrients, such as K, Ca, and Mg, when the concentrations of P and these nutrients in soils are low (Liu et al., 2002).

There is growing interest in inoculating soils with specific types of mycorrhizal fungi to improve crop production. However, all soils contain populations of indigenous mycorrhizal fungi that contribute to crop production. Miller, McGonigle, and Addy (1995) demonstrated that more rapid and vigorous mycorrhizal development in corn occurs with the indigenous fungi already present in no-till soils rather than from fungi introduced to soils through inoculation. Newly introduced fungi must find host roots, colonize them, and finally grow into a mycorrhizal hyphae soil network, whereas indigenous fungi may already exist in the vicinity of the host plant.

In addition, many AM fungi are capable of reducing plant susceptibility to plant-parasitic nematodes and hence improve plant growth and performance (Calvet et al., 2001; Little and Maun, 1996). For example, rhizobacterium (*Rhizobium* spp.) has been demonstrated to induce systemic resistance to infection by the potato cyst nematode *Globodera pallida* (Reitz et al., 2000), and reduce root galls and nematode reproduction (Siddiqui and Mahmood, 2001). The mechanisms that facilitate plant protection from attack by parasitic nematodes through the symbiotic AM fungi are not yet well understood.

Plant-Symbiotic Soil Organisms: N_2-Fixing Bacteria

The conversion of atmospheric dinitrogen (N_2) to ammonia (NH_3) by soil microorganisms is the process of biological N_2 fixation. Biological N_2 fixation produces an estimated 44 to 200 million Mg of fixed N per year. The fertilizer industry fixes about 84 million Mg of N per year, using the Bosch-Haber process that reduces N_2 to NH_3 (Bøckman et al., 1990). Biological N_2 fixation is powered by solar energy, but the industrial production of NH_3 fertilizers accounts for about 1.2% of the world's fossil fuel consumption on an annual basis (Kongshaug, 1998). Farming systems that derive more N from biological fixation than industrial processes may require less fossil fuel consumption to support crop production, and could be more sustainable in the long term.

Biological N_2 fixation by free living, associative and symbiotic organisms has been studied extensively. By far, the most important impact of biologically fixed N to agricultural soils under temperate climates comes from symbiotic associations between *Rhizobium* spp. bacteria and legumes (Weaver and Graham, 1994). The quantity of N_2 fixed in temperate agroecosystems each year varies, but is estimated to range from 164 to 300 kg N ha^{-1} for alfalfa (*Medicago sativa* L.), 57 to 190 kg N ha^{-1} for vetch (*Vicia* sp.) and beans (*Phaseolus* sp.), 46 kg N ha^{-1} for peas (*Pisum sativum* L.), and 17 to 206 kg N ha^{-1} for soybeans (Newton, 1999). Most of the N fixed by the bacteria is transferred to the host plant and used for plant growth, but some of the N fixed is soon released into the soil, probably in root exudates and dead root cells (Hamel, Smith, and Furlan, 1991). Legumes may be more dependent on N_2 fixation to obtain the N needed for growth when they are grown in mixed than pure stands. Grasses, for example, are more competitive for soil N than legumes (Hamel, Furlan, and Smith, 1992). Nitrogenase, the enzyme responsible for converting N_2 to NH_3, is suppressed by NH_4^+ and NO_3^- so legumes will preferentially use N from the soil rather than N_2 fixed from the atmosphere when soil N concentrations are sufficiently high (Paul and Clark, 1996). Therefore, N fertilizer applications will reduce the amount of biological N_2-fixation occurring from the bacterial-legume symbiosis.

Free-Living Soil Organisms: OM Decomposition

Most soil organisms are involved in OM decomposition, the process by which complex organic substrates are physically fragmented and biochemically degraded to produce soluble organic and inorganic molecules that can be assimilated by plants. The end products of this process include CO_2 respired by the soil organisms, NH_4^+, NO_3^-, $H_2PO_4^-$, HPO_4^{2-}, and SO_4^{2-} ions, and stabilized OM (e.g., humus). The rate of decomposition of an organic substrate is affected by the amount, physical size and chemical composition of the organic material, the types of soil organisms present, and the environmental conditions that affect their activity (de Ruiter, Neutal, and Moore, 1994; Lee and Pankhurst, 1992). Soil microorganisms affect decomposition rates by producing the enzymes required to decompose organic substrates and by altering the soil habitat in ways that are beneficial for biological activity. Soil fauna influence decomposition rates by fragmenting organic materials, which increases the surface area available for microbial colonization, grazing on microorganisms, and by altering the soil habitat in ways that are beneficial for soil microorganisms.

The extracellular phosphatase and sulfatase enzymes secreted by plant roots, bacterial and fungal cells hydrolyse ester bonds (C-O-P, C-O-S) to produce $H_2PO_4^-$, HPO_4^{2-}, and SO_4^{2-} ions. Phosphatase and sulfatase enzymes are inhibited by the reaction products, and agricultural soils receiving P and S fertilizers tend to have much lower phosphatase and sulfatase activities than unfertilized soils (Tabatabai, 1994). Enzyme activity within microbial cells generates energy required for microbial growth and respiration and produces NH_4^+ and SO_4^{2-} through the hydrolysis of C-N and C-S bonds. The conversion of organically-bound N, P and S to inorganic forms is known as mineralization. The oxidation of NH_4^+ to NO_3^- in agricultural soils (nitrification) is a two-step process where NH_4^+ is converted to NO_2^- by chemotrophic bacteria such as *Nitrosolobus* spp. and *Nitrosospora* spp., and then from NO_2^- to NO_3^- by *Nitrobacter* spp. (Paul and Clark, 1996). The N, P, and S produced from these enzymatic processes in excess of microbial requirements are released into the soil solution, where they can be absorbed by plant roots. In addition, the C, N, P, and S incorporated into microbial biomass are released when microorganisms die (biomass turnover). The quantities of N, P, and S in microbial biomass can be considerable. Olsson et al. (1999) estimated the annual standing stock biomass of soil microorganisms in a linseed (*Linum usitatissimum* L.) field contained 60 to 110 kg N ha^{-1}, 12 to 22 kg P ha^{-1}, and 1.6 to 3 kg S ha^{-1}. Soil microbial biomass containing 17 to 290 kg P ha^{-1} is estimated to release 11 to 190 kg P ha^{-1} $year^{-1}$ (Frossard et al., 2000). It is evident that soil microorganisms can turn over large quantities of nutrients, but the proportion of these nutrients absorbed by crops under field conditions is not well known and requires further investigation.

Seasonal fluctuations in the size of the microbial biomass C, N, P, and S pools in agricultural soils have been attributed to environmental conditions, particularly soil moisture, and the availability of organic substrates. The quantities of nutrients in the microbial biomass pool may be greatest in the spring (Murphy, Fillery, and Sparling, 1998), summer (Perrott, Sarathchandra, and Dow, 1992) or in the fall after harvest (Joergensen, Meyer, and Mueller, 1994). In a clay loam soil, He et al. (1997) found that seasonal variation in the microbial biomass C pool was related to the timing of OM inputs, but changes in the microbial biomass P were related to soil moisture deficits. Microbial biomass P was lowest when the soil moisture deficit was greatest, perhaps because the reduction in soil moisture limited P diffusion to microorganisms, or because plants and microorganisms were competing for P in soil solution. However, Patra et al. (1990) did not observe seasonal changes in the microbial biomass C, N, and P pools in soils under continuous wheat (*Triticum aestivum* L.) and grass pasture.

Variation in microbial biomass may coincide with changes in the quantities of mineralizable substrates during the decomposition process and from predator-prey interaction (Lee and Pankhurst, 1992). Microbial colonization of organic substrates leads first to the mineralization of labile organic substrates (e.g., carbohydrates) and then to progressively recalcitrant organic substrates such as cellulose and lignin. Microbial growth and respiration attracts microbial predators (e.g., protozoa and nematodes) that graze on the microbial biomass. Ingham et al. (1986) found that the population dynamics of bacteria, fungi, protozoa, and nematodes were explained by simple predator-prey interactions. In the spring, microbial growth appeared to stimulate protozoa and nematode populations, leading to a decline in microbial biomass and an increase in inorganic N in soil. In the fall, protozoa and nematode populations declined, bacterial populations increased due to a reduction in predation, and inorganic N concentrations declined as N was immobilized in bacterial biomass.

Despite their small physical size, the protozoa comprise about 30% of the soil faunal biomass, and account for about two-thirds of soil faunal respiration (Foissner, 1987). Protozoa have relatively high energy conversion efficiency, compared to other soil fauna, and allocate proportionately more of their energy to growth and reproduction than most other soil organisms; the protozoan biomass turns over rapidly (10-12 times per year) compared to other soil fauna (1-2 times per year) (Coleman and Crossley, 1996). The main food source for most protozoa is bacteria, and protozoa have the ability to access and graze upon bacteria in very small pore spaces. Foster and Dormaar (1991) observed amoebae producing elongated and branched pseudopodia to probe micropores less than 1 μm wide within soil microaggregates. Protozoan grazing on bacteria results in a release of CO_2 and mineralizes N, P, and S from the bacterial biomass, contributing to soil respiration and liberating nutrients in forms available for plant uptake. The selective grazing pressure of protozoa on bacteria can alter the composition of soil microbial communities and also keeps bacterial populations physiologically young, which stimulates decomposition and nutrient cycling in soils (Darbyshire, 1994; Gupta and Germida, 1989). Protozoa are responsible for an estimated 14 to 66% of the C mineralized, and 20 to 40% of the N mineralized in soils, hence removal of protozoa from soil food webs can slow residue decomposition rates significantly (Foissner, 1999). Increased N mineralization due to protozoan activity has been linked to improved plant growth and N uptake by plants (Alphei, Bonkowski, and Scheu, 1996; Bonkowski, Griffiths, and Scrimgeour, 2000).

Nematodes are involved in OM decomposition and nutrient mineralization, mainly through their grazing on microorganisms and interactions with plants. Increases in soil available NH_4-N, plant growth and shoot N concentrations in the presence of bacterivorous nematodes have been attributed to N excretion by nematodes and stimulation of bacterial activity by grazing that led to increased N mineralization (Alphei, Bonkowski, and Scheu, 1996; Griffiths, 1994; Ingham et al., 1985). Generally, nematodes excrete about 70% of the microbial biomass N consumed (Bonkowski, Griffiths, and Scrimgeour, 2000; Ferris, Venette, and Lau, 1997). Microfaunal grazing of microorganisms is important for plant growth in temperate agroecosystems. As much as 30% of the N mineralization in agroecosystems has been attributed to the combined microfaunal (protozoa and nematode) activity (Andrén et al., 1990).

Springtails and mites have an important role in OM decomposition and nutrient mineralization through their interactions with microorganisms as well as their effects on soil structure (Coleman et al., 1993; Crossley, Coleman, and Hendrix, 1989). Most of the effects of springtails and mites on nutrient uptake by crop plants in agroecosystems are indirect, since the majority of these organisms do not feed on living plants. Model foodwebs have demonstrated that the feeding activities of springtails and mites can stimulate nutrient turnover from microorganisms and microfauna (de Ruiter, Neutal, and Moore, 1994; Hendrix et al., 1986). Grazing on microorganisms and microfauna is one way that springtails and mites stimulate OM decomposition and N mineralization. Selective feeding on certain fungal species by springtails has been shown to alter the fungal community and hence indirectly affects rates of litter decomposition and nutrient cycling (Moore, Ingham, and Coleman, 1987). Springtails and mites also influence nutrient cycling by fragmenting and redistributing OM in soils, creating new "hot spots" of root, microbial and microfaunal activity (Hansen, 2000).

Although most members of the soil microarthropods do not feed on living plants, the fungal-feeding springtails can alter plant growth and nutrient uptake by selectively feeding on mycorrhizal hyphae. In the laboratory, springtails grazing on mycorrhizal fungi can reduce P uptake and plant yield (Finlay, 1985), but this effect may be due to excessively high numbers of springtails in greenhouse experiments since damage to mycorrhizae by microarthropods is density-dependent (Klironomos and Ursic, 1998). Springtails also feed on several pathogenic fungi that cause take-all and brown foot rot in winter cereal crops and reduce disease severity (Sabatini and Innocenti, 2001). Further investigations will be required to determine whether springtail-mycorrhizal interactions can improve or impair crop performance under field conditions.

The flux of N from earthworm populations in agroecosystems is thought to be considerable, and it is estimated that 10 to 106 kg N ha^{-1} $year^{-1}$ is released from earthworm biomass through mortality and excretion (Andersen, 1983; Schmidt and Curry, 2001; Whalen and Parmelee, 2000). Much of the N released from earthworms through mortality and from excretion products may be readily available for uptake by plants. More than 70% of the N from decomposing earthworms labeled with ^{15}N cycled through the microbial biomass and was recovered in the shoots of ryegrass (*Lolium multiflorum* L.) plants within 16 days after dead earthworms were added to soil (Whalen et al., 1999). In addition, up to 40% of the ^{15}N excreted by earthworms in mucus and urine was recovered in soil NH_4-N and NO_3-N pools (Whalen, Parmelee, and Subler, 2000). Although considerable effort has focused on the role of earthworms in N cycling, there is very little information on how much P, S, and other nutrients are released from earthworms via the direct pathway of mortality and excretion.

Earthworms also influence nutrient cycling indirectly, by modifying soil physical and chemical characteristics and hence altering soil microbial activity, and through grazing on microorganisms. The impact of earthworm species on nutrient cycling via the indirect pathway depends on their feeding habits and life histories, since these factors influence earthworm-microbial interactions. Earthworms consume an estimated 2 to 15 Mg ha^{-1} $year^{-1}$ of OM, including soil OM, surface residues, live and dead roots, mycorrhizae, algae, fungi, bacteria, and protozoa (James, 1991; Whalen and Parmelee, 1999). The portion of ingested organic substrates not assimilated into earthworm tissues is defecated in casts, along with the microorganisms that pass through the earthworm gut. In the short term (e.g., weeks), earthworm casts are a source of plant-available N, P, K, and Ca, and are "hot spots" of bacterial and actinomycete activity compared to bulk soil as these microorganisms further decompose the OM contained in casts (Shipitalo and Protz, 1989; Tiunov and Scheu, 2000). Between 10 and 12% of the N in casts and as much as 50% of the P may be readily-available for plant uptake (James, 1991).

Through time, readily-mineralizable substrates in earthworm casts are utilized and fungi may become relatively more important than bacteria, resulting in a shift from mineralization to immobilization of nutrients in casts and stabilization of cast physical and chemical characteristics (Marinissen and Dexter, 1990; Shipitalo and Protz, 1989). Anecic earthworms also form middens at the top of their burrows, on the soil surface, which contain undecomposed OM and earthworm casts. Middens appear to act as "external rumens," and are "hot spots" of microbial and soil faunal activity compared to bulk soil (Maraun et al., 1999). The drilosphere possesses physical and chemical characteristics that favor the development of bacterial communities, often dominated by gram-negative bacteria; fungal mycelium and the proportion of germinating hyphae are lower in the drilosphere than in bulk soil (Tiunov, Dobrovol'skaya, and Polyanskaya, 1997; Tuinov, Dobrovol'skaya, and Polyanskaya, 2001). Decomposition rates, microbial activity, and the number of protozoa and nematodes are higher in the drilosphere than in bulk soil (Görres, Savin and Amador, 1997; Tuinov et al., 2001).

Free-Living Soil Organisms: Nutrient Solubilization from Soil Minerals

Nutrients can also become available to plants through the dissolution of soil minerals, which releases nutrients into the soil solution. This process contributes to the inherent fertility of a soil, and is generally more important in fine-textured than coarse-textured soils due to the larger proportion of minerals exposed to weathering factors. Plant roots, organic residues and soil microorganisms produce organic acids that react chemically with soil minerals to solubilize nutrients required by plants (Hinsinger, 2001). Soil organic acids include monocarboxylic, dicarboxylic and tricarboxylic acids containing unsaturated C and OH^- groups (Strobel, 2001). Such organic acids have two major functions related to plant nutrition. First, negatively charged organic acids can stimulate the dissolution of soil minerals by chelating metal cations (e.g., Fe, Zn, Cu, and Mn), creating a soluble compound that can be absorbed by plant roots (Jones and Darrah, 1994; Strobel, 2001). For example, citrate dissolves iron phosphate minerals by first forming a ferric hydroxyphosphate complex and then displaces the phosphate ion through chelation reactions (Bolan et al., 1994). Several species of bacteria, yeasts, actinomycetes, ectomycorrhizal and free-living fungi can solubilize P from Ca-P, Fe-P, and Al-P complexes, kaolinite, gibbsite, and goethite minerals (Illmer, Barbato, and Schinner, 1995; Whitelaw, 2000). The ability of organic acids to chelate metal cations is affected by their molecular structure and increases with the number of functional groups, so that tricarboxylic acids are more effective chelation agents than dicarboxylic and monocarboxylic acids (Bolan et al., 1994).

Second, negatively charged organic acids can compete with anions such as HPO_4^{2-}, $H_2PO_4^-$, and SO_4^{2-} for adsorption sites on soil surfaces. Many organic acids are very reactive and will bind strongly to P fixation sites, making these sites unavailable for phosphate ions and hence increasing the concentration of P in the soil solution (Hu et al., 2001; Jones and Kochian, 1996). In addition, inoculation of soil with AM fungi and P solubilizing microorganisms stimulates P uptake and plant growth more than inoculation with either AM fungi or P solubilizing microorganisms alone. The simultaneous inoculation of tomato (*Lycopersicon esculentum* L.) growing in hydroxyapatite-amended substrate with the P solubilizing bacteria *Enterobacter agglomerans* and the mycorrhizal fungi *Glomus etunicatu* improved both N and P uptake in a greenhouse study (Kim, Jordan, and McDonald, 1998). The concentrations of oxalic, 2-keto-P-gluconic and citric acid in the root zone increased most when both *E. agglomerans* and *G. etunicatu* were present, but it was not clear whether some of the organic acids present were exuded from mycorrhizal fungi and plant roots, or whether conditions in the mycorrhiosphere stimulated the production of organic acids by P solubilizing bacteria. Further research is needed to determine the quantities and types of organic acids secreted by plant roots, symbiotic microorganisms and free-living microorganisms in response to P deficiencies in agricultural soils.

Nutrient solubilization by organic acids is well documented in laboratory and greenhouse studies, but information is lacking on how organic acids may contribute to nutrient uptake by plants grown in temperate agricultural soils. The concentration of monocarboxylic acids in soil solution ranges from 0 to 1 mM, whereas dicarboxylic and tricarboxylic are present in lower concentrations, perhaps between 0 and 50 μM (Strobel, 2001). Despite the fact that organic acids can be used as an energy source by microorganisms, appreciable quantities of organic acids likely chelate metals in soil solution or become adsorbed to anion exchange sites in the soil matrix (Jones and Brassington, 1998). Organic acid production varies spatially, and the quantities of organic acids produced by P-solubilizing microorganisms are much higher in the rhizosphere than the surrounding soil (Hu et al., 2001; Whitelaw, 2000). It has been suggested that localized P solubilization by organic acids near plant roots and decomposing organic residues is an important mechanism for increasing P availability to plants and other soil organisms (Whitelaw, 2000). Further research is needed to quantify the contribution of organic acids to nutrient cycling and plant nutrition under field conditions in temperate agricultural soils.

Nutrient Losses and Soil Organisms

Plant roots absorb nutrients from the soil solution, but these plant-available nutrients can also be immobilized in microbial biomass and OM, adsorbed on soil surfaces, or lost from soils via gaseous emissions and leaching. Nutrient immobilization is affected by the C:N:P:S ratios of microbial biomass and the organic substrates available for decomposition. Generally, net immobilization of N occurs when the C:N ratio is greater than 25, and net immobilization of P and S occur when the C:P ratio is greater than 300 and the C:S ratio is greater than 400 (Havlin et al., 1999). Soil organisms are also involved in macroaggregate formation and stabilization, which affects the physical location of organic substrates within the soil matrix and their susceptibility to decomposition. Macroaggregates (> 200 μm diameter) are formed when organic debris from roots, microorganisms, and other sources binds to clay particles and microaggregates. Fungal hyphae and plant roots contribute to macroaggregate formation through the secretion of glycoproteins and polysaccharides that act as cementing agents to stabilize macroaggregates (Beare et al., 1997; Rillig et al., 2001). Macroaggregates are enriched with recently incorporated, relatively undecomposed organic substrates, which can be protected from decomposition for a period of time before it is eventually decomposed and redistributed in smaller macroaggregates or microaggregates (Miller and Jastrow, 1992; Puget, Chenu, and Balesdent, 2000). The quantities of nutrients that are immobilized and mineralized from aggregate fractions by soil microorganisms and other soil fauna is not well known.

Soil organisms likely do not have a large role in the chemical reactions that result in nutrient adsorption on soil surfaces and precipitation in soil minerals. However, they can contribute to nutrient losses from soils by transforming nutrients into forms that can be readily transported from soils, or by altering soil properties in ways that facilitate nutrient losses. Nitrifying bacteria transform NH_4^+ into NO_3^-, a more mobile form of N that is susceptible to leaching from agricultural soils. In tile drains under corn fields, NO_3-N concentrations as high as 120 mg L^{-1} have been reported (Logan, Randall, and Timmons, 1980), but other reports give NO_3-N concentrations of 10-60 mg L^{-1} in drainage waters under agricultural fields (Owens et al., 2000). The quantities of NO_3-N leached through agricultural soils are affected by the form of N at application, rate and timing of nutrient applications, crop residues, irrigation rate and method, precipitation, and soil characteristics such as texture and soil OM (Havlin et al., 1999).

Soil organisms that enhance water movement through the soil profile, such as the anecic earthworms *Lumbricus terrestris* L. and *Aporrectodea longa* L., can increase nutrient losses through leaching. These species form vertical permanent or semi-permanent burrows that may extend several meters in depth, and come to the surface to feed on residues that they then drag into their burrows (Edwards and Bohlen, 1996). Nitrogen mineralization and nitrification rates are higher in the walls of earthworm burrows than bulk soil, which can lead to greater NO_3^- leaching through earthworm burrows than other soil (Görres, Savin, and Amador, 1997; Parkin and Berry, 1999). Agricultural practices that increase the numbers of anecic earthworms are expected to promote NO_3^- leaching through the soil profile (Subler and Kirsch, 1998).

The denitrifying bacteria are facultative or obligate anaerobes that use NO_3^- as an electron acceptor in metabolic processes and produce N_2O and N_2. Other anaerobic bacteria such as *Desulfovibrio* spp. are responsible for converting SO_4^{2-} into SO_2 gas, but the dissimilatory reduction of sulfate to gaseous sulfur is not common in temperate agricultural soils since the process requires low oxidation-reduction potentials that are more typical of flooded soils. As much as 30% of the fertilizer N applied to agricultural soils may be lost via denitrification. Denitrification rates are influenced strongly by temperature and rainfall, as well as soil pH, soil bulk density, and soil OM content (Burton et al., 1997; MacKenzie, Fan, and Cardin, 1997). Earthworm casts and burrows contain higher levels of NO_3, soluble organic C and moisture than the bulk soil, which appears to stimulate denitrification and contributes to higher gaseous N losses from these structures than from bulk soil (Parkin and Berry, 1994; Subler and Kirsch, 1998).

IMPACT OF TILLAGE ON SOIL ORGANISMS

Farming practices are designed to maximize crop yields, but often these practices influence many other components of agroecosystems not considered explicitly, such as weed and crop pest populations, water table levels, soil structure and the activity of soil organisms, and the microclimate of agricultural fields. Fertilization, incorporation of crop residues in the fall, crop rotation and tillage are agricultural practices that induce seasonal variation in microbial processes and nutrients dynamics. The impact of agricultural practices on soil biota and nutrient cycling depends greatly on the cultivation method chosen. Tillage practices alter the quantity and quality of plant residues entering the soil, the seasonal and spatial distribution of these residues, the ratio between inputs from above and belowground and change the quality of nutrient inputs (Kandeler, Tscherko, and Speigel, 1999). Plowing buries OM deeper in the soil profile and, under well-aerated conditions, tends to favor the development of a large bacterial-dominated microbial community, whereas no-till favors the development of a fungal-based microbial community (Beare et al., 1992). The decomposer community in conventional tillage agroecosystems may be more prone to nutrient losses via leaching, whereas no-till systems may conserve more nutrients, due to the dominance of fungi that tend to immobilize rather than mineralize plant-available nutrients (Hendrix et al., 1986). Tillage affects different members of soil food webs in different ways. The numbers, diversity, and activities of soil microorganisms and fauna may all be affected by tillage. Our discussion will focus on how tillage impacts on the key microorganisms and soil fauna.

Effects of Tillage on Soil Microorganisms

Tillage modifies the soil environment by inverting the top 15 to 20 cm of soil, physically fragmenting residues at the soil surface and incorporating them deeper within the soil profile. These alterations to the soil environment affect the distribution, numbers and activity of soil microorganisms significantly. Kabir et al. (1998a) found more fungal hyphae in the top 5 cm of soils under no-tillage than conventional tillage. The abundance and biomass of bacteria, fungi, protozoa, nematodes, and microarthropods are greater on buried litter in conventional tillage systems than surface litter in no-till systems (Beare et al., 1992). However, the response of soil microbial communities to tillage disturbances may depend on soil characteristics. Lupwayi et al. (2001) found that microbial biomass and diversity were lower in tilled than untilled soils that were acidic and contained low OM, but tillage did not affect microbial biomass and diversity significantly in soils with a pH near neutrality and a higher OM content. They concluded that soil microorganisms in acid C-poor soils had less resilience to tillage effects than microorganisms in neutral C-rich soils. Decomposition in no-till systems tends to be slower and more greatly influenced by fungi than in conventionally tilled soils, which have a microbial community dominated by bacteria and bacterivorous soil fauna (Beare et al., 1992).

Tillage can impact AM fungi negatively when tillage operations detach the fungi from their host plant. Colonization of corn roots by AM fungi during the early development of corn was significantly lower in conventionally-tilled than no-tilled soils, but differences in mycorrhizal colonization were not significant after corn had reached the 12 to 14 leaf stage (Kabir et al. 1998b). However, early disruption of the mycorrhizal fungi-corn symbiosis can impact nutrient uptake and yield significantly (Miller, McGonigle, and Addy, 1995; Liu et al., 2000b). Physical disruption of fungal hyphae by fall tillage operations can reduce the survival of AM fungi; the number of fungal hyphae were 20 to 25% lower in soils plowed in the fall than unplowed (Kabir, O'Halloran, and Hamel, 1997). Although tillage can cause a decline in the number of fungal spores/hyphae, the effects may be more pronounced in systems that are tilled more intensely (more often, to a deeper depth) than in those tilled less intensively.

Soil bacteria are generally less impacted by tillage than soil fungi, and bacterial populations may be larger in conventionally tilled than no-tilled soils. Tillage implements often fragment organic residues, increasing their surface area and mixing them more intimately with soil, providing substrates for bacterial colonization and growth. Differences in the distribution and composition of bacterial communities under conventional and no-tillage systems have been documented (Hendrix et al., 1986; Roper and Gupta, 1995). In general, conventionally-tilled soils have higher mineralization and nitrification rates than no-till soils (Aulakh, Rennie, and Paul, 1984; Rice and Smith, 1982). No-till soils tend to have higher soil water contents than conventionally-tilled soils due to the retention of organic residues on the soil surface, and these conditions favor the development of anaerobic microsites. Greater fluxes of N_2O and N_2 from denitrifying bacteria have been documented under no-till than conventionally-tilled soils (Burton et al., 1997; MacKenzie, Fan, and Cardin, 1997).

Effects of Tillage on Soil Micro-, Meso-, and Macro-Fauna

Tillage can affect the numbers, populations and diversity of soil fauna in multiple ways. Leaving crop residues on the soil surface can increase soil moisture and provide a source of food for some of the larger soil fauna. Incorporating residues can permit soils to warm more quickly in the spring and favor the growth of fauna that graze primarily on soil microorganisms, particularly bacteria, and detritus. Some populations of soil fauna decline in tilled systems due to mechanical damage from the tillage implements. The impact of tillage on soil fauna will depend on the frequency and degree of disturbance of the soil physical and chemical properties, as well as the types of organisms present, since some are more susceptible to tillage than others (Kladivko, 2001).

Tillage and other agricultural practices influence protozoan and nematode populations by modifying the soil environment and by altering the quantity of food (bacteria and fungi) available. There have been relatively few studies to assess the effect of tillage practices on protozoa. Bamforth (1997) suggested the number and diversity of protozoa would be greater in no-till than conventional tillage systems because tillage disrupts the continuity of the water-filled pores that are the protozoan habitat. In addition, populations of bacterial-feeding protozoa would be favored in bacterial-dominated conventional tillage than fungal-dominated no-tillage agroecosystems. Roper and Gupta (1995) reported that no-till agroecosystems with retained stubble had 10 to 100 times more fungal-feeding protozoa and 5 to 10 times more bacterial-feeding protozoa than stubble-burnt systems. Beare et al. (1992) assessed protozoan populations in litterbags laid on the soil surface or buried in conventional- and no-tillage agroecosystems eight years after the tillage treatments were established, and found generally no difference in the numbers of amoebae, flagellates and ciliates in litterbags from the two tillage systems.

An increase in tillage intensity can reduce the populations of nematodes susceptible to mechanical damage and alter the vertical distribution of nematodes. Bouwman and Zwart (1994) found greater nematode biomass in integrated management systems with lower agrochemical inputs and reduced tillage than conventional management systems. The dominant microbial-feeding nematodes were bacterivores, and more herbivores and omnivore/predators were found in integrated than conventional management systems. However, burial of crop residues can stimulate microbial activity and provide food for microbiovorous nematodes (Yeates and Bongers, 1999). The effects of tillage on soil microfauna are not entirely clear, and Wardle (1995) contends that there are insufficient data from field studies to fully understand how these soil fauna are affected by tillage. More detailed investigation of protozoan and nematode populations in conventional- and no-tillage agroecosystems under field conditions is warranted.

Springtails and soil mites can be mechanically damaged or disrupted by tillage. The inversion of clods, a common feature in moldboard plowed agroecosystems, can cause individuals to be trapped in soil clods (Wardle, 1995). Springtails and mites of the suborder Mesostigmata that live relatively close to the soil surface tend to decline with increasing tillage intensity (Larink, 1997). However, soil mites vary in their response to tillage events. Cryptostigmatid, mesostignmatid, and astigmatid populations often decline after tillage events, but astigmatid populations appear to recover from tillage disturbances more rapidly (Behan-Pelletier, 1999). Beare et al. (1992) investigated microarthropod populations in litterbags laid on the soil surface or buried in conventional- and no-tillage agroecosystems. There was no difference in the number of microarthropods in surface litterbags from the two tillage systems, but more fungivorous, predatory and total microarthropods in litterbags buried in conventional tillage than no-tillage systems. Beare et al. (1992) suggested that a larger proportion of the springtails and mites in conventional- than no-tillage agroecosystems were omnivores, and shift their feeding habits depending on the availability and biomass of other organisms in soil foodwebs. Although it is difficult to make generalizations about how tillage affects soil mesofauna, Kladivko (2001) reports that the populations of springtails and soil mites are decreased by tillage more often than they are increased.

Tillage affects earthworm populations by changing the amount, quality and location of their food supply, and altering soil physical properties such as soil moisture and temperature. Earthworm populations are also susceptible to mechanical damage from tillage implements, and it has been estimated that rotary cultivation can reduce the biomass of earthworms in a field by up to 68% (Böström, 1995). Inversion of the top 10 to 20 cm of the soil profile during tillage exposes some endogeic earthworm species to avian predators (Giller et al., 1997). Generally, earthworm numbers are higher in no-tillage than conventional tillage agroecosystems (Kladivko, 2001). Higher earthworm numbers and biomass in no-tillage agroecosystems have been attributed to more beneficial soil conditions, including the presence of surface litter, favorable temperature and moisture conditions, and a lack of disturbance. Yet, in some cases, earthworm numbers and biomass may be no different, or slightly lower in no-tillage than conventional-tillage agroecosystems (Kladivko, Alhouri, and Weesies, 1997). After many years of conventional tillage, there may be insufficient numbers of some larger earthworm species, such as *Lumbricus terrestris*, remaining in the area to re-colonize the agroecosystems (Kladivko, 2001). However, earthworm biodiversity and activity generally increase when producers switch from conventional tillage to no-tillage systems (Clapperton et al., 1997; Emmerling, 2001; Parmelee et al., 1990).

CONCLUSIONS AND FUTURE DIRECTIONS

It can be concluded that soil organisms, whether they are associated with plants in a symbiotic relationship or free-living, have a role in nutrient cycling and nutrient uptake by crops. The relationship is much clearer for the symbiotic organisms that are intimately associated with plant roots than the non-symbiotic soil organisms, but there are still many questions to be answered about the interactions among symbiotic organisms, plant roots and free-living soil organisms. Although symbiotic organisms such as mycorrhizal fungi have an important role in nutrient mineralization and uptake, there is growing recognition that mycorrhizal fungi may reduce plant susceptibility to diseases caused by plant parasitic nematodes. Research is needed to better understand these interactions, and to determine their influence on crop production under field conditions.

Many free-living soil organisms, from the smallest soil microorganisms to the largest soil fauna are involved in OM decomposition and nutrient mineralization. Yet, linking the activities of these organisms to nutrient uptake by crops grown in agricultural fields is very difficult. Part of the reason is that most studies showing a linkage between soil organisms and crop production have been conducted in a laboratory setting, and many have focused on the effects of a single, or perhaps a few, soil organisms. Promising techniques that are being used to better understand the interactions among free-living soil organisms and their collective effects on nutrient dynamics and plant growth under field conditions include the use of radioisotopes and stable isotopes (to trace nutrient transformations in the soil-plant system) and manipulation experiments (where organisms are added or removed from the soil-plant system).

The degree of perturbation that occurs in agroecosystems is typically much greater than other terrestrial ecosystems, particularly in those agroecosystems that are continuously cropped and disturbed through cultivation and other agricultural practices. Certain soil organisms are relatively sensitive to tillage and other management practices, but recover rapidly after soils are disturbed, whereas other soil organisms are relatively insensitive to the human perturbations of agroecosystems. These types of soil organisms may have a larger role in nutrient cycling and crop production than those organisms whose populations decline progressively with each disturbance event. There is a growing body of information about the impacts of agricultural practices on the populations and trophic structure of different soil organisms. Future investigations should help us to identify the key organisms or groups of organisms responsible for nutrient cycling in temperate soils under different set of agricultural management practices. Better fertilizer recommendations in temperate cropping systems could be devised if the contributions of soil organisms to crop production were considered. Research is needed to assess the economic and environmental benefits of agricultural practices that consider the role of soil organisms explicitly. This would be an improvement on agricultural management schemes that underrate the complex interactions among the diverse organisms present in the soil environment and largely ignore their role in nutrient cycling and crop production.

REFERENCES

Alexander, D.B. (1998). Bacteria and arachaea. In *Principles and Applications of Soil Microbiology*, eds D.M. Sylvia, J.J. Fuhrmann, P.G. Hartel, and D.A. Zuberer. Upper Saddle River, NJ: Prentice Hall, pp. 44-71.

Alphei J., M. Bonkowski, and S. Scheu. (1996). Protozoa, nematoda and lumbricidae in the rhizosphere of *Hordelymus europaeus* (Poaceae): Faunal interactions, response of microorganisms and effects on plant growth. *Oecologica* 106: 111-126.

Andersen, N.C. (1983). Nitrogen turn over by earthworms in arable plots treated with farmyard manure and slurry. In *Earthworm Ecology: From Darwin to Vermiculture*, ed. J.E. Satchell, London, UK: Chapman and Hall, pp. 139-150.

Andrade, G., R.G. Linderman, and G.J. Bethlenfalvay. (1998). Bacterial associations with the mycorrhizosphere and hyphosphere of the arbuscular mycorrhizal fungus *Glomus mosseae. Plant and Soil* 202:79-87.

Andrén, O., T. Lindberg, U. Bostran, M. Clarholm, A.-C. Hansson, G. Johansson, J. Lagerlöf, K. Paustian, J. Persson, and R. Petterson. (1990). Ecology of arable land: Organisms, carbon and nitrogen cycling. *Ecological Bulletins* 40:85-126.

Aulakh, M.S., D.A. Rennie, and E.A. Paul. (1984). The influence of plant residues on denitrification rates in conventional and zero tilled soils. *Soil Science Society of America Journal* 48:790-794.

Bamforth, S.S. (1997). Protozoa: Recyclers and indicators of agroecosystem quality. In *Fauna in Soil Ecosystems: Recycling Processes, Nutrient Fluxes and Agricultural Production*, ed. G. Benckiser, New York, NY: Marcel Dekker, Inc., pp. 63-84.

Beare, M.H., S. Hus, D.C. Coleman, and P.R. Hendrix. (1997). Influences of mycelial fungi on soil aggregation and organic matter storage in conventional and no-tillage soils. *Applied Soil Ecology* 5:211-219.

Beare, M.H., R.W. Parmelee, P.R. Hendrix, W. Cheng, D.C. Coleman, and D.A. Crossley, Jr. (1992). Microbial and faunal interactions and effects on litter nitrogen and decomposition in agroecosystems. *Ecological Monographs* 62:569-591.

Behan-Pelletier, V.M. (1999). Oribatid mite biodiversity in agroecosystems: Role for bioindication. *Agriculture Ecosystems and Environment* 74:411-423.

Bethlenfalvay, G.J. and R.L. Franson. (1989). Manganese toxicity alleviated by mycorrhizae in soybean. *Journal of Plant Nutrition* 12:953-970.

Blair, J.M., R.W. Parmelee, and P. Lavelle. (1995). Influences of earthworms on biogeochemistry. In *Earthworm Ecology and Biogeography in North America*, ed. P.F. Hendrix, Boca Raton, FL: Lewis Publishers, pp. 127-158.

Bøckman, O.C., O. Kaarstad, O.H. Lie, and I. Richards. (1990). *Agriculture and Fertilizers*. Agricultural Group, Norsk Hydro, Oslo, Norway.

Bolan, N.S. (1991). A critical review on the role of mycorrhizal fungi in the uptake of phosphorus by plants. *Plant and Soil* 134:189-207.

Bolan, N.S., R. Naidu, S. Mahimairaja, and S. Baskaran. (1994). Influence of low-molecular weight organic acids on the solubilization of phosphates. *Biology and Fertility of Soils* 18:311-319.

Bonkowski, M., B. Griffiths, and C. Scrimgeour. (2000). Substrate heterogeneity and microfauna in soil organic 'hotspots' as determinants of nitrogen capture and growth of ryegrass. *Applied Soil Ecology* 14:37-53.

Böström, U. (1995). Earthworm populations (Lumbricidae) in ploughed and undisturbed leys. *Soil and Tillage Research* 35:125-133.

Bouwman, L.A. and K.B. Zwart. (1994). The ecology of bacterivorous protozoans and nematodes in arable soil. *Agriculture Ecosystems and Environment* 51:145-160.

Bowen, G.D. and A.D. Rovira. (1999). The rhizosphere and its management to improve plant growth. *Advances in Agronomy* 66:1-249.

Burton, D.L., D.W. Bergstrom, J.A. Convert, C.Wagner-Riddle, and E.G. Beauchamp. (1997). Three methods to estimate N_2O fluxes as impacted by agricultural management. *Canadian Journal of Soil Science* 77:125-134.

Calvet, C., J. Pinochet, A. Hernandez-Dorrego, V. Estaun, and A. Camprubi. (2001). Field microplot performance of the peach-almond hybrid GF-677 after inoculation with arbuscular mycorrhizal fungi in a replant soil infested with root-knot nematodes. *Mycorrhiza* 10:295-300.

Clapperton, M.J., J.J. Miller, F.J. Larney, and C. W. Lindwall. (1997). Earthworm populations as affected by longterm tillage practices in southern Alberta, Canada. *Soil Biology and Biochemistry* 29:631-633.

Coleman, D.C., P.F. Hendrix, H.M. Beare, W.X. Cheng, and D.A. Crossley, Jr. (1993). Microbial-faunal interactions as they affect soil organic matter dynamics in subtropical agroecosystems. In *Soil Biota, Nutrient Cycling and Farming Systems*, eds. M.G. Paoletti, W. Foissner and D.C. Coleman, Boca Raton, FL: Lewis Publishers, pp. 1-14.

Coleman, D.C. and D.A. Crossley, Jr. (1996). *Fundamentals of Soil Ecology*. San Diego, CA: Academic Press.

Crossley, D.A., Jr., D.C. Coleman, and P.F. Hendrix. (1989). The importance of the fauna in agricultural soils: Research approaches and perspectives. *Agriculture Ecosystems and Environment* 27:47-55.

Curry, J.P. (1998). Factors affecting earthworm abundance in soils. In *Earthworm Ecology*, ed. C.A. Edwards, Boca Raton, FL: CRC Press, pp. 37-64.

Darbyshire, J.F. (1994). *Soil Protozoa*. Wallingford, UK: CAB International.

de Ruiter, P.C., A.M. Neutal, and J.C. Moore. (1994). Modeling food webs and nutrient cycling in agroecosystems. *Trends in Ecology and Evolution* 9:378-383.

Didden, W.A.M., H.-C. Fründ, and U. Graefe. (1997). Enchytraeids. In *Fauna in Soil Ecosystems: Recycling Processes, Nutrient Fluxes and Agricultural Production*, ed. G. Benckiser, New York, NY: Marcel Dekker Inc., pp. 135-172.

Dindal, D.L. (1990). *Soil Biology Guide*. New York, NY: John Wiley and Sons.

Edwards, C.A. and P.J. Bohlen. (1996). *Biology and Ecology of Earthworms*, 3rd edn., London, UK: Chapman and Hall.

Edwards, C.A. (1983). Earthworm ecology in cultivated soils. In *Earthworm Ecology: From Darwin to Vermiculture*, ed. J.E. Satchell, London, UK: Chapman and Hall, pp. 123-137.

Ekelund, F., R. Rønn, and S. Christensen. (2001). Distribution with depth of protozoa, bacteria and fungi in soil profiles from three Danish forest sites. *Soil Biology and Biochemistry* 33:475-481.

Emmerling, C. (2001). Response of earthworm communities to different types of soil tillage. *Applied Soil Ecology* 17:91-96.

Estaun V.C. Calvet, and D.S. Hayman. (1987). Influence of plant genotype on mycorrhizal infection: Response of three pea cultivars. *Plant and Soil* 103:296-298.

Evans, K., D.L. Trudgill, and J.M. Webster (eds). (1993). *Plant Parasitic Nematodes in Temperate Agriculture*. Wallingford, UK: CAB International.

Ferris, H., S.S. Lau, and R.C. Venette. (1995). Population energetics of bacterial feeding nematodes: Respiration and metabolic rates based on carbon-dioxide production. *Soil Biology and Biochemistry* 27:319-330.

Ferris, H., R.C. Venette, and S.S. Lau. (1997). Population energetics of bacterial-feeding nematodes: Carbon and nitrogen budgets. *Soil Biology and Biochemistry* 29: 1183-1194.

Finlay, R. and B. Söderström. (1992). Mycorrhiza and carbon flow to the soil. In *Mycorrhizal Functioning: An Integrative Plant-Fungal Process*, ed. M.F. Allen, New York, NY: Chapman and Hall, pp. 134-160.

Finlay, R.D. (1985). Interactions between soil microarthropods and endomycorrhizal associations of higher plants. In *Ecological Interactions in Soil*, eds. A.H. Fitter, D. Atkinson, and D.J. Read, Oxford, UK: Blackwell, pp. 319-331.

Foissner, W. (1987). Soil protozoa: Fundamental problems, ecological significance, adaptations in ciliates and testaceans, bioindicators and guide to the literature. *Progress in Protistology* 2:69-212.

Foissner, W. (1999). Soil protozoa as bioindicators: Pros and cons, methods, diversity, representative examples. *Agriculture, Ecosystems and Environment* 74:95-112.

Foster, R.C. and J.F. Dormaar. (1991). Bacteria-grazing amoebae in situ in the rhizosphere. *Biology and Fertility of Soils* 11:83-87.

Freckman, D.W. and C.H. Ettema. (1993). Assessing nematode communities in agroecosystems of varying human intervention. *Agriculture Ecosystems and Environment* 45:239-261.

Frossard, E., L.M. Condron, A. Oberson, S. Sinaj, and J.C. Fardeau. (2000). Processes governing phosphorus availability in temperate soils. *Journal of Environmental Quality* 29:15-23.

Garrett, C.J., D.A. Crossley, Jr., D.C. Coleman, P.F. Hendrix, K.W. Kisselle, and R.L. Potter. (2001). Impact of the rhizosphere on soil microarthropods in agroecosystems on the Georgia piedmont. *Applied Soil Ecology* 16:141-148.

Giller, K.E., M.H. Beare, P. Lavelle, A.M.N. Izac, and M.J. Swift. (1997). Agricultural intensification, soil biodiversity and agroecosystem function. *Applied Soil Ecology* 6:3-16.

Görres J.H., M.C. Savin, and J.A. Amador. (1997). Dynamics of carbon and nitrogen mineralization, microbial biomass, and nematode abundance within and outside the burrow walls of anecic earthworms (*Lumbricus terrestris*). *Soil Science* 162: 666-671.

Grayston, S.J., S. Wang, C.D. Campbell, and A.C. Edwards. (1998). Selective influence of plant species on microbial diversity in the rhizosphere. *Soil Biology and Biochemistry* 30:369-378.

Griffiths, B.S. and S. Caul. (1993). Migration of bacterial-feeding nematodes, but not protozoa, to decomposing grass residues. *Biology and Fertility of Soils* 15:201-207.

Griffiths, B.S. (1994). Microbial-feeding nematodes and protozoa in soils: Their effects on microbial activity and nitrogen mineralization in decomposition hotspots and the rhizosphere. *Plant and Soil* 164:25-33.

Gupta, V.V.S.R. and J.J. Germida. (1989). Influence of bacterial-amoebal interactions on sulfur transformations in soil. *Soil Biology and Biochemistry* 21:787-791.

Halaj, J., A.B. Cady, and G.W. Uetz. (2000). Modular habitat refugia enhance generalist predators and lower plant damage in soybeans. *Environmental Entomology* 29:383-393.

Hamel, C., V. Furlan, and D.L. Smith. (1992). Mycorrhizal effects on interspecific plant competition and N transfer in legume-grass forage mixtures. *Crop Science* 32:991-996.

Hamel, C., D.L. Smith, and V. Furlan. (1991). N_2-fixation and transfer in a field grown corn and soybean intercrop. *Plant and Soil* 133:177-185.

Hansen, R.A. (2000). Effects of habitat complexity and composition on a diverse litter microarthropod assemblage. *Ecology* 81:1120-1132.

Harris, K.K. and E.A. Paul. (1987). Carbon requirements of vesicular arbuscular mycorrhizae. In *Ecophysiology of VA Mycorrhizal Plants*, ed. G.R. Safir, Boca Raton, FL: CRC Press, pp. 93-105.

Havlin, J.L., J.D. Beaton, S.L. Tisdale, and W.L. Nelson. (1999). *Soil Fertility and Fertilizers*, 6th edition. Upper Saddle River, NJ: Prentice Hall.

He, Z.L., J. Wu, A.G. O'Donnell, and J.K. Syers. (1997). Seasonal responses in microbial biomass carbon, phosphorous and sulphur in soils under pasture. *Biology and Fertility of Soils* 24:421-428.

Hendrix, P.F., B.R. Mueller, R.R. Bruce, G.W. Langdale, and R.W. Parmelee. (1992). Abundance and distribution of earthworms in relation to landscape factors on the Georgia piedmont, U.S.A. *Soil Biology and Biochemistry* 24:1357-1361.

Hendrix, P.F., R.W. Parmelee, D.A. Crossley, Jr., D.C. Coleman, E.P. Odum, and P.M. Groffman. (1986). Detritus food webs in conventional and no-tillage agroecosystems. *BioScience* 36:374-380.

Hinsinger, P. (2001). Bioavailability of soil inorganic P in the rhizosphere as affected by root-induced chemical changes: A review. *Plant and Soil* 237:173-195.

Hu, H.Q., J.Z. He, X.Y. Li, and F. Liu. (2001). Effect of several organic acids on phosphate adsorption by variable charge soils of central China. *Environment International* 26:353-358.

Illmer, P., A. Barbato, and F. Schinner. (1995). Solubilization of hardly-soluble $AlPO_4$ with P-solubilizing microorganisms. *Soil Biology and Biochemistry* 27:265-270.

Ingham, R.E., J.A.Trofymow, E.R. Ingham, and D.C. Coleman. (1985). Interactions of bacteria, fungi and their nematode grazers: Effects on nutrient cycling and plant growth. *Ecological Monographs* 55:119-140.

Ingham, R.E., J.A. Trofymow, R.N. Ames, H.W. Hunt, C.R. Morley, J.C. Moore, and D.C. Coleman. (1986). Trophic interactions and nitrogen cycling in a semi-arid grassland soil. I. Seasonal dynamics of the natural populations, their interactions and effect on nitrogen cycling. *Journal of Applied Ecology* 23:579-614.

James, S.W. (1991). Soil, nitrogen, phosphorus and organic matter processing by earthworms in a tallgrass prairie. *Ecology* 72:2101-2109.

Joergensen, R.G., B. Meyer, and T. Mueller. (1994). Time-course of the soil microbial biomass under wheat: A one year field study. *Soil Biology and Biochemistry* 26:987-994.

Jones, D.L. and D.S. Brassington. (1998). Sorption of organic acids in acid soils and its implications in the rhizosphere. *European Journal of Soil Science* 49:447-455.

Jones, D.L. and P.R. Darrah. (1994). Role of root derived organic acids in the mobilization of nutrients from the rhizosphere. *Plant and Soil* 166:247-257.

Jones, D.L. and L.V. Kochian. (1996). Aluminum-organic acid interactions in acid soils. 1. Effect of root-derived organic acids on the kinetics of Al dissolution. *Plant and Soil* 182:221-228.

Kabir, Z., I.P. O'Halloran, J.W. Fyles, and C. Hamel. (1997). Seasonal changes of arbuscular mycorrhizal fungi as affected by tillage practices and fertilization: Hyphal density and mycorrhizal root colonization. *Plant and Soil* 192:285-293.

Kabir, Z., I.P. O'Halloran, and C. Hamel. (1997). Overwinter survival of arbuscular mycorrhizal hyphae is favoured by attachment to roots but diminished by disturbance. *Mycorrhiza* 7:197-200.

Kabir, Z., I.P. O'Halloran, P. Widden, and C. Hamel. (1998a). Vertical distribution of arbuscular mycorrhizal fungi under continuous corn in long-term no-till and conventional tillage systems. *Mycorrhiza* 8:53-55.

Kabir, Z., I. P. O'Halloran, J. Fyles, and C. Hamel. (1998b). Dynamics of the mycorrhizal symbiosis of corn: effect of host physiology, tillage practice and fertilization on spatial distribution of extraradical hyphae in the field. *Agriculture Ecosystems and Environment* 68:151-163.

Kalisz, P.J. and H.B. Wood. (1995). Native and exotic earthworms in wildland ecosystems. In *Earthworm Ecology and Biogeography in North America*, ed. P.F. Hendrix, Boca Raton, FL: Lewis Publishers, pp. 117-126.

Kandeler, E., D. Tscherko, and H. Speigel. (1999). Long-term monitoring of microbial biomass, N mineralisation and enzyme activities of a Chernozem under different tillage management. *Biology and Fertility of Soils* 28:343-351.

Kim, K.Y., D. Jordan, and G.A. McDonald. (1998). Effect of phosphate-solubilizing bacteria and vesicular-arbuscular mycorrhizae on tomato growth and soil microbial activity. *Biology and Fertility of Soils* 26:79-87.

Kladivko, E.J., N.M. Alhouri, and G. Weesies. (1997). Earthworm populations and species distribution under no-till and conventional tillage in Indiana and Illinois. *Soil Biology and Biochemistry* 29:613-615.

Kladivko, E.J. (2001). Tillage systems and soil ecology. *Soil and Tillage Research* 61:61-76.

Klironomos J.N. and M. Ursic. (1998). Density-dependent grazing on the extraradical hyphal network of the arbuscular mycorrhizal fungus, *Glomus intraradices*, by the collembolan, *Folsomia candida*. *Biology and Fertility of Soils* 26: 250-253.

Koide, R.T. and M. Li. (1990). On host regulation of the vesicular-arbuscular mycorrhizal symbiosis. *New Phytologist* 114:59-65.

Kongshaug, G. (1998). *Energy Consumption and Greenhouse Gas Emissions in Fertilizer Production.* Proceedings IFA Technical Conference, Marrakesh, Morocco International Fertilizer Industry Association, Paris, pp. 272-289.

Larink, O. (1997). Springtails and mites: important knots in the food web of soils. In *Fauna in Soil Ecosystems: Recycling Processes, Nutrient Fluxes and Agricultural Production*, ed. G. Benckiser, New York, NY: Marcel Dekker, Inc., pp. 225-264.

Lee, K.E. and C.E. Pankhurst. (1992). Soil organisms and sustainable productivity. *Australian Journal of Soil Research* 30: 855-892.

Logan, T.J., G.W. Randall, and D.R. Timmons. (1980). Nutrient content of tile drainage from cropland in the north central region. Wooster, OH: NC Regional Publishers 268, Research Bulletin No. 1119.

Little, L.R. and M.A. Maun. (1996). The 'Ammophila problem' revisited: A role for mycorrhizal fungi. *Journal of Ecology* 84:1-7.

Liu, A., C. Hamel, R.I. Hamilton, and D.L. Smith. (2000a). Mycorrhizal formation and nutrient uptake of new maize (*Zea mays* L.) hybrids with extreme canopy and leaf architecture as influenced by soil N and P levels. *Plant and Soil* 221:157-166.

Liu, A., C. Hamel, R.I. Hamilton, B. Ma, and D.L. Smith. (2000b). Acquisition of Cu, Zn, Mn and Fe by mycorrhizal maize (*Zea mays* L.) grown in soil at different P and micronutrient levels. *Mycorrhiza* 9:331-336.

Liu, A., C. Hamel, A. Elmi, C. Costa, B. Ma, and D.L. Smith. (2002). K, Ca and Mg nutrition of maize colonized by arbuscular mycorrhizal fungi under field conditions. *Canadian Journal of Soil Science* 82:272-278.

Lupwayi, N.Z., M.A. Monreal, G.W. Clayton, C.A. Grant, A.M. Johnston, and W.A. Rice. (2001). Soil microbial biomass and diversity respond to tillage and sulphur fertilizers. *Canadian Journal of Soil Science* 81:577-589.

MacKenzie, A.F., M.X. Fan, and F. Cardin. (1997). Nitrous oxide emissions as affected by tillage, corn-soybean-alfalfa rotations and nitrogen fertilization. *Canadian Journal of Soil Science* 77:145-152.

Maraun, M., J. Alphei, M. Bonkowski, R. Buryn, S. Migge, M. Peter, M. Schaefer, and S. Scheu. (1999). Middens of the earthworm *Lumbricus terrestris* (Lumbricidae): microhabitats for micro- and mesofauna in forest soil. *Pedobiologia* 43:276-287.

Marinissen, J.C.Y. and A.R. Dexter. (1990). Mechanisms of stabilization of earthworm casts and artificial casts. *Biology and Fertility of Soils* 9:163-167.

Marschner, H. (1995). *Mineral Nutrition of Higher Plants.* San Diego, CA: Academic Press Inc.

Miller, M.H., T.P. McGonigle, and H.D. Addy. (1995). Functional ecology of VA mycorrhizas as influenced by P fertilization and tillage in an agricultural ecosystem. *Critical Review of Biotechnology* 15:241-255.

Miller, R.M. and J.D. Jastrow. (1992). The role of mycorrhizal fungi in soil conservation. In *Mycorrhizae in Sustainable Agriculture*, eds. G.J. Bethlenfalvay and R.G. Linderman, Madison, WI: ASA Special Publication Number 54, pp. 29-44.

Moore, J.C., E.R. Ingham, and D.C. Coleman. (1987). Inter- and intraspecific feeding selectivity of *Folsomia candida* (Willem) (Collembola, Isotomidae) on fungi. *Biology and Fertility of Soils* 5:6-12.

Morton, J.B. (1998). Fungi. In *Principles and Applications of Soil Microbiology*, eds. D.M. Sylvia, J.J. Fuhrmann, P.G. Hartel, and D.A. Zuberer. Upper Saddle River, NJ: Prentice Hall, pp. 72-93.

Murphy, D.V., I.R.P. Fillery, and G.P Sparling. (1998). Seasonal fluctuations in gross N mineralisation, ammonium consumption, and microbial biomass in a Western Australian soil under different land uses. *Australian Journal of Agricultural Research* 49:523-535.

Newton, W.E. (1999). Nitrogen fixation and the biosphere. In *Highlights of Nitrogen Fixation Research*, eds. E. Martinez and G. Hernandez, New York, NY: Kluwer Academic/Plenum Publishers, pp. 1-8.

Olsson, P.A., I. Thingstrup, I. Jakobsen, and E. Bååth. (1999). Estimation of the biomass of arbuscular mycorrhizal fungi in a linseed field. *Soil Biology and Biochemistry* 31:1879-1887.

Owens, L.B., R.W. Malone, M.J. Shipitalo, W.M. Edwards, and J.V. Bonta. (2000). Lysimeter study of nitrate leaching from a corn-soybean rotation. *Journal of Environmental Quality* 29:467-474.

Paoletti, M.G. (1999). The role of earthworms for assessment of sustainability and as bioindicators. *Agriculture Ecosystems and Environment* 74:137-155.

Parkin, T.B. and E.C. Berry. (1999). Microbial nitrogen transformations in earthworm burrows. *Soil Biology and Biochemistry* 31:1765-1771.

Parkin, T.B. and E.C. Berry. (1994). Nitrogen transformations associated with earthworm casts. *Soil Biology and Biochemistry* 26:1233-1238.

Parmelee, R.W., M.H. Beare, W. Cheng, P.F. Hendrix, S.J. Rider, D.A. Crossley Jr., and D.C. Coleman. (1990). Earthworms and enchytraeids in conventional and no-tillage agroecosystems: A biocide approach to assess their role in organic matter breakdown. *Biology and Fertility of Soils* 10:1-10.

Patra, D.D., P.C. Brookes, K. Coleman, and D.S. Jenkinson. (1990). Seasonal changes of soil microbial biomass in an arable and a grassland soil which have been under uniform management for many years. *Soil Biology and Biochemistry* 6:739-742.

Paul, E.A., and F.E. Clark. (1996). *Soil Microbiology and Biochemistry*. 2nd ed. San Diego, CA: Academic Press.

Perrott, K.W., S.U. Sarathchandra, and B.W. Dow. (1992). Seasonal and fertilizer effects on the organic cycle and microbial biomass in a hill country soil under pasture. *Australian Journal of Soil Research* 30:383-394.

Peterson, R. L. and S.M. Bradbury. (1995). Use of plant mutants, intraspecific variants and non-hosts in studying mycorrhiza formation and function. In *Mycorrhiza: Structure, Function, Molecular Biology and Biotechnology*, eds. A.K. Varma and B. Hock, Heidelberg, Germany: Springer-Verlag, pp. 157-180.

Poier, K.R. and J. Richter. (1992). Spatial distribution of earthworms and soil properties in an arable loess soil. *Soil Biology and Biochemistry* 24:1601-1608.

Puget, P., C. Chenu, and J. Balesdent. (2000). Dynamics of soil organic matter associated with particle-size fractions of water-stable aggregates. *European Journal of Soil Science* 51:595-605.

Reitz, M., K. Rudolph, I. Schroder, S. Hoffmann-Hergarten, J. Hallmann, and R.A. Sikora. (2000). Lipopolysaccharides of *Rhizobium* etli strain G12 act in potato roots as an inducing agent of systemic resistance to infection by the cyst nematode *Globodera pallida*. *Applied and Environmental Microbiology* 66:3515-3518.

Rice, C.W. and M.S. Smith. (1982). Denitrification in no-till and plowed soils. *Soil Science Society of America Journal* 46:1168-1173.

Rillig, M.C., S.F. Wright, K.A. Nichols, W.F. Schmidt, and M.S. Torn. (2001). Large contribution of arbuscular mycorrhizal fungi to soil carbon pools in tropical forest soils. *Plant and Soil* 233:167-177.

Robertson, G.P. and D.W. Freckman. (1995). The spatial distribution of nematode trophic groups across a cultivated ecosystem. *Ecology* 76:1425-1432.

Roper, M.M. and V.V.S.R. Gupta. (1995). Management practices and soil biota. *Australian Journal of Soil Research* 33:321-329.

Sabatini, M.A. and G. Innocenti. (2001). Effects of collembola on plant-pathogenic fungus interactions in simple experimental systems. *Biology and Fertility of Soils* 33:62-66.

Savin, M.C., J.H. Görres, D.A. Neher, and J.A. Amador. (2001). Uncoupling of carbon and nitrogen mineralization: Role of microbivorous nematodes. *Soil Biology and Biochemistry* 33:1463-1472.

Schmidt, O. and J.P. Curry. (2001). Population dynamics of earthworms (Lumbicidae) and their role in nitrogen turnover in wheat and wheat-clover cropping systems. *Pedobiologia* 45:174-187.

Shipitalo, M.J. and R. Protz. (1989). Chemistry and micromorphology of aggregation in earthworm casts. *Geoderma* 45:357-374.

Siddiqui, Z.A. and T. Mahmood. (2001). Effects of rhizobacteria and root symbionts on the reproduction of *Meloidogyne javanica* and growth of chickpea. *Bioresource Technology* 79:41-45.

Smith, S. E. and D. J. Read. (1997) *Mycorrhizal Symbiosis*. 2nd ed. London, UK: Academic Press.

Sohlenius, B. and S. Boström. (1984). Colonization, population development and metabolic activity of nematodes in buried barley straw. *Pedobiologia* 27:67-78.

St-Arnaud, M., C. Hamel, B. Vimard, M. Caron, and J.A. Fortin. (1995). Altered growth of *Fusarium oxysporum* f.sp. *chrysanthemi* in an in vitro dual culture system with the vesicular arbuscular mycorrhizal fungus *Glomus intraradices* growing on *Daucus carota* transformed roots. *Mycorrhiza* 5:431-438.

Strobel, B.W. (2001). Influence of vegetation on low-molecular-weight carboxylic acids in soil solution–a review. *Geoderma* 99:169-198.

Subler, S. and A.S. Kirsch. (1998). Spring dynamics of soil carbon, nitrogen and microbial activity in earthworm middens in a no-till cornfield. *Biology and Fertility of Soils* 26:243-249.

Tabatabai, M.A. (1994). Soil enzymes. In *Methods of Soil Analysis, Part 2. Microbiological and Biochemical Properties*, eds. R.W. Weaver, S. Angle, P. Bottomley, D. Bezdicek, S. Smith, A. Tabatabai, and A. Wollum, Madison, WI: Soil Science Society of America, Book series, No. 5, pp. 775-883.

Tiunov, A.V., M. Bonkowski, J. Alphei, and S. Scheu. (2001). Microflora, protozoa and nematoda in *Lumbricus terrestris* burrow walls: A laboratory experiment. *Pedobiologia* 45:46-60.

Tiunov, A.V., T.G. Dobrovol'skaya, and L.M. Polyanskaya. (2001). Microbial complexes associated with inhabited and abandoned burrows of *Lumbricus terrestris* earthworm in soddy-podzolic soil. *Eurasian Soil Science* 34:525-529.

Tiunov, A.V., T.G. Dobrovol'skaya, and L.M. Polyanskaya. (1997). Microbial community of the *Lumbricus terrestris* L. earthworm burrow walls. *Microbiology* 66:349-353.

Tiunov, A.V. and S. Scheu. (2000). Microbial biomass, biovolume and respiration in *Lumbricus terrestris* L. cast material of different age. *Soil Biology and Biochemistry* 32:265-275.

Wardle, D.A. (1995). Impacts of disturbance on detritus food webs in agro-ecosystems of contrasting tillage and weed management practices. In *Advances in Ecological Research*, eds. M. Begon and A.H. Fitter, Vol. 26, New York, NY: Academic Press, pp. 105-185.

Weaver, R.W. and P.H. Graham. (1994). Legume nodule symbionts. In *Methods of Soil Analysis, Part 2. Microbiological and Biochemical Properties*, eds. R.W. Weaver, S. Angle, P. Bottomley, D. Bezdicek, S. Smith, A. Tabatabai, and A. Wollum, Madison, WI: Soil Science Society of America, Book series, No. 5, pp. 1019-1045.

Westover, K.M., A.C. Kennedy, and S.E. Kelley. (1997). Patterns of rhizosphere microbial community structure associated with co-occurring plant species. *Journal of Ecology* 85:863-873.

Whalen, J.K. and R.W. Parmelee. (1999). Quantification of nitrogen assimilation efficiencies and their use to estimate organic matter consumption by the earthworms *Aporrectodea tuberculata* (Eisen) and *Lumbricus terrestris* L. *Applied Soil Ecology* 13:199-208.

Whalen, J.K. and R.W. Parmelee. (2000). Earthworm secondary production and determination of N flux through earthworm communities in agroecosystems: Comparison of two approaches. *Oecologia* 124:561-573.

Whalen, J.K., R.W. Parmelee, D.M. McCartney, and J.L. VanArsdale. (1999). Movement of N from decomposing earthworm tissue to soil, microbial and plant N pools. *Soil Biology and Biochemistry* 31:487-492.

Whalen, J.K., R.W. Parmelee, and S. Subler. (2000). Use of ^{15}N to quantify excretion rates of different earthworm species in corn agroecosystems. *Biology and Fertility of Soils* 32:347-352.

Whitelaw, M.A. (2000). Growth promotion of plants inoculated with phosphate-solubilizing fungi. *Advances in Agronomy* 69:99-151.

Wollum, A.G. (1998). Introduction and historical perspective. In *Principles and Applications of Soil Microbiology*, eds. D.M. Sylvia, J.J. Fuhrmann, P.G. Hartel, and D.A. Zuberer. Upper Saddle River, NJ: Prentice Hall, pp. 3-20.

Wolters, V. and K. Ekschmitt. (1997). Gastropods, isopods, diplopods and chilopods: Neglected groups of the decomposer food web. In *Fauna in Soil Ecosystems: Recycling Processes, Nutrient Fluxes and Agricultural Production*, ed. G. Benckiser, New York, NY: Marcel Dekker Inc., pp. 265-306.

Yeates, G.W. and T. Bongers. (1999). Nematode diversity in agroecosystems. *Agriculture Ecosystems and Environment* 74:113-135.

Nutrient Dynamics: Utilizing Biotic-Abiotic Interactions for Improved Management of Agricultural Soils

Chantal Hamel
Christine Landry
Abdirashid Elmi
Aiguo Liu
Timothy Spedding

SUMMARY. Environmental concerns currently trigger the development of more sustainable soil fertility management strategies. It appears that effective sustainable practices are those that enhance natural soil processes. Soil processes include the decomposition of residues and mineralization of organic matter, nitrogen fixation, nitrification, nitrate leaching, denitrification and sulfur reduction. Natural soil processes also include less well-understood interactions, namely, those leading to the dissolution of minerals by organic acids, as well as rhizospheric and mycorrhizospheric interactions. Plants, associated with arbuscular mycor-

---

Chantal Hamel is affiliated with Agriculture and Agri-Food Canada, Swift Current, SK, Canada, S9H 3X2 (E-mail: hamelc@agr.gc.ca).

Christine Landry and Abdirashid Elmi are PhD students, Aiguo Liu is Research Associate, and Timothy Spedding is a MSc student, Department of Natural Resource Science, McGill University, Ste-Anne-de-Bellevue, QB, Canada, H9X 3V9.

[Haworth co-indexing entry note]: "Nutrient Dynamics: Utilizing Biotic-Abiotic Interactions for Improved Management of Agricultural Soils." Hamel, Chantal et al. Co-published simultaneously in *Journal of Crop Improvement* (Food Products Press, an imprint of The Haworth Press, Inc.) Vol. 11, No. 1/2 (#21/22), 2004, pp. 209-248; and: *New Dimensions in Agroecology* (ed: David Clements, and Anil Shrestha) Food Products Press, an imprint of The Haworth Press, Inc., 2004, pp. 209-248. Single or multiple copies of this article are available for a fee from The Haworth Document Delivery Service [1-800-HAWORTH, 9:00 a.m. - 5:00 p.m. (EST). E-mail address: docdelivery@haworthpress.com].

http://www.haworthpress.com/web/JCRIP

Digital Object Identifer: 10.1300/J411v11n01_10

rhizal symbionts, supply and distribute carbon and energy, sustaining most of the biotic mechanisms responsible for nutrient release from soil, and maintaining organic pools of nutrients. Among these pools, the microbial biomass and fine roots pools, with their very fast turnover time, are particularly important as they can maintain large amounts of nutrients in very labile form and, therefore, increase soil fertility. Agricultural soil systems are very dynamic and are characterized by large spatial and temporal variations, which are largely driven by plant development. In addition, nutrient dynamics in agricultural soil systems seem particularly influenced by temperature, moisture, and nitrogen and phosphorus fertilization. Nitrogen losses from soil are reduced in systems where nitrogen release corresponds to plant demand. Biological nitrogen fixation is a sound way to input nitrogen in cropping systems. Phosphorus losses can be reduced through increased reliance on the arbuscular mycorrhizal symbiosis of crops. Soils are diverse and complex systems, which, furthermore, respond to increasingly unpredictable climatic variations. Optimal agricultural soil management is a moving target and, hence, a challenging goal that will never be totally reached.

KEYWORDS. Biogeochemical cycles, soil fertility, soil life, nutrient cycling, soil microorganisms, microbial interactions, soil-plant interactions

INTRODUCTION

Nutrient dynamics in agricultural soil is driven by the same mechanisms as those subtending nutrient dynamics in undisturbed soil ecosystems. In all ecosystems, soil is similarly made of solids and of pores that are occupied by air and by the soil solution. The solid component of soils comprises a mineral fraction intimately bound with humified organic matter (OM), fresh OM and living organisms. Energy flowing within such organized matrix sustains life in natural and agricultural systems too. Most importantly for plants, which need mineral nutrients to build themselves, this energy flow permits transformation of matter. Soil organisms, which act as catalysts of this transformation, helps replenish the soil solution with ions several of which are required by plant. Soils can be seen as the dark interface where the mutual influence of the biotic and abiotic components of Earth results in the dissolution of the very substance of the planet into ions, which cycle intensively before being lost from the soil systems through leaching, volatilization or crop removal. This is largely how plants are fed and how soils evolve.

Mineralization of OM and weathering of mineral soil solids that supplies the soil solution with available mineral nutrients and the cycling of these nutrients through biomass is the natural process that, with biological processes, has maintained life on earth for three billion years (Chernicoff, 1999). The evolution of life within soil inanimate matter has shaped the mechanisms by which living soil organisms, including plants, exploit their environment to better flourish. Today, living soil organisms, even the most immobile ones, are active and important characters in the mobilization of nutrients. The role of soil organisms in nutrient cycling must be understood as their contribution to soil processes and plant growth is the basis of sustainable agricultural production systems.

The role of soil organisms in the fertility of agricultural land in the temperate regions had been somewhat obscured by fertilizer use. The soil is complex and its natural processes are very difficult to predict. In contrast, fertilizers are easy to use. Applications slightly in excess of crop needs were often seen as a low cost measure insuring the production of optimum yields. At this time, however, the environmental impacts of agriculture are being measured on water and air quality and the need for more precise interventions on soil fertility management to insure profitable yields is becoming clearer. More precise and judicious use of fertilizers can only be made with a clear understanding of the natural nutrient processes contributing to plant growth and nutrient loss, which are active in agricultural soils as well as undisturbed soils. This article considers the role of soil organisms in soil processes, nutrient management by plants, and the temporal variation in nutrient release from soil.

SOIL PROCESSES

The taxonomic grouping of soil organisms and the identification of the general characteristics and functions of these groups has helped us understand their role in soil function (Frostegard, Tunlid, and Baath, 1993; Ogram, 2000). However, the prevalence of environmental heterogeneity in soil and the proliferation of microsites with fluctuating conditions (e.g., Pariente, 2002) results in the occurrence of a very large biodiversity of organisms and, consequently, into a high level of complexity and functional redundancy (Wolters, 2001). As many as 170,000 species of soil organisms have been identified (Wall and Virginia, 1999). The number of species of soil prokaryotes was estimated at 30,000; fungi, 1,500,000; algae, 60,000; protozoa, 100,000; and nematodes, 500,000 (Lee and Pankhurst, 1992). Representatives of all these groups may be involved in a single soil process. For example, both heterotrophic microbes and soil animals contribute to OM decomposition. The extreme complexity of the soil system is a major difficulty inherent to any deterministic approach to nutrient dynamics. Hence, soil processes should be considered as a whole to understand soil functioning.

Decomposition of Organic Matter

Paul and Clark (1996) described three pools of soil OM based on their different turnover times. Residues left by crops are decomposed by soil microfauna and flora within hours to days. Decomposers first utilize the more easily degradable portion of these residues. Recalcitrant molecules from plant or secondary production origin, and residual degradation products take more time to decompose. Their turnover rate range from 10 to 100 years. Recalcitrant molecules may accumulate and polymerize into very stable humic substances. This stable pool, which makes up approximately 50% of total soil C, may persist hundreds of years before being decomposed.

Decomposition of organic residues is an important step in nutrient dynamics. Soil organisms of different trophic levels are involved in this activity in which OM changes in both quantity and quality. During decomposition of residues, the soil fauna fragments organic materials. Complex organic molecules are physically and enzymatically dismantled into successively smaller units by a large array of interacting soil organisms and finally, soluble materials are released. Leaching of solubles occur during the process possibly contributing substantially to initial mass losses dependant on the nature of the residues. Concurrently, new simple and complex organic materials are constructed by decomposers from the building blocks released through enzymatic degradation of litter. Fragmented materials and turnover of heterotrophic decomposers replenish the soil OM pool and the decomposition rate is controlled by the rate of inputs of dead plant tissues, quality of the litter input to soil (Paul and Clark, 1996), the structure and complexity of the soil communities (de Ruiter, Neutel, and Moore, 1994; Zheng, Bengtsson, and

Cgren, 1997) and the conditions of the environment influencing growth (Lee and Pankhurst, 1992).

Until recently, the importance of the natural mechanisms underlying nutrient dynamics in field crop production has been underestimated in spite of the large impact natural processes have on crop production. Two approaches to the study of nutrient cycling in soil have been taken. Soil ecologists have modelled decomposition and nutrient flow through the numerous and complex functional groups of soil organisms making up the soil food web (de Ruiter, Neutel, and Moore, 1994; Zheng, Bengtsson, and Cgren, 1997), whereas soil biochemists have investigated the problem through the study of the dynamics of specific nutrient pools (e.g., soil C, N, S, and P) (McGill and Cole, 1981; Sanginga, Mulongoy, and Swift, 1992; Gressel and McColl, 1997). By modelling decomposition and nutrient availability from experimental data, most notably with the Century model, soil biochemists are able to elucidate the problem areas of our understanding of OM decomposition and perhaps make some predictions based upon small perturbations to the cycling of some elements (Paul and Clark, 1996). A general scheme of the decomposition and mineralization-immobilization processes have been drawn from these studies, but more research is needed before we can properly evaluate the impact of different crop management strategies on the soil food web or predict their impact on nutrient release to crops from the soil organic pool.

Soil ecologists concur that most of the mineralization process is conducted by both free-living and insect-gut fungi and bacteria (de Ruiter, Neutel, and Moore, 1994; Zheng, Bengtsson, and Cgren, 1997). The large contribution of fungi, ascomycetes and other bacteria to soil OM cycling is due to their large biomass, rapid turnover and numerous enzymatic capabilities. Bacteria and fungi produce extracellular enzymes to cleave organic polymers into oligomers or monomers that can penetrate microbial cell walls or become immobilized through the action of active transport mechanisms. For microorganisms to be able to use the energy stored within OM, oxidation of organic carbon (C) compounds must take place intracellularly. It is thought that the need for C, and thus, energy drives the mineralization of nitrogen (N); however the fate of mineralized N is often incumbent upon the N economy of the organisms and soil as a whole. Carbon and N are intimately bound within soil organic compounds and N can be released when C is reduced to CO_2 during microbial respiration. Much of soil organic sulphur (S) is also C-bound but 30% to 75% of the total soil organic S exists as sulfate esters (C-O-S) and minor amounts as sulfonate (C-N-S) (Dail and Fitzgerald, 1999).

Because of their link to C, the mineralization rate of these elements depends somewhat on the C:N and C:S ratios of the soil OM being decomposed. Mineralization of C and, hence, of C-bound N and S takes place intracellularly for energy to be recouped by soil organisms through their electron transport chains. Since this process is intracellular, microorganisms may have first access to OM-derived N and S and can immobilize these elements to support growth. Owing to the elemental ratios of substrate and microbial biomass differences, the decomposition of crops residues with C:N ratios above 25:1 and C:S ratios above 400:1 results in no net mineralization of N and S, respectively.

The occurrence of organic S as an ester uncouples to some extent the mineralization of C and that of S (Paul and Clark, 1996; Stevenson, 1999). Phosphorus (P), on the other hand, always occurs in ester form in OM, and is second only to N as an essential nutrient for microbial and plant growth. The mineralization of S and P esters may occur outside of microbial cells through extracellular phosphatases and sulfatases, which are produced by organisms needing these elements. It is important to note that phosphatases and sulfatases are product-repressible enzymes and that their production is reduced or completely inhibited by high levels of available S and P in the soil solution. Consequently, in agricultural soils of temperate regions, phosphatase activity may contribute little to organic P mobilisation because P overfertilization is often practiced, particularly where animal manure is abundant (Coote and Gregorich, 2000).

In general, OM decomposition rates are faster in systems with more trophic levels (Zheng, Bengtsson, and Cgren, 1997). Soil fauna contributes only a small fraction of heterotrophic respiration, although separating faunal and floral decomposition can be difficult because insect gut flora may ultimately be responsible for the decomposition of recalcitrant OM. Hence, the indirect effects of soil fauna on soil OM decomposition, i.e., fragmentation of litter increasing surfaces for microbial attack, improvement of soil structure and control of soil microbial community, and even hosting of digestive track microbial populations, may be more important than their direct contribution. Using ^{15}N as tracer, de Ruter, Neutel, and Moore (1994) estimated that microbial predation by protozoa and bacteriovorous nematodes accounts for approximately 25% to 50% of total N mineralization. Such impacts of predators on N availability is impressive, considering their small biomass in soil and comparatively slow turnover times. The relatively large impact of predators on N release in soil stems from the high C:N ratio of protozoa and nematodes as compared to that of their prey. The N consumed in excess by predators fulfilling their C needs or that lost due to "messy eating" can be released to the soil as available N. Mineralization and immobilization of P are related to processes analogous to that of N (McGill and Cole, 1981). Despite differing enzyme systems making N and P utilizable, there is a highly significant correlation between the rate of N and P conversion to inorganic forms, N mineralization being 8 to 15 times the amount of mineralized P (Alexander, 1977).

Microbial Transformations of Inorganic Nutrients

Although 75% of our atmosphere is comprised of N, this large pool of N is unavailable to most living organisms. Only a few prokaryotes, commonly called N-fixers, possess the enzymatic system necessary to break the triple bond holding the two N atoms of atmospheric dinitrogen and to assimilate N from this source. In the process, N-fixers close the N cycle and bring the inert N_2 into a usable form. Nitrogen is fixed through the enzymatic system of N-fixers, through industrial fixation for fertilizer production, and to a lesser extent, through the action of lightnings. The annual production of biologically fixed N was estimated at 44-200 million tons, while the fertilizer industry produced 84 million tons annually (Brckman et al., 1990). Industrial fixation is a costly process, however, it accounts for 2% of worldwide total energy consumption and also has some negative environmental impacts. The use of N fertilizers can enhance soil acidification and nitrate (NO_3^-) respiration in soils, which is leading to the production of nitrous oxide, a greenhouse gas implicated in ozone layer depletion. Nitrification of ammonium fertilizers can also lead to NO_3^- leaching losses contaminating surface and ground water (Newton, 1999). Biological N fixation is powered by solar energy and believed to be better regulated and synchronized with the N needs of diazotrophs and symbiotic plant associates. Therefore, symbiotic N_2 fixation has fewer environmental impacts than industrial N fertilizer use as it produces moderate amounts of fixed N to fulfil the needs of N-fixers and perhaps that of their symbiotic associates (Triplett, 2000; Abdel-Aziz, 2001). This biologically fixed N, however, will be subjected to the same mineralization and environmental losses as fertilizer-derived N.

Biological N fixation by free living (Kennedy and Islam, 2001), associative (Boddey, 1987; Steenhoudt and Vanderleyden, 2000), and symbiotic organisms (Tajima et al., 2000; Unkovich and Pate, 2000; Wall, 2000) was extensively studied because of the large impact of this process on plant productivity. By far, the most important impact of biological N fixation on agricultural soils under temperate climates comes from symbiotic associations involving legumes. Leguminous crops can favour the input of considerable amounts of N in agricultural soil systems (Table 1). Most of this fixed N is used for host plant growth, but some of it is concurrently released into the soil (Hamel et al., 1991a), through plant root exudates or products of root, plant and microbial turnover. In plant mixtures incorporating legumes, a phenomenon of N sparing occurs in which more N may be available to non-legumes than the amount present in the fraction of the field occupied by them. In pastures, for example, grasses are more competitive for soil N uptake than their leguminous companions and mobilize larger amounts of soil N when they are grown with legumes than when they are grown in monoculture. The interspersed legumes must and do rely more on N_2 for their nutrition when their more competitive companion plants reduce soil available N levels (Hamel, Furlan, and Smith, 1992).

Soil available N (NH_4^+ and NO_3^-) represses the nitrogenase enzyme system responsible for N-fixation (Alexander, 1977) and, consequently, legumes use soil N preferentially to N_2. N fertilizer application reduces the contribution of biological N_2-fixation to crop nutrition (Schubert, 1995). In intensive production systems relying on fertilizers, the contribution of biological N_2-fixation is reduced.

In agricultural soils, chemoautotrophic bacteria oxidize ammonium (NH_4^+) into NO_3^-. These bacteria are collectively called "the nitrifiers" although this term refers to two distinct groups of organisms performing two distinct biochemical reactions. *Nitrosolobus* spp. and *Nitrosospora* spp. are the most common NH_4^+ oxidizers in soil leading to the produc-

TABLE 1. Amount of nitrogen fixed by the symbiosis formed between temperate leguminous plants and diazotrophic bacteria (Bøckman et al., 1990).

Plant	(kg N ha^{-1} yr^{-1})
Clovers	23-620
Alfalfa	164-300
Vetch and Beans	57-190
Peas	46
Lupins	128
Soybean	17-206

tion of nitrite (NO_2^-) while *Nitrobacter* populations may further oxidize NO_2^- into NO_3^- (Paul and Clark, 1996). Reduced S forms from organic or inorganic sources can also be oxidized to sulfate (SO_4^{-2}) via the metabolism of numerous chemoautotrophic and heterotrophic organisms. Nitrifiers enhances plant nutrition much more than do S-oxidizers because most mineral S is already oxidized in aerobic soils and much of the organic-S is already in the SO_4^{-2} form except for sulfhydryl groups of proteins. The N applied to agricultural soils through fertilization is generally a salt of NH_4^+, urea or anhydrous ammonia (NH_3). This N is largely nitrified before plant uptake, dependent on timing of the application, and other N demand. In agricultural systems, the product of N mineralization, NH_3, is also rapidly transformed to NO_3^- after biotic demands for inorganic N is met. Nitrification increases the mobility and hence, the availability of N but, at the same time, renders it more prone to leaching. In Quebec and Ontario, Canada, for example, over 40% of agricultural soils have more than 41 kg ha^{-1} of residual N (N in excess of plant and microbial demand) (Coote and Gregorich, 2000). A survey of 1200 Ontario farms revealed that 14% of wells were contaminated with NO_3^-, an indication of the importance of microbial oxidation of added N (Goss and Barry, 1995).

Nitrate is not only prone to leaching losses but it is also the electron acceptor used by denitrifying bacteria, which use it as a respiratory substitute in the absence of adequate oxygen. The products of NO_3^- reduction by denitrifiers are gases NO, N_2O, and N_2 and are further lost from the soil. Losses of available N from soil through denitrification can be important and commonly range between 0 to 30% of the N fertilizer applied to agricultural soil (Paul and Clark, 1989). Denitrification is influenced by several factors including temperature, rainfall, soil pH, bulk density, OM content, and C availability, in addition to NO_3^-level (Burton et al., 1997; Granli, and Brckman, 1994; MacKenzie, Fan, and Cardin, 1997).

Sulfur can also be lost after dissimilatory reduction of SO_4^{-2} by bacteria in an anaerobic respiration process (Howarth, Stewart, and Ivanov, 1992). This process is in many ways similar to denitrification, but it requires relatively low soil redox potential. Dissimilatory SO_4^{-2} reduction occurs in environments such as lake sediments and flooded soils, and is retarded by the availability of NO_3^-, as the latter is a more favorable electron acceptor wherein more energy is gained using it as a terminal electron dump (Paul and Clark, 1989). Nitrate is also comparatively abundant in agricultural soils and thus amounts of gaseous S lost from agricultural soil by SO_4^{-2} reduction, are likely to be small and unimportant.

Organic Acids and Mineral Dissolution

Plants are also reliant on nutrients that come available via processes other than OM decomposition and fertilization. Biologically important cations may also come from the dissolution of the mineral fraction of soils. In this regard, fine textured soils are more fertile in part because their mineral particles have more surface to volume ratio than coarser soils and therefore, more surfaces exposed to geochemical weathering. An additional important weathering factor is the production of organic acids in soils, which can lead to changes in cation availabilities (Ahumada et al., 2001; Jones and Wilson, 1985; Cornell and Schindles, 1987).

The importance of soil mineral dissolution by organic acids is now being recognized (Whitelaw, 2000). Organic acids are particularly important for the release of P from soil. The reversal of P sorption and the solubilization of stable P complexes by organic acids are particularly important for plant nutrition. Phosphorus, a nutrient required in large amount by plants, exist in soil primarily as insoluble or very poorly soluble inorganic forms including plant phytate (Paul and Clark, 1989). To overcome the P limitation imposed by the low P concentration in soil solutions, three mechanisms have been shaped through evolution: (1) mycorrhizal associations, in which plants benefit from the P-uptake efficient mycelium of their fungal associates that increases the volume of soil exploited by plants, (2) the release by plant roots and microorganisms of phosphatases, which are product repressible enzymes able to cleave phosphate from OM, and (3) the production of organic acids that directly and indirectly affects mineral P solubilities. These mechanisms, which are essential for the P nutrition of plants in undisturbed ecosystems, are also important for crop plants receiving P fertilizers. The phosphate ions released in the soil solution by fertilizers rapidly precipitate with other soil mineral components into insoluble forms (Stevenson and Cole, 1986). Hence, the impact of organic acids on the P fertility of soils is of great agronomic interest, particularly in soils in which P accumulated over decades of fertilization history to become an environmental threat (Coote and Gregorich, 2000).

Organic acids are metabolic products of all organisms. The main sources in soil are plant roots, bacteria, and fungi. They are abundant near decomposing residues and organic amendments such as manure and compost (Table 2). The organic acids found in the soil solution are

TABLE 2. Sources of Organic Acids in Soils

Sources	Organic acids	References
Plant roots	Citrate, malate, oxalate	Jones and Darrah, 1994; Jones and Brassington, 1998
	Acetate, aconitate, fumarate, glycolate, lactate, tartaric acid, succinate	Stevenson 1967; Zhang, Ma, and Cao, 1997
	Lactate, acetate, oxalate, succinate, fumarate, malate, citrate, isocitrate, aconitate,	Jones, 1998
Bacteria	Formic acid, acetate, propionate, butyrate	Stevenson, 1967
	Butyrate, malonate, lactate, succinate, malate, gluconate, acetate, glyconate, fumarate, adipic acid, 2-ketogluconate	Alexander, 1977; Moghimi and Tate, 1978; Moghimi, Tate, and Oades, 1978; Leyval and Berthelin, 1989
Saprophitic and ectomycorrhizal fungi	Citrate, oxalate	Sollins et al., 1981
Decomposing organic matter	-	Evans, 1998
Manure and composted manure	-	Baziramakenga and Simard, 1998

usually mono-, di-, and tricarboxylic acids containing unsaturated C and hydroxy groups (Strobel, 2001). The organic acids most commonly found in soils are pyruvic, citric, aconitic, isocitric, oxalic, ketoglutaric, succinic, fumaric, malic, tartaric, and acetic (Stevenson, 1967; Mo, 1986; Jones and Brassington, 1998). Several of these organic acids can carry a negative charge, which allows them to complex metals in solution and displace anions from the soil matrix (Jones and Brassington, 1998). Organic acids have been implicated in complexation and leaching of metal cations and they may be important in the detoxification of Al and other metallic elements in soil (Strobel, 2001). The release of organic acids from roots is a general mechanism for solubilizing metals from the soil's highly insoluble mineral phase and for mobilizing P, Fe, Zn, Cu, K, and Mn (Römheld and Marschner, 1986; Jones and Darrah, 1994; Strobel, 2001).

The mechanisms underlying nutrient mobilization by organic acids seem unrelated to soil acidification. Several experiments have shown that organic acid excretion by plants has negligible effect on rhizosphere pH (Peterson and Böttger, 1991) and, hence, chelation is more likely responsible for nutrient mobilization (Stevenson, 1967). The ability of organic acids to chelate metal cations is greatly influenced by their molecular structure and increases with the number of functional groups. Hence, tricarboxylic acids are most effective and monocarboxylic acids less effective (Bolan et al., 1994; Hu et al., 2001). Chelation of the metals contained in Ca, Al, and Fe phosphate minerals can results in P release (Kepert, Robson, and Posner, 1979; Coleman, Reid, and Cole, 1983). For example, citrate dissolves Fe phosphates through the formation of a ferric hydroxy-phosphate complex (Gardner, Barber, and Parbery, 1983) and addition of organic acids caused the dissolution of Fe and Al oxides by the complexation of these metal ions (Bolan et al., 1994).

In addition to chelation of P-associated metals, organic acid may improved the solubility of soil P through competition with the later element for adsorption sites. Organic acids are very reactive (Jones and Brassington, 1998) and are strongly bound onto P sorbing sites, making these sites unavailable for reaction with P (Parfitt, 1978; Jones and Kochian, 1996; Hu et al., 2001). Also, organic acids can react with free cations such as Ca^{2+}, Fe^{3+}, and Al^{3+} (Stevenson, 1967), hence competing with P precipitation with these cations.

Several species of bacteria, yeasts, actinomycetes, ectomycorrhizal, and free-living fungi are known for their ability to solubilize P-containing minerals. These P solubilizing activities of microorganisms have been the subject of a number of studies involving P sources such as Ca-phosphate (Kucey, 1983), Fe-phosphate (Jones et al., 1991), and Al-phosphate (Illmer, Barbato, and Schinner, 1995). Some non-mycorrhizal agricultural plant species such as white lupin (*Lupinus* sp.) (Dinkelaker, Römheld, and Marschner, 1989) and rape (*Brassica rapa* L.) (Hoffland, Findenegg, and Nelemans, 1989) are also known to affect P solubility. The impact of P-solubilizing microorganisms was well reviewed by Whitelaw (2000). Microbial solubilization of soil phosphates has been attributed to the excretion of organic acids. For example, Reyes et al. (1999) found that *Penicillium rugulosum*, grown in culture media, was producing gluconic and citric acids. Citric acid, an organic acid produced by fungi and by plants as well (Sollins et al., 1981), can release P from kaolin, geothite, ferric phosphate and Al phosphate (Swenson, Cole, and Sieling, 1949). Citrate, tartrate, oxalate and malate, organic acids produced by plants (Stevenson, 1967; Zhang, Ma, and Cao, 1997; Jones, 1998), fungi (Sollins et al., 1981), and bacteria (Moghimi and Tate, 1978; Moghimi, Tate, and Oades, 1978; Leyval and Berthelin, 1989) are able to desorb P from kaolinite and gibbsite (Nagarajah, Posner, and Quirk, 1970). Some plant species [e.g., alfalfa (*Medicago sativa* L.), rape] increase the amount of organic acid they excrete in the soil or change their composition in response to P deficiency. Zhang, Ma, and Cao (1997) observed that succinic and malic acids production by radish (*Raphanus sativum* L.) was 15 to 60 times larger under P deficient conditions. Different plant species may produce different organic acids to mobilize P from the different types of mineral in the environments to which they are adapted (Ström, Olson, and Tyler, 1994).

Soil microorganisms consume rapidly the organic acids released in the soil solution (Jones, Darah, and Kochian, 1996). Although it is unlikely that organic acids produced in soils would remain long enough to affect the P dynamic of the bulk of soil, organic acids are produced abundantly in localized strategic zones, such as the rhizosphere or the vicinity of decomposing residues, and may accumulate in quantities sufficient to produce functional P-solubilizing microenvironments, or be mobilized by leaching in excess of microbial uptake. High concentrations may be reached in localized zones where biological activity is intense (Stevenson, 1967). The rhizosphere is a zone where P-solubilizing microorganisms are particularly abundant (Whitelaw, 2000) and the amount of organic acids in the immediate vicinity of plant roots have been found to be much higher than that in the bulk of soil (Hu et al., 2001). This strategic localization of P dissolution by small accumulation of organic acids may be another expression of the efficiency of natural processes where little resource investments produce large returns. Furthermore, organic acids are an important C and energy sources for a large number of rhizosphere bacteria and fungi (Sollins et al., 1981; Whitelaw, 2000), which constitutes an important fast-turnover-rate nutrient pool that is particularly important in plant P uptake.

NUTRIENT MANAGEMENT BY PLANTS

Soils are complex, diverse but organized systems shaped by abiotic and biotic influences, plants, fauna, and microorganisms with which co-evolved over the course of millions of years (Cairney, 2000). Plants are far from being passive players that absorb the soil mineral nutrients released through the activity of other soil organisms. Under evolutionary pressures selecting for soil resource use efficiency, plants have become the driving force behind nutrient dynamics in soil. In fact, plant photosynthesis is essentially the basis of energy maintaining the pedosphere populations.

Nutrient Uptake by Plants

Most plants do not solely have roots; rather they have mycorrhizae. Derived from a Greek word 'mycorrhiza' literally means "fungal root." Mycorrhizae are the products of the symbiotic relationship between plants and mycorrhizal fungi to improve a plant's nutrient and water extraction efficiency, while providing the fungi with a consistent and adequate C supply. Most plants possess the endomycorrhizal "arbuscular" type of mycorrhizae. In this association, the fungal symbiont develops an extensive network of microscopic hyphae within the uppermost soil layer (Figure 1). These mycorrhizal fungi establish connections with plant symplast through symbiotic interfaces located at the level of plant cytoplasmic membrane (Strullu, 1986). Plants feed C to their associated fungal symbionts, usually at "tree-like" arbuscules within the plant cell. In return, the enlargement of the root systems' absorptive surfaces enhances plant nutrition and water uptake and plant health is improved

FIGURE 1. Survival of the roots of a range of agricultural species assessed using the mini-rhizotrom method. Reprinted from *Applied Soil Ecology* vol. 15, Anonymous, The beneficial rhizosphere: a dynamic entity, pp. 99-104, 2000, with permission from Elsevier.

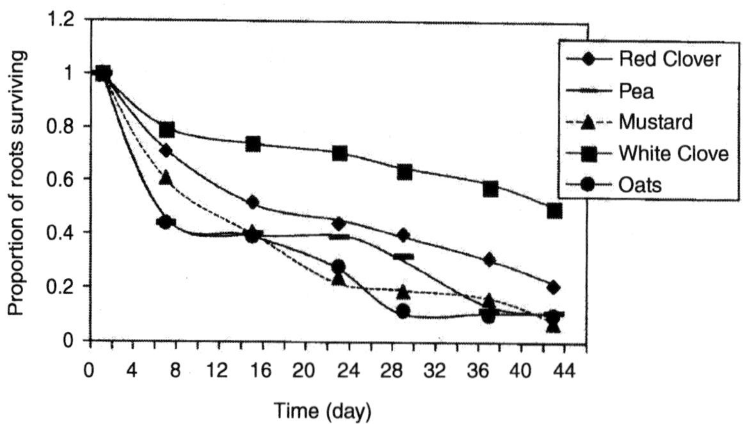

(Smith and Read, 1997). Fungal development in the arbuscular mycorrhizal (AM) symbiosis is quite extensive. Fungal development inside (intraradicular) and outside (extraradicular) the roots is important and was shown to represent 20% of the dry biomass of the mycorrhizae (Bethlenfalvay, Pacovsky, and Brown, 1982).

In the majority of plant species, the extraradical fungal hyphae of the mycorrhizae protruding through the epidermis of fine roots and extending in the soil, may be seen as a last order of root system branching. Numerous greenhouse experiments using inoculated mycorrhizal plants and non-mycorrhizal controls have shown that the contribution of mycorrhizae-mediated plant nutrition improvement is necessary for optimal plant development and yield. Most importantly, mycorrhizae fungal extensions facilitate plant uptake of inorganic P and other plant nutrients of low mobility in soil (Rhodes and Gerdemann, 1975). Root absorption of P is generally faster than P diffusion and replenishment at the root surface, and a P depletion zones rapidly form around absorptive roots. Improved exploitation of the soil volume by mycorrhizal hyphae extending beyond these P depletion zones is probably the major contribution of AM fungi to plant growth (Marschner, 1995). Phosphorus is absorbed from the bulk of soil by mycorrhizal hyphae and translocated within the fungi to plant cytoplasm through a symbiotic interface located at the level of root cortical cell plasmalemma (Smith and Smith, 1990). Hence, the hyphal network of mycorrhizae decreases the distance for diffusion of sparingly mobile nutrients in soil such as P, Cu, and Zn (Bolan, 1991; Liu et al., 2000a). This is important, especially when the availability of these nutrients in soil is low.

Research has shown that the AM symbiosis can also enhance field crop uptake of nutrients carried mainly by mass flow, such as K, Ca, Mg, when the availability of these nutrients and P level are low (Liu et al., 2002). The extent of mycorrhizae development is regulated by P availability (Koide and Li, 1990). Plants invest less C in the maintenance of fungal symbiont under condition of P abundance such that hyphal extensions are not required for P uptake as was the case for leguminous plants under N-fertilization. However, when available P is abundant, other mycorrhizae-derived benefits, such as uptake of other plant nutrients or water, are not expressed. Mycorrhizal development also varies with crop species and genotypes (Daft, 1991; Estaun, Calvet, and Hayman, 1987; Smith, Robson, and Abbott, 1992; Liu et al., 2000b). Plants with extensive root development would rely less on hyphal extension for uptake.

The great importance of mycorrhizae to crop production is not always fully appreciated. The value of the AM symbiosis is sometimes underestimated because field-grown plants often do not respond to mycorrhizal inoculation. The absence of a response often indicates that an indigenous mycorrhizal flora already contributes successfully to plant growth and that inoculation is not necessary (Hamel et al., 1997). Arbuscular mycorrhizal hyphae are not only abundant in soil of undisturbed ecosystems, but they also dominate agricultural soils (Hamel et al., 1991a; Olsson et al., 1999). Mycorrhizal fungi not only improve crop nutrition, but also interact in numerous ways with soils and soil microbes in the soil zone termed 'mycorrhizosphere *sensus*' (Linderman, 1992). Mycorrhizae and soil microbes in general may increase soil aggregate stability through the production of exogenous compounds that glue soil particles together. The abundance of mycorrhizal fungi in soil and their impact on other soil inhabitants within the mycorrhizosphere is another expression of plant influence on the soil system.

In the light of all this, it appears that the management of the mycorrhizal symbiosis in crop production is possible. It would allow optimal crop production with lesser amounts of fertilizers and hence improve the sustainability of cropping systems.

Nature of the Mycorrhizosphere

Plants mobilize nutrients from the mineral and the organic phases of the soil through the production of organic acids and phosphatase enzymes, and they store nutrients in above and below-ground biomass. In association with their mycorrhizal fungal associates, plants also have a large impact on soil organisms and influence soil processes at both the macro- and microscale. They also profoundly influence soil conditions through respiration and exudation and, in doing so, influence the activity of other soil organisms. Plants stimulate soil organisms and, in turn, benefit from the activity of these organisms that impact upon the release of plant available nutrients into the soil solution. The mycorrhizosphere is dominated by interactions.

Some of the C and energy invested into the soil milieu by plants may either be released to soil or transferred through symbiotic interfaces to microbial partners. Plants commonly loose 10% to 30% or their photosynthate-C into the soil through passive and active root excretions, root cap cells sloughing off, and epidermal cell senescence (Bowen and Rovira, 1999). This constitutes considerable C input to soil. In corn (*Zea mays* L.), for example, 24.4% of total fixed C that is lost to soil translated into 9.91 Mg ha^{-1} dry mass rhizodeposition in a corn crop producing 12.74 Mg ha^{-1} above ground dry mass and 1.64 Mg ha^{-1} below ground dry mass (Molina et al., 2001). This input of C in soil stimulates the proliferation of soil organisms in the vicinity of roots and creates what is known as the "rhizosphere effect."

Some more intimate root-microorganism associations have developed highly regulated and efficient mechanisms through which plant photosynthate is channelled directly to specific microbial associates. This is the case with the AM symbiosis (Nakano, Takahashi, and Kimura, 1999), the most widespread of these associations. The amount of photosynthesis-derived C channelled to AM fungi could be as high as 40% to 50% of net plant production (Harris and Paul, 1987). Part of the C invested in mycorrhizal biomass is made available to the microbial community not only upon senescence of these fungi, but also by way of these "living interactions." For example, the presence of actively growing mycorrhizal hyphae enhanced *Fusarium* growth and reproduction on gellifyed water, suggesting that this fungus could utilize exudates from mycorrhizal hyphae (St-Arnaud et al., 1995).

The soil zone under the influence of plants is significantly enlarged by the fungal extensions of mycorrhizal plant roots systems. These hyphae can extend into the soil several centimetres beyond the root surface. Since the AM fungi are normally abundant in soil and are tapping directly plant-fixed C, they constitute a major input of C and energy to soil. This C and energy is distributed throughout the soil of the rooting zones (Jakobsen and Rosendahl, 1990; Finlay and Söderström, 1992) where it is used by soil animals and microbes, including fungal-feeding and predatory mites and nematodes (Smith, Hartnett, and Rice, 2000). Arbuscular mycorrhizal hyphae appear to have a complex interaction with other soil organisms. They can influence soil organisms, including those involved in OM mineralization in the vicinity of plant roots and soil-born pathogens. Arbuscular mycorrhizal fungi are not the only plant root symbionts influencing soil organisms. For example, *Rhizobium* sp. has induced systemic resistance to infection by the potato cyst nematode (Reitz et al., 2000), and reduced root galls and nematode reproduction (Siddiqui and Mahmood, 2001).

The large amount of organic material released by plant root systems and their associated mycorrhizal fungi is used by soil organisms, which recycle the nutrients they contain (Bowen and Rovira, 1999; Linderman, 1988). This C stimulates preferentially the proliferation of certain organisms in the mycorrhizosphere (Filion, St-Arnaud, and Fortin, 1999; Green et al., 1999). The specific microbial communities developing in the mycorrhizospheres appears to be both plant (Westover, Kennedy, and Kelley, 1997) and AM fungi specific (Andrade, Linderman, and Bethlenfalvay, 1998). In turn, rhizosphere organisms influence mycorrhizal roots and whole plant development (Kim, Jordan, and McDonald, 1998). They also stimulate the release of carbonaceous compound to the soil by roots.

The multiple interactions taking place between soil organisms, including mycorrhizae, are extremely complex but the high level of redundancy found in soil communities allows us to draw generalizations. It appears that the mechanisms subtending the impacts of rhizosphere organisms on plant growth are primarily indirect and relate to nutrient mobilisation while others, like antibiotic or plant growth factor production are more direct (Marschner, 1998).

Nutrient Mobilization in the Mycorrhizosphere

Considering the large amounts of nutrient efflux from plant roots to the mycorhizosphere, reabsorption is clearly an important step in the nutrient cycle in this soil zone. Retrieval of amino acids seems to be important, revising our previous notions that plant uptake be limited to inorganic nutrients (Marschner, 1998). Nutrients are also retrieved by plants root systems after rhizosphere organisms have consumed the C and released mineral constituents from various organic molecules making up root exudates, secretions and lysates (Bowen and Rovira, 1999). The involvement of mycorrhizal fungi in the retrieval of N previously lost by roots was demonstrated using ^{15}N (Hamel et al., 1991b). The relative importance of direct reabsorption of amino or organic acids contained in rhizodeposition by roots as compared to the amount of rhizodeposition-derived N reabsorbed by roots, however, is still uncertain.

Plant roots, mycorrhizal fungi as well as other soil microorganisms are involved in the uptake and mobilization of P bound to the mineral fraction of the soil solids, or in their organic fraction. The contribution of AM fungi on plant P nutrition is thought to be more of a physical nature than through phosphatases and organic acid release in soil. Plants are considered to draw on the same available P pool whether or not they are mycorrhizal (Sanders and Tinker, 1971; Barea, 1991). No published study has demonstrated that AM fungi may exude organic acids to increase P availability. However, recent and yet unpublished results suggest that significant mycorrhizae-induced modifications in organic acid release, in the root zone of field-grown corn, is a component of the mycorrhizal effect (Landry, 2002).

The direct or indirect involvement of AM in the solubilization of mineral would explained the observation that AM plants can respond readily to additions of sparingly soluble P sources such as rock phosphate (Tinker, 1980; Barea, Azcon, and Hayman, 1975; Manjunath, Hue, and Habte, 1989). Further evidence for a synergistic effect of AM fungi on the solubilization of P-containing minerals come from experiments where dual inoculation with AM fungi and phosphate solubilizing bacteria or fungi stimulated plant growth better than inoculation with either microorganisms alone (Azcon, Barea, and Hayman, 1976; Raj, Bagyaraj, and Manjunath, 1981; Piccini and Azcon, 1987). For example, the mixed inoculation with *Aspergilus niger*, *Bacillus mobilis*, and *Glomus fasciculatus* increased the P content of onion (*Allium cepa* L.) to a greater extent than inoculation with either microorganism alone (Manjunath, Mohan, and Bagyaraj, 1981). In another greenhouse experiment, simultaneous inoculation of tomato (*Lycopersicon esculentum* L.) growing in hydroxyapatite-amended substrate with a phosphate-solubilizing bacteria, *Enterobacter agglomerans*, and by a mycorrhizal fungi, *G. etunicatum*, produced a synergistic effect on plant N and P uptake (Kim, Jordan, and McDonald, 1998). In this case, dual inoculation effect on plant nutrient uptake was concurrent with a marked increase in oxalic, 2-keto-P-gluconic, and citric acid concentrations in the plant root zone. It is unclear whether the changes in organic acid release can be attributed to the mycorrhizal fungus, to the modification of plant root excretions by mycorrhizal colonization, or to the stimulation of those soil microorganisms, which are particularly effective in P solubilization. However, it is likely that mycorrhizal fungi improve the recovery of the P solubilized by organic acids.

The interaction between AM fungi, plants and other soil organisms is further complicated by the level of soil available P. Mycorrhizae development, the release of organic acids and the release of phosphatase enzymes by plant roots and microorganisms of the mycorrhizosphere are reduced by high levels of available P (Marschner, 1998). Resource investment by plants and other living organisms is proportional to the extent of their needs, i.e., plants with better P nutrition excrete less organic acids and phosphatases. This effect of mycorrhizae-mediated improved plant phosphorus nutrition may reduce the uptake of metallic elements. The uptake of Mn was reduced in mycorrhizal soybean (*Glycine max* L. Merr.), which in this way, alleviated Mn toxicity (Bethlenfalvay and Franson, 1989). The reduced uptake of Mn by soybean was attributed to reduced root excretion by mycorrhizal plant in this study. Liu et al. (2000a) also observed reduced uptake of Mn and Fe by mycorrhizae in corn grown under high P availability. High level of P fertilizer increased the uptake of these metallic nutrients suggesting the occurrence of an uptake-selective mechanism for mycorrhizal plant absorption of Mn and Fe when the availability of these nutrients is high. Hence, it appears that the mycorrhizal symbiosis not only increases plant nutrient uptake but also contributes to the maintenance of optimal plant nutrient balance.

Interactions in the mycorrhizosphere are numerous and can also be influenced by N availability. For example, Reyes et al. (1999) found in a petri dish experiment, that the N nutritional economy of *Penicillium rugulosum* affects its phosphate solubilization activity. *Penicillium rugulosum* grew better on NO_3^- than on NH_4^+ when relatively insoluble P sources were used. Similar results have been reported for other fungi like the ectomycorrhizal fungus *Hebeloma cylindrosporum* (Scheromm, Plassard, and Salsac, 1990). Cunningham and Kuiack (1992) also reported that oxalic and citric acids were the major organic acids produced by *P. bilaii* when the N source was NO_3^-, although neither was detected when the N source was NH_4^+. Hence, it seems that the form of available N influences organic acid production by soil organisms.

Biological N fixation constitutes a very important input of N to temperate crops. The AM fungi also positively influence biological N fixation. The mycorrhizal symbiosis is well known to increase the amount of N fixed in nodulating legumes (Barea and Azcón-Aguilar, 1983; Haselwandter and Bowen, 1996), nodule number, nodule size (Koffa and De la Cruz, 1995; Zhang et al., 1995), plant N content and protein yield (Hamel, Smith, and Furlan, 1991). The stimulation of N fixation in mycorrhizal legumes is generally attributed to improved P nutrition of the host plant leading to increased growth and higher N-demand concurrent with improved photosynthate supply to nodules.

Mycorrhizal fungi have also been observed to stimulate non-symbiotic N fixers such as *Azospirillum* (Biró et al., 2000). The basis for mycorrhizal enhancement of N fixation in associative systems is not well understood. It may be related to the fact that some soil organisms are favoured by the presence of AM fungi while other are negatively affected (Andrade, Linderman, and Bethlenfalvay, 1998). Arbuscular mycorrhizal fungi could give a competitive advantage to diazotrophes or relieve energy constraints upon N-fixation. This differential effect of mycorrhizal fungi on soil organisms could be directly related to change in the quality and quantity of root exudates (Schwab, Menge, and Leonard, 1983), or to other change in the soil environment under the influence of mycorrhizal hyphae (Li, Marschner, and George, 1991). The production of substances by AM fungi was found to stimulate the germination of *Trichoderma harzianum* conidia and the growth of *Pseudomonas chlororaphis* spores, to inhibit conidial germination of *Fusarium oxysporum* f. sp. *chrysanthemi*, while they had no effect on *Clavibacter michiganensis* subsp. *michiganensis* (Filion, St-Arnaud, and Fortin, 1999).

Microorganisms as a Labile Nutrient Pool

Non-biomass soil OM is a large pool of plant nutrients. Another and, perhaps, more important nutrient source is the biomass pool. Using microbial biomass (SMB) data reported by Olsson et al. (1999) and based on the fact that there is 7.5% of N, 1.5% of P, and 0.2% of S in SMB (Paul and Clark, 1996), it can be estimated that there was 60 to 110 kg N, 12 to 22 kg P, and 1.6 to 3 kg S stored in the SMB ha^{-1} of a linseed (*Linum usitatissimum* L.) field in the Olsson et al. (1999) study. This figure suggests that the SMB pool is a pool that can fulfil an important part of crop nutrient needs, and in addition it must be noted that this soil "nutrient" pool has a comparatively rapid turn over rate. For example, Frossard et al. (2000) reported fluxes of 11 to 190 kg P ha^{-1} yr^{-1} through a SMB pool that contained 17 to 290 kg P ha^{-1}, in fields where herbage exportation amounts 2 to 11 kg P ha^{-1} yr^{-1}. This very dynamic SMB-P pool is considered particularly important to crop nutrition because, as mentioned earlier, phosphate is somewhat insoluble and has a strong tendency to bind chemically to soil mineral matter (Gressel and McColl, 1997). These chemisorption losses are reduced when P is held in biomass pools. Ultimately, the contribution of the SMB pool to crop nutrition will depend on how well nutrient release from SMB corresponds to crop uptake of nutrients, and is an area of continued study.

Organic P accumulates in soil in the form of phytate, a plant-P storage molecule mainly found in seeds, which is somewhat resistant to mineralization (Taranto, Adams, and Polglase, 2000). It is known that mineral P is released from phytate through the enzymatic action of phytases (Bishop, Chang, and Lee, 1994). But since phytate reacts with Fe to form Fe-phytate in soil, it appears that organic acids could facilitate the enzymatic release of P from phytate in soil. Corn-soybean meal-based pig feed pretreatment including citric acid and microbial phytase did not improved P availability to the animals more than did phytase pretreatment alone (Radcliffe, Zhang, and Kornegay, 1998). The absence of synergistic impact of organic acid and phytases on phytate-P mineralization in animal guts, however, does not mean that there is no such impact on phytate-P mineralization in soil. Further studies are needed to clarify the question.

Maintenance of a Favorable Soil Environment

Soil processes and nutrient cycling are largely controlled by the conditions of the soil environment. Microorganisms modify their environment particularly through their involvement in soil macroaggregate formation and stabilization, and in turn are influenced by crops and climate factors. Large contributions in biomass, microbial exudates, and rapid turnover rates of microorganisms influence biochemical transformations in soil and, hence, nutrient availabilities to plants. According to Puget, Chenu, and Balesdent (2000), soil magroaggregates (aggregates larger than 200 μm diameter) are formed from the clumping of microaggregates around organic debris from roots and other sources and their associated decomposers, which all act as cementing agents. Arbuscular mycorrhizal hyphal network also contribute to macroaggregate stabilization (Degens, Sparling, and Abbot, 1996) through the enmeshment of microaggregates, hence, creating a "skeletal-like" structure to macroaggregates, and by the creation, along with plant root excretions, of the conditions conducive to the formation of macroaggregates (Miller and Jastrow, 1992). Small and often ignored mycorrhizal hyphae are very abundant in soil and can account for 25% of total soil biomass of agricultural soils (Hamel et al., 1991b; Olsson et al., 1999). These hyphae have a fast turnover in the order of days to weeks and constitute the organic matrix around which macroaggregates can form. Arbuscular mycorrhizal fungi have been shown to produce in abundance a glycoprotein recalcitrant to decomposition that has been well correlated to soil aggregate stability (Rillig et al., 2001).

Macroaggregates are enriched in young organic C (Angers and Giroux, 1996) and the process of packaging soil particles into aggregates is sometimes seen as a means by which soil OM can be protected from oxidation (Miller and Jastrow, 1992). While it is not certain whether the enrichment of macroaggregates can be interpreted as evidence for OM preservation in macroaggregates, it seems that OM in the first stage of decomposition holds macroaggregates together before being redistributed to finer non-aggregated material (Puget, Chenu, and Balesdent, 2000). It is quite clear, in contrast, that the maintenance of a good soil structure favours diffusion of O_2 in soil and, hence, microorganisms and plant roots growth. In a positive feedback loop, this enhanced biological activity favors soil structure stabilisation. Well-aerated and biologically active soil conditions, in turn, favor the most plant available states of nutrient such as N, S and most micronutrients. This favors uptake of N and S by plants and maintains the availability of metallic micronutrients below toxic levels. Good soil structure also favors plant biomass production and enhances OM decomposition.

SPATIAL AND TEMPORAL VARIATION IN NUTRIENT RELEASE

Nutrient Cycling in the Mycorrhizosphere

Rhizosphere resources are transient both in time and space. Rhizodeposition are localized near the root tips. Because of constant root elongation and rapid turnover, which changes continuously the point of C release in soil, and because the factors triggering plant response are sometimes transient, the mycorhizosphere is a very dynamic environment. Rhizosphere microbial populations along wheat (*Triticum aestivum* L.) roots have been described as "travelling waves" responding to the forward movement of root tips, which can also be described as soil C sources on the move (Zelenev, van Bruggen, and Semenov, 2000). In the model system of Zelenev, van Bruggen, and Semenov, (2000), the first bacterial wave peaked at 15 cm from the root tip. Successions of bacterial populations then come to bloom and use, in turn, the resource freed in soil by the death of the previous cohorts, up to exhaustion of the plant-derived C input. Fast growing bacteria, stimulated by the abundance of available C behind the root apex, colonize the zone of new growth (Lee and Pankhurst, 1992). They multiply as long as N does not become limiting. Predators are attracted by the CO_2 released by bacterial activity. Because predator's respiration requirement impose the release of a continuous flow of CO_2 and since their C:N ratio is about the same as bacteria, they excrete close to two third of the N contained in their prey, creating a "wave" of N at a more distal zone of the root. The root tip is also free from endomycorrhizal fungi, which commonly enter the root at the root hair level (Smith and Read, 1997). Mycorrhizal hyphae running along plant roots (Friese and Allen, 1991) could help retrieve the mineral nutrients liberated by successive bacterial waves in distal zones of the rhizosphere.

Mycorrhizal hyphae and crop roots are also transient in nature. A mini-rhizotron technique has revealed short root longevity, the span of which varied substantially with crop species (Anonymous, 2000). Often, root buds start to develop then rapidly disappear. This could be an important source of "rhizodeposition." Eighty percent of white clover (*Trifolium repens* L.) roots survived longer than seven days while in contrast, over 50% of pea (*Pisum sativum* L.) and oats (*Avena sativa* L.) roots had senesced during this period (Figure 1). Associated fine roots in forests may be much longer lived (Gaudinski et al., 2000). The minirhizotron technique was used to assess the longevity of fungal hyphae under mycorrhizal forest plants and which were presumed to be mycorrhizal. The majority of these hyphae survived less than seven days (Atkinson and Dawson, 2000). It can be inferred that microbial nutrient recycling intensity is likely crop-dependant. Recycling activities occurs preferentially in the vicinity of root and mycorrhizal hyphae, and at locations where root and mycorrhizal hyphae have existed in the recent past.

Seasonal Variation in Nutrient Uptake by Crop Plants

The production of annual crops imposes particularly marked seasonal variation within soils. In the spring, there are large amounts of decaying roots and aboveground plant residues in soil, but very few living roots. The nutrient uptake capacity of crops increases with mycorrhizal development. Kabir (1997) investigated the spatial distribution of mycorrhizal hyphae in two Canadian agricultural soils; in a corn field, most hyphae were located in the top 25 cm of the soil surface (Figure 2; Kabir et al., 1998b). Hyphal abundance followed not only the spatial distribution of corn plants but, as might be expected, also varied with plant development stage, being less abundant in the spring, maximum in mid summer and reduced at harvest (Figure 3; Kabir et al., 1998a). Maximum internal and external fungal development occurred at mid-season when plant biomass accumulation and, hence, nutrient needs are larger. It can be concluded from these studies that the capacity of crop nutrient extraction systems varies spatially, being more intense in the rows than between the rows, and seasonally, being most important at the reproductive stage of corn.

Dynamics of Nutrient Release from Soil

Nutrient release from soil also varies seasonally. Soil conditions are never static and soil biological activity is continuously fluctuating in response to changing soil and climatic conditions. The abundance and ac-

FIGURE 2. Variation in the abundance of fungal hyphae associated with corn, in a St-Benoît sandy loam under conventional tillage practice. Light shade, metabolically active (iodonitrotetrazolium reduction) hyphae; Dark, non-active hyphae. Reprinted from *Mycorrhiza* vol. 8, Kabir et al. Vertical distribution . . ., pp. 53-55, 1998 with permission from Springer-Verlag.

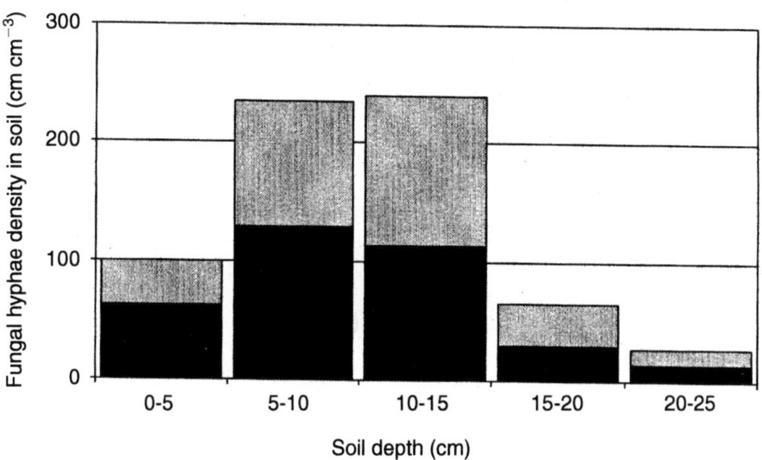

FIGURE 3. Variation in the abundance of metabolically active (iodonitrotetrazolium reduction) hyphae with time in the growing season and with placement relative to corn rows, in a St-Benoît sandy loam. Values are averages of 2 years of measurement. Reprinted from *Agriculture, Ecosystems and Environment*, vol. 68, Kabir et al. Dynamics of mycorrhizal . . ., pp. 151-163, 1998 with permission from Elsevier.

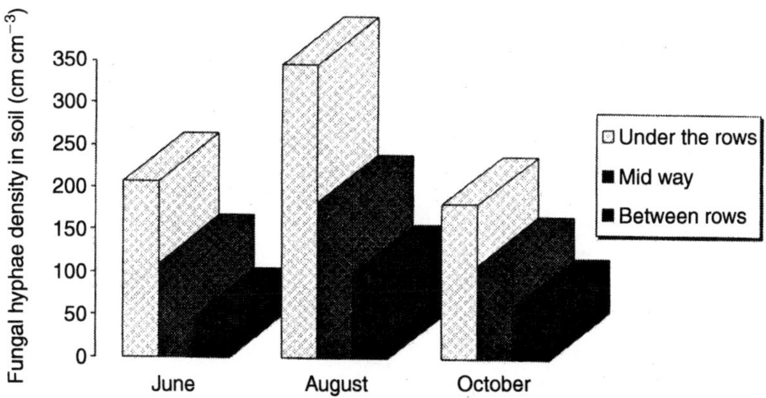

tivities of soil organisms, which are fundamental to the breakdown of OM and to the mineralization and immobilization of nutrients, are influenced by various environmental factors (e.g., soil type, temperature, pH, moisture, salinity, dissolved nutrients, predation and competition) and plant factors (species, age, biomass and health) (Grayston et al., 1998).

Nutrient dynamics in temperate crop production systems is influenced by seasonal fluctuations in SMB, which is directly involved in the turnover of OM and nutrient cycling according to specific climatic conditions (McGill et al., 1986). Ultimately, nature provides seasonal fluctuations in temperature and moisture, govern microbial activity, and thus the rates of decomposition and nutrient availability. Soil microbial biomass, which is sometimes used as a proxy for the nutrient cycling capacity of a soil, can increase in summer when conditions are favorable to biological activity. For example, Lynch and Panting (1980, 1982) found that at 0-15 cm depth in wheat fields, SMB-C remained constant from autumn to spring, increased to a maximum in the summer and then declined with declining temperatures. Murphy, Fillery, and Sparling (1998) found that drought modified the effect of temperature on SMB. Microbial biomass-N and mineralizable N under wheat fluctuated during the growing season, with biomass N levels at the end of the season being half of that found at the beginning of the growing season. SMB was lower during the summer when compared to winter, when desiccation was high. It can be concluded from these studies that fluctuations in SMB pools are driven by changes in the availability of mineralizable substrates and soil conditions, especially soil water content, and that the soil cycling of C and N may be somewhat uncoupled.

Joergensen, Meyer, and Mueller (1994) found that SMB-C and SMB-N increased from a December minimum to a peak in October, with biomass N showing more temporal fluctuation during the season and slight variations in SMB-C to N ratio. The decomposition of organic materials in soil is considered the result of successional populations of soil organisms (Lee and Pankhurst, 1992). Primary saprophytes would first utilize readily available, soluble constituents of plant residues. Cellulolitic organisms and those able to degrade complex organic molecules such as lignin would then appear, followed by microbial predators, amoeba and nematodes, which are attracted by microbial CO_2 release or other cues. Successional populations may differ in elemental composition, and this may explain some of the seasonal variation in the ratio of C to other elements in SMB.

It seems that the supply of P and perhaps that of S to crops under temperate climate can be sometimes buffered by SMB. While Patra et al. (1990) found no unequivocal evidence for seasonal changes in biomass C, N, and P within continuous wheat and grass fields, on a clay loam soil, He et al. (1997) found that SMB-C, SMB-S and SMB-P varied considerably with season (March to December), reflecting the potential for biomass to release or immobilize S and P at different times of the year. The authors concluded that biomass C was closely related to inputs of fresh OM while biomass P was more sensitive to soil moisture deficits, markedly decreasing between July and October mirroring a decline in soil moisture content. It was suggested that while biomass was turning over, it was not decreasing in size, and that moisture deficits may reduce the diffusion and, hence, the availability of phosphate to microorganisms. This favors microorganisms that have a limited capacity for immobilizing P. When the moisture deficit was eliminated later in the growing season (October), biomass P decline ceased in all experimental plots.

According to He et al. (1997), the impact of low rainfall in mid summer would be compounded by the vigorous crop growth at this time that would be expected to exacerbate the soil moisture deficit by evapotranspiration, but also compete with soil microorganisms for P. When this occurs, SMB turnover would lead to a lower biomass P level. This is in agreement with Perrott, Sarathchandra, and Dow (1992) who found that SMB-P was lowest in the spring and autumn, and significantly correlated with soil moisture. Release of P from OM and microbial biomass occurred during periods of rapid pasture growth and could account for P uptake by the pasture at those times.

Reduction in SMB-P could result in high available P flux in soil. Simard, Cambouris, and Nolin (2000) studied the seasonal variation in available P in a fine-textured soil under corn in Eastern Canada, using anion exchange membranes (AEM), and found that AEM-P was low in spring, peaked in mid summer (July) and decreased to a minimum in fall (September) (Figure 4). These authors did not report soil moisture conditions or soil water flux, but concluded that the variation in available P observed followed soil temperature. Similar results were previously reported for other soil inorganic P fractions (Magid and Nielsen, 1992).

Ingham et al. (1986) explained the observed seasonal changes in available N level and in the population dynamics of various groups of soil organisms living in a non-cultivated soil, by simple predator-prey interactions. In the spring, populations of bacteria and fungi increased as did those of their consumers, protozoa and nematodes, and then the

FIGURE 4. Seasonal variation in anion exchange membrane P in a fine-textured Humaquept soil growing maize in St-Marc (Qc) Canada, from May to the end of September. (Modified from Simard, Cambouris, and Nolin, 2000.)

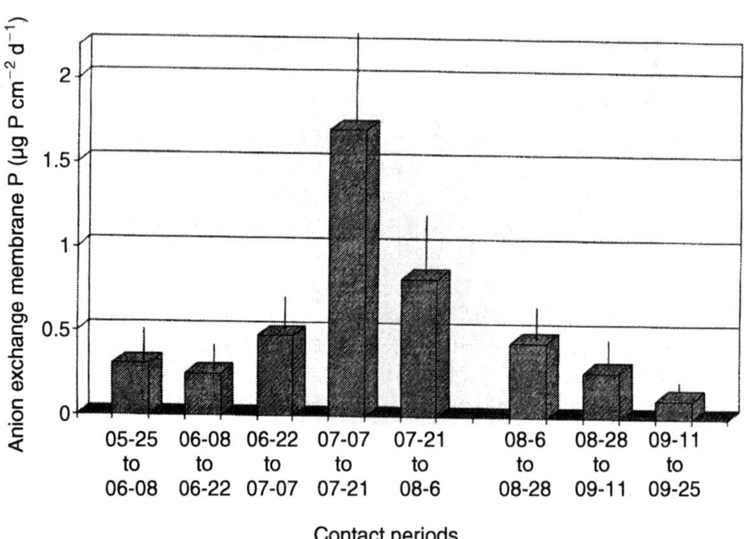

biomass of the microflora decreased and inorganic N levels increased. In the fall, bacterial population size increased while inorganic N levels decreased. Similarly, the addition of amoebae and bacteria resulted in greater release of available N in soil than bacteria alone.

Agricultural soils receive abundant N fertilization. For example, in Eastern Canada, 120 to 170 kg ha^{-1} of fertilizer-N is recommended to grow a corn crop (CPVQ, 2000). In comparison, a soil with 3% OM would release 33 to 132 kg of N to the crop through N mineralization, as per Alexander's (1977) estimates for temperate latitudes. In accordance with this figure, Tran and Giroux (1998) found that half of the N taken up by corn grown in Eastern Canada was derived from fertilizer. In this study, the recovery of N determined from ^{15}N-labelled fertilizer recovery by corn, varied from 47 to 51% of the amount applied.

The seasonal variation in available N in agricultural soils is dominated by the effect of N fertilization and timing thereof. This is partly due to the high reliance on N fertilizer for crop production and because, in contrast to phosphate, the N brought in the soil solution by fertilization reacts very little with soil constituents and largely remains in soluble forms in soil, mainly as NO_3^-. This soluble N form can be subjected to leaching and denitrification losses or living organisms can immobilize it. Hence, soil available N level rises in soil after fertilization in spring and slowly decreases afterwards (Figure 5) due to crop uptake, leaching and denitrification losses. The fate of soil N is highly dependent on moisture. Denitrification requires reducing conditions and is highly dependent on soil moisture. Typically, denitrification is highest in early summer because of high NO_3^- levels and moisture in the soil (Figure 6). Transpiration by annual crops is very low and rainfalls are more frequent in spring. Denitrification rates usually decrease in mid summer with increasing evapotranspiration and decreasing soil moisture, especially when precipitation events are scarce, and increase slightly in fall along with soil moisture. Tran and Giroux (1998) measured annual losses of N-derived fertilizers ranging from 13.6% in a soil with high capacity for N immobilization to 44.1%, in a silt loam. Fertilizer-derived N remaining in the soil in fall varied from 1.4% of the total N applied in a wet growing season, to 20.6% in a dry growing season. Immobilization of N fertilizer in the soil organic pool amounted to 16.3%, 24.6%, and 38.9% in sandy soils, in a silty loam and a clay loam.

CONCLUSIONS

The fact that the dynamics of soil nutrients under temperate climates is largely driven by soil organisms and plant roots cannot be ignored.

FIGURE 5. Seasonal variation in nitrate concentration in the top 0-20 cm of a sandy loam soil in St-Emmanuel (Qc) Canada, from May to October 1997. Nitrogen fertilizer was applied on May 23 at a rate of 160 kg ha^{-1}.

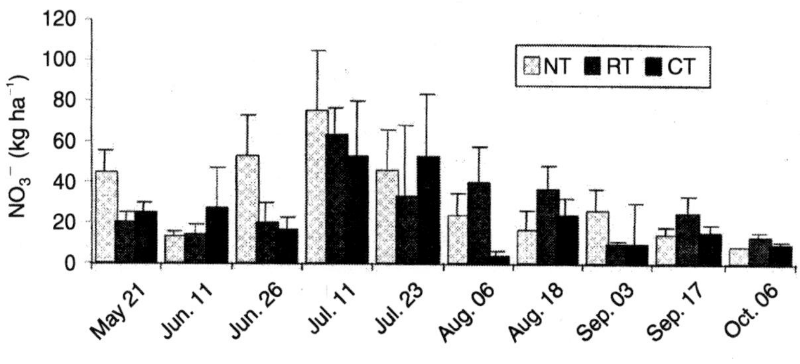

FIGURE 6. Seasonal variation in nitrous oxide production in the top 0-20 cm of a sandy loam soil in St-Emmanuel (Qc) Canada, from May to October 1997. Nitrogen fertilizer was applied on May 23 at a rate of 160 kg ha^{-1}.

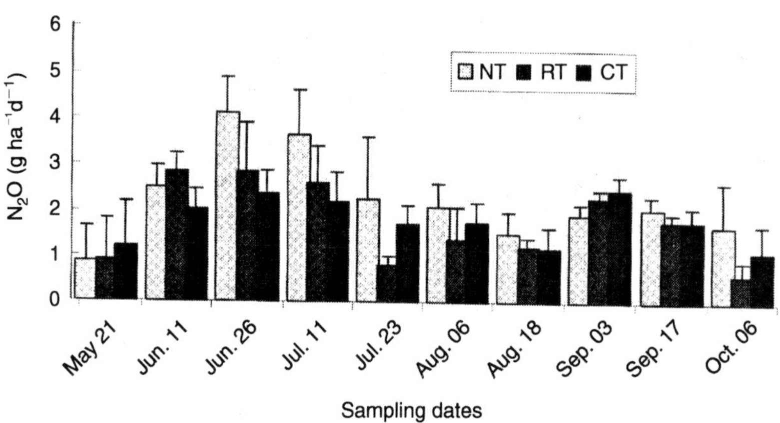

These involve the release of nutrients from both the mineral and organic fractions of soils, and the uptake of nutrients by plants. Living organisms also maintain in their biomass a pool of labile nutrients, which could otherwise be lost through leaching or fixation onto soil solids. The recycling of the SMB nutrient pool is very active in the mycorrhizosphere, which corresponds to the site of nutrient uptake by plant. Nutrients are lost in cropped soils mainly because nutrient release from soil is not synchronized with plant nutrient uptake, particularly in annual crop production, which leaves the soil bare for a significant part of the year. The design of efficient sustainable cropping systems requires a good understanding of the natural processes that have shaped the interactions between the biotic and the abiotic soil components through evolution. It appears that the P economy of cropping systems can be improved through a larger reliance on the AM symbiosis, which is enhanced in no-tilled systems and under conditions of reduced P fertilization. The maintenance of a large soil microbial community would maintain a high level of nutrient availability to crops, especially in the case of P. The N economy of cropping systems is more difficult to manage, particularly in non-leguminous crops. Most crops depend on abundant N fertilizer applications much of which is lost through denitrification and leaching. However, the synchrony between crop nutrient uptake and soil mineral N availability in fertilized crops can be improved by the storage of fertilizer-derived N into biomass and OM, which are slow release N sources. It is clear that the most efficient and sustainable cropping systems will be those in which the soil and crop management strategies are selected to support and enhance natural soil processes. Each soil is unique, complex and sensitive to environmental conditions such as temperature and moisture, which are increasingly difficult to predict in this period of climate change. The optimal management of crop-soil systems, therefore, is likely to be a continuing challenge.

REFERENCES

Abdel-Aziz, R.A. (2001). Effect of rhizobial inoculants and soil-available nitrogen on strain effectiveness and competition for nodulation in some legumes. In *Plant Nutrition: Food Security and Sustainability of Agro-Ecosystems Through Basic and Applied Research.* Fourteenth International Plant Nutrition Colloquium, Hannover, Germany. Dordrecht, Netherlands: Kluwer Academic Publishers. pp. 658-659.

Ahumada, I., J. Mendoza, P. Escudero, K. Mossert, and L. Ascar (2001). Determination of organic acids of low molecular weight and phosphate in soils by capillary electrophoresis. *Journal of the Association of Official Analytical Chemist*, 84: 1057-1064.

Alexander, M. (1977). *Introduction to Soil Microbiology.* 2nd ed. New York, NY: John Wiley & Sons, Inc.

Andrade, G., R.G. Linderman, and G.J. Bethlenfalvay. (1998). Bacterial associations with the mycorrhizosphere and hyphosphere of the arbuscular mycorrhizal fungus *Glomus mosseae. Plant and Soil* 202: 79-87.

Angers, D.A. and M. Giroux. (1996). Recently-deposited organic matter in soil water-stable aggregates. *Soil Science Society of America Journal* 60:1547-1551.

Anonymous. (2000). The beneficial rhizosphere: A dynamic entity. *Applied Soil Ecology* 15:99-104.

Atkinson, D. and L.A. Dawson. (2000). Root growth. Methods of measurement. In *Soil and Environmental Analysis: Physical Methods*, eds. K.A. Smith and C.E. Mullins, 2nd edition, New York, NY: Marcel Dekker Inc. pp. 431-493.

Azcon, R., J.M. Barea, and D.S. Hayman. (1976). Utilization of rock phosphate in alkaline soils by plants inoculated with mycorrhizal fungi and phosphate-solubilizing bacteria. *Soil Biology and Biochemistry* 8:135-138.

Barea, J.M. (1991). Vesicular-arbuscular mycorrhizae as modifiers of soil fertility. In *Advances in Soil Science.* New York, NY: Springer-Verlag. 15:1-40.

Barea, J.M., R. Azcon, and D.S. Hayman (1975). Possible synergistic interactions between endogone and phosphate-solubilizing bacteria in low-phosphate soils. In *Endomycorrhizas*, eds., F.E. Sanders, B. Mosse, and P.B. Tinker, New York, NY: Academic Press, pp. 409-417.

Barea, J.M. and C. Azcón-Aguilar. (1983). Mycorrhizas and their significance in nodulating nitrogen fixing plants. *Advances in Agronomy* 36:1-39.

Baziramakenga, R. and R.R. Simard. (1998). Low molecular weight aliphatic acid contents of composted manures. *Journal of Environmental Quality* 27:557-561.

Bethlenfalvay, G.J. and R.L. Franson. (1989). Manganese toxicity alleviated by mycorrhizae in soybean. *Journal of Plant Nutrition* 12:953-970.

Bethlenfalvay, G.J., R.S. Pacovsky, and M.S. Brown. (1982). Parasitic and mutualistic associations between a mycorrrhizal fungus and soybean: Development of the endophyte. *Phytopathology*. 72:894-897.

Biró, B., K. Köves-Péchy, I. Vörös, T. Takács, P. Eggenberger, and R.J. Strasser. (2000). Interrelations between *Azospirillum* and *Rhizobium* nitrogen-fixers and arbuscular mycorrhizal fungi in the rhizosphere of alfalfa in sterile, AMF-free or normal soil conditions. *Applied Soil Ecology* 15:159-168.

Bishop, M.L., A.C. Chang, and R.W.K. Lee. (1994). Enzymatic mineralization of organic phosphorus in a volcanic soil in chile. *Soil Science* 157:238-243.

Boddey, R.M. (1987). Methods for quantification of nitrogen fixation associated with gramineae. *Critical Review in Plant Science* 6:209-266.

Bolan, N.S. (1991) A critical review on the role of mycorrhizal fungi in the uptake of phosphorus by plants. *Plant and Soil* 134:189-207.

Bolan, N.S., R. Naidu, S. Mahimairaja, and S. Baskaran. (1994). Influence of low-molecular-weight organic acids on the solubilization of phosphates. *Biology and Fertility of Soils* 18: 311-319.

Bowen, G.D., and A.D. Rovira. (1999). The rhizosphere and its management to improve plant growth. *Advances in Agronomy* 66:1-249.

Burton, D.L., D.W. Bergstrom, J.A. Convert, C. Wagner-Riddle, and E.G. Beauchamp. (1997). Three methods to estimate N_2O fluxes as impacted by agricultural management. *Canadian Journal of Soil Science* 77:125-134.

Bøckman, O.C., O. Kaarstad, O.H. Lie, and I. Richards. (1990). *Agriculture and Fertilizers*. Agricultural Group, Norsk Hydro, Oslo, Norway.

CPVQ. (2000). Fertilization recommendations. Agdex 540. Conseil des production végétales du Québec Inc. Québec, Canada.

Cairney, J.W.G. (2000). Evolution of mycorrhiza systems. *Naturwissenschaften* 87: 467-475.

Chernicoff, S. (1999). *Geology: An Introduction to Physical Geology*. 2nd ed., Boston, MA: Houghton Mifflin Company, 596 p.

Coleman, D.C., C.P.P. Reid, and C.V. Cole (1983). Biological strategies of nutrient cycling in soil systems. *Advances in Ecological Research* 13:1-55.

Coote, D.R. and L.J. Gregorich. (2000). *The Health of Our Water: Toward Sustainable Agriculture in Canada*. Agriculture and Agro-Food Canada Research Branch, Publication 2020/E.

Cornell, R.M. and P.W. Schindler. (1987). Photochemical dissolution of geothite in acid oxalate solution. *Clay Minerals* 35:347-352.

Cunningham, J.E. and C. Kuiack. (1992). Production of citric and oxalic acids and solubilization of calcium phosphate by *Penicillium bilaii*. *Applied and Environmental Microbiology* 58:1451-1458.

Daft, M.J. (1991) Influences of genotypes, rock phosphate and plant densities on mycorrhizal development and the growth responses of five different crops. *Agriculture, Ecosystems and Environment* 35:151-169.

Dail, D.B. and J.W. Fitzgerald. (1999). S cycling in soil and stream sediment: Influence of season and in situ concentrations of carbon, nitrogen and sulfur. *Soil Biology and Biochemistry* 31:1395-1404.

de Ruiter, P.C., A.M. Neutel, and J.C. Moore. (1994). Modeling food webs and nutrient cycling in agroecosystems. *Trends in Ecology and Evolution* 9:378-383.

Degens, B.P., G.P. Sparling, and L.K. Abbot. (1996). Increasing the length of hyphae in a sandy soil increases the amount of water-stable aggregates. *Applied Soil Ecology* 3:149-159.

Dinkelaker, B., V. Römheld, and H. Marschner (1989). Citric-acid excretion and precipitation of calcium citrate in the rhizosphere of white lupin (*Lupinus albus* L.). *Plant Cell Environment* 12:285-292.

Estaun, V., C. Calvet, and D.S. Hayman. (1987). Influence of plant genotype on mycorrhizal infection: Response of three pea cultivars. *Plant and Soil* 103:296-298.

Evans, A.J. (1998). Biodegradation of ^{14}C-labeled low molecular organic acids using three biometer methods. *Journal of Geochemical Exploration* 65:17-25.

Filion, M., M. St-Arnaud, and J.A. Fortin. (1999). Direct interaction between the arbuscular mycorrhizal fungus *Glomus intraradices* and different rhizosphere microorganisms. *New Phytologist* 141:525-533.

Finlay, R. and B. Söderström. (1992). Mycorrhiza and carbon flow to the soil. In *Mycorrhizal Functioning. An Integrative Plant-Fungal Process*, ed. M.F. Allen, New York, London: Chapman and Hall. pp. 134-160.

Friese, C.F. and Allen, M.F. (1991). The spread of VA mycorrhizal fungal hyphae in the soil: Inoculum types and external hyphal architecture. *Mycologia* 83:409-418.

Frossard, E., L.M. Condron, A. Oberson, S. Sinaj, and J.C. Fardeau. (2000). Processes governing phosphorus availability in temperate soils. *Journal of Environmental Quality* 29: 15-23.

Frostegard, C., A. Tunlid, and E. Baath. (1993). Phospholipid fatty acid composition, biomass, and activity of microbial communities from two soil types experimentally exposed to different heavy metals. *Applied and Environmental Microbiology* 59: 3605-3617.

Gardner, W.K., D.A. Barber, and D.G. Parbery. (1983). The acquisition of phosphorus by *Lupinus albus* L. III The probable mechanism by which phosphorus movement in the soil/root interface is enhanced. *Plant and Soil* 70:107-124.

Gaudinski, J.B., S.E. Trumbore, E.A. Davidson, and S.H. Zheng. (2000). Soil carbon cycling in a temperate forest: Radiocarbon-based estimates of residence times, sequestration rates and partitioning of fluxes. *Biogeochemistry* 51:33-69.

Goss, M.J. and D.A.J. Barry. (1995). Groundwater Quality: Responsible agriculture and public perceptions. *Journal of Agricultural and Environmental Ethics*. 8:52-64.

Granli,T.and O. Brckman. (1994). Nitrous oxide from agriculture. *Norwegian Journal of Agricultural Sciences. Supplement No.* 12.

Grayston, S.J., S. Wang, C.D. Campbell, and A.C. Edwards. (1998). Selective influence of plant species on microbial diversity in the rhizosphere. *Soil Biology and Biochemistry* 30: 369-378.

Green H., J. Larsen, P.A. Olsson, D.F. Jensen, and I. Jakobsen. (1999). Suppression of the biocontrol agent *Trichoderma harzianum* by mycelium of the arbuscular mycorrhizal fungus *Glomus intraradices* in root-free soil. *Applied and Environmental Microbiology.* 65:1428-1434.

Gressel, N. and J.G. McColl. (1997). Phosphorus mineralization and organic matter decomposition: A critical review. In *Driven by Nature: Plant Litter Quality and Decomposition*, eds., G. Cadisch and K.E. Giller, Wallingford, UK: CAB International, pp. 297-309.

Hamel, C., C. Neeser, U. Barrantes-Cartín, and D.L. Smith. (1991b). Endomycorrhizal fungal species mediate ^{15}N-transfer from soybean to corn in non-fumigated soil. *Plant and Soil* 138:41-47.

Hamel, C., Y. Dalpé, S. Parent, and V. Furlan. (1997). Indigenous populations of arbuscular mycorrhizal fungi and soil aggregate stability are major determinants of leek (*Allium porrum* L.) response to inoculation with *Glomus intraradices* Schenck & Smith or *Glomus versiforme* (Karsten) Berch. *Mycorrhiza* 7:187-196.

Hamel, C., Smith, D.L., and V. Furlan. (1991). N_2-fixation and transfer in a field grown corn and soybean intercrop. *Plant and Soil* 133:177-185.

Hamel, C., U. Barrantes-Cartín, V. Furlan, and D.L. Smith. (1991a). Endomycorrhizal fungi in nitrogen transfer from soybean to corn. *Plant and Soil* 138:33-40.

Hamel, C., V. Furlan, and D.L. Smith. (1992). Mycorrhizal effects on interspecific plant competition and N transfer in legume-grass forage mixtures. *Crop Science* 32:991-996.

Harris, K.K. and E.A. Paul. (1987). Carbon requirements of vesicular arbuscular mycorrhizae. In *Ecophysiology of VA Mycorrhizal Plants*, ed., G.R. Safir, Boca Raton, FL: CRC Press, pp. 93-105.

He, Z.L., J. Wu, A.G. O'Donnell, and J.K. Syers. (1997). Seasonal responses in microbial biomass carbon, phosphorous and sulphur in soils under pasture. *Biology and Fertility of Soils* 24:421-428.

Haselwandter, K. and G.D. Bowen. (1996). Mycorrhizal relations in trees for agroforestry and land rehabilitation. *Forest Ecology and Management* 81:1-17.

Hoffland, E., G.R. Findenegg, and J.A. Nelemans (1989). Solubilization of rock phosphate by rape. II. Local root exudation of organic acids as a response to P-starvation. *Plant Soil* 113:161-165.

Howarth, R.W., J.W.B. Stewart, and M.V. Ivanov. (1992). *Sulphur Cycling on the Continents: Wetlands, Terrestrial Ecosystems and Associated Water Bodies.* Chichester, UK: John Wiley & Sons, Ltd. 350 pp.

Hu, H. Q., J.Z. He, X.Y. Li, and F. Liu (2001). Effect of several organic acids on phosphate adsorption by variable charge soils of central China. *Environment International* 26:353-358.

Illmer, P., A. Barbato, and F. Schinner (1995). Solubilization of hardly-soluble $AlPO_4$ with P-solubilizing microorganisms. *Soil Biology and Biochemistry* 27:265-270.

Ingham, R.E., J.A. Trofymow, R.N. Ames, H.W. Hunt, C.R. Morley, J.C. Moore, and D.C. Coleman. (1986). Trophic interactions and nitrogen cycling in a semi-arid grassland soil. I. Seasonal dynamics of the natural populations, their interactions and effect on nitrogen cycling. *Journal of Applied Ecology* 23:579-614.

Jakobsen, I. and L. Rosendahl. (1990). Carbon flow into soil and external hyphae from roots of mycorrhizal cucumber plants. *New Phytologist* 115:77-83.

Joergensen, R.G., B. Meyer, and T. Mueller. (1994). Time-course of the soil microbial biomass under wheat: A one year field study. *Soil Biology and Biochemistry* 26:987-994

Jones, D. and M.J. Wilson. (1985). Chemical activity of lichens on mineral surfaces–a review. *International Biodeterioration* 21:99-104.

Jones, D.L. (1998). Organic acids in the rhizosphere–a critical review. *Plant and Soil* 205:25-44.

Jones, D.L. and D.S. Brassington. (1998). Sorption of organic acids in acid soils and its implications in the rhizosphere. *European Journal of Soil Science* 49:447-455.

Jones, D.L. and L.V. Kochian. (1996). Aluminium organic-acid interactions in acid soils. I. Effect of root-derived organic acids on the kinetics of Al dissolution. *Plant and Soil* 182:221-228.

Jones, D.L. and P.R. Darrah. (1994). Role of root derived organic acids in the mobilization of nutrients from the rhizosphere. *Plant and Soil* 166:247-257.

Jones, D.L., P.R. Darrah, and L.V. Kochian. (1996). Critical evaluation of organic acid mediated iron dissolution in the rhizosphere and its potential role in root iron uptake. *Plant and Soil* 180:57-66.

Jones, D., B.F.L. Smith, M.J. Wilson, and B.S. Goodman. (1991). Phosphate solubilizing fungi in a Scottish upland soil. *Mycological Research* 95:1090-1093.

Kabir, Z. (1997). *Dynamics of Mycorrhizal Association in Corn (Zea mays* L.*): Influence of Tillage and Manure*. PhD thesis. Faculty of Agricultural and Environmental Sciences, McGill University, Canada.

Kabir, Z., I.P. O'Halloran, J. Fyles, and C. Hamel. (1998a). Dynamics of the mycorrhizal symbiosis of corn: Effect of host physiology, tillage practice and fertilization on spatial distribution of extraradical hyphae in the field. *Agriculture, Ecosystems and Environment* 68:151-163.

Kabir, Z., I.P. O'Halloran, P. Widden, and C. Hamel. (1998b). Vertical distribution of arbuscular mycorrhizal fungi under continuous corn in long-term no-till and conventional tillage systems. *Mycorrhiza* 8:53-55.

Kennedy I.R. and N. Islam. (2001). The current and potential contribution of asymbiotic nitrogen fixation to nitrogen requirements on farms: A review. *Australian Journal of Experimental Agriculture* 41:447-457.

Kepert, D.G., A.D. Robson, and A.M. Posner. (1979). The effect of organic root products on the availability of phosphorus to plants. In *The Soil-Root Interface*, eds. J.L. Harley and R. Scott-Russell, London, UK: Academic Press. pp. 115-124.

Kim, K.Y., D. Jordan, and G.A. McDonald. (1998). Effect of phosphate-solubilizing bacteria and vesicular-arbuscular mycorrhizae on tomato growth and soil microbial activity. *Biology and Fertility of Soils* 26:79-87.

Koffa, S.N. and R.E. De la Cruz. (1995). Screenhouse performance of VAM-inoculated seedlings of *Leucaena leucocephala* (Lam) De wit. in a phosphorus-deficient and aluminum sulfate-treated medium. *New Forests* 9:273-279.

Koide, R.T. and M. Li. (1990). On host regulation of the vesicular-arbuscular mycorrhizal symbiosis. *New Phytologist* 114:59-65.

Kucey, R.M.N. (1983). Phosphate-solubilising bacteria and fungi in various cultivated and virgin Alberta soils. *Canadian Journal of Soil Science* 63:671-678.

Landry, C. (2002). *Phosphorus Fertilization and Soil Biology Inputs in Soil Phosphorus Dynamics, Corn Nutrition and Growth Under Reduced-Tillage Practices*. PhD thesis. Faculty of Agricultural and Environmental Sciences, McGill University, Canada.

Lee, K.E. and C.E. Pankhurst. (1992). Soil organisms and sustainable productivity. *Australian Journal of Soil Research* 30:855-892.

Leyval, C. and J. Berthelin. (1989). Interactions between *Laccaria laccata, Agrobacterium radiobacter* and beech roots: Influence on P, K, Mg, and Fe mobilization from minerals and plant growth. *Plant and Soil* 117:103-110.

Li, X.-L., H. Marschner, and E. George. (1991). Phosphorus depletion and pH decrease at the root-soil and hyphae-soil interfaces of VA mycorrhizal white clover fertilized with ammonium. *New Phytologist* 119:397-404.

Linderman, R.G. (1988). Mycorrhizal interactions with the rhizosphere microflora: The mycorrhizosphere effect. *Phytopathology*. 78:366-371.

Linderman, R.G. (1992). Vesicular-arbuscular mycorrhizae and soil microbial interactions. In *Mycorrhizae in Sustainable Agriculture*, Special publication No. 54, eds. G.J. Bethlenfalvay and R.G. Linderman. Madison, WI: American Society of Agronomy, pp. 1-27.

Liu, A., C. Hamel, R.I. Hamilton, and D.L. Smith. (2000b). Mycorrhizal formation and nutrient uptake of new maize (*Zea mays* L.) hybrids with extreme canopy and leaf architecture as influenced by soil N and P levels. *Plant and Soil* 221:157-166.

Liu, A., C. Hamel, R. I. Hamilton, B. Ma, and D. L. Smith. (2000a). Acquisition of Cu, Zn, Mn and Fe by mycorrhizal maize (*Zea mays* L.) grown in soil at different P and micronutrient levels. *Mycorrhiza* 9:331-336.

Liu, A., C. Hamel, A, Elmi, C. Costa, B. Ma, and D.L. Smith. (2002). K, Ca and Mg nutrition of maize colonized by arbuscular mycorrhizal fungi under field conditions. *Canadian Journal of Soil Science* 82:271-278.

Lynch, J.M. and L.M. Panting. (1980). Cultivation and the soil biomass. *Soil Biology and Biochemistry* 12:29-33.

Lynch, J.M. and L.M. Panting. (1982). Effects of season, cultivation and nitrogen fertilizer on the size of microbial biomass. *Journal of the Science of Food and Agriculture* 33:249-252.

MacKenzie, A.F., M.X. Fan, and F. Cardin. (1997). Nitrous oxide emissions as affected by tillage, corn-soybean-alfalfa rotations and nitrogen fertilization. *Canadian Journal of Soil Science* 77:145-152.

Magid, J. and N.R. Nielsen. (1992). Seasonal variation in organic and inorganic phosphorus fractions of temperate-climate sandy soils. *Plant and Soil* 144:155-165.

Manjunath, A., N.V. Hue, and M. Habte (1989). Response of *Leucaena leucocephala* to vesicular-arbuscular mycorrhizal colonization and rock phosphate fertilization in an oxisol. *Plant and Soil* 114:127-133.

Manjunath, A., R. Mohan, and D.J. Bagyaraj. (1981). Interaction between *Beijerinckia mobilis, Aspergillus niger* and *Glomus fasciculatus* and their effects on growth of onion. *New Phytologist* 87:723-727.

Marschner, H. (1995). *Mineral Nutrition of Higher Plants*. San Diego, CA: Academic Press Inc.

Marschner, H. (1998). Role of root growth, arbuscular mycorrhiza, and root exudates for the efficiency in nutrient acquisition. *Field Crop Research* 56:203-207.

McGill, W.B., K.R. Cannon, J.A. Roberstson, and F.D. Cook. (1986). Dynamics of soil microbial biomass and water soluble organic C in Breton L after 50 years of cropping to two rotations. *Canadian Journal of Soil Science* 66:1-19.

McGill, W.B. and C.V. Cole. (1981). Comparative aspects of cycling of organic C, N, S and P through soil organic matter. *Geoderma* 26:267-286.

Miller, R.M. and J.D. Jastrow. (1992). The role of mycorrhizal fungi in soil conservation. In *Mycorrhizae In Sustainable Agriculture*, eds. G.J. Bethlenfalvay and R.G. Linderman, ASA Special Publication No. 54, Madison, WI: American Society of America and Soil Science Society of America. pp. 29-44.

Mo, S.X. (1986). Production and transformation of soil organic acids and their significance on soil fertility. *Advances in Soil Science* 4:9-11.

Moghimi, A. and M.E. Tate (1978). Does 2-ketogluconate chelate calcium in the pH range 2.4 to 6.4? *Soil Biology and Biochemistry* 10:289-292.

Moghimi, A., M.E. Tate, and J.M. Oades (1978). Characterization of rhizosphere products especially 2-ketogluconic acid. *Soil Biology and Biochemistry* 10:283-287.

Molina, J.A.E., C.E. Clapp, D.R. Linden, R.R. Allmaras, M.F. Layese, R.H. Dowdy, and H.H. Cheng. (2001). Modeling the incorporation of corn (*Zea mays* L.) carbon from roots and rhizodeposition into soil organic matter. *Soil Biology and Biochemistry* 33:83-92.

Murphy, D.V., I.R.P. Fillery, and G.P. Sparling. (1998). Seasonal fluctuations in gross N mineralisation, ammonium consumption, and microbial biomass in a Western Australian soil under different land uses. *Australian Journal of Agricultural Research* 49:523-535.

Nagarajah, S., A.M. Posner, and J.P. Quirk. (1970). Competitive adsorption of phosphate with polygalacturonate and other organic anions on kaolinite and oxide surfaces. *Nature* 228:83-85.

Nakano, A., K. Takahashi, and M. Kimura. (1999). The carbon origin of arbuscular mycorrhizal fungi estimated from $\delta^{13}C$ values of individual spores. *Mycorrhiza* 9:41-47.

Newton, W.E. (1999). Nitrogen fixation and the biosphere. In *Highlights of Nitrogen Fixation Research*, eds. E. Martinez and G. Hernandez, New York, NY: Kluwer Academic/Plenum Publishers, pp. 1-8.

Ogram, A. (2000). A soil molecular microbial ecology at age 20: Methodological challenges for the future. *Soil Biology and Biochemistry* 32:1499-1504.

Olsson, P.A., I. Thingstrup, I. Jakobsen, and E. Bååth. (1999). Estimation of the biomass of arbuscular mycorrhizal fungi in a linseed field. *Soil Biology and Biochemistry* 31:1879-1887.

Parfitt, R.L. (1978). Anion adsorption by soils and soil materials. *Advances in Agronomy* 30:1-50.

Pariente, S. (2002). Spatial patterns of soil moisture as affected by shrubs, in different climatic conditions. *Environmental Monitoring and Assessment* 73:237-251.

Patra, D.D., P.C. Brookes, K. Coleman, and D.S. Jenkinson. (1990). Seasonal changes of soil microbial biomass in an arable and a grassland soil which have been under uniform management for many years. *Soil Biology and Biochemistry* 6:739-742.

Paul, E.A. and F.E. Clark. (1989). *Soil Microbiology and Biochemistry*, New York, NY: Academic Press.

Paul, E.A. and F.E. Clark. (1996). *Soil Microbiology and Biochemistry*. 2nd ed. New York, NY: Academic Press.

Perrott, K.W., S.U. Sarathchandra, and B.W. Dow. (1992). Seasonal and fertilizer effects on the organic cycle and microbial biomass in a hill country soil under pasture. *Australian Journal of Soil Research* 30:383-394.

Peterson, W. and M. Böttger. (1991). Contribution of organic acids to the acidification of the rhizosphere of maize seedling. *Plant and Soil* 132:159-163.

Piccini, D. and R. Azcon. (1987). Effect of phosphate-solubilizing bacteria and vesicular-arbuscular mycorrhizal fungi on the utilization of Bayover rock phosphate by alfalfa plants using a sand-vermiculite medium. *Plant and Soil* 101:45-50.

Puget, P., C. Chenu, and J. Balesdent. (2000). Dynamics of soil organic matter associated with particle-size fractions of water-stable aggregates. *European Journal of Soil Science* 51:595-605.

Radcliffe, J.S., Z. Zhang, and E.T. Kornegay. (1998). The effects of microbial phytase, citric acid and their interaction in a corn-soybean meal-based diet for weanling pigs. *Journal of Animal Science*. 76:1880-1886.

Raj, J., D.J. Bagyaraj, and A. Manjunath. (1981). Influence of soil inoculation with vesicular-arbuscular mycorrhiza and a phosphate dissolving bacterium on plant growth and ^{32}P uptake. *Soil Biology and Biochemistry* 13:105-108.

Reitz, M., K. Rudolph, I. Schroder, S. Hoffmann-Hergarten, J. Hallmann, and R.A. Sikora. (2000). Lipopolysaccharides of *Rhizobium etli* strain G12 act in potato roots as an inducing agent of systemic resistance to infection by the cyst nematode *Globodera pallida*. *Applied and Environmental Microbiology* 66:3515-3518.

Reyes, I., L. Bernier, R.R. Simard, and H. Antoun. (1999). Effect of nitrogen source on the solubilization of different inorganic phosphates by an isolate of *Penicillium rugulosum* and two UV-induced mutants. *FEMS Microbiology Ecology* 28:281-290.

Rhodes, L.H. and J.W. Gerdemann. (1975). Phosphate uptake zones of mycorrhizal and nonmycorrhizal onions. *New Phytologist* 75:555-561.

Rillig, M.C., S.F. Wright, K.A. Nichols, W.F. Schmidt, and M.S. Torn. (2001). Large contribution of arbuscular mycorrhizal fungi to soil carbon pools in tropical forest soils. *Plant and Soil* 233:167-177.

Römheld, V. and H. Marschner. (1986). Evidence for a specific uptake system for iron phytosiderophores in roots of grasses. *Plant Physiology* 80:175-180.

Sanders, F.E. and P.B. Tinker. (1971). Mechanism of absorption of phosphate from soil by Endogone mycorrhizas. *Nature* 233:278-279.

Sanginga N., K. Mulongoy, and M.J. Swift. (1992). Contribution of soil organisms to the sustainability and productivity cropping systems in the tropics. *Agriculture, Ecosystems and Environment* 41:135-152.

Scheromm, P., C. Plassard, and L. Salsac. (1990). Effect of nitrate and ammonium nutrition on the metabolism of the ectomycorrhizal basidiomycete, *Hebeloma cylindrosporum* Romagn. *New Phytologist* 114:227-234.

Schubert, S. (1995). Nitrogen assimilation by legumes–processes and ecological limitations. *Fertilizer Research* 42:99-107.

Schwab, S.M., J.A. Menge, and R.T. Leonard. (1983). Quantitative and qualitative effects of phosphorus on extracts and exudates of Sudangrass roots in relation to vesicular-arbuscular mycorrhiza formation. *Plant Physiology* 73:761-765.

Siddiqui, Z.A. and T. Mahmood. (2001). Effects of rhizobacteria and root symbionts on the reproduction of *Meloidogyne javanica* and growth of chickpea. *Bioresource Technology* 79:41-45.

Simard, R.R., A.N. Cambouris, and M.C. Nolin. (2000). Spatio-temporal variation of anion exchange membrane P in a corn field. *Proceedings of the Fifth International Conference on Precision Agriculture*, Madison, WI: ASA-CSSA-SSSA.

Smith, M.D., D.C. Hartnett, and C.W. Rice. (2000). Effects of long-term fungicide applications on microbial properties in tallgrass prairie soil. *Soil Biology and Biochemistry*. 32:935-946.

Smith, S.E. and D.J. Read. (1997). *Mycorrhizal Symbiosis*. 2nd eds. London, UK: Academic Press.

Smith, S.E., A.D. Robson, and L.K. Abbott. (1992). The involvement of mycorrhizas in assessment of genetically dependent efficiency of nutrient uptake and use. *Plant and Soil* 146:169-179.

Smith, S.E. and F.A. Smith. (1990). Structure and function of the interfaces in biotrophic symbioses as they relate to nutrient transport. *New Phytologist* 114:1-38.

Sollins, P., K. Cromack, R. Fogel, and C.Y. Li (1981). Role of low-molecular-weight organic acids in the inorganic nutrition of fungi and higher plants. In *The Fungal Community: Its Organization and Role in the Ecosystem*, eds. D.T. Wicklow and G.C. Carroll, New York, NY: Marcel Dekker, pp. 607-619.

St-Arnaud, M., C. Hamel, B. Vimard, M. Caron, and J.A. Fortin. (1995). Altered growth of *Fusarium oxysporum* f.sp. *chrysanthemi* in an *in vitro* dual culture system with the vesicular arbuscular mycorrhizal fungus *Glomus intraradices* growing on *Daucus carota* transformed roots. *Mycorrhiza* 5:431-438.

Steenhoudt, O. and Vanderleyden, J. (2000). *Azospirillum*, a free-living nitrogen-fixing bacterium closely associated with grasses: Genetic, biochemical and ecological aspects. *FEMS Microbiology Reviews* 24:487-506.

Stevenson, F.J. (1967). Organic acids in soil. In *Soil Biochemistry*, eds. A.D. McLaren and G.H. Peterson, New York, NY: Marcel Dekker, pp. 119-146.

Stevenson, F.J. and M.A. Cole (1986). The phosphorus cycle. In *Cycles of Soil: Carbon, Nitrogen, Phosphorus, Sulfur, Micronutrients*. New York, NY: John Wiley and Sons, pp. 231-284.

Strobel, B.W. (2001). Influence of vegetation on low-molecular-weight carboxylic acids in soil solution–a review. *Geoderma* 99:169-198.

Ström, L., T. Olsson, and G. Tyler. (1994). Differences between calcifuge and acidifuge plants in root exudation of low-molecular organic acids. *Plant and Soil* 167: 239-245.

Strullu, D.G. (1986). Vers un modPle unifié de fonctionnement des symbioses mycorhiziennes. *Physiologie végétale* 24:219-225.

Swenson, R.M., C.V. Cole, and D.H. Sieling. (1949). Fixation of phosphate by iron and aluminium and replacement by organic and inorganic ions. *Soil Science* 67:3-22.

Tajima, S., K. Takane, M. Nomura, and H. Kouchi. (2000). Symbiotic nitrogen fixation at the late stage of nodule formation in *Lotus japonicus* and other legume plants. *Journal of Plant Research* 113:467-473.

Taranto, M.T., M.A. Adams, and P.J. Polglase. (2000). Sequential fractionation and characterisation (P-31-NMR) of phosphorus-amended soils in *Banksia integrifolia* (L.f.) woodland and adjacent pasture. *Soil Biology and Biochemistry* 32:169-177.

Tinker, P.B. (1980). The role of rhizosphere microorganisms in phosphorus uptake by plants. In *The Role of Phosphorus in Agriculture*, eds. F. Kwasaneh and E. Sample, Madison, WI: American Society of Agronomy. pp. 617-654.

Tran, T.S. and M. Giroux. (1998). Fate of 15N-labelled fertilizer applied to corn grown on different soil types. *Canadian Journal of Soil Science*. 78:597-605.

Triplett, E.W. (2000). *Prokaryotic Nitrogen Fixation: A Model System for the Analysis of a Biological Process*. Wymondham, UK: Horizon Scientific Press. 800 pp.

Unkovich, M.J.S. Pate. (2000). An appraisal of recent field measurements of symbiotic N_2 fixation by annual legumes. *Field Crops Research* 65:211-228.

Wall, D.H. and R.A. Virginia. (1999). The world beneath our feet: Soil biodiversity and ecosystem functioning. In *Nature and Human Society: The Quest for a Sustainable World*, eds. P.R. Raven and T. Williams, Washington, DC: National Academy Press.

Wall, L.G. (2000). The actinorhizal symbiosis. *Journal of Plant Growth Regulation* 19:167-182.

Westover, K.M., A.C. Kennedy, and S.E. Kelley. (1997). Patterns of rhizosphere microbial community structure associated with co-occurring plant species. *Journal of Ecology* 85:863-873.

Whitelaw, M.A. (2000). Growth promotion of plants inoculated with phosphates-solubilizing fungi. *Advance in Agronomy* 69:99-178.

Wolters, V. (2001). Biodiversity of soil animals and its function. *European Journal of Soil Biology* 37:221-227.

Zelenev, V.V., A.H.C. van Bruggen, and A.M. Semenov. (2000). "BACWAVE," a spatial-temporal model for traveling waves of bacterial populations in response to a moving carbon source in soil. *Microbial Ecology* 40:260-272.

Zhang, F. S., J. Ma, and Y.P. Cao. (1997). Phosphorus deficiency enhances root exudation of low-molecular weight organic acids and utilization of sparingly soluble inorganic phosphates by radish (*Raghanus sativus* L.) and rape (*Brassica napus* L.) plants. *Plant and Soil* 196:261-264.

Zhang, F., C. Hamel, H. Kianmehr, and D.L. Smith. (1995). Root-zone temperature and soybean *Glycine max* (L.) Merr. vesicular-mycorrhizal mycorrhizae: Development and interactions with the nitrogen fixing symbiosis. *Environmental and Experimental Botany* 35:287-298.

Zheng, D.W., J. Bengtsson, and G.I. Cgren. (1997). Soil food webs and ecosystem processes: Decomposition in donor-control and Lotka-Volterra systems. *The American Naturalist* 145:125-148.

Microbial and Genetic Diversity in Soil Environments

Ping Wang
Warren A. Dick

SUMMARY. Soil microorganisms play a key role in soil biogeochemical cycling processes, in sustaining natural soil ecosystems and in agricultural production. Microbial diversity is an index of the community that takes into account both species richness and the relative abundance of species. This paper discusses concepts of microbial diversity from various points of view and extends the concepts to include genetic diversity. Advantages and disadvantages of various research methods (i.e., methods involving culturable microorganisms, biochemical markers and indirect methods, microscopy and other direct methods, and nucleic acid methods) to study microbial diversity are listed to help match research goals with the appropriate research method. Recent progress has yielded a better understanding of microbial diversity in selected environments and these results are reviewed with emphasis on the plant root rhizosphere and some typical agroecosystems. The study of microbial diversity and how it relates to sustainable agriculture and environmental protection is an emerging focus area that has great potential. We believe

---

Ping Wang is Professor, Department of Microbiology, Huazhong Agricultural University, Wuhan, 430070, People's Republic of China.

Warren A. Dick is Professor, School of Natural Resources, The Ohio State University, Wooster, OH 44691 USA.

Address correspondence to: Warren A. Dick at the above address (E-mail: dick.5@osu.edu).

[Haworth co-indexing entry note]: "Microbial and Genetic Diversity in Soil Environments." Wang, Ping, and Warren A. Dick. Co-published simultaneously in *Journal of Crop Improvement* (Food Products Press, an imprint of The Haworth Press, Inc.) Vol. 12, No. 1/2 (#23/24), 2004, pp. 249-287; and: *New Dimensions in Agroecology* (ed: David Clements, and Anil Shrestha) Food Products Press, an imprint of The Haworth Press, Inc., 2004, pp. 249-287. Single or multiple copies of this article are available for a fee from The Haworth Document Delivery Service [1-800-HAWORTH, 9:00 a.m. - 5:00 p.m. (EST). E-mail address: docdelivery@haworthpress.com].

http://www.haworthpress.com/web/JCRIP

Digital Object Identifer: 10.1300/J411v12n01_01

that advances will continue to occur as we apply new and reliable research methods both in the laboratory and in the field.

KEYWORDS. Microbial diversity, genetic diversity, functional diversity, bacterial community, soil nucleic acids, 16S rDNA, PCR, rhizosphere, soil ecology, sustainable agriculture, rice soils

INTRODUCTION

Soil is recognized as being formed as a result of five factors. These 'soil forming' factors are parent material, climate, organisms, relief (topography) and time. Although generally considered as operating independently they, in fact, interact to create a very diverse and complex habitat. The soil forming factors set the boundaries of the potential microbial diversity in soil. Within these boundaries, an almost endless variety of microniches exist that serve as unique environments where microbial diversity can develop and thrive. Thus, microbial diversity in soil results from the complex nature of soil.

Ribosomal DNA (rDNA) and molecular phylogenetic methods have provided means for identifying the types of organisms that occur in microbial communities without the need for cultivation (Amann, Ludwig, and Schleifer, 1995; Hugenholtz and Pace, 1996). Comparative analyses of small-subunit rDNA (16S or 18S rDNA) and other gene sequences show that life falls into three primary domains–*Bacteria*, *Eucarya*, and *Archaea* (Woese, Kandler, and Wheelis, 1990). The general properties of representatives of the three domains indicate that the earliest life was based on inorganic nutrition and that photosynthesis and use of organic compounds for carbon (C) and energy metabolism came comparatively later (Pace, 1997). Based on rDNA trees, the main extent of earth's biodiversity is microbial (Hugenholtz, Goebel, and Pace, 1998).

DNA and RNA analyses imply prokaryotic diversity far greater than was expected, and are beginning to hint at the role of bacterial and viral diversity in global ecological cycles (Torsvik, Øvreås, and Thingstad, 2002). A single gram of soil may contain 10^{10} bacterial cells (Fagri, Torsvik, and Goksoyr, 1977) and represent more than 13,000 species of bacteria as revealed by DNA analysis (Torsvik, Goksøyr, and Daae, 1990).

The genome size (complexity) of prokaryotic organisms can be calculated from the reassociation rate of denatured (i.e., single stranded) DNA, which depends on the amount of homogeneous DNA present in a sample. This method can be used to estimate prokaryotic genomes in a community, which can then be used as a measure of the total genomic diversity in a community. A high degree of genomic diversity in prokaryotic communities in pristine soil and sediments with high organic content was estimated through this method.

The DNA diversity seen in 30- to 100-cm^3 samples corresponds to about 3,000 to 11,000 different genomes. By using the DNA-based species definition, and assuming that strains with $\geq 70\%$ DNA homology belong to the same species, it has been estimated that these samples contain ~10^4 different prokaryotic species (Torsvik, Øvreås, and Thingstad, 2002).

To date, only a fraction of the various kinds of microorganisms in soil have been identified and studied (Amann, Ludwig, and Schleifer, 1995). The functional and genetic potential of the sum total of the microbial community may exceed that of higher organisms and provide a valuable source for novel products (e.g., new drugs) and technologies (Handelsman et al., 1998; Osburne et al., 2000). Microorganisms in soil probably represent more undiscovered biological diversity than that found in any other environment on earth (Tiedje et al., 1999).

The Convention on Biodiversity, agreed to in 1992 and now ratified by more than 140 countries (CIESIN, 1992) has stimulated the study of microbial diversity. Extensive efforts are currently being made to determine whether microbial species are uniquely placed and isolated geographically, the current evidence is that microbial communities are relatively similar among widely varying geographic regions and agricultural practices. For example, the same species of fungi were found in newly cultivated soils and soils where crop production was well-established (Supardiyono and Smith, 1997).

The findings that similar microorganisms can be basically found anywhere in the world is not surprising given that recent evidence has clearly shown a global mixing of fine particles in the atmosphere that are eventually deposited in locations far removed from their source of origin (Prospero, 2001). However, the relative distribution or abundance of members of a microbial community may be unique and dependent on the origin of the sample, even if the individual members themselves are not unique (Rondon, Goodman, and Handelsman, 1999). Also, the fact that there is potential for individual organisms to reside in an environment does not necessarily ensure the presence of that organism.

Generally, reductions in soil microbial diversity are thought to cause reductions in the functional capability of soils (Giller et al., 1997). This does not always occur because the high degree of functional redundancy in soil microbial communities may result in no changes even though changes in the microbial diversity of the community can be clearly perceived. In the case where a large number of species conduct similar functions, a reduction in any single or small numbers of species has little effect on overall soil processes since other organisms fill the functional role (Andren, Bengtsson, and Clarholm, 1995; Yang et al., 2000). However, where a function relies on only a few or one species to be accomplished, removing that species or narrowing the diversity of species will also impact the function. The identification of such key species is an important area of study.

Knowledge about microbial community structure and diversity is essential if we are to understand the relationship between environmental factors and ecosystem functions (Atlas, 1984; Pankhurst, 1997). Soil microorganisms play a key role in many processes that are essential to proper functioning of soil. In agricultural soils, microniches are created or altered by the type of crops grown, tillage, drainage, fertilizer and lime additions, pesticide additions, and mechanical traffic. Plants themselves can create unique environments, by creating microniches in the rhizoplane and rhizosphere of plant roots available to diverse microorganisms.

It is in this context that we strive to better understand the microbial diversity and the unique function of each component of the soil microbial community as they contribute to functions such as the recycling of plant nutrients, maintenance of soil structure, degradation of agro-chemicals and pollutants, and the control of plant and animal pests. In this review, we will restrict ourselves to discussion of diversity of the soil bacteria. The discussion on genetic diversity of specific plant associated bacteria such as rhizobia and *Pseudomonas fluorescens* in the soil are not included in this paper. Recent reviews and articles by Bridge and Spooner (2001), Drijber et al. (2000), Douds and Millner (1999), and Smit et al. (1999) provide useful information on fungal diversity in soils.

Our objectives in writing this review are to: (1) define a framework from which discussion of microbial diversity can take place, (2) briefly review methods used to study microbial diversity, and (3) review microbial diversity in selected environments and agroecosystems. Obviously such a review must be highly selective in the material included. For additional detail, the reader is advised to consult the original papers cited in this review.

WHAT IS MICROBIAL DIVERSITY

Ecology began with natural history observations of plants and animals (Ford, 2000). Microbial ecology has borrowed many concepts used in classical ecology to help characterize microbial diversity and microbial communities. A 'community' is a group of interacting populations of different species. These interactions can be direct or indirect. How these species interact greatly influences the diversity of the community. Species that make up a large (or largest) proportion of the community are called the 'dominant' species. In more general terms, dominance is the influence that a single (or defined number) of species exerts over a community. Dominance can be measured in several ways including density, biomass, and the frequency of occurrence of a specific function or a species. Species 'richness' is the number of species in a community and species 'evenness' describes how evenly distributed individuals are spread among species. Communities with more evenness are more diverse, even if they have the same richness.

The soil microbial community is a composite of all of the different microorganisms in the soil. The diversity of the microbial community can be characterized in several different ways (Figure 1).

Species diversity is an index of community diversity that takes into account both species richness and the relative abundance of species. In a traditional sense, such an approach would involve defining a soil microbial community in terms of the numbers of different known bacterial, archaeal or fungal species and the relative contribution of each of these species in terms of the total number of individual organisms in the population. Such diversity is traditionally determined by culturability and isolation of species, but these species may represent 10% or less of the total bacteria present and active in soil. Microbial diversity, however, can also be defined in molecular terms as the number of different nucleic acid sequence types present in a habitat. Relatedness of each species is not expressed in terms of morphological or physiological characteristics but rather genetic characteristics. Thus genetic diversity may be defined as the relative abundance of sequence, gene or genome variance in a population. Extending this definition to community structure is to include quantitative information on the number of individuals of different

FIGURE 1. Conceptual classification of microbial diversity in soils and how diversity relates to the total soil microbial community.

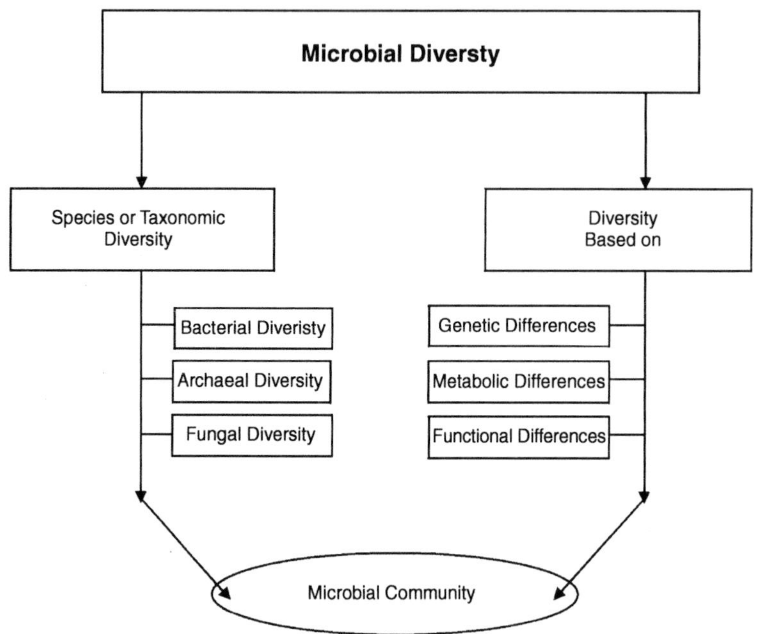

taxa or physiological groups (Liesack et al., 1997) that fall within a predefined set of genetic boundaries.

Genetic diversity, however, is also limited in describing the diversity of a soil microbial community because it cannot, as yet, be used to predict functional or metabolic diversity. Functional diversity includes the magnitude and capacity of soil inhabitants that are involved in key roles, such as nutrient cycling, decomposition of various compounds and other transformations (Zak et al., 1994). Functional diversity can be reduced, for example, by intensive land-use. This may have implications regarding the resistance of the soils to stress or disturbance (Degens et al., 2001). The term, metabolic diversity, implies a potential capability whereas functional diversity actually suggests an active expression of the capability (Degens, 1998a, 1998b).

Approximations of metabolic diversity using profile tests such as BIOLOG provide only a glimpse of the functional diversity in the soil microbial community at any single moment (Tate, 2000). This is because the tested substrates do not accurately represent the types of substrates present in the ecosystems, and the metabolic redundancy of species implies that changes in the response may only crudely represent the actual microbial community population dynamics (Konopka, Oliver, and Turco, 1998).

Describing the sum total of the potential metabolic diversity or expressed functional diversity in an agricultural soil is an ideal, but in reality impossible given our current limitations. Therefore, it is important to clearly define the type of information required of the soil microbial community and then focus on obtaining diversity information that will best meet that need. For example, if one is only interested in the capabilities of a soil to reduce nitrate to dinitrogen gas via denitrification, a functional measurement approach may be the best approach. One could simply incubate the soil and measure the amount of dinitrogen gas that evolves. However, if one wants to know how contamination of a heavy metal may potentially impact the denitrification function, then it may be much more informative to measure changes in the metabolic capacity of the soil or the narrowing of the genetic basis (i.e., the genetic diversity) upon which denitrification depends.

METHODS TO MEASURE SOIL MICROBIAL DIVERSITY

Techniques to Measure Species Diversity in Soil

Soil differs from most other microbial habitats in that it is dominated by a solid phase consisting of particles of different sizes and containing different pore sizes that are filled with either water or air. The very fine clay particles impute a large surface area to which microorganisms can attach. In addition, most soils contain organic matter (OM) in various levels of decomposition. Some of the OM is tightly bound to clays. The interaction of solid particles (especially clay), OM, water, and microorganisms creates problems in extracting microorganisms or cell components of microorganisms (e.g., cell wall components and nucleic acids). Also, the complex natures of the energy sources and microenvironments that can be found in soil are very difficult to reproduce in the laboratory and limits the species that we can actually isolate based on culture methods. Humified OM and other compounds in soil often are potent inhibitors of enzymes used in nucleic acid techniques such as polymerases and restriction enzymes.

The scale at which biological reactions take place in soil and the scale at which attempts are made to measure these reactions are also often grossly different. Microorganisms live and function in microenvironments that often are at a nanometer scale, but the methods we used to study these microorganisms are often at a scale hundreds or thousands of times larger. The concentration of a bacterial substrate may be very high in a localized microenvironment, which may or may not be replicated in a culture plate or media solution and certainly not at the same scale across the spatial boundaries in the soil.

Given this dichotomy, what are the best methods to study microbial diversity in soil and how can we make sense of our data? What level of confidence can we apply to the numbers we generate in our studies? It is wise to realize that all of our measurements are merely a snapshot of what is actually occurring in soil. This is because if we choose a very high-resolution measurement such as that for a single cell within a soil microenvironment, we are limited by the sheer number of sites that must be measured to actually gain a real world representation of the total microbial diversity. If we choose a more integrative approach, such as extraction of total phospholipid fatty acid, then we must be satisfied with the knowledge that we may not be able to understand the actual interactions and rates and workings of microbial cells at the individual community level.

Measuring species diversity of a microbial community implies we have a means of accurately and precisely measuring the different kinds of species in soil and their numerical (or relative) abundance. This assumption is obviously very often violated because many of the resident species simply cannot be identified and counted using classical approaches. However, culture based techniques do provide some very real advantages and cannot simply be abandoned. Other methods have been developed and have proved useful in studying soil microbial diversity (Ward, Weller, and Bateson, 1990; Ward et al., 1992; Amann, Ludwig, and Schleifer, 1995; Torsvik, Sorheim, and GoksØyr, 1996). A summary of the four basic types of approaches–culture based techniques, biochemical markers and other indirect techniques, microscopy and direct measurement techniques, and nucleic acid based techniques is provided in Table 1. In practice, these approaches are often combined to create techniques that offer unique advantages in answering specific questions. A nucleic acid probe containing a chromophore that is visually observable in a microscopic field is an example where a nucleic acid based approach is tied to a microscopic approach.

TABLE 1. Summary of approaches used to measure microbial diversity in soil.

Approach Used	Advantages and Disadvantages of Approach	Examples of Specific Techniques
Culturable Microorganism	<u>Advantages</u> In some cases it is possible to isolate individual species. Most microbiology laboratories are equipped to conduct studies using culture-based methods. Results obtained can be used in defining diversity indices and other metabolic properties. <u>Disadvantages</u> Many species still cannot be cultured. Thus the information gained is limited. Techniques can be labor extensive, although the reagent cost may be very reasonable. Reproducibility is often poor from one sample to another.	(1) Plate counts and most probable number (MPN) techniques. (2) Use of selective media such as specific growth compounds, antibiotics and/or choice of physical conditions such as temperature, solid supports and pressure. (3) Substrate utilization (e.g., BIOLOG).
Biochemical Markers and Other Indirect Measurements	<u>Advantages</u> Provides data useful in calculating diversity indices for community analyses. Can be applied to whole soils or to cultured microbial species. Some markers are directly proportional to microbial biomass size in soil. Sensitive to changes in the microbial habitat caused by soil disturbance, amendments and climatic variables. <u>Disadvantages</u> For some methods (e.g., FAME) it is not possible to distinguish between live and dead cells. Often requires specialized equipment and reproducibility is highly dependent on technical skill. Incomplete extraction of markers may limit the quality of quantitative information.	(1) Cell wall membranes components including fatty acid methyl esters (FAME) and phospholipid fatty acids (PFLAs) (2) ATP, enzymes or other specific biochemical markers (e.g., ergosterol as a marker for fungi) (3) Microbial biomass analyses using extraction or respiration based techniques

TABLE 1 (continued)

Approach Used	Advantages and Disadvantages of Approach	Examples of Specific Techniques
Microscopy and Other Direct Measurement Techniques	Advantages Reporter systems can be made very specific. Many of these techniques do not require culturing of the microorganism and can be applied to whole soil samples. Sensitivity can be very high. High resolution and ability to discern detail can be achieved. Disadvantages Requires specialized equipment, some of which can be very expensive. Techniques can be tedious and some involve a large amount of subjective interpretation of what is a cell and what is merely nonspecific staining. Cross reactivity may be a problem. For some microscopic and flow cytometry techniques, the small size of the microbial cell in soil is a limitation.	(1) All types of light-based microscopy (2) Electron microscopy (3) Combinations of nucleic acid hybridization probes and microscopy (e.g., fluorescence *in situ* hydridization-FISH, and reporter systems such as the use of the *gus*A gene, *lux* gene, *gfp* gene and *fhu*A gene) (4) Immunological techniques (5) Flow cytometry
Nucleic Acid Approaches	Advantages Selective amplification of marker genes including 16s rRNA genes used to classify microbial species. Reverse transcription measures the active individuals of a microbial population. Does not require culturing microorganisms. Can be used to calculate diversity indices and phylogenetic trees. Measures relative diversity of communities and especially useful in comparing communities in different samples. Disadvantages Amplification may be selective and can also create chimeric products. Nucleic acids may be incompletely extracted and thus may not be representative of the total microbial community. The procedures can be long, expensive and generate toxic chemicals. Gel matrix effects can influence results. Data analyses can be complicated and requires specialized software.	(1) Polymerase chain reaction (PCR) amplification of nucleic acids. (2) Real-Time PCR (3) Reverse transcription (RT) of RNA to DNA followed by PCR amplification (4) Denaturing gradient electrophoresis techniques (e.g., DGGE, TGGE) (5) Single-Strand Conformational Polymorphism (SSCP) (6) Terminal Restriction fragment length polymorphism (T-RFLP) and variations.

No matter what approach is used to study microbial diversity, there will be advantages and disadvantages (Table 1). However, the disadvantages can mostly be classified into four main areas–sample bias, cultural bias, extraction bias or artifacts, and nucleic acid amplification bias.

Because microorganisms in soil are so strongly tied to a microenvironment, studies that are based upon a single sample are probably suspect from the very beginning. Like a nutrient fertility test, soils must be sampled randomly from more than one location in a field or plot and then the samples combined. Sample bias is not only related to whether the sample for analyses is representative of the experimental unit being investigated, but also relates to sample preparation prior to analyses. Storage of samples should always be a cause for concern as it has great potential to introduce numerous artifacts. Miller et al. (1999), in a study of storage effects, found that DNA yields decreased rapidly for refrigerated soil samples and decreased much more slowly over several weeks for frozen samples. Soils are commonly frozen for storage and then all samples from an experiment or a replicate are analyzed at one time to avoid day-to-day laboratory variations. Freezing, however, causes changes in the physical, chemical, and biological properties of the soil. It is always best to put samples on ice, conduct the analyses as soon as possible after sampling, and use proper controls to try and account for variations due to operator bias and normal laboratory procedures. If this cannot be done, quick-freezing in liquid nitrogen and then storage at $-20°C$ is probably the best option.

Cultural bias is commonly thought to be restricted to classical studies where microorganisms are grown in a synthetic medium, which is not representative of the natural habitat. This is certainly a problem, but culturing is also a part of other methods to study microbial diversity such as the use of phospholipid fatty acids (PLFAs) to identify microbial species. Microorganisms synthesize a variety of PLFAs through various biochemical pathways. Only under very controlled conditions are results reproducible (Morgan and Winstanley, 1997; Kozdroj and van Elsas, 2001a).

Extraction of DNA or PLFAs from whole soils is not always quantitative. For example, there have been almost no systematic studies, which report the relative extraction efficiency of one type of PLFA versus another type in various soils. Likewise, extracted microbial cells, used later as a source of DNA, are assumed to be representative of the original soil population even though this is probably not true. If attempts are first made to break the cells before extracting the DNA, then there are problems associated with some cells being protected in soil and thus not available to the lysing agents. Even if one assumes all of the cells are broken, the DNA can be tightly sorbed onto clays, trapped in humus complexes or degraded, so that it quickly becomes less than an accurate representation of the original microbial community (Zhou, Bruns, and Tiedje, 1996; Duarte et al., 1998; Kuske et al., 1998; Griffiths et al., 2000; Martin-Laurent et al., 2001; Roose-Amsaleg, Garnier-Sillam, and Harry, 2001; Courtois et al., 2001; Hurt et al., 2001).

Nucleic acid approaches suffer additional problems related to amplification bias (Suzuki and Giovannoni 1996; Frostegard et al., 1999; Becker et al., 2000; Qiu et al., 2001; Schmalenberger , Schwieger, and Tebbe, 2001; Ishi and Fukui, 2001). If cloning is involved, there may be bias due to inappropriate choice of vectors and hosts. Finally, many methods require the use of specific probes and these may be designed inappropriately or used in a manner that does not provide reliable data (Wintzingerode, Gobel, and Stackebrandt, 1997; Ranjard et al., 2001; Stach et al., 2001).

The choice of approach to be used in studying the diversity of microbial communities will be dependent on the type of question being addressed, the resources available and the level of expertise found in a particular laboratory. The methods of statistics analysis and interpretation of the data are also very important for reaching a conclusion and they are dependent on the approach and the goal (Dollhopf, Hashsham, and Tiedje, 2001).

Several excellent books have been recently published that provide detailed procedures of the techniques that can generate data useful for the characterization of microbial diversity (Akkermans, van Elsas, and de Bruijn, 1996; Burlage et al., 1998). Microarray technology is a very promising new technique that can provide a vast amount of nucleic acid-based data from complex systems. It is primarily being used in the medical field, but work has already been initiated that applies microarray techniques to characterize nucleic acids extracted from soils and other environmental samples (Guschin et al., 1997; Wu et al., 2001; Small et al., 200, Lucchini, Thompson, and Hinton, 2001; Ye et al., 2001).

Statistical Methods

After sampling and analyses of the samples has been completed, we are still left with sets of data that require some sort of interpretation and analyses. The most common methods are statistical approaches and they describe results based on (1) central tendency, (2) scatter about a central tendency, and (3) limits based on some prejudged amount of scatter that is considered acceptable to provide some level of confidence in the result obtained (McSpadden Gardener and Lilley, 1997).

Richness is simply the value obtained by counting the total of species (*n*) in a community. Diversity indices combine into a single value both the measures of the number of species (richness) and the distribution of the total number of individuals among the species (evenness). There are several different diversity indices commonly used in microbial ecology. The Shannon-Wiener Diversity Index and the Simpson Diversity Index both lend themselves to easy statistical comparisons of microbial diversity in different soils.

Shannon-Wiener Diversity Index: $H' = -\sum (p_i)(\ln p_i)$

Simpson Diversity Index: $D = 1/\sum p_i^2$

where p_i is equal to the proportion of species to the total number of species. This information can then be used to calculate maximum diversity and evenness as follows:

Maximum Diversity: $H'_{max} = \ln S$

Evenness: $E = H'/H'_{max}$

where S is the total number of species in the community (richness). Maximum evenness occurs when individuals are spaced equally among species.

MICROBIAL DIVERSITY IN THE RHIZOSPHERE

There is probably no portion of the soil richer in microbial abundance and diversity than the plant rhizosphere. The rhizosphere is the volume of soil adjacent to and influenced by the plant root and is directly impacted by the type of crop grown. Sloughed off root materials or actively secreted root exudates are responsible for creating the rhizosphere effect. The amounts and composition of these organic materials vary among plant species and cultivars, stages of plant development, and among old and young components of the root system (Bowen and Rovira,

1999). As a result, bacterial communities will differ in composition, density and physiological state depending on their usage of specific substrates of the rhizosphere (Goodman et al., 1997). This results in the buildup of a diverse microflora in the rhizosphere.

A general description of the microbial community in the rhizosphere can be made (Bolton, Frederickson, and Elliot, 1992). Microorganisms that are primary root colonizers occur in the zone of elongation behind the root tips because this is where easily degradable sugars and organic acids are exuded. Fungi and bacteria adapted to crowded, oligotrophic conditions are found in the older root zones where carbon is deposited primarily as sloughed cells containing more recalcitrant cellulose, hemicellulose and lignified materials. Other mature microbial communities can be found at nutritionally rich and distinct sites such as where lateral roots emerge and at non-growing root tips.

Rhizosphere microorganisms have an important role in plant nutrition and growth promotion, biocontrol of soil-borne diseases, bioremediation of contaminated soil, drought resistance of plants through transformation of soil organic and inorganic nutrients, increasing uptake of nutrients to plants, production of phytohormones, antibiotics and siderophores, inducing systemic resistance in plants, and biodegradation of pollutants in soil (Pinton, Varanini, and Nannipieri, 2001). Regulating the rhizosphere microflora by genetically modifying plants and their rhizosphere environment to stimulate specific beneficial microorganisms has been proposed (Phillips and Streit, 1998).

That spatial variation in available C along the root coincides with different physiological groups of bacteria was clearly evident in studies by Maloney, van Bruggen, and Hu (1997). Rhizosphere bacteria were isolated from different locations along the taproot of lettuce (*Lactuca sativa* L., cv. Salinas) and tomato (*Lycopersicon esculentum* L., cv. Castlepeel II.) plants grown under greenhouse and field conditions. Overall numbers of both oligotrophs and copiotrophs were high at the upper portions of the root and lower at tip locations and in the bulk soil environment. Competitive interactions between microorganisms and physiological constraints with respect to substrate affinity were considered the key factors influencing bacterial populations and diversity of the rhizosphere communities.

Nutritional changes affecting the plant will also eventually be reflected in changes in the rhizosphere microbial community. Microbial communities in the rhizosphere of 30-day-old barley (*Hordeum vulgare* L.) plants grown under iron-limiting and iron-sufficient conditions were studied using 16S rDNA fingerprints generated by PCR-denaturing gradient gel electrophoresis (DGGE) (Yang and Crowley, 2000). Although the microbial communities associated with the different root locations produced many common 16S rDNA bands, communities associated with the two differently grown plants could be distinguished. Approximately 40% of the variation between communities could be attributed to plant iron nutritional status confirming that a rhizosphere community can be altered by changes in root exudate composition caused, in turn, by changes in plant iron nutritional status.

The knowledge obtained concerning microbial diversity in the plant rhizosphere is highly dependent on the study methods used. For example, Kaiser, Püler, and Selbitschka (2001) studied the microbial community of oilseed rape (*Brassica napus* cv. Westar) in the root rhizoplane (i.e., the surface area of the root directly in contact with the soil) using both a culture-dependent as well as a culture-independent approach based on 16S rDNA amplification. Phylogenetic 16S rDNA sequence analysis of 103 clones of this library revealed considerable differences from the corresponding nucleotide sequences of 111 cultured bacteria. Whereas the 16S rDNA clone library was dominated by α-*Proteobacteria* (51% of total clones) and bacteria of the *Cytophaga–Flavobacterium–Bacteroides* phylum (30% of total clones), less than 17% of the cultured bacteria belonged to these two groups. In contrast, more than 64% of the cultivated isolates were allocated to the β- and γ-subclasses of the *Proteobacteria*, which were present in the clone library at about 14% abundance. Most of the clones of the α-*Proteobacteria* of the library showed highest similarity to *Bradyrhizobium* species. No such bacteria were found in the culture collection. Similarly, the second dominant group of the clone library comprising members of the *Cytophaga–Flavobacterium–Bacteroides* phylum was represented in the culture collection by a single isolate. The phylogenetic analysis of isolates of the culture collection also clearly emphasized the need to use different growth media for recovery of rhizoplane bacteria. Whereas most of the α-*Proteobacteria* was recovered on complex medium, most of the β-*Proteobacteria* was isolated on minimal media.

Plant Effects Dominate the Rhizosphere Community

Microbial communities in the rhizosphere of four week old plants of wheat (*Triticum aestivum* L.), ryegrass (*Lolium perenne* L.), bentgrass (*Agrostis* L.), and clover (*Trifolium* L.) were studied by Grayston et al. (1998) using a BIOLOG approach. Canonical variety analysis (CVA) could not differentiate between microbial communities from the two different soil types but could clearly discern significant clustering of microbial communities related to plant species. Thus plants exert a much greater influence on microbial communities than soil in the soil-plant root interface called the rhizosphere.

Similarly, a greenhouse study using soil-plant microcosms was conducted to investigate the effects of plant [alfalfa (*Medicago sativa* L.) and rye (*Secale cereale* L.)] and soil origin (different locations and with differing histories of leguminous crop rotation) on microbial rhizosphere communities (Miethling et al., 2000). Three community-level targeting approaches–community level physiological profiles (CLPP), fatty acid methyl ester analysis (FAME), and temperature gradient gel electrophoresis (TGGE)–identified the crop species as the major determinant of microbial community characteristics. The influence of soil was of minor importance.

Characteristics of fluorescent pseudomonad populations associated with the roots of two plant species [flax (*Linum usitatissinum* L.) and tomato (cv. H63-5)] were compared with those from an uncultivated soil (Lemanceau et al., 1995). The results indicated that bacterial populations were not randomly distributed along the roots and that the plant species clearly was the most important selector for specific populations of fluorescent pseudomonads in the rhizosphere. Similar results were reported by Latour et al. (1996).

Plant-dependent shifts in the relative abundance of bacterial populations in the rhizosphere of field-grown strawberry (*Fragaria ananassa* Duch.), oilseed rape, and potato (*Solanum tuberosum* L.) were also observed using a DGGE approach (Smalla, Bucher, and Wieland, 2001). All these plants were hosts of the pathogenic fungus *Verticillium dahliae*. The population shifts became more pronounced in the second year.

Effect of Plant Cultivars

Not only different plant species, but also different cultivars of the same species can select for a distinct microbial population in the rhizosphere. Phospholipid fatty acids from rhizosphere bacteria of two different wheat cultivars, grown under controlled environmental conditions, were extracted and analyzed by gas chromatography (Diab El Arab, Vilich, and Sikora, 2001). The results consistently demonstrated the clustering of the samples into two distinct groups, each group belonging specifically to one of the two wheat cultivars. Profiles of one cultivar showed higher amounts of cyclopropane acid 19:0cy and Sif 7 and higher bacterial counts (cfu per g fresh root weight) of Gram-negative bacteria than what were observed in the rhizosphere of the other cultivar.

Isolates of the newly described *Pseudomonas brassicacearum* populations were obtained from the rhizosphere of two ecotypes of *Arabidopsis thaliana* (Wassilewskija and Columbia), a mutant of Columbia impaired in starch metabolism (pgm mutant), and a genetically distant plant (wheat) grown in a French eutric cambisol (Mereville) (Fromin et al., 2001). The isolates were obtained using semi-selective iron-deficient media. Their diversity was assessed using repetitive extragenic palindromic (REP)-PCR profiling and their affiliation to the *P. brassicacearum* species was measured using ARDRA and siderotyping. A total of 379 isolates from two experiments were clustered into 68 REP-genotypes. Statistical analysis showed that the genetic structure of the *P. brassicacearum* populations was homogeneous for isolates obtained from different plants of the same genotype within the same experiment, but significantly differed across the four tested plant genotypes.

A comparison of PCR-TGGE patterns for DNA obtained from the rhizosphere of two corn (*Zea mays* L.) cultivars was conducted. The rhizosphere effect was much more pronounced for young roots, which exhibited a reduced diversity (i.e., approximately five dominating TGGE bands) compared to that observed for mature corn plants that yielded TGGE patterns similar to those of soil (Gomes et al., 2001). Sequencing of dominant clones indicated that the dominant population found at all plant growth stages could be assigned to *Arthrobacter* populations.

Effect of Transgenic Plants

Because of the sensitivity of the microbial community to plant properties, it is not surprising that questions have been asked concerning whether transgenic plants affect rhizosphere microbial diversity. Rhizosphere bacterial communities between parental and two transgenic alfalfa lines of isogenic background were compared based on metabolic fingerprinting using BIOLOG GN microplates and DNA fingerprinting (using Enterobacterial Repetitive Intergenic Consensus Sequence-PCR or ERIC-PCR) (Di Giovanni et al., 1999). The two transgenic alfalfa expressed either bacterial (*Bacillus licheniformis*) genes for alpha-amylase or fungal (*Phanerochaete chrysosporium*) genes for Mn-dependent lignin peroxidase. Cluster analysis and PCA of the BIOLOG fingerprints indicated consistent differences in lignin peroxidase activity between the parental and transgenic alfalfa rhizosphere bacterial communities. Cluster analysis of ERIC-PCR fingerprints of the bacterial communities confirmed the differences in the types of bacteria in the rhizosphere of each alfalfa genotype. Comparison of ERIC-PCR fingerprints suggested that a limited number of populations were responsible for substrate oxidation in these wells. This is evidence that transgenic plant genotypes can affect rhizosphere microorganisms.

The indigenous bacterial communities of three soil plots located within an agricultural field were compared with T-RFLP analysis (Lukow, Dunfield, and Liesack, 2000). The first site was planted with non-transgenic potato plants, while the other two were planted with transgenic GUS and Barnase/Barstar potato plants, respectively. Barnase, a ribonuclease, when coupled with the *gst1* promoter, should lead to suicide of *Phytophthora infestans* infected cells and thus prevent spreading of the pathogen. The *barstar* gene inhibits Barnase synthesis and was inserted to minimize the detrimental effects of background activity of the *barnase* gene in non-infected tissue. Instead of the *barnase/barstar* gene construct, the transgenic GUS potato plants carried a *gus* gene (*uidA* gene) coding for β-glucuronidase. Statistical analyses clearly revealed spatial (i.e., potato treatment) and temporal effects, as well as potato treatment by time interaction effects, on the structural composition of the bacterial communities. Terminal restriction fragments (T-RFs), which showed the highest correlation to the discriminant factors, were different from those fragments which showed the largest variations between the individual plots.

Effects of Introduced Bacteria

Overlapping nutritional requirements of inoculated bacteria and indigenous microorganisms results in competition for niche colonization (Fukui et al., 1994). As a consequence, both the survival of inoculated bacteria and the population and diversity of the indigenous microorganisms may be reduced (Phillips and Streit, 1998; Streeter, 1994). This imposes an ecological risk if it affects important functions such as nutrient cycling and plant disease control (Leung et al., 1994; Jones, Broder, and Stotzky, 1991; Brimecombe, De Leij, and Lynch, 1999; Kennedy, 1999). Thus, an understanding of the effect of inoculated bacteria on the resident microbial community is important.

For example, fluorescent *Pseudomonas* spp. that produce DAPG (2,4-diacetylphloroglucinol) are known to have biocontrol activity against damping-off, root rot, and wilt disease caused by soil-borne fungal pathogens. The impact of a DAPG-producing biocontrol *Pseudomonas* agent on the non-target resident populations associated with the roots of field-grown sugar beet seedlings was evaluated by Moënne-Loccoz et al. (2001). Nineteen days after sowing, the introduced strain had replaced some of the resident culturable fluorescent pseudomonads at the rhizoplane but had no effect on the number of these bacteria in the rhizosphere. The introduced pseudomonad also induced a major shift in the composition of the resident culturable fluorescent *Pseudomonas* community. The percentage of rhizoplane isolates capable of growing on three different carbon substrates (erythritoll, adonitol, and L-trytophan) was increased from less than 10% to more than 40%. The shift in the resident populations was spatially limited to the surface of the root (i.e., the rhizoplane) and did not affect the relative proportions of phylogentic groups or the high level of strain diversity of the resident culturable fluorescent *Pseudomonas* community.

A luciferase gene-tagged *Sinorhizobium meliloti* L33, released in field plots seeded with alfalfa, was found 14 weeks later in bulk soil from noninoculated control plots (Schwieger and Tebbe, 2000). In the rhizosphere of alfalfa, *S. meliloti* L33 could be detected in noninoculated plots 12 weeks after inoculation, indicating that growth in the rhizosphere preceded spread into bulk soil. To determine whether inoculation affected bacterial diversity, 1,119 bacteria were isolated from the rhizosphere of alfalfa and common lambsquarters (*Chenopodium album* L.), which was the dominant weed in the field plots. In the rhizosphere of alfalfa, inoculation reduced the numbers of cells of *A. calcoaceticus* and members of the genus *Pseudomonas* and increased the number of rhizobia. Cultivation-independent PCR-single strand conformation polymorphism (SSCP) profiles of a 16S rRNA gene region confirmed the effect of the inoculant.

The parent strains and two derivatives of *Pseudomonas putida* WCS358r, genetically modified to constitutively produce the antifungal compound, phenazine-1-carboxylic acid, were inoculated onto wheat for two consecutive years (Glandorf et al., 2001). The numbers, composition, and activities of the rhizosphere microbial populations were then measured. During both growing seasons, all three bacterial strains decreased from 10^7 CFU per g of rhizosphere sample to below the limit of detection (10^2 CFU per g) one month after harvest. None of the strains, however, affected the metabolic activity of the soil microbial population (substrate-induced respiration), soil nitrification potential, cellulose decomposition, plant height, or plant yield. Thus, the non-target impact of the engineered strains seems to be limited.

MICROBIAL DIVERSITY AND AGRICULTURAL PRACTICES

Microbial diversity in an agricultural soil is a function of the combination of climate and landscape factors and anthropogenic factors (Figure 2). We can control anthropogenic practices such as crop rotations, tillage, weed control methods, fertilizer additions, etc. These practices interact with each other and with the climatic and landscape factors to define the microbial diversity in the soil.

As soils are changed from their natural state into a crop production state, there is generally a decline in soil OM and, presumably, a narrowing of microbial diversity. However, there is a paucity of knowledge on microbial community diversity and naturally occurred seasonal variations in agricultural soil as impacted by crop production practices. The remainder of this section briefly reviews climatic, tillage, soil properties, and agricultural chemical impacts on soil microbial diversity. The last part of this section focuses on a single crop, rice (*Oryza sativa* L.), and how microbial diversity in a rice field can be impacted by a variety of factors.

Temporal and Tillage Effects on Field Level Diversity

The soil microbial community of a wheat field on an experimental farm in The Netherlands was studied using both cultivation-based and molecule-based methods (Smit et al., 2001). Samples were taken in the different seasons over a 1-year period. Some genera, such as *Micrococcus*, *Arthrobacter*, and *Corynebacterium* were detected throughout the year,

FIGURE 2. Climatic and landscape factors and anthropogenic factors that influence microbial diversity in a dryland agricultural soil (adapted from Kennedy and Gewin, 1997).

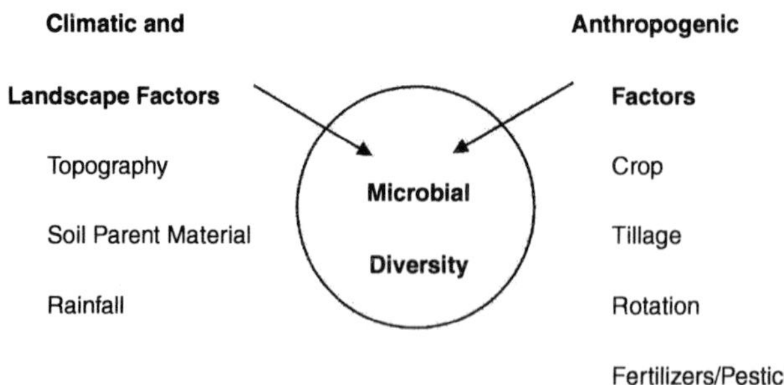

while *Bacillus* was found only in July. Isolate diversity was lowest in July. Denaturing gradient gel electrophoresis (DGGE) similarly revealed significant differences between samples taken in different seasons. The bacterial community in July clearly differed from those in the other months. Analysis of 16S rDNA sequences of these same soils showed that diversity of cloned 16S ribosomal DNA (rDNA) sequences was greater than the diversity among cultured isolates but that the seasonal trends were similar.

A BIOLOG™ system was used to detect specific patterns of substrate utilization by bacteria as affected by tillage (Lupwayi, Rice, and Clayton, 1998). Surface soil (0-7.5 cm) from a no-tillage or a conventional tillage wheat field on a Gray Luvisol in Northern Alberta, Canada showed that tillage significantly reduced the diversity of bacteria in one year out of two by reducing both richness and evenness. The effect of tillage on diversity was more prominent at the flag leafstage than at planting time and more prominent in bulk soil than in the root rhizosphere. The substrate utilization patterns also revealed greater diversity of the microbial community under no-tillage than conventional tillage. These results contrasted with those of Coutinho et al. (1999) who found that biodiversity did not differ between tilled and untilled soybean (*Glycine max* L. Merr.) plots. However, rhizobial strain composition differed between tilled and no-till fields.

DNA was extracted from six soils from agroecosystems in Norway and the USA under different agronomic treatments (crops, rotation, and tillage) and one soil contaminated with polyaromatic hydrocarbons (PAH, 700 mg kg^{-1}). Reproducible, characteristic profiles of the communities were obtained by DGGE separation of the PCR amplification products. The number of fragments resolved by DGGE indicated bacterial diversity was far greater than that of the Archaea in the agricultural soils. The soil contaminated with PAHs had reduced bacterial diversity, as evidenced by a distinct DGGE profile (Nakatsu, Torsvik, and Øvreas, 2000).

Tillage significantly affects microbial diversity because it has such a large influence on soil structure, which in turn, is a result of the association between mineral soil particles (sand, silt, and clay) and OM. Our own preliminary results of T-RFLP analysis of bacterial community structure in the surface layer (0-12 cm) of soil indicated bacterial diversity (species evenness) was far greater and the richness of domain strains was far less, in plow tilled soil than in long-term no-tillage soil (Wang and Dick, 2001).

Soil Properties

Little information is available on the structure of microbial populations in microhabitats associated with various sized soil particles. Sessitsch, Gerzabek, and Weilharter (2001) recently fractionated soil according to particle size and studied microbial diversity in these fractions using T-RFLP analysis and cloning and sequencing of 16S rRNA genes. The microbial community structure was significantly affected by particle size, yielding higher diversity of microbes in small size fractions than in coarse size fractions. Greater amounts of biomass commonly found in silt and clay fractions were attributed to higher diversity rather than to better colonization of particular species. Low nutrient availability, protozoan grazing, and competition with fungal organisms may have been responsible for reduced diversities in larger size fractions. Addition of different types of organic amendments (green manure, animal manure, sewage sludge, and peat) caused less variation in microbial diversity than what was observed among particle size fractions.

The rRNA intergenic spacer analysis (RISA) was used to study the genetic structure of the bacterial pools present in various microenvironments of a silt loam soil (Ranjard et al., 2000). Different fingerprints and consequently different genetic structures were observed between the unfractionated soil and the microenvironments, and also among the various microenvironments, giving evidence that some populations were specific to a given location in addition to common populations being present in all microenvironments. Cluster and multivariate analysis of RISA profiles showed the weak contribution of the pools located in the macroaggregate fractions to the whole soil community structure, as well as the clear distinction between the pool associated to the macroaggregate fractions and the pools associated to the microaggregate ones.

Two different agricultural soils in Norway were investigated: one organic soil and one sandy soil (Øvreås and Torsvik, 1998). The sandy soil was a field frequently tilled and subjected to crop rotations. The organic soil was permanent grazing land, infrequently tilled. The diversity of 200 cultivable bacteria strains was described at phenotypic, phylogenetic, and genetic levels. The total bacterial diversity was determined by reassociation analysis of DNA isolated from the bacterial fraction of environmental samples, combined with ARDRA and DGGE analysis. Organic soil exhibited a higher diversity for all analyses performed than the sandy soil. Analysis of cultivable bacteria resulted in different resolution levels and revealed a high biodiversity within the population of cultured isolates. The difference between the two agricultural soils was significantly higher for the total bacterial population than for the cultivable population.

Agricultural Chemicals

The impact of agricultural chemicals on soil microbial communities has been studied extensively by conventional methods using cultivation of the microbial communities and by measurements of their metabolic activities (Bromilow et al., 1996; Somerville and Greaves, 1987; Wardle and Parkinson, 1990). This information has led to our current understanding of the topic. The new molecular methods do not negate the previous findings but enrich our knowledge of effects of agricultural chemicals on soil microbial community (Desaint et al., 2000). A number of studies have been done on the effects of heavy metals contamination on the soil microbial community structure and function (Sandaa et al. 1999; KozdrÓj and van Elsas, 2000, 2001b; Müller et al. 2001, 2002; Shi et al. 2002). Some review papers on this topic have also been recently published (Giller, Witter, and Mcgrath, 1998; Kozdrój and van Elsas, 2001a). We will only focus on the effects of organic agricultural chemicals in this paper.

The DNA sequence diversities of microbial communities in soils treated with the herbicide (triadimefon) or fertilizer (ammonium bicarbonate) were evaluated by Random Amplified Polymorphic DNA (RAPD) analysis (Yang et al., 2000) and by measuring microbial biomass content. The richness, modified richness, Shannon-Weiner index, and a DNA similarity coefficient were calculated to quantify the diversity in the microbial communities of treated and untreated soil. The results showed microbial communities were clearly distinguishable among the differently treated soils. The herbicide treatment caused a decrease in the soil microbial biomass but maintained a high degree of diversity at the DNA level compared to the control soil (i.e., the untreated soil). In contrast, the fertilizer treated soil showed an increase in the soil biomass but a decrease in DNA diversity.

Polychlorinated biphenyl (PCB)-polluted soil exhibited bacterial diversity that included representatives of two novel lineages (Nogales et al., 2001). The apparent abundance of bacteria affiliated to the beta-subclass of the *Proteeobacteria*, and to the genus *Burkholderia* in particular, was confirmed by FISH analysis. The possible influence on apparent diversity caused by low template concentrations was assessed by dilution of the RNA template prior to amplification by reverse transcription-PCR.

An Aroclor 1260 (polychlorinated biphenyl, PCB)-laden soil and one heavily contaminated with polycyclic aromatic hydrocarbons (PAHs) from a secure, engineered landfill site in Québec were analyzed for microbial diversity using a clone library of the 16S rDNA sequences (Lloyd-Jones and Lau, 1998). Phylogenetic analysis revealed that three phyla and their major subdivisions of the domain *Bacteria* were highly represented in these samples despite the high pollution, particularly by PAHs. None of the 16S rDNA sequences obtained matched known sequences from cultivated bacterial species or from 16S rDNA sequences amplified directly from other environmental samples.

The effect of three phenyl urea herbicides (diuron, linuron, and chlorotoluron) on microbial communities in soil with 10-year histories of treatment was studied by Saïd et al. (1999). Both the structure and metabolic potential of soil microbial communities were significantly affected by the long-term application of these urea herbicides. Analyses of the 16S rRNA genes (16S rDNA) using DGGE revealed that the most affected species in the soils treated with diuron and linuron belonged to an uncultivated bacterial group. Substrate utilization (BIOLOG) patterns of the microbial communities from the same soils were also compared using principal component analysis. In all cases, the functional abilities of the soil microbial communities were altered by the application of the herbicides.

Application of methyl bromide (MeBr) to agricultural soils before planting of high-value cash crops has, for many years, been the major method for controlling nematodes, soilborne pathogens, and weeds in warm regions of the United States. The ability of soil microbial communities to recover after treatment with fumigants is very important if the soil is to maintain a healthy ecological state. Changes in soil microbial community structure were examined in a microcosm experiment following the application of methyl bromide (MeBr), methyl isothiocyanate, 1,3-dichloropropene, and chloropicrin (Ibekwe et al., 2001). The effect of MeBr on heterotrophic microbial activities was assessed using a BIOLOG approach and was found to be most severe in the first week after treatment. Thereafter the effects of MeBr and the other fumigants were much less. Microbial communities in the same soil were also analyzed using phospholipid fatty acid (PLFA) analysis and clearly demonstrated a community shift in all treatments to one dominated by gram-positive bacteria. In addition, different 16S rDNA profiles from fumigated soils were quantified by DGGE band patterns. High Shannon-Weiner diversity indices were maintained for the control soil and the fumigant-treated soils except in MeBr treated soil (the diversity index decreased from 1.4 to 0.13). Sequence analysis of clones generated from unique bands showed taxonomically unique clones had emerged from the MeBr-treated samples. Generally, MeBr had the greatest impact on soil microbial community diversity and 1,3-dichloropropene had the least impact.

Denaturing gradient gel electrophoresis (DGGE) of 16S rRNA genes showed that inoculation of soil with a *Pseudomonas* strain that could act as a plasmid donor of genes required for 2,4-dichlorophenoyacetic acid (2,4-D) degradation was followed by a shift in the microbial community structure of B-horizon soils (Dejonghe et al., 2000). The observed change in the community was due to proliferation of transconjugants formed in soil, especially when additional nutrients were added to the nutrient poor B-horizon soil. Affecting the soil community in this way clearly has implications related to decontamination of soils, which are poor in nutrients and low in microbial activity.

Microbial Diversity in Rice Field Soil

Rice is the world's most important agronomic plant, with 143 million ha under cultivation globally. Because of the importance of rice to the well being of so many people in the world, the last section of this paper reviews microbial diversity in rice soil agroecosystems. Rice soil also represents a suitable model system to study fundamental aspects of microbial ecology, such as diversity, structure, and dynamics of microbial communities as well as structure-function relationships between microbial groups.

On about 75% of the land where it is grown, rice grows under flooded conditions (Roger, Zimmerman, and Lumpkin, 1993). Flooded rice paddy soil can be considered as a system with three compartments (oxic surface soil, anoxic bulk soil, and rhizosphere) characterized by different physiochemical conditions. In the surface soil and the rhizosphere, the oxygen profile seems to govern gradients of other electron acceptors (e.g., nitrate, iron (III), and sulfate) and reduced compounds (e.g., ammonium, iron (II), and sulfide). Functional groups of microorganisms consist of metabolically related organisms, e.g., oxygen-respiring bacteria, nitrate reducers, iron reducers, sulfate reducers, but also fermenting bacteria and CH_4-producing archaea. The concerted action of all functional groups of microorganisms drives the carbon, nitrogen, sulfur, and iron cycle in rice paddies.

Besides the spatial distribution of functional groups of microorganisms, there is also a temporal development that occurs as the field goes through a flooding cycle followed by drainage at harvest. When the rice field is flooded, aerobic bacteria and chemical oxidation reactions consume oxygen, and oxygen is rapidly depleted in most regions of the soil. In the anoxic zone alternative electron acceptors are used with nitrate being the first electron acceptor reduced after oxygen depletion followed by Mn(IV), Fe(III), SO_4^{-2}, and CO_2 (Liesack, Schnell, and Revsbech, 2000). Along oxic-anoxic interfaces that occur directly beneath the flood water-soil interface and in the rhizosphere/rhizoplane soil, alternative electron acceptors will be regenerated because of the reoxidation of reduced compounds.

Finally, microbial diversity is also controlled by an input of OM into the rice soil. After rice harvest, the remaining plant material is plowed under the soil. In principle, the pathway of anaerobic polysaccharide degradation is well known and has been elucidated experimentally for anoxic rice soil (Chin and Conrad, 1995; Chin et al., 1998; Chin et al., 1999; Chin, Lukow, and Conrad, 1999). It primarily begins with the activity of fermenting bacteria that excrete enzymes that hydrolyze the polysaccharides. The resulting sugar monomers are then converted to alcohol, fatty acids, and H_2. In the presence of alternative electron acceptors, these substrates are degraded completely to CO_2. Under limitation of external electron acceptors, syntrophic bacteria further degrade the alcohol and fatty acids to acetate, formate, and CO_2. Acetate and formate finally serve as substrates for methanogenic archaea. An alternative route is the direct conversion of monomers (e.g., sugars) to acetate by homoacetogenic bacteria. Acetate then serves as substrate for acetotrophic methanogens converting it to CH_4 and CO_2 (Liesack, Schnell, and Revsbech, 2000).

The archaeal community of the anoxic rice soil slurry has been found to be dominated by members of the families Methanosarcinaceae and Methanosaetaceae, the kingdom Crenarchaeota, and a novel, deeply branching lineage of the (probably methanogenic) kingdom Euryarchaeota that has recently been detected on rice roots (Grosskopf, Stubner, and Liesack, 1998). The population dynamics of *Archaea* after flooding of an Italian rice field soil were also studied using a dual approach involving molecular sequence retrieval and fingerprinting of small-subunit (SSU) rRNA genes. Archaeal sequences were retrieved from four clone libraries (30 each) constructed to represent different time points (days 0, 1, 8, and 17) after flooding of the soil. Significant changes were detected in the frequency of individual clones, especially those affiliated with the Methanosaetaceae and Methanobacteriaceae. However, these findings could not be confirmed by T-RFLP analysis of SSU rDNA amplicons. Direct T-RFLP analysis of archaeal populations in DNA obtained directly from rice field soil slurries did reveal the presence of Methanosaetaceae and Methanosarcinacea. Only the relative abundance of Methanosarcinaceae was increased, roughly doubling to 29% of total archaeal gene frequency within the first 11 days. A functionally dynamic flooded rice field agroecosystem thus seems to be linked to a relatively stable archaeal community structure (Lurdes and Friedrich, 2000).

Because rice straw is a major substrate in paddy soils, it plays a major role in controlling changes in redox potential and the production of methane in flooded rice fields. The bacterial community involved in degrading rice straw under anoxic conditions was investigated by incubating straw in paddy soil for 71 days (Weber, Stubner, and Conrad, 2001). DGGE analyses of the amplified bacterial 16S rRNA genes showed that the composition of the bacterial community changed during the first 15 days but then was stable until the end of incubation. DNA from fifteen DGGE bands and from straw incubated in the soil for 1 and 29 days was amplified and cloned. From these efforts a total of 31 clones were obtained. Approximately two-thirds (20 clones) belonged to different phylogenetic clusters of the *Clostridia*. The relative abundance of various phylogenetic groups in the rice straw-colonizing community was determined by FISH. Most (55%) of the active cells again were found to belong to the genus *Clostridium*. The bacterial community colonizing and decomposing rice straw thus seems to rapidly develop and is dominated by members of different clostridia.

Microbial biomass and community structure, in paddy rice soil during the vegetation period of rice, were estimated by analysis of their fatty acid components directly extracted from the soil (Bai, Gattinger, and Zelles, 2000). A clear change in the composition of the community structure at different sampling periods was observed as indicated by the principal component analyses. A dramatic decline of ester-linked PLFA was observed while the non-ester-linked PLFA composition did not change. The hydroxy fatty acids of lipopolysaccharides as well as ether lipids continuously decreased during the observation period. Total microbial abundance was estimated to be $4.1\text{-}7.3 \times 10^9$ cells g^{-1} soil (dry weight) with an overall population average described as being 44% aerobic, 32% facultative anaerobic, and 24% archaea. The paddy soil at the August sampling period contained more facultative anaerobic bacteria (ca. 36%) and archaea (ca. 37%), but the total microbial biomass was significantly lower than in the remaining sampling periods. As the plant approached maturity, the microbial community structure in the soil changed to contain more Gram-negative bacteria and methanotrophs.

Specific PCR assays were used to amplify the 16S rDNA genes of the Desulfobacteriaceae and the Desulfovibrionaceae from DNA extracted from rice roots and bulk soils and were then analyzed using T-RFLP analysis (Scheid and Stubner, 2001). Results indicated that Desulfobacteriaceae and the Desulfovibrionaceae represent predominant Gram-negative sulfate reducers in rice fields. Quantification of the bacterial abundances was accomplished by rDNA dot blot hybridization. The Gram-negative sulfate reducers accounted for approximately 2-3% of the total rDNA content with the relative rDNA abundance of the Desulfobacteriaceae (at 1.4%) being higher than that of the Desulfovibrionaceae (0.5%).

Molecular ecology techniques were applied to assess changes in the bacterial community structure along a vertical oxygen gradient in flooded paddy soil cores (Lüdemann, Arth, and Liesack, 2000). Microsensor measurements showed that oxygen was rapidly depleted at a depth of approximately 2.0 mm beneath the floodwater/soil interface. Bacterial 16S rRNA gene (rDNA)-based community fingerprint patterns were obtained from 200-μm-thick soil slices of both the oxic and anoxic zones by using the T-RFLP technique. An easily identifiable shift in the community patterns was observed in the different zones and this shift correlated with oxygen depletion measured with depth. Comparative sequence analysis of 16S rDNA clones identified members of the subclasses of *Proteobacteria* as the most abundant microorganisms in the oxic zone. In contrast, members of *Clostridia* were determined to be the predominant bacterial group in the oxygen-depleted soil. There was agreement in the results obtained by the extraction of total RNA followed by reverse transcription-PCR of the bacterial 16S rRNA and by T-RFLP analysis for both oxic and anoxic zones of flooded soil. This finding suggests that the microbial groups detected on the rDNA level are the metabolically active populations within these zones.

CONCLUSIONS

One of the continuing challenges facing environmental microbiologists is to improve existing methods or develop new ones for describing microbial diversity. These methods should allow meaningful comparison of samples from different sites and should be sufficiently rapid to allow the monitoring of time-dependent changes. The main objective for developing such methods is not to study diversity for diversity's sake, but to gain an understanding of the role of microorganisms that comprise the microbial community at a particular site (Voordouw, 1998).

There are now many major groups of bacteria and archaea known only from molecular sequences and, until these organisms can be cultivated, the only means of understanding their role in the environment is through culture-independent characterization linked to determination of *in situ* metabolic activity (Gray and Head, 2001).

In recent years, a number of methodologies have been developed that allow us to directly correlate a specific metabolic activity with phylogenetically identifiable units (single cells or DNA sequences) in natural environments. These techniques not only allow us to identify specific metabolic activities of defined phylogentic entities within a complex community but, by determining the response of specific metabolic properties to natural or experimentally imposed perturbations, it is possible to determine environmental factors that regulate the activity.

Nonetheless, researchers studying microbial diversity are quick to point out molecular probes and genetic analyses have supplemented, not replaced, culturing techniques. Molecular approaches can give us excellent insights into what microorganisms are present in a habitat. Classical approaches can still provide understanding of what an organism might be doing in that habitat (Stencel, 2000).

Advances in genomics and bioinformatics have made it possible to integrate genomics into contemporary microbial systematics, evolution, and ecology (Check, 2002). A more complete description of microbial diversity and the environmental processes controlled by that microbial diversity will require a deeper understanding of the basic units of organization and their interactions. Communities, not total biomass, control process that drive the biogeochemical cycles sustaining the biosphere. Descriptions of the temporal and spatial dimensions of the microbial community structure and the complex gene expression patterns of microbial communities provide meaningful explanation of microbial diversity in soil (American Academy of Microbiology, 2001).

Improved methods to study microbial diversity in soil remains an on-going goal of soil microbial ecologists and advances are being made. Studies are yielding an increased understanding of the microbial community as to their identity, their relative abundance, and the functions in which they are involved. Information is also accumulating regarding how diverse members respond to each other and to various types of natural and anthropogenic factors. This can only help in creating more sustainable agricultural systems and in remediating soil already negatively impacted by past activities.

REFERENCES

Akkermans, A. D. L., J. D. van Elsas, and F. J. de Bruijn. (1996). *Molecular Microbial Ecology*. Dordrecht/Boston/London: Kluwer Academic Publishers.

Amann, R. I., W. Ludwig, and K-H. Schleifer. (1995). Phylogenetic identification and *in situ* detection of individual microbial cells without cultivation. *Microbiological Reviews* 59: 143-169.

American Academy of Microbiology. (2001). *Ecology and Genomics: A Crossroads of Opportunity*. http://www.asmusa.org/acasrc/pdfs/MicroEcoReportBW.pdf

Andren, O., J. Bengtsson, and M. Clarholm. (1995). Biodiversity and species redundancy among litter decomposers. In *The Significance and Regulation of Soil Biodiversity*, eds. H. P. Collins, G. P. Robertson, and M. J. Klug, Dordrecht: Kluwer Academic Publishers. pp. 141-151.

Atlas, R. M. (1984). Use of bacterial diversity measurements to assess environmental stress. In *Current Perspectives in Bacterial Ecology*, eds. M. J. Klug and C. A. Reddy, Washington DC: American Society for Microbiology. pp. 540-545.

Bai, Q., A. Gattinger, and L. Zelles. (2000). Characterization of microbial consortia in paddy rice soil by phospholipid analysis. *Microbial Ecology* 39: 273-281.

Becker, S., P. Boger, R. Oehlmann, and A. Ernst. (2000). PCR bias in ecological analysis: A case study for quantitative Taq nuclease assays in analyses of microbial communities. *Applied and Environmental Microbiology* 66: 4945-4953.

Bolton, H., J. K. Frederickson, and L. F. Elliot. (1992). Microbial ecology of the rhizosphere. In *Soil Microbial Ecology*, ed. F. B. Metting, New York, NY: Marcel Dekker. pp. 27-36.

Bowen, G. D. and A. D. Rovira. (1999). The rhizosphere and its management to improve plant growth. *Advances in Agronomy* 66: 1-101.

Bridge, P. and B. Spooner. (2001). Soil fungi: Diversity and detection. *Plant and Soil* 232: 147-154.

Brimecombe, M. J., F. A. A. M. De Leij, and J. M. Lynch. (1999). Effect of introduced *Pseudomonas fluorescens* strains on the uptake of nitrogen by wheat from N-15-enriched organic residues. *World Journal of Microbiology and Biotechnology* 15: 417-423.

Bromilow, R. H., A. A. Evans, P. H. Nicholls, A. D. Todd, and G. G. Briggs. (1996). The effect on soil fertility of repeated applications of pesticides over 20 years. *Pesticide Science* 48:63-72.

Buckley, D.H. and T.M. Schmidt. (2001). The structure of microbial communities in soil and the lasting impact of cultivation. *Microbial Ecology* 42: 11-21.

Burlage, R. S., R. Atlas, D. Stahl, G. Geesey, and G. Sayler. (1998). *Techniques in Microbial Ecology*. New York, NY: Oxford University Press.

Center for International Earth Science Information Network (CIESIN). 1992. Full text of treaty accessed at *<http://sedac.ciesin.org/entri/register/reg-170.rrr.html>*. Site confirmed February 15, 2002.

Check, B. (2002). AAM sees bright progress for microbial ecology research in the genomic era. *ASM News* 68: 427-431.

Chin, K-J. and R. Conrad. (1995). Intermediary metabolism in methanogenic paddy soil and the influence of temperature. *FEMS Microbiology Ecology* 18: 85-102.

Chin, K-J., T. Lukow, and R. Conrad. (1999). Effect of temperature on structure and function of the methanogenic archaeal community in an anoxic rice field soil. *Applied and Environmental Microbiology* 65: 2341-2349.

Chin, K-J., T. Lukow, S. Stubner, and R. Conrad. (1999). Structure and function of the methanogenic archaeal community in stable cellulose-degrading enrichment cultures at two different temperatures (15 and 30°C). *FEMS Microbiology Ecology* 30: 313-326.

Chin, K-J., F. A. Rainey, P. H. Janssen, and R. Conrad. (1998). Methanogenic degradation of polysaccharides and the characterization of polysaccharolytic clostridia from anoxic rice field soil. *Systemic and Applied Microbiology* 21: 185-200.

Courtois, S., A. Frostegard, P. Goransson, G. Depret, P. Jeannin, and P. Simonet. (2001). Quantification of bacterial subgroups in soil: Comparison of DNA extracted directly from soil or from cells previously released by density gradient centrifugation. *Environmental Microbiology* 3: 431-439.

Coutinho, H. L. C., V. M. Oliveira, A. Lovato, A. H. N. Maia, and G. P. Manfio. (1999). Evaluation of the diversity of rhizobia in Brazilian agricultural soils cultivated with soybeans. *Applied Soil Ecology* 13: 159-167.

Degens, B. P. (1998a). Decreases in microbial functional diversity do not result in corresponding changes in decomposition under different moisture conditions. *Soil Biology and Biochemistry* 30: 1989-2000.

Degens, B. P. (1998b). Microbial functional diversity can be influenced by the addition of simple organic substrates to soil. *Soil Biology and Biochemistry* 30: 1981-1988.

Degens, B. P., L. A. Schipper, G. P. Sparling, and L. C. Duncan. (2001). Is the microbial community in a soil with reduced catabolic diversity less resistant to stress or disturbance? *Soil Biology and Biochemistry* 33: 1143-1153.

Dejonghe, W., J. Goris, Saïd El Fantroussi, M. Höfte, P. De Vos, W. Verstraete, and E. M. Topp. (2000). Effect of dissemination of 2,4-dichlorophenoxyacetic acid (2,4-D) degradation plasmids on 2,4-D degradation and on bacterial community structure in two different soil horizons. *Applied and Environmental Microbiology* 66: 3297-3304.

Desaint, S., A. Hartmann, N. R. Parekh, and J-C Fournier. (2000). Genetic diversity of carbofuran-degrading soil bacteria. *FEMS Microbiology Ecology* 34: 173-180.

Di Giovanni, G. D., L. S. Watrud, R. J. Seidler, and F. Widmer. (1999). Comparison of parental and transgenic alfalfa rhizosphere bacterial communities using BIOLOG GN metabolic fingerprinting and enter repetitive intergenic consensus sequence-PCR (ERIC-PCR). *Microbial Ecology* 37: 129-139.

Diab El Arab, H. G., V. Vilich, and R. A. Sikora. (2001). The use of phospholipid acids (PLFA) in the determination of rhizosphere specific microbial communities (RSMC) of two wheat cultivars. *Plant and Soil* 228: 291-297.

Dollhopf, S. L., S. A. Hashsham, and J. M. Tiedje. (2001). Interpreting 16S rDNA T-RFLP data: Application of self-organizing maps and principal component analysis to describe community dynamics and convergence. *Microbial Ecology* 42: 495-505.

Douds Jr., D. D. and P. D. Millner. (1999). Biodiversity of arbuscular mycorrhizal fungi in agroecosystems. *Agriculture, Ecosystems and Environment* 74: 77-93.

Drijber, R. A., J. W. Doran, A. M. Parkhurst, and D. J. Lyon. (2000). Changes in soil microbial community structure with tillage under long-term wheat-fallow management. *Soil Biology and Biochemistry* 32: 1419-1430.

Duarte, G. F., A. S. Rosado, L. Seldin, A. C. Keijzer- Wolters, and J. D. van Elsas. (1998). Extraction of ribosomal RNA and genomic DNA from soil for studying the diversity of the indigenous bacterial community. *Journal of Microbiological Methods* 32: 21-29.

Fagri, A., V. L. Torsvik, and J. Goksoyr. (1977). Bacterial and fungal activities in soil: Separation of bacteria and fungi by a rapid fractionated centrifugation technique. *Soil Biology Biochemistry* 9: 105-112.

Ford, E.D. (2000). *Scientific Methods for Ecological Research.* Cambridge, UK: Cambridge University Press.

Fromin, N., W. Achouak, J. M. Thiery, and T. Heulin. (2001). The genotypic diversity of *Pseudomonas brassicacearum* populations isolated from roots of *Arabidopsis thaliana*: Influence of plant genotype. *FEMS Microbiology Ecology* 37: 21-29.

Frostegard, A., S. Courtois, V. Ramisse, S. Clerc, D. Bernillon, F. L. Gall, P. Jeannin, X. Nesme, and P. Simonet. (1999). Quantification of bias related to the extraction of DNA directly from soils. *Applied and Environmental Microbiology* 65: 5409-5420.

Fukui, R., M. N. Schroth, M. Hendson, and J. G. Hancock. (1994). Interaction between strains of pseudomonads in sugar-beet spermospheres and their relationship to pericarp colonization by *Pythium ultimum* in soil. *Phytopathology* 84: 1322-1330.

Giller, K. E., M. H. Beare, P. Lavelle, A. M. N. lzac, and M. J. Swift. (1997). Agricultural intensification, soil biodiversity and agroecosystem function. *Applied Soil Ecology* 6: 3-16.

Giller, K.E., E. Witter, and S. P. Mcgrath. (1998). Toxicity of heavy metals to microorganisms and microbial processes in agricultural soils: A review. *Soil Biology and Biochemistry* 30: 1389-1414.

Glandorf, D. C. M., P. Verheggen, T. Jansen, Jan-Willem Jorritsma, E. Smit, P. Leeflang, K. Wernars, L. S. Thomashow, E. Laureijs, J. E. Thomas-Oates, P. A. H. M. Bakker, and L. C. van Loon. (2001). Effect of genetically modified *Pseudomonas putida* WCS358r on the fungal rhizosphere microflora of field-grown wheat. *Applied and Environmental Microbiology*, 67: 3371-3378.

Gomes, N. C. M., H. Heur, J. Schonfeld, R. Costa, L. Mendonca-Hagler, and K. Smalla. (2001). Bacterial diversity of the rhizosphere of maize (*Zea mays*) grown in tropical soil studied by temperature gradient gel electrophoresis. *Plant and Soil* 232: 167-180.

Goodman, R. M., S. B. Bintrim, J. Handelsman, B. F. Quirino, J. C. Rosas, H. M. Simon, and K. P. Smith. (1997). A dirty look: Soil microflora and rhizosphere microbiology. In *Advances and Perspectives on the Function of Plant Roots*, eds. H. E.

Flores, J. P. Lynch, D. Eissenstat, Rockville, MD: American Society of Plant Physiologists. pp. 219-231.

Gray, N. D. and I. M. Head. (2001). Linking genetic identity and function in communities of uncultured bacteria. *Environmental Microbiology* 3: 481-492.

Grayston, S. J., S. Wang, C. D. Campbell, and A. C. Edwards. (1998). Selective influence of plant species on microbial diversity in the rhizosphere. *Soil Biology and Biochemistry* 30: 369-378.

Griffiths, R. I., A. S. Whiteley, A. G. O'Donnell, and M. J. Bailey. (2000). Rapid method for coextraxtion of DNA and RNA from natural environments for analysis of ribosomal DNA-and RNA-based microbial community composition. *Applied and Environmental Microbiology* 66: 5488-5491.

Grosskopf, R., S. Stubner, and W. Liesack. (1998). Novel eyryarchaeotal lineages detected on rice roots and in the anoxic bulk soil of flooded rice microcosms. *Applied and Environmental Microbiology* 64: 4983-4989.

Guschin, D. Y., B. K. Mobarry, D. Proudnikov, D. A. Stahl, B. E. Rittmann, and A. D. Mirzabekov. (1997). Oligonucleotide microchips as genosensors for determinative and environmental studies in microbiology. *Applied and Environmental Microbiology* 63: 2397-2402.

Handelsman, J., M. R. Rondon, S. B. Bintrim, and R. M. Goodman. (1998). The agroecosystem as a source of microbial metabolites. *Development in Industrial Microbiology* 37: 49-52.

Hugenholtz, P. and N. R. Pace. (1996). Identifying microbial diversity in the natural environment: A molecular phylogenetic approach. *Trends in Biotechnology* 14: 190-197.

Hugenholtz, P., B. M. Goebel, and N. R. Pace. (1998). Impact of culture-independent studies on the emerging phylogenetic view of bacteria diversity. *Journal of Bacteriology* 180: 4765-4774.

Hurt, R., X. Qiu, L. Wu, Y. Roh, A. V. Palumbo, J. M. Tiedje, and J. Zhou. (2001). Simultaneous recovery of RNA and DNA from soils and sediments. *Applied and Environmental Microbiology* 67: 4495-4503.

Ibekwe, A. M., S. K. Papiernik, J -Y Gan, S. R. Yates, C -H Yang, and D. E. Crowley. (2001). Impact of fumigants on soil microbial communities. *Applied and Environmental Microbiology* 67: 3245-3257.

Ishi, K. and M. Fukui. (2001). Optimization of annealing temperature to reduce bias caused by a primer mismatch in multitemplate PCR. *Applied and Environmental Microbiology* 67(8): 3753-3755.

Jones, R. A., M. W. Broder, and G. Stotzky. (1991). Effects of genetically engineered microorganisms on nitrogen transformations and nitrogen-transforming microbial populations in soil. *Applied and Environmental Microbiology* 57: 3212-3219.

Kaiser, O., A. Pühler, and W. Selbitschka. (2001). Phylogenetic analysis of microbial diversity in the rhizoplane of oilseed rape (*Brassica napus* cv. Westar) employing cultivation-dependent and cultivation-independent approaches. *Microbial Ecology* 42(2): 136-149.

Kennedy, A. C. (1999). Bacterial diversity in agroecosystems. *Agriculture, Ecosystem and Environment* 74: 65-76.

Kennedy, A. C. and V. L. Gewin. (1997). Soil microbial diversity: Present and future considerations. *Soil Science* 162: 607-617.

Konopka, A., L. Oliver, and R. F. Turco. (1998). The use of carbon substrate utilization patterns in environmental and ecological microbiology. *Microbial Ecology* 35: 103-115.

Kozdroj, J. and J. D. van Elsas. (2000). Response of the bacterial community to root exudates in soil polluted with heavy metals assessed by molecular and cultural approaches. *Soil Biology and Biochemistry* 43: 197-212.

Kozdroj, J. and J. D. van Elsas. (2001a). Structural diversity of microorganisms in chemically perturbed soil assessed by molecular and cytochemical approaches. *Journal of Microbiological Methods* 43: 197-212.

Kozdroj, J. and J. D. van Elsas. (2001b). Structural diversity of microbial communities in arable coils of a heavily industrialized area determined by PCR-DGGE fingerprinting and FAME profiling. *Applied Soil Ecology* 17: 31-42.

Kuske, C. R., K. L. Banton, D. L. Adorada, P. C. Stark, K. K. Hill, and P. J. Jackson. (1998). Small-scale DN sample preparation method for field PCR detection of microbial cells and spores in soil. *Applied and Environmental Microbiology* 64: 2463-2472.

Latour, X., T. Corberand, G. Laguerre, F. Allard, and P. Lemanceau. (1996). The composition of fluorescent pseudomonad populations associated with roots is influenced by plant and soil type. *Applied and Environmental Microbiology* 62: 2449-2456.

Lemanceau, P., T. Corberand, L. Gardan, X. Latour, G. Laguerre, J. Boeufgras, and C. Alabouvette. (1995). Effect of two plant species, flax (*Linum usitatissinum* L.) and tomato (*Lycopersicon esculentum* Mill.), on the diversity of soilborne populations of fluorescent pseudomonads. *Applied and Environmental Microbiology* 61: 1004-1012.

Leung, K., L. S. England, M. B. Cassidy, J. T. Trevors, and S. Weir. (1994). Microbial diversity in soil: Effect of releasing genetically engineered microorganisms. *Molecular Ecology* 3: 413-422.

Liesack, W., P. H. Janssen, F. A. Rainey, N. L. Ward-Rainey, and E. Stackebrandt. (1997). Microbial diversity in soil: The need for a combined approach using molecular and cultivation techniques. In Modern Soil Microbiology, eds. J. D. van Elsas, J. T. Trevors and E. M. H. Wellington, New York, NY: Marcel Dekker Inc. pp. 375-439.

Liesack, W., S. Schnell, and N. P. Revsbech. (2000). Microbiology of flooded rice paddies. *FEMS Microbiology Reviews* 24: 625-645.

Lloyd-Jones, G. and P. C. K. Lau. (1998). A molecular view of microbial diversity in a dynamic landfill in Quebec. *FEMS Microbial Letters* 162: 219-226.

Lucchini, S., A. Thompson, and J. C. D. Hinton. (2001). Microarrays for microbiologists. *Microbiology* 147: 1403-1414.

Lüdemann, H., I. Arth, and W. Liesack. (2000). Spatial changes in the bacterial community structure along a vertical oxygen gradient in flooded paddy soil cores. *Applied and Environmental Microbiology* 66: 754-762.

Lukow, T., P. F. Dunfield, and W. Liesack. (2000). Use of the T-RFLP technique to assess spatial and temporal changes in the bacterial community structure within an agricultural soil planted with transgenic and non-transgenic potato plants. *FEMS Microbiology Ecology* 32: 241-247.

Lupwayi, N. Z., W. A. Rice, and G. W. Clayton. (1998). Soil microbial diversity and community structure under wheat as influenced by tillage and crop rotation. *Soil Biology and Biochemistry* 30: 1733-1741.

Lurdes, T. and M. Friedrich. (2000). Archaeal population dynamics during sequential reduction processes in rice field soil. *Applied and Environmental Microbiology* 66: 2732-2742.

Maloney, P. E., A. H. C. van Bruggen, and S. Hu. (1997). Bacterial community structure in relation to the carbon environments in lettuce and tomato rhizospheres and in bulk soil. *Microbial Ecology* 34: 109-117.

Martin-Laurent, F., L. Philippot, S. Hallet, R. Chaussod, J. C. Germon, G. Soulas, and G. Catroux. (2001). DNA extraction from soils: old bias for new microbial diversity analysis methods. *Applied and Environmental Microbiology* 67: 2354-2359.

Martin-Laurent, F., L. Philippot, S. Hallet, R. Chaussod, J. C. Germon, G. Soulas, and G. Catroux. (2001). DNA extraction from soils: old bias from new microbial diversity analysis methods. *Applied and Environmental Microbiology* 67: 2354-2359.

McSpadden Gardener, B. B. and A. K. Lilley. (1997). Application of common statistical tools. In *Modern Soil Microbiology*, eds. J. D. van Elsas, J. T. Trevors and E. M. H. Wellington, New York, NY: Marcel Dekker Inc. pp. 501-523.

Miethling, R., G. Wieland, H. Backhaus, and C. C. Tebbe. (2000). Variation of microbial rhizosphere communities in response to crop species, soil origin, and inoculation with *Sinorhizobium meliloti* L33. *Microbial Ecology* 41: 43-56.

Miller, D. N., J. E. Bryant, E. L. Madsen, and W.C. Ghiorse. (1999). Evaluation and optimization of DNA extraction and purification procedures for soil and sediment samples. *Applied and Environmental Microbiology* 65: 4715-4724.

Moënne-Loccoz, Y., Hans-Volker Tichy, A. O'Donnell, R. Simon, and F. O'Gara. (2001). Impact of 2,4-diacetylphloroglucinol-producing biocontrol strain *Pseudomonas fluorescens* F113 on intraspecific diversity of resident culturable fluorescent pseudomonads associated with the roots of field-grown sugar beet seedlings. *Applied and Environmental Microbiology* 67: 3418-3425.

Morgan, J. A. W. and C. Winstanley. (1997). Microbial biomarkers. In *Modern Soil Microbiology*, eds. J. D. van Elsas, J. T. Trevors and E. M. H. Wellington, New York, NY: Marcel Dekker Inc. pp. 331-352.

Müller, A. K., K. Westergaard, S. Christensen, and S. J. Sorensen. (2001). The effect of long-term mercury pollution on the soil microbial community. *FEMS Microbiology Ecology* 36: 11-19.

Müller, A. K., K. Westergaard, S. Christensen, and S. J. Sorensen. (2002). The diversity and function of soil microbial communities exposed to different disturbances. *Microbial Ecology* 44: 49-58.

Nakatsu, C. H., V. Torsvik, and L. Øvreås. (2000). Soil community analysis using DGGE of 16S rDNA polymerase chain reaction products. *Soil Science Society of American Journal* 64: 1382-1388.

Nogales, B., E. R. B. Moore, E. Llobet-Brossa, R. Rossello-Mora, R. Amann, and K. N. Timmis. (2001). Combined use of 16S ribosomal DNA and 16S rRNA to study the bacterial community of polychlorinated biphenyl-polluted soil. *Applied and Environmental Microbiology* 67: 1874-1884.

Osburne, M. S., T. H. Grossman, P. R. August, and I. A. MacNeil. (2000). Tapping into microbial diversity for natural products drug discovery. *ASM News* 66: 411-417.

Øvreås, L. and V. Torsvik. (1998). Microbial diversity and community structure in two different agricultural soil communities, *Microbial Ecology* 36: 303 -315.

Pace, N. R. (1997). A molecular view of microbial diversity and the biosphere. *Science* 276: 734-740.

Pankhurst, C. E. (1997). Biodiversity of soil organisms as an indicator of soil health. In *Biological Indicators of Soil Health*, eds. C. E. Pankhurst, B. M. Doube, and V. V. S. R. Gupta, Wallingford, UK: CAB International. pp. 197-324.

Phillips, D. A. and W. R. Streit. (1998). Modifying rhizosphere microbial communities to enhance nutrient availability in cropping systems. *Field Crop Research* 56: 217-221.

Pinton, R., Z. Varanini, and P. Nannipieri. (2001). The rhizosphere as a site of biochemical interactions among soil components, plants, and microorganisms. In *The Rhizosphere*, eds. R. Pinton, Z. Varanini, and P. Nannipieri, New York, NY: Marcel Dekker Inc. pp. 1-18.

Prospero, J. M. (2001). African dust in America. In *Geotimes*, Alexandria, VA: American Geological Institute.

Qiu, X., L.Wu, H. Huang, P. E. McDonel, A. V. Palumbo, J. M. Tiedje, and J. Zhou. (2001). Evaluation of PCR-generated chimeras, mutations, and heteroduplexes with 16S rRNA gene-based cloning. *Applied and Environmental Microbiology* 62: 880-887.

Ranjard, L., F. Poly, J. Combrisson, A. Richaume, F. Gourbiere, J. Thioulouse, and S. Nazaret. (2000). Heterogeneous cell density and genetic structure of bacterial pools associated with various soil microenvironments as determined by enumeration and DNA fingerprinting approach (RISA). *Microbial Ecology* 39: 263-272.

Ranjard, L., F. Poly, J.-C. Lata, C. Mougel, J. Thioulouse, and S. Nazaret. (2001). Characterization of bacterial and fungal soil communities by automated ribosomal intergenic spacer analysis fingerprints: biological and methodological variability. *Applied and Environmental Microbiology* 67: 4479-4487.

Roger, P. A., W. J. Zimmerman, and T.A. Lumpkin. (1993). Microbiological management of wetland rice fields. In *Soil Microbial Ecology: Applications in Agricultural and Environmental Management*, ed., F. B. Metting Jr., New York, NY: Marcel Dekker Inc. pp. 417-455.

Rondon, M. R., R. M. Goodman, and J. Handelsman. (1999). The Earth's bounty: Assessing and accessing soil microbial diversity. *Trends in Biotechnology* 17:403-409.

Roose-Amsaleg C.L., E. Garnier-Sillam, and M. Harry. (2001). Extraction and purification of microbial DNA from soil and sediment samples. *Applied Soil Ecology* 18: 47-60.

Saïd. F., L. Verschuere, W. Verstraete, and E. M. Topp. (1999). Effect of phenylurea herbicides on soil microbial communities estimated by analysis of 16S rRNA gene fingerprints and community-level physiological profiles. *Applied and Environmental Microbiology* 65: 982-988.

Sandaa, R.-A., V. Torsvik, Ø. Enger, F. L. Daae, T. Castberg, and D. Hahn. (1999). Analysis of bacterial communities in heavy metal-contaminated soils at different levels of resolution. *FEMS Microbiological Ecology* 30: 237-251.

Scheid, D. and S. Stubner. (2000). Structure and diversity of Gram-negative sulfate-reducing bacteria on rice roots. *FEMS Microbiology Ecology* 36: 175-183.

Schmalenberger, A., F. Schwieger, and C. C. Tebbe. (2001). Effect of primers hybridizing to different evolutionarily conserved regions of the small-subunit rRNA gene in PCR-based microbial community analyses and genetic profiling. *Applied and Environmental Microbiology* 67(8): 3557-3563.

Schwieger, F. and C. C. Tebbe. (2000). Effect of field inoculation with *Sinorhizobium meliloti* L33 on the composition of bacterial communities in rhizospheres of a target plant (*Medicago sativa*) and a non-target plant (*Chenopodium album*) linking of 16S rRNA gene-based single-strand conformation polymorphism community profiles to the diversity of cultivated bacteria. *Applied and Environmental Microbiology* 66: 3556-3565.

Sessitsch, A., M. H. Gerzabek, and A. Weilharter. (2001). Microbial population structures in soil particle size fractions of a long-term fertilizer field experiment. *Applied and Environmental Microbiology* 67: 4215-4224.

Shi, W., J. Becker, M. Bischoff, R. F. Turco, and A. E. Konopka. (2002). Association of microbial community composition and activity with lead, chromium, and hydrocarbon contamination. *Applied and Environmental Microbiology* 68: 3859-3866.

Small, J., D. R. Call, F. J. Brockman, T. M. Straub, and D. P. Chandler. (2001). Direct detection of 16S rRNA in soil extracts by using oligonucleotide microarrays. *Applied and Environmental Microbiology* 67: 4708-4716.

Smalla, K., A. Buchner, and G. Wieland. (2001). Bulk and rhizosphere soil bacterial communities studied by denaturing gradient gel electrophoresis: Plant-dependent enrichment and seasonal shifts revealed. *Applied and Environmental Microbiology* 67(10): 4742-4751.

Smit, E., P. Leeflang, B. Glandorf, J. D. van Elsas, and K. Wernars. (1999). Analysis of fungal diversity in the wheat rhizosphere by sequencing of cloned PCR-amplified genes encoding 18S rRNA and temperature gradient Gel electrophoresis. *Applied and Environmental Microbiology* 65: 2614-2621.

Smit, E., P. Leeflang, S. Gommans, J. van den Broek, S. van Mil, and K. Wernars. (2001). Diversity and seasonal fluctuations of the dominant members of the bacterial soil community in a wheat field as determined by cultivation and molecular methods. *Applied and Environmental Microbiology* 67(5): 2284-2291.

Somerville, L. and M. P. Greaves. (1987). *Pesticide Effects on Soil Microflora*. New York, NY: Taylor and Francis.

Stach, J. E. M., S. Bathe, J. P. Clapp, and R. G. Burns. (2001). PCR-SSCP comparison of 16S rDNA sequence diversity in soil DNA obtained using different isolation and purification methods. *FEMS Microbiology Ecology* 36: 139-151.

Stencel, C. (2000). Microbial diversity: Eyeing the big picture. *ASM News* 66: 142-146.

Streeter, J. G. (1994). Failure of inoculant rhizobia to overcome the dominance of indigenous strains for nodule formation. *Canadian Journal of Microbiology* 40: 513-522.

Supardiyono, E. K. and D. Smith. (1997) Technical report–Microbial diversity: *ex situ* conservation of Indonesian microorganisms. *World Journal of Microbiological Biotechnology*. 13: 359-361.

Suzuki, M. T. and S. J. Giovannoni. (1996). Bias caused by template annealing in the amplification of mixtures of 16S rRNA genes by PCR. *Applied and Environmental Microbiology* 62: 625-630.

Tate III, R. L. (2000). *Soil Microbiology*, 2nd edition, New York, NY: John Wiley & Sons Inc.

Tiedje, J. M., S. Asuming-Brempong, K. Nusslein, T. L. Marsh, and S. J. Flynn. (1999). Opening the black box of soil microbial diversity. *Applied Soil Ecology* 13: 109-122.

Torsvik, V., J. GoksØyr, F., and F. L. Daae. (1990). High diversity in DNA of soil bacteria. *Applied and Environmental Microbiology* 56: 782-787.

Torsvik, V., R. Sorheim, and J. Goksøyr. (1996). Total bacterial diversity in soil and sediment communities-a review. *Journal of Industrial Microbiology* 17: 170-178.

Torsvik, V., L. Øvreås, and T. F. Thingstad. (2002). Prokaryotic diversity–Magnitude, dynamics, and controlling factors. *Science* 296: 1064-1066.

Voordouw, G. (1998). Reverse sample genome probing of microbial community dynamics. *ASM News* 64: 627-633.

Wang, P. and W. A. Dick. (2001). Bacterial community assessment in soil and corn roots under long-term no-tillage soil management. Abstract from the 2001 Annual Meetings of Soil Science Society of America. Published as CD-ROM, Soil Science Society of America, Madison, WI.

Ward, D., M. Bateson, R. Weller, and A. Ruffroberts. (1992). Ribosomal RNA analysis of microorganisms as they occur in nature. *Advance in Microbial Ecology* 12: 219-286.

Ward, D. M., R. Weller, and M. M. Bateson. (1990). 16S rRNA sequences reveal numerous uncultured microorganisms in a natural community. *Nature* 344: 63-65.

Wardle, D. A. and D. Parkinson. (1990). Effects of three herbicides on soil microbial biomass and activity. *Plant and Soil* 122: 21-28.

Weber, S., S. Stubner, and R. Conrad. (2001). Bacterial populations colonizing and degrading rice straw in anoxic paddy soil. *Applied and Environmental Microbiology* 67: 1318-1327.

Wintzingerode, F. V., U. B. Gobel, and E. Stackebrandt. (1997). Determination of microbial diversity in environmental samples: Pitfalls of PCR-based rRNA analysis. *FEMS Microbiology Reviews* 21: 213-229.

Woese, C. R., O. Kandler, and M. L. Wheelis. (1990). Towards a naturally system of organisms: Proposal for the domains archaea, bacteria, eucarya. *Proceedings of National Academy of Science of America* 87: 4576-4579.

Wu L., D. K. Thompson, G. Li, R. A. Hurt, J. A. Tiedje, and J. Zhou. (2001). Development and evaluation of functional gene arrays for detection of selected genes in the environment. *Applied and Environmental Microbiology* 67: 5780-5790.

Yang, Y., H. J. Yao, S. Hu, and Y. Qi. (2000). Effects of agricultural chemicals on DNA sequence diversity of soil microbial community: A study with RAPD marker. *Microbial Ecology* 39: 72-79.

Yang, C-H. and D. E. Crowley (2000). Rhizosphere microbial community structure in relation to root location and plant iron nutritional status. *Applied and Environmental Microbiology* 66: 345-351.

Ye, R. W., T. Wang, L. Bedzyk, and K. M. Croker. (2001). Applications of DNA microarrays in microbial systems. *Journal of Microbiological Methods* 47: 257-272.

Yin, B., D. Crowley, G. Sparovek, W. J. De Melo, and J. Borneman. (2000). Bacterial functional redundancy along a soil reclamation gradient. *Applied and Environmental Microbiology* 66: 4361-4365.

Zak, J. C., M. R. Willig, D. L. Moorhead, and H. G. Wildman. (1994). Functional diversity of bacterial communities: A quantitative approach. *Soil Biology and Biochemistry* 26: 1101-1108.

Zhou, J., M. A. Bruns, and J. M. Tiedje. (1996). DNA recovery from soils of diverse composition. *Applied and Environmental Microbiology* 62: 316-322.

Impact of Global Change on Biological Processes in Soil: Implications for Agroecosystem Management

Shuijin Hu
Weijian Zhang

SUMMARY. The Earth is undergoing rapid environmental changes due to human activities. Three components of the ongoing global change, elevated atmospheric CO_2, N deposition, and global warming, may significantly impact soil biota directly through modifying the physical and chemical environment, and indirectly through altering aboveground plant growth and community composition. The biomass, community structure, and activities of microbes and animals in soil as well as their interactions will likely be affected, leading to changes in ecological processes and functions. Biological processes that may be modified by

Shuijin Hu is affiliated with the Department of Plant Pathology, North Carolina State University, Raleigh, NC 27607 (E-mail: shuijin_hu@ncsu.edu).

Weijian Zhang is affiliated with the Department of Agronomy, Nanjing Agricultural University, Nanjing 210095, China.

The authors express special thanks to D. R. Clements and two anonymous reviewers for their valuable comments.

This work was partially supported by the grants from National Science Foundation (DEB-96-27368; DEB-00-01686) to Shuijin Hu and a fellowship from the Chinese Education Commission to Weijian Zhang.

[Haworth co-indexing entry note]: "Impact of Global Change on Biological Processes in Soil: Implications for Agroecosystem Management." Hu, Shuijin, and Weijian Zhang. Co-published simultaneously in *Journal of Crop Improvement* (Food Products Press, an imprint of The Haworth Press, Inc.) Vol. 12, No. 1/2 (#23/24), 2004, pp. 289-314; and: *New Dimensions in Agroecology* (ed: David Clements, and Anil Shrestha) Food Products Press, an imprint of The Haworth Press, Inc., 2004, pp. 289-314. Single or multiple copies of this article are available for a fee from The Haworth Document Delivery Service [1-800-HAWORTH, 9:00 a.m. - 5:00 p.m. (EST). E-mail address: docdelivery@haworthpress.com].

http://www.haworthpress.com/web/JCRIP

Digital Object Identifer: 10.1300/J411v12n01_02

global change include organic matter decomposition, N mineralization, food web interaction, and biotic N fixation. Lack of the complexity in agroecosystems may amplify the effects of global change on many biological processes in agricultural soils. However, minimizing human disturbance and thus increasing the complexity of agroecosystems may enhance the potential of C sequestration in agricultural soils and the stability of belowground systems, thereby contributing to ecological sustainability.

KEYWORDS. Elevated CO_2, nitrogen deposition, global warming, decomposition, N mineralization, soil C sequestration, food web interactions, agroecosystems

INTRODUCTION

The Earth is undergoing rapid changes due to human disturbance, especially in the concentrations of the trace gases in its atmosphere and global land cover (Jenkinson, 2001; Vitousek, 1994; Wuebbles et al., 1999). The annually averaged CO_2 concentration in the atmosphere has increased by almost one third since the beginning of the Industrial Revolution (Whorf and Keeling, 1998), and is projected to double by the end of the 21st century due to increased combustion of fossil fuels and changes in land use (IPCC, 1996). At the same time, soluble nitrogen (N)-deposition has increased two to several fold, as a result of combustion processes and fertilizer inputs in agriculture (Körner, 2000). During the past four decades, we have witnessed historically unprecedented additions of reactive N to terrestrial systems (Vitousek et al., 1997): anthropogenic N production has increased from about 40 Tg N $year^{-1}$ in 1961 to about 160 Tg N $year^{-1}$ in 1995 (Smil, 1999). Moreover, increasing evidence supports the notion that the Earth's average surface temperature has increased by about 0.5°C over the past century (Gaffen et al., 2000; Santer et al., 2000) and is predicted to increase by 1.4-5.8°C within this century (Houghton, 2001).

For the last two decades or so, great efforts have been directed to evaluate the effects of atmospheric CO_2 enrichment (Delucia et al., 1999; Reich et al., 2001), N deposition (Vitousek et al., 1997), and climatic warming (Rustad et al., 2001; Shaver et al., 2000) on terrestrial ecosystems. Although a vast majority of the studies have been focused on aboveground processes, an increasing number of the studies examined changes in belowground biological processes caused by global change (Hodge, 1996; Hu, Firestone, and Chapin, 1999; Oren et al., 2001). This paper attempts to synthesize the results obtained from various studies, and to examine the implications of these results for agriculture in a changing global environment.

IMPACT OF ELEVATED CO_2, N DEPOSITION, AND GLOBAL WARMING ON PLANT PRIMARY PRODUCTION

Elevated CO_2, N deposition, and global warming can significantly impact aboveground net primary production (NPP) (DeLucia et al., 1999; Reich et al., 2001; Sturm, Racine, and Tape, 2001), tissue composition of plant materials (Booker and Maier, 2001; Estiarte et al., 1999), and plant community structure (Harte and Shaw, 1995; Meier and Fuhrer, 1997; Wedin and Tilman, 1996).

Over the short term (weeks to years), CO_2 enrichment generally enhances plant growth through stimulating photosynthesis (Drake, 1992; Jablonski, Wang, and Curtis, 2002; Pooter, Roumet, and Campbell, 1996; Strain, 1987), even in the face of significant resource limitations and environmental stresses (Idso and Idso, 2001). CO_2 stimulation of plant growth can also occur by increasing plant nutrient acquisition and water use efficiency (Owensby, Coyne, and Auen, 1993) (Figure 1). Over the long term, CO_2 effects on plants decrease as other factors (e.g., nutrients) become more limiting (Oren et al., 2001). N deposition can enhance NPP as NPP in many terrestrial ecosystems is N-limited (Padgett, Allen, and Bytnerowicz, 1999; Verburg et al., 1998) (Figure 2).

Warming may increase plant production through stimulating N mineralization and lengthening the growing season for some plants (Rustad et al., 2001; Kirschbaum, 2000), but reduce production through exacerbating water limitations (De Valpine and Harte, 2001) (Figure 2). Global warming is more likely to affect the productivity of high-latitude ecosystems since the magnitude of warming is predicted to be more significant in mid- to high-latitude regions than low-latitude ones (IPCC, 1996). Various experiments have now recorded different responses of specific plant species to elevated CO_2 (Edwards, Clark, and Newton, 2001; Meier and Fuhrer, 1997), N deposition (Bowman and Steltzer, 1998; Heijmans et al., 2001), and climatic warming (Kellomaki et al., 2001; Meier and Fuhrer, 1997). Different responses of plant species to

FIGURE 1. Over the short term, elevated CO_2 in the atmosphere increases plant photosynthesis, enhancing plant biomass, water use efficiency and nutrient uptake. These changes will likely modify plant-microbial interactions and biological processes in soil.

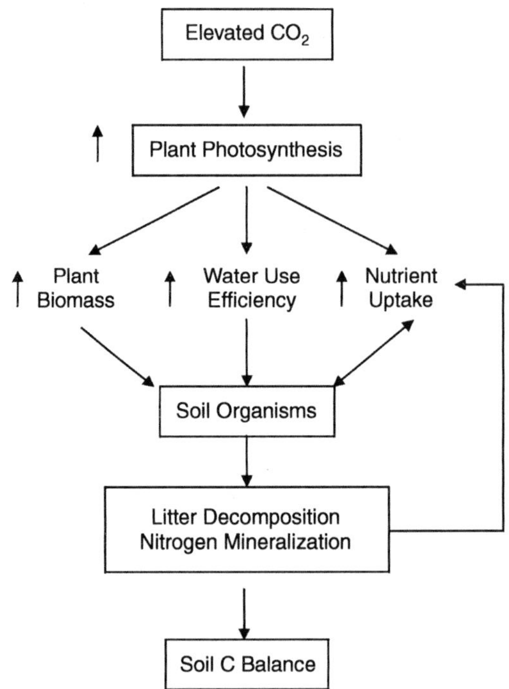

CO_2 enrichment, N deposition, and global warming may thus alter plant community structure. In general, photosynthesis of C_3 plants is more limited by CO_2 concentration than C_4 plants, thereby being more responsive to CO_2 enrichment (Pooter, Roumet, and Campell, 1996).

Increasing deposition of anthropogenic N may impact the plant community structure of terrestrial ecosystems by reducing the competitive advantage of many native species that have adapted to low fertility soils (Wedin and Tilman, 1996). Bowman and Steltzer (1998) showed that increases in N availability altered plant species composition as a result of competitive displacement in alpine tundra of the southern Rocky Mountains. Paschke, McLendon, and Redente (2000) found that the addition of N generally resulted in increased abundance of annual forbs and grasses relative to perennials at the previously cul-

FIGURE 2. Climatic warming and nitrogen deposition impact plants and soil organisms, modifying their interactions and biological processes in soil.

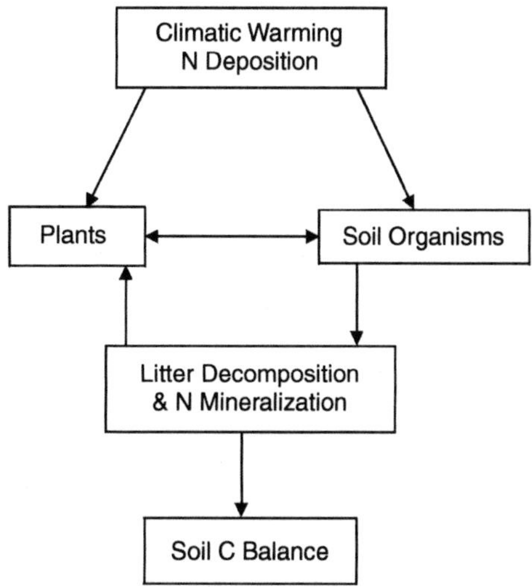

tivated sites. Experimental reduction of N availability generally increased the relative abundance of perennials in old-fields in shortgrass steppe of Colorado.

A few experimental warming studies have demonstrated changes in plant community structure (Harte and Shaw, 1995; Sternberg et al., 1999). In experimentally heated plots in a Rocky Mountain meadow aboveground biomass of sagebrush (*Artemisia tridentata*) increased in the drier habitat, and that of a shrub cinquefoil (*Pentaphylloides floribunda*) increased in the wetter habitat, but aboveground forb biomass decreased in the heated plots (Harte and Shaw, 1995). Warmer winters can provide sites for colonization of annuals, as demonstrated in a calcareous grassland in Oxfordshire, UK (Sternberg et al., 1999).

It is critically important to note that elevated CO_2, N deposition, and warming may interactively impact plants. For example, the combination of elevated CO_2 and global warming would increase plant growth if N deposition mitigated N limitation (Kirschbaum, 2000). Thus, in a future climate with higher CO_2, N deposition, and temperature, plant production may increase, particularly in ecosystems with few nutrient and soil moisture limitations.

EFFECTS OF GLOBAL CHANGE-INDUCED ALTERATION IN PLANT GROWTH ON SOIL C AND N AVAILABILITY

Elevated CO_2, N deposition, and global warming can alter the quantity and quality of carbon (C) inputs to soil and energy supply to soil organisms (Cannell and Thornley, 1998; King et al., 2001; Wiemken et al., 2001), thereby modifying belowground biological processes (Coûteaux and Bolger, 2000; Hu, Firestone, and Chapin, 1999).

The quantity of C inputs to soil generally increases in response to elevated CO_2 owing to increased NPP (Paterson et al., 1997; Zak et al., 2000a). Even when there is no significant stimulation of aboveground biomass, CO_2-stimulation of photosynthesis often results in increased allocation of C belowground (Körner and Arnone, 1992). Quantitative alterations in C inputs to the soil include increase in root growth, rhizodeposition and root exudates (Paterson et al., 1997; Rouhier et al., 1994; Wiemken et al., 2001), and C supply to symbionts and total litter inputs (Soussana and Hartwig, 1996).

The magnitude of CO_2-induced increase of C inputs to soil is likely constrained by soil nutrient availability, N in particular. Rates of root production and turnover were significantly greater at elevated CO_2 when trees grew in a high-N soil, but were unchanged in a low-N soil (Zak et al., 2000b). N deposition might favor biomass allocation to the aboveground plant parts (Hattenschwiler and Körner, 2000) and/or decrease fine-root biomass, but fine-root turnover and root exudation could significantly increase (Nadelhoffer, 2000). Thus, enhanced N availability may likely increase litterfall and total belowground C (Ceulemans, Janssens, and Jach, 1999; Holland et al., 1997).

Global change components may also alter the quality of C inputs to soil. Plant materials grown in high CO_2 environments often have higher C:N or lignin:N ratios (Baron et al., 2000; Bazzaz, 1990; Jablonski, Wang, and Curtis, 2002; Norby et al., 2001), nonstructural carbohydrates (Coley, Bryant, and Chapin, 1985; Herms and Mattson, 1992; Kinney et al., 1997; Serraj, Sinclair, and Allen, 1998) and C-based secondary compounds, such as phenolics and tannins (Booker and Maier, 2001; Gebauer, Strain, and Reynolds, 1998; Penuelas and Estiarte, 1998). In contrast, N addition often leads to low C:N ratios of plant materials, especially in grasses and green materials of trees. Additionally, changes in plant community structure can alter litter composition as described in the previous section. Warming impacts soil C availability directly through enhancing litter decomposition and indirectly by altering plant growth. Rustad et al. (2001) analyzed results from ecosystem warming experiments at 32 sites and found that on average warming increased soil respiration by 20% and plant productivity by 19%. Pregitzer et al. (2000) suggested that warming might increase the rates of root growth and root turnover if soil moisture and nutrient availability are adequate. Based on results from incubations and field measurements, Kirschbaum (2000) has suggested that warming will reduce soil organic C because decomposition will be stimulated more than NPP. Soil available N may decrease because of increased plant N uptake efficiency under elevated CO_2 and warming conditions due to enhanced root growth and mycorrhizal colonization (Constable et al., 2001; Hungate et al., 2000; Rillig, Hernandez, and Newton, 2000). Enhanced fine root production and mycorrhizal colonization allow plants to better exploit and capture available N.

CONSEQUENCES OF ALTERED C AND N AVAILABILITY AND TEMPERATURE ON BELOWGROUND PROCESSES

While elevated CO_2 indirectly affects soil microbes through altering plant water and nutrient utilization and C inputs (Figure 1), N deposition and warming can directly affect microbes and modify their activities (Figure 2).

Soil Microbial Biomass, Activity, and Community Structure

Enhanced C inputs under elevated CO_2 can stimulate microbial biomass and activities because soil microbes are usually C-limited (Paterson et al., 1997; Paul and Clark, 1996). Higher microbial biomass has been documented in grasslands (Hungate et al., 2000), forests (Zak et al., 1993), and rice (*Oryza sativa* L.) paddies (Inubushi et al., 2001). However, other experiments showed no significant effects of elevated CO_2 on soil microbial biomass (Allen et al., 2000; Berntson and Bazzaz, 1998; Wiemken et al., 2001). Microbial biomass is controlled by resource availability (the "bottom up" control) and faunal predation (the "top-down" control) (Hu, Firestone, and Chapin, 1999) and combining microbial biomass and activities may better reflect any effects of global change components. Enhanced C but decreased N availability under elevated CO_2 may favor fungi over bacteria (Hu et al., 2001), thereby altering the microbial community structure (Rillig, Field, and Allen, 1999a,b). Dhillion, Roy, and Abrams (1996) documented a significant increase in hyphal length (predominantly of saprophytic fungi), and xylanase, cellulase, and phosphatase activities in elevated CO_2 under low nutrient conditions. Grayston et al. (1998) reported a preferential stimulation of fungal growth under elevated CO_2 conditions in a native C_3 Australian grassland. In addition, there is evidence to suggest that the arbuscular mycorrhizal community composition could be changed by N enrichment (Egerton-Warburton and Allen, 2000; Rillig, Field, and Allen, 1999b). Over the long term, soil N limitation may inhibit microbial response to elevated CO_2 because enhanced plant N uptake and N sequestration in dead or alive plant materials may further constrain microbes (Hu et al., 2001). Thus, N deposition may benefit microbial activity under the elevated CO_2 conditions (Klironomos, Rillig, and Allen, 1996; Körner, 2000). Rice et al. (1994) reported that adding N to the soil stimulated the microbial activity in tallgrass prairie under elevated CO_2 levels. Warming may also affect the structure and activities of the soil microbial community (Christensen et al., 1997; Ruess et al., 1999).

Soil Fauna Dynamics and Food Web Interactions

Soil animals such as protozoa and nematodes in particular feed on soil microbes and play a critical role in nutrient cycling (Coleman and Crossley, 1996). In some systems, effects of global change on soil fauna may potentially alter ecosystem processes and overall functioning. Considerable efforts have recently been directed to quantify the effects of global change components on soil faunal dynamics (Briones, Ineson, and Piearce, 1997; Coûteaux and Bolger, 2000; Sulkava, Huhta, and Laakso, 1996). Most studies predict that warming will increase faunal biomass, abundance, and diversity if there are no soil moisture limitations (Cannon, 1998; Harte, Rawa, and Price, 1996). Briones, Ineson, and Piearce (1997) discovered that various species of enchytraeids responded to warming differently. Ruess et al. (1999) showed that warming and nutrient applications for eight years doubled nematode density in a subarctic dwarf shrub. Changes in temperature also influence the life cycles of soil fauna (David et al., 1999; Tsukamoto and Watanabe, 1997) and spatial distribution (Briones, Ineson, and Piearce, 1997).

Alterations in faunal dynamics and microbial biomass and community structure may modify the food web interactions since soil microbes serve as a major food source for many soil fauna. For example, Jones et al. (1998) showed that doubling atmospheric CO_2 altered the composition of fungi and some fungal-feeding collembola in model terrestrial ecosystems. These effects of global change on food web components may profoundly influence structure and function of ecosystems (Petchey et al., 1999).

Organic Matter Decomposition and Ecosystem C Balance

Changes in CO_2 concentration, N deposition and temperature are expected to have significant effects on soil C dynamics. Terrestrial ecosystems contain nearly three times more C (~2,060 Gt) than the atmosphere (~735 Gt C) with about 65% in the soil (Schlesinger, 1996). Whether soil will be a C sink or source in the future climate will depend on the net balance between soil C input and output.

Elevated CO_2 may enhance soil C sequestration by increasing C inputs and/or reducing decomposition. Evidence of CO_2 enhancement of C inputs is plentiful as described in the previous section. Reduction of decomposition can occur due to reduced litter decomposability (Frederiksen, Ronn, and Christensen, 2001) or to inhibition of microbial activities (Hu et al., 2001; Nijs, Salager, and Fabreguettes, 2000). For grasses, negative effects of elevated CO_2 on litter decomposability seems to be species-dependent and for naturally senesced tree leaves they may not be as important as previously suggested (Norby et al., 2001). Nijs, Salager, and Fabreguettes (2000) reported that CO_2 enrichment inhibited root respiration or microbial C-utilization in early successional Mediterranean ecosystems. Elevated CO_2 may favor plants over microbes for N utilization in grassland ecosystems, thereby slowing microbial decomposition and increasing C sequestration (Hu et al., 2001).

Elevated CO_2 and N deposition may interactively affect soil C dynamics. Oren et al. (2001) found that low soil fertility limited C sequestration by a plantation forest exposed to elevated atmospheric CO_2, but that N addition to this system stimulated C accumulation. Results from limited experiments show variable effects of global warming on soil C dynamics. On the one hand, warming may increase NPP and C inputs (Rustad et al., 2001). On the other hand, warming often stimulates decomposition and microbial activities, leading to an increase in CO_2 release from soil (Grogan and Chapin, 2000; Mertens et al., 2001). Most estimates of warming effects on soil C are based on incubation experiments. However, recent field studies seem to suggest that this effect may have been overestimated as old soil organic matter (OM) was less sensitive to temperature (Liski et al., 1999) and soil respiration may acclimate to warming (Luo et al., 2001).

Nitrogen Mineralization

The vast majority of soil N exists in the form of OM and only a very small part of the soil N is readily available for soil organisms and plants (Paul and Clark, 1996). The quantity and quality of soil OM, soil organisms and environmental factors (moisture and temperature in particular) exert critical control over transformations of soil N. It is obvious that N deposition may increase N availability in soil and the focus of this section will be on elevated CO_2 and warming effects on N availability.

There is some evidence showing that elevated CO_2 decreases soil N turnover rates, because in the presence of plants elevated CO_2 can exacerbate N constraints on microbes in N-limited ecosystems (Hu et al., 2001). CO_2-induced increase in litter C:N ratios may also reduce N mineralization. Hungate et al. (1999) reported that elevated CO_2 reduced the rate of gross N mineralization in soil, and resulted in lower recovery of nitrate on resin lysimeters. Over the short term, warming can stimulate N mineralization, as demonstrated by various experimental warming studies (Rustad et al., 2001; Shaver et al., 2000; Shaw and Harte, 2001).

Elevated CO_2 and warming will likely exert interactive controls over N cycling. Loiseau and Soussana (2000) found that CO_2 enrichment slowed N cycling down at ambient temperatures, but an increase of 3°C alleviated this effect by increasing soil N mineralization and N derived from the soil. Thus, a temperature increase under elevated CO_2 may stimulate N cycling and facilitate N transfer from soil to vegetation, but any long-term consequences are unclear and warrant further investigation.

Plant-Microbial Symbiotic Interactions

Rhizobia and mycorrhizal fungi are most likely to be affected by any changes in plant C budgets as they directly depend on photosynthates from their hosts (Hodge, 1996; Soussana and Hartwig, 1996). First, increased C availability under elevated CO_2 may enhance symbiotic N fixation and plant P acquisition through mycorrhizae and improved P nutrition may further enhance N_2 fixation (Stocklin and Körner, 1999). Experimental results have shown that elevated CO_2 increases the number and mass of nodules in various plants, including soybean (*Glycine max* L. Merr.), alfalfa (*Medicago sativa* L.), and white clover (*Trifolium repens* L.) (DeLuis et al., 1999; Hungate et al., 1999; Nakamura et al., 1999; Serraj, Sinclair, and Allen, 1998; Zanetti et al., 1996). At the same time, elevated CO_2 may directly impact rhizobial population composition. Montealegre et al. (2000) reported that the isolates of *Rhizobium leguminosarum bv. trifolii* from white clover nodules grown under elevated CO_2 were genetically different from those isolates obtained from plants grown under ambient conditions. Moreover, elevated CO_2 may improve N-fixation conditions by increasing water utilization efficiency and soil moisture, which has been shown to be responsible for enhancing soybean N-fixation under drought conditions (Serraj, Sinclair, and Allen, 1998).

In contrast to CO_2 enrichment, N deposition may reduce symbiotic N-fixation as high N availability can inhibit N_2 fixation (Paul and Clark, 1996). Thomas, Bashkin, and Richter (2000) recently examined the impact of N deposition and elevated atmospheric CO_2, and soil N availability on the N_2 fixation rate of *Gliricidia sepium* (a tropical leguminous tree species). They found that addition of 10 mM N of NH_4NO_3 solution significantly reduced nodule number and mass, specific nitrogenase activity, the specific rate of N-fixation, and the proportion of plant N derived from N_2 fixation, but elevated CO_2 strongly counteracted the inhibitory effects of N fertilization. These results suggest that CO_2-induced enhancement in N fixation may partially offset the inhibition of N_2 fixation caused by N deposition. Elevated CO_2 and N deposition cause similar effects on plant-rhizobial and plant-mycorrhizal symbioses. Under elevated CO_2, both ecto- and endo-mycorrhizae often increase due to higher fine root biomass, greater colonization, and hyphal length in soil (Hodge, 1996; Rillig, Hernandez, and Newton, 2000). The effects of N on mycorrhizae varied greatly, probably depending on mycorrhizal general and soil nutrient status (Nilsen et al., 1998; Treseder and Allen, 2002). In addition, limited studies suggest that N deposition might increase turnover rates of fungal tissue and negate CO_2 effects on hyphal biomass (Treseder and Allen, 2000).

Global warming may exacerbate soil moisture deficiency in arid ecosystems, which may negatively affect symbiotic N-fixation. Lilley et al. (2001) studied herbage symbiotic N fixation in pure and mixed swards of subterranean clover (*Trifolium subterraneum* L.) and phalaris (*Phalaris aquatica* L.) in field tunnels at ambient and elevated CO_2 (690 PPM) as well as at ambient and warmed air temperature (+3.4°C). They found that the proportion of clover-N derived from N_2 fixation increased by 12% under elevated CO_2, but decreased by 6% under warming. In a review of N fixation in the tropics, Hungria and Vargas (2000) concluded that high temperature and moisture deficiency were major causes of nodulation failure, adversely affecting all stages of the symbiosis and rhizobial growth and survival in soil. Thus, the net effect of elevated CO_2 and warming on rhizobia may depend on water availability.

IMPLICATIONS FOR THE MANAGEMENT OF AGROECOSYSTEMS

Human activities such as tillage, fertilization and cropping sequences dominate the disturbance regimes in many cropping systems. Intensive human intervention makes it difficult to predict global change effects on agroecosystems, but also provides opportunities for taking advantage of any potential beneficial impact of increased CO_2, N deposition, and temperature. Conventional agriculture is characterized by low diversity of plant species and high inorganic inputs of synthetic chemicals, such as fertilizers and pesticides. Recently, alternative agricultural practices such as no-tillage and organic farming, which significantly reduce disturbance frequency and inorganic inputs, have been increasingly adopted. In other less-disturbed agroecosystems such as pastures and forage lands, biological interactions and processes play a more important role in determining their structure and functioning. In this section, we will focus on some effects caused by global change that may be of direct implications for agroecosystem management.

Crop Community Structure and Crop-Weed Interactions

Increasing CO_2 concentration in the atmosphere will stimulate crop growth but have differential effects on C_3 and C_4 crops or weeds (Kimball, Kobayashi, and Bindi, 2002). C_3 plants may become more abundant in pasture and forage lands in a CO_2-enriched environment regardless of water availability (Gavazzi et al., 2000). Potvin and Vasseur (1997) demonstrated that CO_2 enrichment in a pasture for three years stimulated early-successional species and prevented complete dominance by late-successional species. This effect can be amplified over time as CO_2-enhanced growth may also enhance seed production (Edwards, Clark, and Newton, 2001), leading to alteration in dominant species. However, a temperature increase can theoretically cause changes in the grass community in favor of C_4 species over C_3 species as the former is more adapted to high temperatures (White et al., 2000). The community structure in intensive agroecosystems depends primarily on human intervention. New crop species or cultivars may emerge as warming changes growing seasons and planting times (Mahmood, 1997).

Weed composition may change due to differential effects of elevated CO_2 and warming on C_3 and C_4 weeds, modifying their competitive relationships with crops. For example, Ziska, Teasdale, and Bunce (1999) demonstrated that elevated CO_2 enhanced the tolerance of C_3 weed species to glyphosate, a widely used, phloem mobile, herbicide.

Elevated CO_2 benefits plant-microbial symbioses, thereby enhancing the competitive ability of some legume plants (Stocklin and Körner, 1999). The vast majority of agricultural plants form symbioses with arbuscular mycorrhizal fungi (AMF), and over the long term, CO_2-stimulation of AMF may alter the competitive ability of different plants. These effects should be taken into account in future grassland management because possible changes in the plant community affect the nutritional quality of grasses. Chronic N addition may enhance the competitive ability of invasive species over many native ones (Wedin and Tilman, 1996; Vitousek et al., 1997) and warming may allow the northward range expansion of other exotic noxious weeds in North America (Patterson, 1995). Together, N deposition and warming may enforce the dominance of a small number of invasive species.

Host-Parasite Interactions

Most crop insects reside in the soil for at least part of their life cycle and many pathogens are soilborne. Global changes and their associated soil alterations may modify the life cycles and activities of many parasites as well as the physiology of their hosts.

Global change may cause major changes in the geographical distribution of pests and their hosts (Coakley, Scherm, and Chakraborty, 1999). A warmer climate may generally promote the northward expansion of insect and pathogen distributions (Cannon, 1998). First, increases in food sources and changes in soil environmental conditions might affect insect activities and populations. Collier et al. (1991) reported that an increase of 3°C in the mean daily temperature caused the cabbage root fly (*Delia radicum* L.) to become about one month earlier in the year. Similarly, Bezemer, Jones, and Knight (1998) found that both elevated CO_2 (ambient + 200 mu mol/mol) and temperature (ambient + 2.0°C) increased aphid abundance in model terrestrial ecosystems. Secondly, elevated CO_2 and warming may alter insect community compositions and spatial distributions, leading to shifts in the dominant species in a certain region (Coviella and Trumble, 1999). In addition, CO_2 enrichment and warming might modify interactions between hosts and pests by altering crop tissue chemical composition, and insect life cycles and activity (Coviella and Trumble, 2000; Marks and Lincoln, 1996). Under elevated CO_2, consumption rates of insect herbivores generally increase, but this does not necessarily compensate fully for reduced leaf N (Cannon, 1998). Over the long term, it is not clear whether this reduction in tissue quality will result in more herbivorous consumption or reduced insect reproduction.

A few studies have examined the potential impact of global changes on crop diseases (Coakley, Scherm, and Chakraborty, 1999; Jwa and Walling, 2001; Patterson et al., 1999; Scherm et al., 2000). Increased C:N ratios in plant green leaves under elevated CO_2 may in general reduce pathogen infection. Elevated CO_2 also has the potential to alter plant physiology and response to pathogens. Tomato (*Lycopersicon esculentum* L.) plants showed a degree of tolerance against *Phytophthora parasitica* (root rot) at elevated CO_2 levels, probably due to the effect of CO_2 concentration on the transcription or post-translational turnover of pathogenesis-related proteins, or through increased photosynthesis and water use efficiency (Jwa and Walling, 2001).

Potential alterations in the relationships between pests and their natural enemies can pose serious challenges to pest control and changes in pest and crop physiology might affect pesticide uptake rates of both crops and insects (Coakley, Scherm, and Chakraborty, 1999; Patterson et al., 1999; Scherm et al., 2000). This risk could be significant because of the lack of complexity in many agroecosystems. Agroecosystem diversity could be critical in maintaining populations of natural enemies to reduce the possibility of pest outbreaks.

One likely scenario of the future climate under global change is increasing intra- and inter-annual variability in surface temperature, which may increase the uncertainty of pathogen or pest outbreaks. Dry, hot summers generally reduce infestations of most fungal diseases. Cannon (1998) suggested that marked changes in the distribution of well-documented species, e.g., Odonata in northwestern Europe, in response to unusually hot summers, provide useful indications of the potential effects of climate change.

Soil Organic Matter Preservation and C Storage in Agricultural Soils

CO_2 emissions caused by transformations of forest- and grass-lands for agriculture are the second most important source after fuel combustion. Cultivation for agriculture generally reduces soil OM and the fertility and sustainability of soils. Several factors need to be considered while estimating effects of global change on C balance of agricultural soils. First, many agricultural practices disturb soils, directly exposing soil OM to microbial attack. Second, many agricultural soils have been managed with high N loads over the years or decades, leading to relative N excess over C in soil. Third, residues in agricultural soils are highly decomposable as they consist predominantly of less recalcitrant materials with low lignin content. Finally, intensive disturbance and simplicity of agroecosystems further add to the vulnerability of organic materials to any change in climatic conditions.

In the last several decades, soil C in US agricultural lands has increased due to the adoption of conservation tillage and removal of lands from crop production (Eve et al., 2002) and C sequestration in agricultural soils has continuously been promoted to partially mitigate further CO_2 increase (Robertson, Paul, and Harwood, 2000). Enhanced residue quantity with higher C:N ratio under elevated CO_2 may be beneficial to C accumulation in agricultural soils. Direct evidence demonstrating significant CO_2-induced C sequestration in agricultural soils is still lacking (Kimball, Kobayashi, and Bindi, 2002). However, any beneficial effects from decreased residue quality may be of minor importance compared to the effect of varying crops, agricultural practices or land use (Frederiksen, Ronn, and Christensen, 2001).

Soil N status may be important in determining elevated CO_2 effects on soil OM dynamics in crop fields (Cheng and Johnson, 1998; Kimball, Kobayashi, and Bindi, 2002; Torbert et al., 2000). Nitrogen availability affects decomposition rates because microbes may be N-limited, especially when significant amounts of new residues are added (Wang and Bakken, 1997). Cheng and Johnson (1998) found that elevated CO_2 in wheat (*Triticum aestivum* L.) fields increased soil OM decomposition (22%) in the N-added treatment, but decreased soil OM decomposition (18%) in the treatment without N addition. These results clearly demonstrate the importance of integrative management of N fertilizers and residues.

Use Efficiency of Nitrogen Fertilizers

The use efficiency of N fertilizers is low and active N is highly mobile in the environment, which poses some great challenges both to agricultural production and the environment (Vitousek et al., 1997).

CO_2 enrichment has frequently resulted in increased efficiency of N fertilizer utilization (Mosier, 1998), probably due to enhanced plant N captures. For example, CO_2 enrichment increased total N uptake by accelerating cotton (*Gossypium hirsutum* L.) growth, though the concentrations of N-nutrient in cotton tissues decreased (Owensby, Auen, and Coyne, 1994; Prior et al., 1998). The increase in total N accumulation may result from physiological changes in plant N use efficiency rather than simply a size-dependent phenomenon due to accelerated dry matter production (Harmens et al., 2001). At the same time, CO_2-induced decrease in soil extractable N can reduce nitrogenous gas emissions and leaching from soil (Xu, Shaffer, and Al-kasisi, 1998). For instance, Hungate et al. (1997) showed that elevated CO_2 increased microbial N immobilization during wet-up, leading to a 43% reduction in gross nitrification and a 55% reduction in NO emission from soil.

Warming may lengthen plant growing period and total plant N uptake, reducing N leaching and N_2O emissions. N_2O emission rates are higher from soils in warm climates than from soils in temperate climates, because warm and moist conditions promote microbial processes that generate N_2O (Dobbie and Smith, 2001). However, warming might decrease N_2O emissions by decreasing the freezing-thawing frequency, as N_2O emissions are strongly affected by the freezing-thawing cycle (Hantschel, Kamp, and Bease, 1995; Ruser et al., 2001). Warmer temperatures during the early spring may stimulate microbial activities and N release, underlining the importance of effective residue management for synchronizing N release.

Nitrogen deposition in some regions adjacent to intensive agriculture fields and major urban areas has significantly altered N inputs (Vitousek et al., 1997). For example, N deposition may bring in N inputs as high as 60 kg $ha^{-1}year^{-1}$ in Germany (Weigel, Russow, and Korschens, 2000). Therefore, N inputs through this pathway constitute a significant source and should be taken into account in the design of fertilization strategies.

CONCLUDING REMARKS

Increasing evidence has demonstrated that global change can profoundly affect the population dynamics and activities of soil organisms and their interactions, modifying the structure and functioning of global terrestrial ecosystems. The alterations induced by elevated CO_2, N deposition, and global warming are particularly important in understanding effects of global change on agriculture as well as the roles of agroecosystems in terrestrial feedbacks to global change.

Most studies have predicted that the impact on food and forage crops of mid- to high-latitude regions will be less severe than in low-latitude regions, and possibly even beneficial, although the influence of pests and diseases is rarely taken into account (Cannon, 1998). However, changes in pest and disease outbreaks that may devastate crop production warrant more attention in future studies. Studies integrating effects of climatic changes, precipitation, and temperature with biological processes, are urgently needed. The continuing decline in complexity of many agroecosystems across the world poses great risks to food production in a changing climate. Uncertainty about future climates further underlines the need for complex agricultural systems that are more dependable in production and more sustainable in terms of resource conservation than simple ones.

The roles of agroecosystems in mitigating further air CO_2 increase are unclear and the potential of C sequestration in agricultural soils requires further investigation (Kimball, Kobayashi, and Bindi, 2002). In the last several decades, C sequestration in agricultural soils of North America and Western Europe has likely occurred mainly due to conservation of farming lands and introduction of alternative farming practices. However, the potential for C sequestration in the vast developing world is uncertain.

Humans will continue to face the dilemma in managing N cycling: Low N availability limits the crop production in many agricultural soils and the relative N excess causes undesirable changes in the composition and functioning of many natural ecosystems. This conflict highlights the needs for better N management strategies that are based on a sound understanding of N cycling in nature. It would be useful to focus attention on transformation and the fate of N fertilizers in future research.

REFERENCES

Allen, A. S., J. A. Andrews, A. C. Finzi, R. Matamala, D. D. Richter, and W. H. Schlesinger. (2000). Effects of free-air CO_2 enrichment (FACE) on belowground processes in a *Pinus taeda* forest. *Ecological Applications* 10:437-448.

Baron, J. S., H. M. Rueth, A. M. Wolfe, K. R. Ndick, E. J. Allstott, J. T. Minear, and B. Moraska. (2000). Ecosystem responses to nitrogen deposition in the Colorado Front Range. *Ecosystems* 3:352-368.

Bazzaz, F. F. (1990). The response of natural ecosystems to the rising global CO_2 levels. *Annual Review of Ecology and Systematics* 21:167-196.

Berntson, G. M. and F. A. Bazzaz. (1998). Regenerating temperate forest mesocosms in elevated CO_2: Belowground growth and nitrogen cycling. *Oecologia* 113:115-125.

Bezemer, T. M., J. H. Jones, and K. J. Knight. (1998). Long-term effects of elevated CO_2 and temperature on populations of the peach potato aphid *Myzus persicae* and its parasitoid *Aphidius matricariae*. *Oecologia* 116:128-135.
Booker, F. L. and C. A. Maier. (2001). Atmospheric carbon dioxide, irrigation, and fertilization effects on phenolic and nitrogen concentrations in loblolly pine (*Pinus taeda*) needles. *Tree Physiology* 21:609-616.
Bowman, W. D. and H. Steltzer. (1998). Positive feedbacks to anthropogenic nitrogen deposition in Rocky Mountain Alpine tundra. *AMBIO* 27:514-517.
Briones, M. J. I., P. Ineson, and T. G. Piearce. (1997). Effects of climate change on soil fauna; Responses of enchytraeids, Diptera larvae and tardigrades in a transplant experiment. *Applied Soil Ecology* 6: 117-134.
Cannell, M. G. R., and J. H. M. Thornley. (1998). N-poor ecosystems may respond more to elevated [CO_2] than N-rich ones in the long term. A model analysis of grassland. *Global Change Biology* 4:431-442.
Cannon, R. J. C. (1998). The implications of predicted climate change for insect pests in the UK, with emphasis on non-indigenous species. *Global Change Biology* 4:785-796.
Ceulemans, R, I. A. Janssens, and M. E. Jach. (1999). Effects of CO_2 enrichment on trees and forests: Lessons to be learned in view of future ecosystem studies. *Annals of Botany* 8:577-590.
Cheng, W. X. and D. W. Johnson. (1998). Elevated CO_2, rhizosphere processes, and soil organic matter decomposition. *Plant and Soil* 202:167-174.
Christensen, T. R., A. Michelsen, S. Jonasson, and I. K. Schmidt. (1997). Carbon dioxide and methane exchange of a subarctic heath in response to climate change related environmental manipulations. *Oikos* 79:34-44.
Coakley, S. M., H. Scherm, and S. Chakraborty. (1999). Climate change and plant disease management. *Annual Review of Phytopathology* 37:399-426.
Coleman, D. C. and D. A. Crossley. (1996). *Fundamentals of Soil Ecology*. San Diego, CA: Academic Press, Inc. pp. 51-106.
Coley, P. D., J. P. Bryant, and F. S. Chapin, III. (1985). Resource availability and plant antiherbivore defense. *Science* 230:895-899.
Collier, R. H., S. Finch, K. Phelps, and A. R. Thompson. (1991). Possible impact of global warming on Cabbage root fly (*Delia-radicum*) activity in the UK. *Annals of Applied Biology* 118:261-271.
Constable, J. V. H., H. Bassirirad, J. Lussenhop, and A. Zerihun. (2001). Influence of elevated CO_2 and mycorrhizae on nitrogen acquisition: Contrasting responses in *Pinus taeda* and *Liquidambar styraciflua*. *Tree Physiology* 21:83-91.
Coûteaux, M. M. and T. Bolger. (2000). Interactions between atmospheric CO_2 enrichment and soil fauna. *Plant and Soil* 224:123-134.
Coviella, C. E. and J. T. Trumble. (1999). Effects of elevated atmospheric carbon dioxide on insect-plant interactions. *Conservation Biology* 13:700-712.
Coviella, C. E. and J. T. Trumble. (2000). Effect of elevated atmospheric carbon dioxide on the use of foliar application of *Bacillus thuringiensis*. *Biocontrol* 45:325-336.
David, J. F., J. F. Ponge, P. Arpin, and G. Vannier. (1999). Periods of dormancy and cohort-splitting in the millipede *Polydesmus angustus* (Diplopoda, Polydesmidae). *European Journal of Entomology* 96:111-116.

De Luis, I., J. J. Irigoyen, and M. Sanchez-Diaz. (1999). Elevated CO_2 enhances plant growth in droughted N_2-fixing alfalfa without improving water status. *Physiologia Plantarum* 107:84-89.

De Valpine, P. and J. Harte. (2001). Plant responses to experimental warming in a montane meadow. *Ecology* 82:637-648.

DeLucia, E. H., J. G. Hamilton, S. L. Naidu, R. B. Thomas, J. A. Andrews, A. Finzi, M. Lavine, R. Matamala, J. E. Mohan, G. R. Hendrey, and W. H. Schlesinger. (1999). Net primary production of a forest ecosystem with experimental CO_2 enrichment. *Science* 284:1177-1179.

Dhillion, S. S., J. Roy, and M. Abrams. (1996). Assessing the impact of elevated CO_2 on soil microbial activity in a Mediterranean model ecosystem. *Plant and Soil* 187:333-342.

Dobbie, K. E. and K. A. Smith. (2001). The effects of temperature, water-filled pore space and land use on N_2O emissions from an imperfectly drained gleysol. *European Journal of Soil Science* 52:667-673.

Drake, B. G. (1992). The impact of rising CO_2 on ecosystem production. *Water Air Soil Pollution* 64:25-44.

Edwards, G. R., H. Clark, and P. C. D. Newton. (2001). The effects of elevated CO_2 on seed production and seedling recruitment in a sheep-grazed pasture. *Oecologia* 127:383-394.

Egerton-Warburton, L. M. and E. B. Allen. (2000). Shifts in arbuscular mycorrhizal communities along an anthropogenic nitrogen deposition gradient. *Ecological Applications* 10:484-496.

Estiarte, M., J. Penuelas, B. A. Kimball, D. L. Hendrix, P. J. Pinter, G. W. Wall, R. L. LaMorte, and D. J. Hunsaker. (1999). Free-air CO_2 enrichment of wheat: Leaf flavonoid concentration throughout the growth cycle. *Physiologia Plantarum* 105: 423-433.

Eve, M. D., M. Sperow, K. Paustian, and R. F. Follett. (2002). National-scale estimation of changes in soil carbon stocks on agricultural lands. *Environmental Pollution* 116:431-438.

Frederiksen, H. B., R. Ronn, and S. Christensen. (2001). Effect of elevated atmospheric CO_2 and vegetation type on microbiota associated with decomposing straw. *Global Change Biology* 7:313-321.

Gaffen, D. J., B. D. Santer, J. S. Boyle, J. R. Christy, N. E. Graham, and R. J. Ross. (2000). Multidecadal changes in the vertical temperature structure of the tropical troposphere. *Science* 287:1242-1245.

Gavazzi, M., J. Seiler, W. Aust, and S. Zedaker. (2000). The influence of elevated carbon dioxide and water availability on herbaceous weed development and growth of transplanted loblolly pine (*Pinus taeda*). *Environmental and Experimental Botany* 44:185-194.

Gebauer, R. L. E., B. R. Strain, and J. F. Reynolds. (1998). The effect of elevated CO_2 and N availability on tissue concentrations and whole plant pools of carbon-based secondary compounds in loblolly pine (*Pinus taeda*). *Oecologia* 113:29-36.

Grayston, S. J., C. D. Campbell, J. L. Lutze, and R. M. Gifford. (1998). Impact of elevated CO_2 on the metabolic diversity of microbial communities in N-limited grass swards. *Plant and Soil* 203:289-300.

Grogan, P. and F. S. Chapin. (2000). Initial effects of experimental warming on above- and belowground components of net ecosystem CO_2 exchange in arctic tundra. *Oecologia* 125:512-520.

Hantschel, R. E., T. Kamp, and F. Beese. (1995). Increasing the soil temperature to study global warming effects on the soil nitrogen cycle in agroecosystems. *Journal of Biogeography* 22:375-380.

Harmens, H., C. Marshall, C. M. Stirling, and J. F. Farrar. (2001). Partitioning and efficiency of use of N in *Dactylis glomerata* as affected by elevated CO_2: Interaction with N supply. *International Journal of Plant Sciences* 162:1267-1274.

Harte, J., A. Rawa, and V. Price. (1996). Effects of manipulated soil microclimate on mesofaunal biomass and diversity. *Soil Biology & Biochemistry* 28:313-322.

Harte, J. and R. Shaw. (1995). Shifting dominance within a mountain vegetation community-results of a climate-warming experiment. *Science*. 267:876-880.

Hattenschwiler, S. and C. Körner. (2000). Biomass allocation and canopy development in spruce model ecosystems under elevated CO_2 and increased N deposition. *Oecologia*. 106:172-180.

Heijmans, M. M. P. D., F. Berendse, W. J. Arp, A. K. Masselink, H. Klees, W. de Visser, and N. van Breemen.(2001). Effects of elevated carbon dioxide and increased nitrogen deposition on bog vegetation in the Netherlands. *Journal of Ecology* 89:268-279.

Herms, D. A. and W. J. Mattson. (1992). The dilemma of plants: To grow or defend. *Quarterly Review of Biology* 67: 283-335.

Hodge, A. (1996). Impact of elevated CO_2 on mycorrhizal associations and implications for plant growth. *Biology and Fertility of Soils* 23:388-398.

Holland, E. A., B. H. Braswell, J. F. Lanmarque, A. Townsend, J. Suylzman, J. F. Mueller, F. Dentener, G. Brasseur, H. II Levy, J. E. Penner, and G. Roelofs. (1997). Variations in the predicted spatial distribution of atmospheric nitrogen deposition and their impact on carbon uptake by terrestrial ecosystems. *Journal of Geophysical Research (Atmospheres)* 102:15849-15866.

Houghton, J. (2001). The science of global warming. *Interdisciplinary Science Reviews* 26: 247-257.

Hu, S. J., M. K. Firestone, and F. S. Chapin, III. (1999). Soil microbial feedbacks to atmospheric CO_2 enrichment. *Trends in Ecology & Evolution* 14:432-437.

Hu, S., F. S. Chapin, III., M. K. Firestone, C. B. Field, and N. R. Chiariello. (2001). Nitrogen limitation of microbial decomposition in a grassland under elevated CO_2. *Nature* 409:188-191.

Hungate, B. A., C. H. Jaeger, G. Gamara, F. S. Chapin, and C. B. Field. (2000). Soil microbiota in two annual grasslands: Responses to elevated atmospheric CO_2. *Oecologia* 124:589-598.

Hungate, B. A., E. A. Holland, R. B. Jackson, F. S. Chapin, H. A. Mooney, and C. B. Field. (1997). The fate of carbon in grasslands under carbon dioxide enrichment. *Nature* 388:576-579.

Hungate, B. A., P. Dijkstra, D. W. Johnson, C. R. Hinkle, and B. G. Drake. (1999). Elevated CO_2 increases nitrogen fixation and decreases soil nitrogen mineralization in Florida scrub oak. *Global Change Biology* 5:781-789.

Hungria, M. and M. A. T. Vargas. (2000). Environmental factors affecting N_2 fixation in grain legumes in the tropics, with an emphasis on Brazil. *Field Crops Research* 65:151-164.

Idso, S. B. and K. E. Idso. (2001). Effects of atmospheric CO_2 enrichment on plant constituents related to animal and human health. *Environmental and Experimental Botany* 45:179-199.

Inubushi, K., M. Hoque, S. Miura, K. Kobayashi, H. Y. Kim, M. Okada, and S. Yabashi. (2001). Effect of free-air CO_2 enrichment (FACE) on microbial biomass in paddy field soil. *Soil Science and Plant Nutrition* 47:737-745.

IPCC (Intergovernmental Panel for Climate Change). (1996). *Climate Change. The Second Assessment Report of IPCC Working Group I.* Cambridge, UK: Cambridge University Press.

Jablonski, L.M., X. Z.Wang, and P. S. Curtis (2002). Plant reproduction under elevated CO_2 conditions: A meta-analysis of reports on 79 crop and wild species. *New Phytologist* 156:9-26.

Jenkinson, D. S. (2001). The impact of humans on the nitrogen cycle, with focus on temperate arable agriculture. *Plant and Soil* 228:3-15.

Jones, T. H., L. J. Thompson, J. H. Lawton, T. M. Bezemer, R. D. Bardgett, T. M. Blackburn, K. D. Bruce, P. F. Cannon, G. S. Hall, S. E. Hartley, G. Howson, C. G. Jones, C. Kampichler, E. Kandeler, and D. A. Ritchie. (1998). Impacts of rising atmospheric carbon dioxide on model terrestrial ecosystems. *Science* 280:441-443.

Jwa, N. S. and L. L. Walling. (2001). Influence of elevated CO_2 concentration on disease development in tomato. *New Phytologist.*149:509-518.

Kellomaki, S., I. Rouvinen, H. Peltola, H. Strandman, and R. Steinbrecher. (2001). Impact of global warming on the tree species composition of boreal forests in Finland and effects on emissions of isoprenoids. *Global Change Biology* 7:531-544.

Kimball, B.A., Kobayashi, K., and Bindi, M. (2002). Responses of agricultural crops to free-air CO_2 enrichment. *Advances in Agronomy* 77:293-368.

King, J. S., K. S. Pregitzer, D. R. Zak, J. Sober, J. G. Isebrands, R. E. Dickson, G. R. Hendrey, and D. F. Karnosky. (2001). Fine-root biomass and fluxes of soil carbon in young stands of paper birch and trembling aspen as affected by elevated atmospheric CO_2 and tropospheric O_3. *Oecologia* 128:237-250.

Kinney, K. K., R. L. Lindroth, S. M. Jung, and E. V. Nordheim. (1997). Effects of CO_2 and NO_3-availability on deciduous trees: Phytochemistry and insect performance. *Ecology* 78:215-230.

Kirschbaum, M. U. F. (2000). Will changes in soil organic carbon act as a positive or negative feedback on global warming? *Biogeochemistry* 48:21-51.

Klironomos, J. N., M. C. Rillig, and M. F. Allen. (1996). Below-ground microbial and microfaunal responses to *Artemisia tridentata* grown under elevated atmospheric CO_2. *Functional Ecology* 10:527-534.

Körner, C. (2000). Biosphere responses to CO_2 enrichment. *Ecological Applications* 10:1590-1619.

Körner, C. and J. A. Arnone. (1992). Responses to elevated carbon dioxide in artificial tropical ecosystems. *Science* 257:1672-1675.

Lilley, J. M., T. P. Bolger, M. B. Peoples, and R. M. Gifford. (2001). Nutritive value and the nitrogen dynamics of *Trifolium subterraneum* and *Phalaris aquatica* under warmer, high CO_2 conditions. *New Phytologist* 150:385-395.

Liski, J., H. Ilvesniemi, A. Makela, and C. J. Westman. (1999). CO_2 emission from soil in response to climatic warming are overestimated-The decomposition of old soil organic matter is tolerant of temperature. *Ambio* 28:171-174.

Loiseau, P. and J. F. Soussana. (2000). Effects of elevated CO_2, temperature and N fertilization on nitrogen fluxes in a temperate grassland ecosystem. *Global Change Biology* 6:953-965.

Luo, Y., S. Wan, D. Hui, and L. L. Wallace. (2001). Acclimatization of soil respiration to warming in a tall grass prairie. *Nature* 413:622-625.

Mahmood, R. (1997). Impacts of air temperature variations on the bore rice phenology in Bangladesh: Implications for irrigation requirements. *Agricultural and Forest Meteorology* 84:233-247.

Marks, S. and D. E. Lincoln. (1996). Antiherbivore defense mutualism under elevated carbon dioxide levels: A fungal endophyte and grass. *Environmental Entomology* 25:618-623.

Meier, M. and J. Fuhrer. (1997). Effect of elevated CO_2 on orchard grass and red clover grown in mixture at two levels of nitrogen or water supply. *Environmental and Experimental Botany* 38:251-262.

Mertens, S., I. Nijs, M. Heuer, F. Kockelbergh, L. Beyens, A. Van Kerckvoorde, and I. Impens. (2001). Influence of high temperature on end-of-season tundra CO_2 exchange. *Ecosystems* 4:226-236.

Montealegre, C. M., C. van Kessel, J. M. Blumenthal, H. G. Hur, U. A. Hartwig, and M. J. Sadowsky. (2000). Elevated atmospheric CO_2 alters microbial population structure in a pasture ecosystem. *Global Change Biology* 6:475-482.

Mosier, A. R. (1998). Soil processes and global change. *Biology and Fertility of Soils* 27:221-229.

Nadelhoffer, K. J. (2000). The potential effects of nitrogen deposition on fine-root production in forest ecosystems. *New Phytologist* 147:131-139.

Nakamura, T., T. Koike, T. Lei, K. Ohashi, T. Shinano, and T. Tadano. (1999). The effect of CO_2 enrichment on the growth of nodulated and non-nodulated isogenic types of soybean raised under two nitrogen concentrations. *Photosynthetica* 37: 61-70.

Nijs, I., J. Roy, J. L. Salager, and J. Fabreguettes. (2000). Elevated CO_2 alters carbon fluxes in early successional Mediterranean ecosystems. *Global Change Biology* 6:981-994.

Nilsen, P., I. Borja, H. Knutsen, and R. Brean. (1998). Nitrogen and drought effects on ectomycorrhizae of Norway spruce [*Picea abies* L. (Karst.)]. *Plant and Soil* 198: 179-184.

Norby, R. J., M. F. Cotrufo, P. Ineson, E. G. O'Neill, and J. G. Canadell. (2001). Elevated CO_2, litter chemistry, and decomposition: A synthesis. *Oecologia* 127:153-165.

Oren, R., D. S. Ellsworth, K. H. Johnsen, N. Phillips, B. E. Ewers, C. Maier, K. V. R. Schafer, H. McCarthy, G. Hendrey, S. G. McNulty, and G. G. Katul. (2001). Soil fertility limits carbon sequestration by forest ecosystems in a CO_2-enriched atmosphere. *Nature* 411:469-472.

Owensby, C. E., L. M. Auen, and P. I. Coyne. (1994). Biomass production in a nitrogen-fertilized, tallgrass prairie ecosystem exposed to ambient and elevated levels of CO_2. *Plant and Soil* 165:105-113.

Owensby, C. E., P. I. Coyne, and L. M. Auen. (1993). Nitrogen and phosphorus dynamics of a tallgrass prairie ecosystem exposed to elevated carbon dioxide, *Plant Cell and Environment* 16:843-850.

Padgett, P. E., E. B. Allen, and R. A. Bytnerowicz. (1999). Changes in soil inorganic nitrogen as related to atmospheric nitrogenous pollutants in southern California. *Atmospheric Environment* 33:769-781.

Paschke, M. W., T. McLendon, and E. F. Redente. (2000). Nitrogen availability and old-field succession in a shortgrass steppe. *Ecosystems* 3:144-158.

Paterson, E., J. M. Hall, E. A. S. Rattray, B. S. Griffiths, K. Ritz, and K. Killham. (1997). Effect of elevated CO_2 on rhizosphere carbon flow and soil microbial processes. *Global Change Biology* 3:363-377.

Patterson, D. T. (1995). Weeds in a changing climate. *Weed Science* 43:685-701.

Patterson, D. T., J. K. Westbrook, R. J.V. Joyce, P. D. Lingren, and J. Rogasik. (1999). Weeds, insects and diseases. *Climatic Change* 43:711-727.

Paul, E. A. and F. E. Clark. (1996). *Soil Microbiology and Biochemistry.* San Diego, CA: Academic Press.

Penuelas, J. and M. Estiarte. (1998). Can elevated CO_2 affect secondary metabolism and ecosystem function? *Trends in Ecology & Evolution* 13:20-24.

Petchey, O. L., P. T. McPhearson, T. M. Casey, and P. J. Morin. (1999). Environmental warming alters food-web structure and ecosystem function. *Nature* 402:69-72.

Pooter, H., C. Roumet, and B. D. Campbell. (1996). Interspecific variation in the growth response of plants to elevated CO_2: A search for functional types. In *Carbon dioxide, Populations, and Communities*, eds. C. Körner, and F. A. Bazzaz, New York, NY: Academic Press, Inc. pp. 375-412.

Potvin, C. and L. Vasseur. (1997). Long-term CO_2 enrichment of a pasture community: Species richness, dominance, and succession. *Ecology* 78:666-677.

Pregitzer, K. S., J. S. King, A. J. Burton, and S. S. Brown. (2000). Responses of tree fine roots to temperature. *New Phytologist* 147:105-115.

Prior, S. A., H. A. Torbert, G. B. Runion, G. L. Mullins, H. H. Rogers, and J. R. Mauney. (1998). Effects of carbon dioxide enrichment on cotton nutrient dynamics. *Journal of Plant Nutrition* 21:1407-1426.

Reich, P. B., J. Knops, D. Tilman, J. Craine, D. Ellsworth, M. Tjoelker, T. Lee, D. Wedin, S. Naeem, D. Bahauddin, G. Hendrey, S. Jose, K. Wrage, J. Goth, and W. Bengston. (2001). Plant diversity enhances ecosystem responses to elevated CO_2 and nitrogen deposition. *Nature* 410:809-812.

Rice, C. W., F. O. Garcia, C. O. Hampton, and C. E. Owensby. (1994). Soil microbial response in tallgrass prairie to elevated CO_2. *Plant and Soil* 165:67-74.

Rillig, M. C., C. B. Field, and M. F. Allen. (1999a). Soil biota responses to long-term atmospheric CO_2 enrichment in two California annual grasslands. *Oecologia* 119: 572-577.

Rillig, M. C., C. B. Field, and M. F. Allen. (1999b). Fungal root colonization responses in natural grasslands after long-term exposure to elevated atmospheric CO_2. *Global Change Biology* 5:577-585.

Rillig, M. C., G. Y. Hernandez, and P. C. D. Newton. (2000). Arbuscular mycorrhizae respond to elevated atmospheric CO_2 after long-term exposure: Evidence from a

CO_2 spring in New Zealand supports the resource balance model. *Ecology Letters* 3:475-478.

Robertson, G. P., E. A. Paul, and R. R. Harwood. (2000). Greenhouse gases in intensive agriculture: Contributions of individual gases to the radiative forcing of the atmosphere. *Science* 289:1922-1925.

Rouhier, H., G. Billes, A. Elkohen, M. Mousseau, and P. Bottner. (1994). Effect of elevated CO_2 on carbon and nitrogen distribution within a tree (*Castanea sativa* Mill.) soil system. *Plant and Soil* 162:281-292.

Ruess, L., A. Michelsen, I. K. Schmidt, and S. Jonasson. (1999). Simulated climate change affecting microorganisms, nematode density and biodiversity in subarctic soils. *Plant and Soil* 212:63-73.

Ruser, R., H. Flessa, R. Schilling, F. Beese, and J. C. Munch. (2001). Effect of crop-specific field management and N fertilization on N_2O emissions from a fine-loamy soil. *Nutrient Cycling in Agroecosystems* 59:177-191.

Rustad, L. E., J. L. Campbell, G. M. Marion, R. J. Norby, M. J. Mitchell, A. E. Hartley, J. H. C. Cornelissen, and J. Gurevitch. (2001). A meta-analysis of the response of soil respiration, net nitrogen mineralization, and aboveground plant growth to experimental ecosystem warming. *Oecologia* 126:543-562.

Santer, B. D., T. M. L. Wigley, D. J. Gaffen, L. Bengtsson, C. Doutriaux, J. S. Boyle, M. Esch, J. J. Hnilo, P. D. Jones, G. A. Meehl, E. Roeckner, K. E. Taylor, and M. F. Wehner. (2000). Interpreting differential temperature trends at the surface and in the lower troposphere. *Science* 287:1227-1232.

Scherm, H., R.W. Sutherst, R. Harrington, and J. S. I. Ingram. (2000). Global networking for assessment of impacts of global change on plant pests. *Environmental Pollution* 108:333-341.

Schlesinger, W. H. (1996). *Biogeochemistry: An Analysis of Global Change*. San Diego, CA: Academic Press.

Serraj, R., T. R. Sinclair, and L. H. Allen. (1998). Soybean nodulation and N_2 fixation response to drought under carbon dioxide enrichment. *Plant Cell and Environment* 21:491-500.

Shaver, G. R., J. Canadell, F. S. Chapin, J. Gurevitch, J. Harte, G. Henry, P. Ineson, S. Jonasson, J. Melillo, L. Pitelka, and L. Rustad. (2000). Global warming and terrestrial ecosystems: A conceptual framework for analysis. *Bioscience* 50:871-882.

Shaw, M. R. and J. Harte. (2001). Control of litter decomposition in a subalpine meadow-sagebrush steppe ecotone under climate change. *Ecological Applications* 11:1206-1223.

Smil, V. (1999). Nitrogen in crop production: An account of global flows. *Global Biogeochemical Cycles* 13:647-662.

Soussana, J. F. and U. A. Hartwig. (1996). The effects of elevated CO_2 on symbiotic N_2 fixation: A link between the carbon and nitrogen cycles in grassland ecosystems. *Plant and Soil* 187:321-332

Sternberg, M., V. K. Brown, G. J. Masters, and I. P. Clarke. (1999). Plant community dynamics in a calcareous grassland under climate change manipulations. *Plant Ecology* 143:29-37.

Stocklin, J. and C. Körner. (1999). Interactive effects of elevated CO_2, P availability and legume presence on calcareous grassland: Results of a glasshouse experiment. *Functional Ecology* 13:200-209.

Strain, B. R. (1987). Direct effects of increasing atmospheric CO_2 on plants and ecosystems. *Trends in Ecology & Evolution* 2:18-21.

Sturm, M., C. Racine, and K. Tape. (2001). Climate change- Increasing shrub abundance in the Arctic. *Nature*, 411:546-547.

Sulkava, P., V. Huhta, and J. Laakso. (1996). Impact of soil faunal structure on decomposition and N-mineralisation in relation to temperature and moisture in forest soil. *Pedobiologia* 40:505-513.

Thomas, R. G. B., M. A. Bashkin, and D. D. Richter. (2000). Nitrogen inhibition of nodulation and N_2 fixation of a tropical N_2-fixing tree (*Gliricidia sepium*) grown in elevated atmospheric CO_2. *New Phytologist* 145:233-243.

Torbert, H. A., S. A. Prior, H. H. Rogers, and C. W. Wood. (2000). Review of elevated atmospheric CO_2 effects on agro-ecosystems: residue decomposition processes and soil C storage. *Plant and Soil* 224:59-73.

Treseder, K. K. and M. F. Allen. (2000). Mycorrhizal fungi have a potential role in soil carbon storage under elevated CO_2 and nitrogen deposition. *New Phytologist* 147:189-200.

Treseder, K.K. and M. F. Allen. (2002). Direct nitrogen and phosphorus limitation of arbuscular mycorrhizal fungi: a model and field test. *New Phytologist* 155:507-515.

Tsukamoto, J., and H. Watanabe. (1997). Influence of temperature on hatching and growth of *Eisenia foetida* (oligochaeta, Lumbricidae). *Pedobiologia* 17:338-342.

Verburg, P. S. J., A. Gorissen, and W. J. Arp. (1998). Carbon allocation and decomposition of root-derived organic matter in a plant-soil system of *Calluna vulgaris* as affected by elevated CO_2. *Soil Biology & Biochemistry* 30:1251-1258.

Vitousek, P. M. (1994). Beyond global warming-ecology and global change. *Ecology* 75:1861-1876.

Vitousek, P. M., J. D. Aber, R. W. Howarth, G. E. Likens, P. A. Matson, D. W. Schindler, W. H. Schlesinger, and D. G. Tilman. (1997). Human alteration of the global nitrogen cycle: Sources and consequences. *Ecological Applications* 7:737-750

Wang, J. G., and L. R. Bakken. (1997). Competition for nitrogen during mineralization of plant residues in soil: Microbial response to C and N. *Soil Biology & Biochemistry* 29:163-170.

Wedin, D. A. and D. Tilman. (1996). Influence of nitrogen loading and species composition on the carbon balance of grasslands. *Science* 274:1720-1723.

Weigel, A., R. Russow, and M. Korschens. (2000). Quantification of airborne N-input in long-term field experiments and its validation through measurements using N-15 isotope dilution. *Journal of Plant Nutrition and Soil Science* 163:261-265.

White, T. A., B. D. Campbell, P. D. Kemp, and C. L. Hunt. (2000). Sensitivity of three grassland communities to simulated extreme temperature and rainfall events. *Global Change Biology* 6:671-684.

Whorf, T. and C. D. Keeling. (1998). Rising carbon. *New Scientist* 157:54-54.

Wiemken, V., E. Laczko, K. Ineichen, and T. Boller. (2001). Effects of elevated carbon dioxide and nitrogen fertilization on mycorrhizal fine roots and the soil microbial community in beech-spruce ecosystems on siliceous and calcareous soil. *Microbial Ecology* 42:126-135.

Wuebbles, D. J., A. Jain, J. Edmonds, D. Harvey, and K. Hayhoe. (1999). Global change: State of the Science. *Environmental Pollution* 100:57-86.

Xu, C., M. J. Shaffer, and M. Al-kasisi. (1998). Simulating the impact of management practices on nitrous oxide emissions. *Soil Science Society of America Journal* 62:736-742.

Zak, D. R., K. S. Pregitzer, J. S. King, and W. E. Holmes. (2000b). Elevated atmospheric CO_2, fine roots and the response of soil microorganisms: A review and hypothesis. *New Phytologist* 147:201-222.

Zak, D. R., K. S. Pregitzer, P. S. Curtis, and W. E. Holmes. (2000a). Atmospheric CO_2 and the composition and function of soil microbial communities. *Ecological Applications* 10:47-59.

Zak, D.R., K. S. Pregitzer, P. S. Curtis, J. A. Teeri, R. Fogel, and D. L. Randlett. (1993). Elevated CO_2 and feedback between carbon and nitrogen cycles. *Plant and Soil* 151:105-117.

Zanetti, S., U.A. Hartwig, A. Luscher, T. Hebeisen, M. Frehner, B. U. Fischer, G. R. Hendrey, H. Blum, and J. Nosberger. (1996). Stimulation of symbiotic N_2 fixation in *Trifolium repens* L. under elevated atmospheric pCO_2 in a grassland ecosystem. *Plant Physiology* 112:575-583.

Ziska, L. H., J. R. Teasdale, and J. A. Bunce. (1999). Future atmospheric carbon dioxide may increase tolerance to glyphosate. *Weed Science* 47:608-615.

The Importance of Biodiversity in Agroecosystems

Lori Ann Thrupp

SUMMARY. This article identifies the important functions and benefits of biodiversity in agriculture, for sustainable crop production and food security. After clarifying impacts from the decline of biodiversity in agriculture, the article summarizes principles, practices and policies for biodiversity conservation and enhancement in farming systems and in landscapes. Some of the strategies identified build upon valuable local experiences and knowledge in traditional farming practices, while others take advantage of recent scientific findings in agroecology and ecosystem health. The analysis suggests the value of adopting an agroecosystems approach, beyond a focus on genetic resource conservation alone–to implement other biodiversity-enhancing methods in farms,

---

Lori Ann Thrupp is former Director of Sustainable Agriculture, World Watch Institute, 1776 Massachusetts Avenue N.W., Washington, DC 20036-1904. Current address: 5930 Fern Street, El Cerrito, CA 94530 (E-mail: athrupp@igc.org).

The author is grateful to research assistants Rita Banerji, Dina Matthews, and Nabiha Megateli, and to the following colleagues at the World Resources Institute for sharing their expertise: Walt Reid, Kenton Miller, Thomas Fox, Robert Blake, Arthur Getz, Nels Johnson, Paul Faeth, and Consuelo Holguin. The author also appreciates the comments from the World Bank Environment and Agriculture Departments, particularly Jitendra Srivastava, Lars Vidaeus, and John Kellenberg, and also the input of Calestous Juma, Miguel Altieri, Kristin Schafer, Daniel Debouck, and David Williams. The author appreciates support from the World Bank and Swedish International Development Cooperation Agency for this project.

[Haworth co-indexing entry note]: "The Importance of Biodiversity in Agroecosystems." Thrupp, Lori Ann. Co-published simultaneously in *Journal of Crop Improvement* (Food Products Press, an imprint of The Haworth Press, Inc.) Vol. 12, No. 1/2 (#23/24), 2004, pp. 315-337; and: *New Dimensions in Agroecology* (ed: David Clements, and Anil Shrestha) Food Products Press, an imprint of The Haworth Press, Inc., 2004, pp. 315-337. Single or multiple copies of this article are available for a fee from The Haworth Document Delivery Service [1-800-HAWORTH, 9:00 a.m. - 5:00 p.m. (EST). E-mail address: docdelivery@haworthpress.com].

http://www.haworthpress.com/web/JCRIP

Digital Object Identifer: 10.1300/J411v12n01_03

such as integrated ecological pest and soil management. Attention is also given to the challenges by conflicting agricultural policies that contribute to the decline of biodiversity in farming systems, and potential of reforming such policies. The practices and approaches reviewed in this analysis show effective ways to link biodiversity conservation and sustainable agriculture.

KEYWORDS. Agrobiodiversity, agroecology, biological diversity, genetic diversity, participatory approach, resource conservation, sustainable agriculture

INTRODUCTION

The functions of biodiversity have been recognized in recent studies of agriculture and farming systems. Scientists and farmers have given increasing attention to agricultural biodiversity (called "agrobiodiversity") in research projects and scientific literature, and policy implications of agrobiodiversity losses have been discussed in international conventions and policy documents (CBD, 2001; FAO, 1998, 1999; Forno and Smith, 1996; Kellenberg and Vidaeus, 1997; Shand, 1996). The incorporation of biodiversity in indigenous crop management practices on traditional small farms has proven to be beneficial for ecosystem health and contributes to productivity in the broader farming system. At the same time, farmers who are currently pursuing sustainable farming have also reaped practical values from conserving and enhancing biodiversity in both large- and small-scale crop production.

Growing numbers of studies and farmers' initiatives reveal that biodiversity is fundamental to agricultural production, food technology innovations, and food security, as well as being an ingredient of environmental conservation (Brookfield and Padoch, 1994; CBD, 2001; FAO, 1998, 1999; Forno and Smith, 1996). Yet, studies have shown biodiversity in agroecosystems has been declining and has become eroded at the genetic, species, and ecosystem level–in plants, animals, and soils. This decline of agricultural biodiversity has caused economic losses, jeopardizing productivity and food security, and leading to broader social costs. Also of concern is the loss of biodiversity in "natural" habitats from the expansion of agricultural production to frontier areas.

In spite of emerging evidence and scientific literature in this field by selected authors and farmers, there is still a great lack of understanding and appreciation of agricultural biodiversity (Thrupp, 1998; 2000). Only a minimal number of scientists in agriculture universities are addressing this matter, and few organizations or programs are taking this issue seriously as part of research and development efforts. The trend of agricultural biodiversity erosion is continuing at a rapid pace in many places–as monocultural systems and uniform technology models tend to predominate in industrial agricultural production in the Western World, and increasingly in developing countries as well.

This paper identifies the main features, functions and benefits of biodiversity in agriculture, including ecosystem services and other roles in a broad farming systems perspective. It summarizes indicators of decline and erosion of such agrobiodiversity, and identifies practices and approaches that conserve and/or enhance biodiversity in agroecosystems. The evidence reveals that practices for conserving, sustainably using and enhancing biodiversity are important for maintaining the health of agricultural ecosystems, to ensure and/or increase agricultural production, and to contribute to food security.

AGROBIODIVERSITY AS A BASIS FOR SUSTAINABLE CROP PRODUCTION

Biodiversity and detailed knowledge about it have allowed farming systems to evolve since agriculture began some 12,000 years ago (GRAIN, 1994). Although sometimes perceived as an enemy of biodiversity, agriculture is actually based on richly diverse biological resources. Likewise, agriculture comprises a variety of managed ecosystems or "agroecosystems" that benefit from resources in natural habitats.

Agricultural biodiversity or "agrobiodiversity" is a fundamental feature of farming systems around the world. It encompasses many types of biological resources tied to agriculture, including:

- genetic resources–the essential living materials of plants and animals;
- edible plant and crop species, including traditional varieties, cultivars, hybrids, and other genetic material developed by breeders;
- livestock species (small and large, lineal breeds or thoroughbreds) and freshwater fish;

- soil organisms vital to soil fertility, structure, quality, and soil health;
- insects, bacteria, and fungi (including types that live in the soil);
- agroecosystem components and types (polycultural/monocultural, small/large scale, rainfed/irrigated, etc.) indispensable for nutrient cycling, stability, and productivity; and
- "wild" resources (species and other elements) of natural habitats and landscapes that can provide ecosystem functions and services (for example, pest control and stability) to agriculture.

Agrobiodiversity therefore includes not only a wide variety of species and genetic resources, but also refers to the ways in which farmers can use biological diversity to produce and manage crops, land, water, insects, and biota (Altieri, 1991; Brookfield and Padoch, 1994). The concept also includes habitats and species outside of farming systems that benefit agriculture and/or enhance agroecosystem services, such as host plants that harbor natural enemies and predators of agricultural pests.

Many early agricultural settlements made use of diverse plants, livestock, and agroecosystems. Examples have been recorded in Egyptian, Mesopotamian, Chinese, and Andean civilizations. Over many centuries, farmers have employed numerous practices to use, enhance, and conserve this diversity in traditional farming systems. Many such practices continue today: the use of diverse species for pest control and the integration of trees and woody shrubs into farming systems are two examples. Wild plant and animal species in surrounding habitats also provide services and value to the farming system. Such practices are a basis of crop production and well-being for millions of people (Altieri, 1991; Bookfield and Padoch, 1994; FAO, 1999; Forno and Smith, 1996).

The majority of staple crops consumed globally originated from a few areas, mostly in Asia, Africa, and Latin America, and are often called "megadiversity" centers. Although many of these crops have been cultivated by farmers around the world, crop diversity is still most concentrated in these regions. The rich diversity of crops and plant varieties in these locations has also served as a basis for the growth of important civilizations (Greenburg, 1994). From ancient times to the present day, plant collecting has also enhanced agrobiodiversity. Throughout the colonial period, the search for and collection of diverse plants and foods was a driving interest of European explorers and played an important role in colonial expansion.

Examples of these traditional farming systems that maximize diversity are small polycultural farms, sometimes called home gardens, which are still found in many regions today, including Central America, Southeast Asia, sub-Saharan Africa, and even Europe. Other similar examples are traditional agroforestry systems, such as the shaded coffee (*Coffea* sp.) plantations common throughout Central and South America. These traditional agroforestry systems commonly contain well over 100 annual and perennial plant species per field (Altieri, 1991). Farmers often integrate leguminous trees, fruit trees, trees for fuelwood, and types that provide fodder on their coffee farms. The trees also provide habitat for birds and animals that benefit the farms. For example, a shaded coffee plantation in Mexico supports up to 180 species of birds that help control insect pests and disperse seeds (Greenburg, 1994). Ethnobotanical studies show that the Tzeltal Mayans of Mexico can recognize more than 1,200 species of plants, while the Purepechas recognize more than 900 species and Yucatan Mayans some 500 (Altieri, 1991). Such knowledge is used to make production decisions.

Numerous studies show that shifting cultivation systems, especially in traditional forms, are agroecologically diverse and contain numerous plant species. These can also be relatively sustainable in certain areas of the world, especially where economic and demographic pressures for growth are low (Brookfield and Padoch, 1994; Thrupp, 1996). Another dimension of traditional agrobiodiversity is the use of so-called folk varieties, also known as landraces. Defined as "geographically or ecologically distinctive populations [of plants and animals] which are conspicuously diverse in their genetic composition . . ." (Brown, 1978), landraces are products selected by local farmers over time, for their various production benefits (Cleveland, Soleri, and Smith, 1994; NRC, 1984). In some areas in the Andean region for example, farmers have developed complex techniques to select, store, and propagate the seeds of landraces.

Many of these practices for conserving and enhancing biodiversity are tied to the rich cultural diversity and local knowledge that are valuable elements for community livelihood. Rural women are particularly knowledgeable about diverse plants and tree species and their uses for health care, fuel, and fodder, as well as food (Thrupp, 1984). Many principles, as well as intuitive knowledge, from traditional systems are applied today in both large- and small-scale production. In fact, "traditional multiple cropping systems still provide as much as 20 percent of the world food supply" (UNDP, 1995).

Starting in the late 19th and early 20th centuries, scientists who recognized the value of diverse crop varieties discovered plant breeding methods that have boosted crop productivity. The innovative uses of plant genetic resources has continued to be valuable for scientific advances in plant and livestock breeding and seed improvements up to the present day. Agrobiodiversity is therefore important to industrial agribusiness as well as traditional small-scale farming and livelihoods. Access to germplasm is vital for modern agriculture, and for the development of medicinal products, fibers and foods. For example, in the U.S. [for two major crops soybean (*Glycine max* L. Merr.) and corn (*Zea mays* L.)] exotic germplasm "adds a value of $3.2 billion to the nation's $1 billion annual soybean production and $7 billion to its $18 billion annual corn crop" (Shand, 1996). By incorporating genes from wild tomatoes (*Lycopersicon* spp.), a potential increase in $8 million-a-year tomato sales was realized from a one-time $42 investment (Iltis, 1988).

In summary, the many components of agrobiodiversity yield an array of benefits. They reduce risk and contribute to resilience, and sustainability. Agricultural biodiversity also provides ecosystem services, by enhancing or improving the health of soils and they also benefit nutrition and productivity, and in many cases, income generation.

DECLINE OF AGROBIODIVERSITY

Agricultural Production Trends in Relation to Natural Resources

The links between agriculture and biodiversity have changed over time and should be viewed in the context of global agricultural development trends. Agricultural production and productivity have increased greatly in the last 30 years, and the international trade and exchange of seeds and food products have also expanded. This growth stems from both the expansion of cultivated area (extensification) and the increased output per unit of land (intensification) through technological inputs, improved varieties, and the management of biological resources, such as soil and water. Although agricultural development has enabled significant yield increases in many areas, these development trends have also involved problems: for example, hunger and malnutrition persist, food security has declined in many developing countries, the benefits and costs of production are distributed inequitably, and biological resources have been degraded.

These general trends in agriculture and biodiversity have been shaped by demographic pressures, including high population growth rates, the migration of people into frontier areas, and imbalances in population distribution. Additional influential forces are the predominant paradigms of industrial agriculture and the Green Revolution. These paradigms generally emphasize maximizing yield per unit of land, uniform varieties, reduction of multiple cropping, standardized farming systems (particularly generation and promotion of high-yielding varieties), and the standardized applications of agrochemicals. Seed and agrochemical companies have also influenced these trends. High yielding varieties (HYVs) or "miracle seeds," are planted on high percentages of agricultural land [52% for wheat (*Triticum aestivum* L.), 54% for rice (*Oryza sativa* L.), and 51% for corn]. Use of HYVs has increased production in many regions and sometimes reduced pressure on habitats by curbing the need to farm new lands. However, the widespread adoption of these varieties and other technologies has led to reduction of biodiversity. In turn, the ecosystem services provided by agricultural biodiversity have declined, and can therefore undermine ecosystem health. These problems pose tremendous challenges to meet production needs while conserving resources.

Erosion of Crop and Livestock Diversity

Although people consume approximately 7,000 species of plants, only 150 species are commercially important, and about 103 species account for 90% of the world's food crops. Just three crops (rice, wheat, and maize) account for about 60% of the calories and 56% of the protein people derive from plants (Raeburn, 1995). Along with this trend towards uniform monocropping, the dependence on high levels of inputs such as irrigation, fertilizers, and pesticides has increased worldwide. The reduction in diversity often increases vulnerability to climate and other stresses, raises risks for individual farmers, and can undermine the stability of agriculture. In Bangladesh, for example, "promotion of HYV rice monoculture has decreased diversity, including nearly 7,000 traditional rice varieties and many fish species. The production of HYV rice per acre in 1986 dropped by 10% from 1972, in spite of a 300% increase in agrochemical use per acre" (Hussein, 1994). In the Philippines, HYVs have displaced more than 300 traditional rice varieties that had been the principal source of food for generations. In India, by 1968, the so-called "miracle" HYV seed had replaced half of the native varieties, but these seeds were not high-yielding unless cultivated on irrigated land with high rates of fertilizer, which is often unaffordable to poor farmers (Shiva, 1991). Thus, the expected production increases were not realized in many areas, while genetic uniformity of crops has clearly increased (Table 1).

In Africa, transfer of the Green Revolution model has also reduced diversity. In Senegal, for example, a traditional cereal called fonio (*Panicum laetum*) (which is highly nutritious as well as robust in lateritic soils) has been threatened by extinction because of its replacement by modern crop varieties (IFOAM, 1994). In the Sahel, reports also confirm that traditional systems of polyculture are being replaced with monocultures that cause further food instability (Mann, 1994 or 1995).

Homogenization also occurs in high-value crops. Nearly all the coffee trees in South America, for example, descended from a single tree in a botanical garden in Holland. *Coffea arabica* was first obtained from forests of southwest Ethiopia that have virtually disappeared (Table 2). Uniform varieties are also common in export crops of bananas (*Musa* sp.), cacao (*Theobroma cacao* L.), and cotton (*Gossypim hirsutum* L.), replacing traditional diverse varieties. Such changes have increased productivity, but the risks of narrowing varietal selection have become clear over time.

In the North, similar losses in crop diversity are occurring. Many fruit and vegetable varieties listed by the USDA in 1903 are now extinct. Of more than 7,000 apple (*Malus* sp.) varieties grown in the U.S. between

TABLE 1. Extent of genetic uniformity in selected crops, based on Goombridge (1992)

Crop	Country	Number of Varieties
Rice	Sri Lanka	From 2,000 varieties in 1959 to less than 100 today 75% descend from a common stock
Rice	Bangladesh	62% of varieties descend from a common stock
Rice	Indonesia	74% of varieties descend from a common stock
Wheat	USA	50% of crop in 9 varieties
Potato	USA	75% of crop in 4 varieties
Soybeans	USA	50% of crop in 6 varieties

1804 and 1904, 86% are no longer cultivated, and of 2,683 pear (*Pyrus* sp.) varieties, 88% are no longer available (Fowler and Mooney, 1990) (See Table 2). Evidence from Europe shows similar trends; thousands of varieties of flax (*Linum usitatissimum* L.) and wheat vanished after HYVs were introduced (Mooney, 1979). Similarly, varieties of oats (*Avena sativa* L.) and rye (*Secale cereale* L.) are also declining in Europe (Vallve, 1993). In Spain and Portugal, various legumes that had been an important part of the local diet are being replaced by homogeneous crops, and in the Netherlands, four crops are grown on 80% of Dutch farmlands (Vallve, 1993).

Livestock is also suffering genetic erosion; the FAO estimates that somewhere in the world at least one breed of traditional livestock dies out every week. Many traditional breeds have disappeared as farmers focus on new breeds of cattle, pigs, sheep, and chickens (Plucknett and Horne, 1992). Of the 3,831 breeds of cattle, water buffalo, goats, pigs, sheep, horses, and donkeys believed to have existed in this century, 16% have become extinct, and a further 15% are rare (Hall and Ruanne, 1993). Among extant livestock breeds, 474 can be regarded as rare and

TABLE 2. Reduction of diversity in fruits and vegetables, Varieties in NSSL Collection, 1903 to 1983, based on Fowler and Mooney (1990)

Vegetable	Taxonomic Name	Number in 1903	Number in 1983	Loss (percent)
Asparagus	*Asparagus officinalis*	46	1	97.8
Bean	*Phaseolus vulgaris*	578	32	94.5
Beet	*Beta vulgaris*	288	17	94.1
Carrot	*Daucus carota*	287	21	92.7
Leek	*Allium ampeloprasum*	39	5	87.2
Lettuce	*Lactuca sativa*	497	36	92.8
Onion	*Allium cepa*	357	21	94.1
Parsnip	*Pastinaca sativa*	75	5	93.3
Pea	*Pisum sativum*	408	25	93.9
Radish	*Raphanus sativus*	463	27	94.2
Spinach	*Spinacia oleracea*	109	7	93.6
Squash	*Cucurbita* spp.	341	40	88.3
Turnip	*Brassica rapa*	237	24	89.9

a further 617 have become extinct since 1892 (Hall and Ruanne, 1995). Over 80 breeds of cattle are found in Africa, and some are being replaced by exotic breeds (Rege, 1994). These losses weaken breeding programs that could improve hardiness of livestock.

As these forms of biodiversity are eroded, their ecosystem services also decline, and food security may also be reduced in many areas. Evidence indicates that such changes can decrease sustainability and productivity in farming systems (Tillman, Wedline, and Knops, 1996). Loss of diversity also reduces the resources available for future adaptation.

Increased Vulnerability to Insect Pests and Diseases

Homogenization of varieties increases vulnerability to insect pests and diseases, which can devastate a uniform crop, especially on large plantations. History has shown serious economic losses and suffering from relying on monocultural uniform varieties (Table 3). Renowned examples include: the potato (*Solanum tuberosum* L.) famine of Ireland during the 19th century, a winegrape (*Vitis vinifera* L.) blight that wiped out valuable vines in both France and the U.S., and a virulent disease *(Sigatoka)* that damaged extensive banana plantations in Central America in recent decades and devastating mold that infested hybrid corn in Zambia.

TABLE 3. Past crop failures due to genetic uniformity, based on Groombridge (1992)

Date	Location	Crop	Effects
1846	Ireland	Potato	Potato famine
1800s	Sri Lanka	Coffee	Farms destroyed
1940s	USA	U.S. crops	Crop loss to insects doubled
1943	India	Rice	Great famine
1960s	USA	Wheat	Rust epidemic
1970	USA	Maize	$1 billion loss
1970	Philippines, Indonesia	Rice	Tungo virus epidemic
1974	Indonesia	Rice	3 million tons destroyed
1984	USA (Florida)	Citrus	18 million trees destroyed

In addition, there has been a serious decline in soil organisms and soil nutrients. Beneficial insects and fungi also suffer under agriculture that involves heavy pesticide inputs and uniform stock–making crops a more susceptible to pest problems. These losses, along with fewer types of agroecosystems, also increase risks and can reduce productivity. In addition, many insects and fungi commonly seen as enemies of food production are actually valuable. Benefits of insects include pollination, contributions to biomass, natural nutrient production and cycling, and their roles as natural enemies to insect pests and crop diseases. Mycorrhizae, the fungi that live in symbiosis with plant roots, are essential for nutrient and water uptake.

The global proliferation of modern agricultural systems has eroded the range of insects and fungi, a trend that lowers productivity. Dependence on agrochemicals, and particularly the heavy use or misuse of pesticides, is largely responsible. Agrochemicals generally kill natural enemies and beneficial insects, as well as the "target" pest. "Pesticides [especially when overused] destroy a wide array of susceptible species in the ecosystem while also changing the normal structure and function of the ecosystem" (Pimentel, 1992).

This disruption in the agroecosystem balance can lead to perpetual resurgence of pests and outbreaks of new pests and can provoke resistance to pesticides. This disrupted cycle often leads farmers to apply increasing amounts of pesticides or to change products–a strategy that is not only ineffective, but that also further disrupts the ecosystem services and elevates costs. This "pesticide treadmill" has occurred in countless locations. Reliance on monocultural species and the decline of natural habitat around farms also cuts beneficial insects out of the agricultural ecosystem.

Additional Losses in Habitats, Nutrition, and Knowledge

Agricultural expansion has also reduced the diversity of natural habitats, including tropical forests, grasslands, and wetland areas. Projections of food needs in the coming decades indicate probable further expansion of cropland, which could add to this degradation. Modifying natural systems is necessary to fulfill the food needs of growing populations, but many conventional forms of agricultural development, particularly large-scale conversion of forests or other natural habitats to monocultural farming systems, erode the biodiversity of flora and fauna. Intensive use of pesticides and fertilizers can also disrupt and erode biodiversity in natural habitats and ecosystem services that surround agricultural areas, particularly when these inputs are used inappropriately.

Other documented direct effects of reduced diversity of crops and varieties include:

- Decline in the variety of foods adversely affects nutrition (IIED, 1995);
- High-protein legumes have often been replaced by less nutritious cereals (Shiva, 1991);
- Local knowledge about diversity is lost as uniform agricultural technologies predominate (Altieri, 1987); and
- Institutions and companies in industrial countries often have unfair advantages in exploiting the diverse biological resources from the tropics (Shand, 1996).

In sum, the loss of agrobiodiversity has immediate costs to producers, social costs to communities and nations, and long-term effects on agricultural productivity.

Causes of Agrobiodiversity Losses

The causes of agrobiodiversity losses are complex, and they obviously vary under different conditions. However, in general, the proximate causes have often been associated with the use of unsustainable agriculture methods and degrading land-use practices, as well as reliance on monocultural uniform varieties. More deeply, the roots underlying the erosion of agricultural biodiversity are tied to demographic pressures, and the dominance of industrial agricultural policies, institutions, and commercial interested that support and contribute to use of monocultural industrial agriculture paradigm that eliminates diversity by design, and instead promotes the reliance on uniform monocultures and chemicals. Other contributing factors are the depreciation and devaluation of diversity and accumulated local knowledge, and market and consumer demands for standardized products, disparities in farmers' access to information and resources (Raeburn, 1995; Shand, 1996).

ENHANCING AGROBIODIVERSITY

Humanity faces a major challenge to prevent and mitigate these losses, and to instead integrate biodiversity in agriculture. Meeting this challenge requires addressing root causes of agrobiodiversity loss, and therefore involves changing practices, paradigms, and policies, as well as commitments by governments and institutions. The conservation and enhancement of agrobiodiversity have been supported and mandated in major international conventions, particularly by the Convention on Biological Diversity (CBD, 2001), and have been encouraged by institutions such as the World Bank and the Food and Agriculture Organization of the U.N. (FAO, 1998, 1999; Forno and Smith, 1996; Kellenberg and Vidaeus, 1997). The Convention on Biological Diversity establishes a global commitment to biodiversity conversation for all signatory nations, and also includes specific measures and mandates for implementing agrobiodiversity conservation, sustainable use, and benefit-sharing of plant and animal genetic resources. While these kinds of global policies can help establish broad support for agrobiodiversity, actions also are being developed at the national and local levels, where experience provides lessons and promising opportunities for integrating biodiversity in agriculture.

Diversity Through Sustainable Agriculture Principles and Practices

Practical and effective approaches to conserve and enhance agrobiodiversity already exist and have been documented in literature. Many of these methods can be fit within a general framework of sustainable agriculture (Figure 1). This framework or paradigm merges the goals of productivity, food security, and social equity, and ecological soundness. A shift to sustainable agriculture requires changes in production methods, models, and policies, as well as the full participation of local people. In this approach, scientific advancements in genetics and improved varieties also can have significant roles, but need to be reoriented towards conserving and using diversity in farming systems.

To achieve such transformations for the conservation and enhancement of agricultural biodiversity, the following main approaches are important.

- *Application of agroecological principles* helps conserve, use, and enhance biodiversity on farms and can increase sustainable productivity and intensification, which avoids extensification, thereby reducing pressure on off-farm biodiversity.

- *Participation and empowerment of farmers*, and protection of their rights, are important means of conserving agrobiodiversity in research and development.
- *Adaptation of methods* to local agroecological and socioeconomic conditions, building upon existing successful methods and local knowledge, is needed to link biodiversity and agriculture.
- *Conservation* of plant and animal genetic resources–especially *in situ* efforts–help protect biodiversity for current livelihood security as well as future needs and ecosystem functions.
- *Reforming genetic research and breeding programs* for agrobiodiversity enhancement is essential and can also have production benefits.
- Creating a *supportive policy environment*–including eliminating incentives for uniform varieties and for pesticides, and implementing policies for secure tenure and local rights to plant genetic resources–is vital for agricultural biodiversity enhancement and for food security.

Applying these basic principles can generate considerable public and private benefits that were described earlier. Specific practices that have proven effective for this purpose have been discovered and adapted in many areas of the world. Building upon the knowledge of rural people has also proven to be effective in many contexts to make scientific advancements and to help ensure adoption and spread the benefits of agrobiodiversity innovations (Altieri, 1991; Brookfield and Padoch,

FIGURE 1. General Elements of Sustainable Agriculture

Economic
(Profitable;
Productive)
Ecological
(Environmentally
Sound)
SUSTAINABLE
AGRICULTURE
Social
(Food Security,
Health,
& Safety)

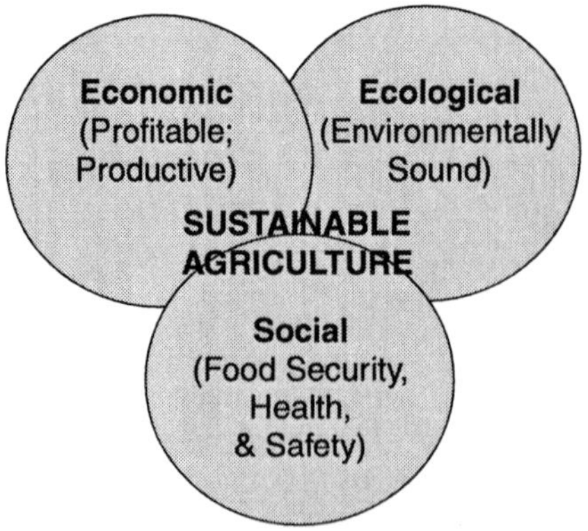

1994; Shand, 1996; Thrupp, 1996). The use of such principles and practices has resulted in production increases, in both small and large-scale farms. Additional advantages include improvement of soil nutrient cycles and soil quality; added economic value; increase in sustainability and stability of systems; and alleviation of pressures on habitats.

Integrated pest management (IPM) methods illustrate well the use and benefits of biodiversity. Ecologically-based IPM approaches usually highlight diversity as a key feature. Examples of effective IPM practices include the following:

- multiple cropping and/or crop rotations, used to prevent build-up of pests;
- plants that are grown within crops or around farm fields to house predators of insect pests;
- other intercropped plants that act as alternative host plants for pests (and provide soil benefits), such as the use of cover crops for vineyards and orchards in California;
- use of certain plants as natural pesticides: for example, in Ecuador, castor (*Ricinus communis* L.) leaves that contain a paralyzing agent are used to control the tenebronid beetle;
- weeds that are used to repel insects: for example in Colombia, grassweeds are grown around bean (*Phaseolus* sp.) fields to repel leafhoppers, and in Southern Chile, the shrub *Cestrum parqui* is used to repel beetles in potato fields;
- integration of biocontrol agents, including various parasites, animals, and fish that consume insect pests (such as use of ducks and fish in rice paddies in Asia);
- elimination or reduction of pesticide use to avoid adverse agroecological effects on the insect diversity in agroecosystems;
- effective disease-management practices using agrobiodiversity include:
  - mixed crop stands that slow the spread of diseases by altering the micro-environment; for example, in Central America, cowpea (*Vigna unguiculata* L.) grown with corn are less susceptible to the fungus *Ascochyta phaselolorum*, and to the cowpea mosaic virus.
  - use of non-host plants as "decoy" crops, to attract fungus (or nematodes) (Altieri, 1987; Thrupp, 1996; UNDP, 1992, 1995).

Successful IPM programs in Asia illustrate that building agrobiodiversity–particularly using diverse beneficial insects–is a key ingredient of effective pest management in rice production. These initiatives, coordinated by the Food and Agriculture Organization, along with government and non-governmental organizations, have resulted in remarkable reductions of pesticide use and increased rice yields (examples in Thrupp, 1996).

In Bangladesh, thousands of farmers involved in IPM projects have also integrated fish into rice paddies and have adopted agroecological methods to restore the natural balance between insects and other fauna, and have planted vegetables on the dikes around the edges. This approach has increased rice yields, provided new sources of nutrition and has made hazardous chemical use unnecessary. For example, farmers in the pilot IPM program achieved an 11% increase in rice production while eliminating pesticides (Thrupp, 1996).

Practices for soil fertility/health and nutrient cycling also make use of agrobiodiversity. Good examples include the following:

- compost from crop residues, tree litter, and other plant/organic residues;
- intercropping and cover crops, particularly legumes, which add nutrients, fix nitrogen, and "pump" nutrients to the soil surface;
- use of mulch and green manures (through collection and spread of crop residues, litter from surrounding areas, and organic materials, and/or under crop);
- integration of earthworms (vermiculture) or other beneficial organisms and biota into the soil to enhance fertility, organic matter, and nutrient recycling; and
- elimination or reduction of agrochemicals–especially toxic nematicides–that destroy diverse soil biota, organic material, and valuable soil organisms (Altieri, 1991, 1987; Lee, 1990; UNDP, 1992).

These kinds of soil-management practices have proven effective and profitable in a variety of farming systems. Agroforestry illustrates an effective practice of using agrobiodiversity that also generates multiple benefits (Michon and de Foresta, 1990). In many contexts, the integration of trees into farming systems is highly efficient, and the trees have multiple functions, such as providing fuel, fodder, shade, nutrients, timber for construction, and aiding soil conservation and water retention. In West Sumatra, agroforestry gardens occupy 50 to 85% of the total agricultural land. Complex forms of agroforestry exhibit forest-like structures, as well as a remarkable degree of plant and animal diversity, combining conservation and natural resource use. In Indonesia, for example, small-holder "jungle rubber" gardens incorporate numerous tree species. Agroforestry systems in traditional forms also shelter hundreds of plant species, constituting valuable forms of *in situ* conservation (Michon and de Foresta, 1990). Many of the practices noted here serve multiple purposes. For example, intercropping provides pest and soil management as well as enhanced income. Similarly, about 70-90% of beans, and 60% of corn in South America are intercropped. Farmers in many other parts of the world have recognized such diversity as valuable sources of soil nutrients, nutrition, and risk reduction–essential for production and other economic values (UNDP, 1995).

A common misperception is that agrobiodiversity enhancement is feasible only in small-scale farms. In fact, experience shows that large production systems also benefit from incorporating these principles and practices. Crop rotations, intercropping, cover crops, IPM techniques, and green manures are the most common methods being used profitably in larger commercial systems, in both industrialized countries and developing countries. Examples of agrobiodiversity enhancement in large scale commercial systems are found in tea (*Camellia sinensis* L. Kuntze) and coffee plantations in the tropics, and in vineyards and orchards in temperate zones. For instance, large scale producers in California, have reported valuable results from using diverse cover crops (including improved soil health, soil conservation, and improved insect management), as well as advantages from planting more diverse varieties in their farms (UNDP, 1995). In some large-scale farms that are changing from monocultural to diverse systems, this transition involves added costs during the first two or three years. However, after the initial transition, producers have found that the incorporation of biodiversity is profitable as well as ecologically sound for commercial production and that they present new valuable opportunities.

Using Participatory Approaches

The incorporation of farmers' local knowledge, practices, and experimentation is advantageous in such efforts to conserve and enhance agrobiodiversity. Experiences have shown that full involvement of local farming practices in agricultural research and development (R&D)–through participation and leadership of local people–has had beneficial outcomes. In other words, an understanding of farmers' knowledge and incorporation of their strategies for agrobiodiversity enhancement increases the chances of success (Rajasekaran, Martin, and Warren, 1993; Thrupp, 1996). At the same time, the involvement of farmers as partners in research and development helps to ensure adoption of sustainable farming methods and can help to empower local people.

In Mexico, for example, researchers worked with the local people to re-create *chinampas*–multicropped, species-diverse gardens developed from reclaimed lakes–which were native to the Tabasco region and part of Mexico's pre-Hispanic tradition. A similar project conducted in Veracruz also incorporated the traditional Asiatic system of mixed farming, mixing *chinampas* with animal husbandry, and aquaculture. These gardens also made more productive use of local resources, and integrated from plant and animal waste, as fertilizers. Yields of such systems equaled or surpassed these of conventional systems.

In Burkina Faso, on the other hand, a soil-conservation and integrated cropping project in Yatenga province was based largely on an indigenous technology of Dogon farmers in Mali–building rock bunds to prevent water run-off. The project added innovation, bunds along contour lines, and revived an indigenous technique called "*zai*," which is adding compost to holes in which seeds of millet, sorghum (*Sorghum bicolor* L. Moench), and peanut (*Arachis hypogaea* L.) are planted. These crops are in a multicropping system. Animals are incorporated for their manure. In the fields using these techniques, yields were consistently higher than in fields using conventional practices, ranging from 12% higher in 1982, to 91% in 1984. Yields in the *zai* method reached 1,000-1,200 kg ha^{-1}, compared to conventional yields of 700 kg ha^{-1}. Water management was enhanced, and food security, a priority concern of local people, was also improved through this approach. The techniques have been widely adopted, covering 3,500 ha by the end of 1988 (UNDP, 1995).

In such efforts, the full participation of women has significant benefits. As managers of biodiversity in and around farming systems in many areas of the world, women can make important contributions and have a promising role in research, development, and conservation of agrobiodiversity. In Rwanda, for example, in a plant-breeding project of CIAT (International Center for Tropical Agriculture), scientists worked with women farmers from the early stages of a project on breeding new varieties of beans to suit local peoples' needs (CGIAR, 1994). Together they identified the characteristics desired to improve beans, run experiments, manage, and evaluate trials, and make decisions on the trial results. The experiments resulted in stunning outcomes: the varieties selected and tested by women farmers over four seasons "performed better than the scientists' own local mixtures 64-89% of the time"

(CGIAR, 1994). The women's selections also produced substantially more beans, with average production increases as high as 38%.

The development of participatory approaches requires deliberate measures, training, and time to change the conventional approaches of agricultural R&D. The application of such two-way approaches improves the likelihood of adoption and success of agrobiodiversity efforts. Basic principles of participatory rural appraisal in agroecological R&D include:

- Joint problem-solving among farmers and scientists, and responsiveness to local need;
- Mutual listening/learning between farmers and scientists;
- Understanding of complexity;
- Triangulation (investigate a theme in different ways);
- Flexibility in selecting methods, adjusting to timing;
- Interdisciplinary and holistic perspective; and
- Inclusive and equitable representation (Chambers, Pacey, and Thrupp, 1987; Thrupp, 1996).

In sum, the use of these participatory approaches can help planners and communities to identify and develop "best practices" in sustainable production, i.e., practices that are adapted to diverse local conditions and that build convergence between agriculture and biodiversity, as well as create socioeconomic opportunities.

Merging Agrobiodiversity and Habitat Conservation

Efforts to conserve and enhance agrobiodiversity must also address the underlying policies that accelerate its loss. Broader policies and institutional structures focused on agrobiodiversity conservation drive practical, field-level changes. Many policy initiatives and institutions have already been established to address these issues. For example, several international institutions influence and regulate the use of plant genetic resources. Among the key players are the Consultative Group on International Agricultural Research, the International Plant Genetic Resources Institute, the Food and Agriculture Organization, the Commission on Plant Genetic Resources, and the World Intellectual Properties Organization. Recent important international conventions and agreements, particularly the Convention on Biological Diversity and the General Agreement on Tariffs and Trade, are also influential in setting guidelines that affect agrobiodiversity and use of genetic resources.

Concerns about the control of plant genetic resources have led to many intellectual property regulations that govern the activities of public institutions and private companies and that are intended to protect farmers' legal access to genetic resources. Gene banks conserve a remarkable diversity of plant genetic resources, and increasing numbers of agricultural research institutes have begun *in situ* conservation projects as well. Along with these large formal institutions, many NGOs and local organizations are also increasingly involved in promoting the conservation and equitable distribution of benefits from agrobiodiversity.

Policy and Institutional Changes

Although many institutions are already actively involved, more coordination and work is needed at all levels to ensure effective reforms and agrobiodiversity-conservation policies that benefit the public, especially the poor. Policy changes that attack the root causes of the problems and that benefit agriculture and the broader society. Ideas needing further attention include:

- Ensuring public participation in the development of agricultural and resource use policies;
- Eliminating subsidies and credit policies for HYVs;
- Fertilizers, and pesticides to encourage the use of more diverse seed types and farming methods;
- Policy support and incentives for effective agroecological methods that make sustainable intensification possible;
- Reform of tenure and property systems that affect the use of biological resources to ensure that local people have rights and access to necessary resources;
- Regulations and incentives to make seed and agrochemical industries socially responsible;
- Development of markets and business opportunities for diverse organic agricultural products; and
- Changing consumer demand to favor diverse varieties instead of uniform products.

Building complementarities between agriculture and biodiversity will also require changes in agricultural R&D, land use, and breeding approaches. The types of practices and policies outlined here constitute potential solutions and promising opportunities. Such changes are urgently needed to overcome threats from the ongoing erosion of genetic resources and biodiversity. Experience shows that enhancing agrobiodiversity economically benefits both small- and large-scale farmers, while at the same time serving the broader social interests of ecosystem health and food security. Implementing the changes and policies suggested in this paper will support agrobiodiversity and lead to wide-ranging socioeconomic and ecological gains.

REFERENCES

Altieri, M. (1991). Traditional farming in Latin America. *The Ecologist* 21(2): 93-96.

Altieri, M. (1987). *Agroecology: The Scientific Basis of Sustainable Agriculture*. Boulder, CO: Westview Press.

Altieri, M. and L. Merrick. (1988). Agroecology and *in situ* conservation of native crop diversity in the Third World. In *Biodiversity*, eds. E. Wilson and F. Peter, Washington, DC: National Academy of Sciences.

Brookfield, H. and C. Padoch.(1994). Appreciating agrodiversity: A look at the dynamism and diversity of indigenous farming practices. *Environment* 36(5): 7-44.

Brown, A.H.D. (1978). Isozymes, plant population genetic structure, and genetic conservation. *Theoretical Applied Genetics* 52: 145-57.

CBD. (2001). Convention on Biological Diversity, Secretariat for the Convention on Biological Diversity, UN Environment Programme. (www.biodiv.org).

CGIAR. (1994). *Partners in Selection*. Washington, DC: Consultative Group on International Agricultural Research.

Chambers, R., A. Pacey, and L.A. Thrupp. (1987). *Farmer First: Farmer Innovation and Agricultural Research*. London, UK: IT Publications

Cleveland, D., D. Soleri, and S. E. Smith. (1994). Do folk crop varieties have a role in sustainable agriculture? *Bioscience* 44(11): 740-51.

FAO. (1998). Agricultural biodiversity: Assessment of ongoing activities and priorities for a programme of work. UNEP/CMD/SBSTTA/5/10, Rome: Food and Agriculture Organization of the U.N. (www.fao.org/biodiversity/).

FAO. (1999). Background paper on agricultural biodiversity, Rome: Food and Agriculture Organization of the UN. (www.fao.org/biodiversity/).

Forno, D. and N. Smith. (1996). *Biodiversity and Agriculture: Implications for Conservation and Development*. World Bank Technical Paper #321. Washington, DC: World Bank.

Fowler, C. and P. Mooney. (1990). *Shattering: Food, Politics, and the Loss of Genetic Diversity*. Tucson, AZ: University of Arizona Press.

GRAIN. (1994). Biodiversity in agriculture: Some policy issues. *IFOAM Ecology and Farming*. January: 14.

Goombridge, B. (1992). *Global Biodiversity: Status of the Earth's Living Resources*, World Conservation Monitoring Center. London, UK: Chapman and Hall.

Greenburg, R. (1994). Phenomena, comment and notes. *Smithsonian* 25(8): 24-7.

Hall, S.J.G. and J. Ruane. (1993). Livestock breeds and their conservation: A global overview. *Conservation Biology*. 7(4): 815-825, cited in Smith, N. 1996. "The Impact of Land Use Systems on the Use and Conservation of Biodiversity." Draft paper, Washington, DC: World Bank, p. 43.

Hussein, M. (1994). Regional Focus News–Bangladesh. *Ecology and Farming: Global Monitor*, IFOAM. January: 20.

IFOAM. (1994). Biodiversity: Crop resources at risk in Africa. *Ecology and Farming-Global Monitor*. January: 5.

IIED. (1995). Hidden Harvests Project Overview. Sustainable Agriculture Program. London, UK: International Institute for Environment and Development.

Iltis, H. H. (1988). Serendipity in the exploration of biodiversity: What good are weedy tomatoes? In *Biodiversity*, ed., E.O. Wilson, Washington, DC: National Academy Press. Pp. 98-105.

Kellenberg, J. and L. Vidaeus. (1997). Mainstreaming biodiversity in agricultural development: Toward good practice. World Bank Environment Paper #15. Washington, DC: World Bank.

NRC. (1989). *Alternative Agriculture*. Washington, DC: National Academy Press.

Lee, K.E. (1990). The diversity of soil organisms. In *The Biodiversity of Microorganisms and Invertebrates: Its Role in Sustainable Agriculture*, ed. D.K. Hawksworth, London, UK: CAB International.

Mann, R.D. (1994). Time running out: The urgent need for tree planting in Africa. *The Ecologist* 20(2): 48-53.

Michon, G. and H. de Foresta. (1990). Complex agroforestry systems and the conservation of biological diversity. In *Harmony with Nature, Proceedings of International Conference on Tropical Biodiversity*. Kuala Lumur, Malaysia: SEAMEO-BIOTROP.

Mooney, P. (1979). *Seeds of the Earth: A Private or Public Resource?* Ann Arbor, MI: Canadian Council for International Cooperation.

Morales, H.L. 1984. Chinampas and integrated farms: Learning from the rural traditional experience. In *Ecology and Practice*: Vol. 1, eds. F. De Castri, G. Baker and M. Hadley, Ecosystem Management. Dublin, Ireland: Tycooly.

Pimentel, D., D.A. Takacs, H.W. Brubaker, A.R. Dumas, J.J. Meaney, J.A.S. O'Neill, D.E. Onsi, and D.B. Corzilius. (1992). Conserving biological diversity in agricultural/forestry systems. *Bioscience* 42: 354-362.

Plucknett, D. and M.E. Horne. (1992). Conservation of genetic resources. *Agriculture, Ecosystems, and the Environment*. 42: 75-92, cited in Smith, N. 1996. The Impact of Land Use Systems on the Use and Conservation of Biodiversity. draft paper, Washington, DC: World Bank, p. 23 .

Rajasekaran, B., R.A. Martin, and D.M. Warren. (1993). Framework for incorporating indigenous knowledge systems into agricultural extension. *Indigeneous Knowledge and Development Monitor* 1(3): 21-24.

Raeburn, P. (1995). *The Last Harvest: The Genetic Gamble That Threatens to Destroy American Agriculture*. New York, NY: Simon and Schuster, p. 40.

Rege, J.E.O. (1994). International livestock center preserves Africa's declining wealth of animal biodiversity. *Diversity* 10(3): 21-5, cited in Smith, N. 1996. p. 43.

Shand, H. (1996). *Human Nature: Agricultural Biodiversity and Farm-Based Security.* Ottawa: RAFI.

Shiva, V. (1991). The Green Revolution in the Punjab. *The Ecologist.* 21(2): 57-60.

Thrupp, L.A., ed. (1996). *New Partnerships for Sustainable Agriculture.* Washington DC: World Resources Institute.

Tillman, D., D. Wedline, and J. Knops. (1996). Productivity and sustainability influenced by biodiversity in grassland ecosystems. *Nature* 379(22): 718-720.

UNDP. (1995). *Agroecology: Creating the Synergisms for Sustainable Agriculture.* New York, NY: United Nations.

UNDP. (1992). *Benefits of Diversity: An Incentive Toward Sustainable Agriculture.* New York, NY: UN Development Programme.

UNDP. (1995). *Agroecology: Creating the Synergism for a Sustainable Agriculture.* New York, NY: United Nations Development Programme.

UNEP, (1995). *Global Biodiversity Assessment.* New York, NY: United Nations Environmental Programme.

Vallve, R. (1993). The decline of diversity in European agriculture. *The Ecologist* 23(2): 64-69.

World Bank. (1996). *Integrated Pest Management: Strategy and Options for Promoting Effective Implementation.* Draft document, Washington, DC: World Bank.

Biophysical and Ecological Interactions in a Temperate Tree-Based Intercropping System

N. V. Thevathasan
A. M. Gordon
J. A. Simpson
P. E. Reynolds
G. Price
P. Zhang

SUMMARY. Tree-based intercropping is considered an excellent farming system and can contribute much to our understanding of sustainable agriculture practices. Our current research goals are to address and quantify the numerous biophysical interactions that occur at the tree-crop interface in order to enhance our understanding of the ecology of tree-based intercropping (a form of agroforestry).

N. V. Thevathasan, A. M. Gordon, J. A. Simpson, and P. Zhang are affiliated with the Department of Environmental Biology, University of Guelph, Guelph, Ontario, Canada N1G 2W1 (E-mail: nthevath@uoguelph.ca).

P. E. Reynolds is affiliated with Natural Resources Canada, Canadian Forest Service, 1219 Queen Street East, Sault Ste. Marie, Ontario, Canada P6A 5M7.

G. Price is affiliated with the Department of Land Resource Science, University of Guelph, Guelph, Ontario, Canada N1G 2W1.

The authors wish to thank Peter Williams for granting permission to utilize results stemming from his research on intercropping microclimates.

Financial assistance received from the Ontario Ministry of Food and Agriculture (OMAF) is also gratefully acknowledged.

[Haworth co-indexing entry note]: "Biophysical and Ecological Interactions in a Temperate Tree-Based Intercropping System." Thevathasan, N. V. et al. Co-published simultaneously in *Journal of Crop Improvement* (Food Products Press, an imprint of The Haworth Press, Inc.) Vol. 12, No. 1/2 (#23/24), 2004, pp. 339-363; and: *New Dimensions in Agroecology* (ed: David Clements, and Anil Shrestha) Food Products Press, an imprint of The Haworth Press, Inc., 2004, pp. 339-363. Single or multiple copies of this article are available for a fee from The Haworth Document Delivery Service [1-800-HAWORTH, 9:00 a.m. - 5:00 p.m. (EST). E-mail address: docdelivery@haworthpress.com].

http://www.haworthpress.com/web/JCRIP

Digital Object Identifer: 10.1300/J411v12n01_04

In 1987, the University of Guelph established a large field experiment on 30 ha of prime agricultural land in Wellington county southern Ontario, Canada to investigate various aspects of intercropping trees with agricultural crops. A variety of spacing, crop compatibility and tree growth, and survival experiments were initiated at that time, utilizing 10 tree species within the genera *Picea*, *Thuja*, *Pinus*, *Juglans*, *Quercus*, *Fraxinus*, *Acer*, and *Populus*. Two between row-spacings (12.5 m or 15 m) and two within row-spacings (3 m, or 6 m) were utilized in conjunction with all possible combinations of three agricultural crops (soybean, corn, and either winter wheat or barley).

Investigations over the last decade have documented several complementary biophysical interactions. Nitrogen (N) transfer from fall-shed leaves to adjacent crops with enhanced soil nitrification as the proposed mechanism was estimated to be 5 kg N ha^{-1}. Soil organic carbon (C) adjacent to tree rows has increased by over 1%, largely as a result of tree litterfall inputs and fine root turnover. It is estimated that intercropping has reduced nitrate loading to adjacent waterways by 50%, a hypothesized function of deep percolate interception by tree roots. We have also noticed increased bird diversity and usage within the intercropped area as compared to mono-cropped adjacent agricultural areas, and have recorded increases in small mammal populations. Earthworm distribution and abundance was also found to be higher closer to the tree rows when compared to earthworm numbers in the crop alleys. We speculate that these are indicative of major changes in the flow of energy within the trophic structure identified with intercropping systems.

In light of climate change mitigation processes, C sequestration and NO_2 reduction potentials in tree-based intercropping systems were studied and compared to conventional agricultural systems. The results suggest that sequestration of C was 5 times more in the former system than in the latter. Competitive interactions between trees and crops for nutrients, moisture and light were also studied.

The tangible benefits that are derived from properly designed and managed tree-based intercropping systems place this land management option above conventional agriculture in terms of long-term productivity and sustainability.

KEYWORDS. Agroforestry, sustainable agriculture, carbon sequestration, biodiversity, agroecosystem, biophysical interactions

INTRODUCTION

Agroforestry is an approach to landuse that incorporates trees into farming systems, and allows for the production of trees and crops or livestock from the same piece of land in order to obtain economic, ecological, environmental, and cultural benefits (Gordon and Newman, 1997). Agroforestry has its roots in the developing world, where lack of land resources in the presence of high population growth necessitated the development of novel and simultaneous wood and food production systems by indigenous peoples. In North America, many different types of agroforestry have been employed historically (Gordon, Newman, and Williams, 1997), but the vast potential for economic and environmental benefits attributed to agroforestry have yet to be realized on a large scale. The main types of agroforestry practices currently being researched in many areas of North America are windbreaks and shelterbelts, silvipastoral systems (animals, pasture and trees), integrated riparian forest systems, forest farming systems, and tree-based intercropping systems (crops grown between widely spaced tree rows) (Gordon and Newman, 1997; Garrett et al., 2000). These systems have also been extensively researched in southern Ontario, Canada and are summarized in Table 1.

When trees and annual crops are properly combined in a tree-based intercropping system, they create a dynamic agroecosystem that can in-

TABLE 1. Summary of research on various agroforestry systems undertaken in southern Ontario, Canada.

Agroforestry system	Reference
Windbreaks and shelterbelts	Kenney, 1987; Loeffler, Gordon, and Gillespie, 1992
Integrated riparian forest systems	O'Neill and Gordon, 1994; Oelbermann and Gordon, 2000, 2001
Forest farming systems	Matthews et al., 1993; Christrup, 1993; Williams et al., 1997
Tree-based intercropping systems	McLean, 1990; Ball, 1991; Gordon and Williams, 1991; Williams and Gordon, 1992, 1994, 1995; Ntayombya, 1993; Ntayombya and Gordon, 1995; Kotey, 1996; Thevathasan and Gordon, 1995, 1997; Thevathasan, 1998; Price and Gordon, 1999; Dyack, Rollins, and Gordon, 1999; Price, 1999; Simpson, 1999; Zhang, 1999; Gray, 2000; Howell, 2001
Silvipastoral systems	Bezkorowajnyj, Gordon, and McBride, 1993

crease and diversify farm income (Dyack, Rollins, and Gordon, 1999), enhance wildlife habitat, abate soil erosion, and lower nutrient loading to waterways (Williams et al., 1997). By properly designing and managing a tree-based intercropping system, diversified short- and long-term economies are possible and commercial markets for certain products (e.g., nut crops, Christmas trees, etc.) from intercropping systems may be realized (Dyack, Rollins, and Gordon, 1999). Further, these systems can also potentially address societal concerns about climate change, by sequestering carbon (C) in both above- and below-ground components (Kort and Turnock, 1999).

Interaction in agroforestry is defined as the effect of one component of the system on the performance of another component and/or the overall system (Nair, 1993). Rao, Nair, and Ong (1998) have indicated that the study of interactions in agroforestry systems requires the examination of a number of complex processes, including processes related to soil fertility (which includes soil chemical, physical and biological interactions), competition (which includes competitive interactions for soil, water and nutrients, and solar radiation), microclimate, pest and diseases (which include interactions related to weeds, insects, and diseases), soil conservation (which includes reduction in soil erosion and nutrient leaching) and allelopathy. Exploitation of positive interactions between the woody (tree) and non-woody (agricultural or annual crop) components and the minimization of negative interactions are the key to the success of tree-based intercropping systems. A better understanding of these types of interactions will provide a strong scientific basis for improvement and adoption of tree-based intercropping systems.

In 1987, a long-term tree-based intercropping research experiment was initiated at the University of Guelph Agroforestry Research Station, Ontario, Canada (43° 32′ 28″ N, 80° 12′ 32″ W) using 10 different species of hardwood and coniferous trees which were annually intercropped with corn (*Zea mays* L.), soybean (*Glycine max* L. Merr.), and either winter wheat (*Triticum aestivum* L.) or barley (*Hordeum vulgare* L.). Between tree-row spacings of 12.5 and 15 m and within-row spacings of either 3 m or 6 m were also incorporated into the design, which covers 30 ha. The soil type was sandy loam (Order-Alfisols, group-Typic Hapludalf). Conventional standard cultural practices were implemented annually for the respective crops grown between the widely spaced tree rows (Figure 1 A and B).

This contribution will deal mainly with the last 10 years of the study. Identifying and quantifying biophysical interactions are important so that management strategies can be established that promote comple-

FIGURE 1. (A) Conventional harvest of soybeans grown in between widely spaced tree rows. (B) Winter wheat crop grown between widely spaced white ash trees, at the University of Guelph Agroforestry Research Station, Guelph, ON, Canada.

mentary interactions and reduce or eliminate negative competitive interactions. The establishment of suitable management strategies will facilitate the adoption of tree-based land-use practices in southern Ontario and other suitable geographical regions. In this contribution, we discuss results from a series of studies that addressed these biophysical interactions. The specifics of the materials and methods of each substudy is beyond the scope of this contribution. The reader is, therefore, advised to refer to the cited literature for appropriate materials and methods.

INTERACTIONS RELATED TO SOIL FERTILITY

The effects of poplar (hybrid clone DN 177; *Populus deltoides* × *Populus nigra* 177) leaf biomass distribution (leaf fall) on soil nitrogen (N) transformations and soil organic carbon (SOC) was studied from 1993 to 1995. Two experiments were conducted simultaneously and in both experiments the main treatment was distance from the tree row. In field experiment 1, poplar leaves were removed after leaf senescence in 1993 and 1994; in experiment 2 leaves were not removed (see Thevathasan and Gordon, 1997). The study commenced in the summer of 1993 and the fall-shed leaves were removed from the experimental plots in fall 1993 and in fall 1994 (experiment 1). In experiment 2, the leaves were not removed during fall 1993 and fall 1994. Poplar leaf biomass distribution showed a distinct pattern, with almost 80% of the leaves falling within 2.5 m of the poplar tree row (Table 2).

Differing rates of poplar leaf biomass input across the field created distinct zones of soil N and C accumulation. Based on these inputs, the intercropped alley was divided into three zones: the area close to the

TABLE 2. Poplar leaf-biomass (leaf fall) distribution in a poplar-barley intercropping system during the 1993 and 1994 growing seasons, University of Guelph Agroforestry Research Station, Guelph, ON, Canada (adapted from Thevathasan and Gordon, 1997).

Zone	Leaf biomass ($Mg \cdot ha^{-1}$)	
(Distance from the poplar tree row (m))	1993	1994
0-2.5	2.67 ± 0.04	2.76 ± 0.14
2.5-6.0	0.52 ± 0.05	0.61 ± 0.06

poplar tree row (0 to 2.5 m on either side of the tree row), the middle of the crop alley (2.5 to 8 m from the tree row), and the area furthest away from the tree row (8.0 to 15.0 m). Observed mean soil nitrate (NO_3) production in the above zones during 1993 (June to August) was 73.1, 41.0, and 34.0 μg 100 g^{-1} dry soil day^{-1}, respectively. In 1995, as a result of the removal of poplar leaves from the field for two consecutive years (fall 1993 and fall 1994), NO_3 production values were decreased to 17.6, −2.8, and −1.7 μg 100 g^{-1} dry soil day^{-1} in the same zones, respectively. However, in experiment 2 (leaves not removed) mean NO_3 production (June to August 1995) in the same zones was 109.4, 15.4, and 5.7 μg 100 g^{-1} dry soil day^{-1}, respectively. It appears that the addition of poplar leaves significantly ($p < 0.05$) affected NO_3 production rates, especially in the zones close to the tree row and in the middle of the crop alley. It also appears that the major portion of NO_3 was released from the labile organic pool (recently added poplar leaf biomass) rather than from the recalcitrant organic pool, since the removal of poplar leaves from the field did not significantly change the SOC pool over the three-year period.

The results also suggest that differences in resource pools (e.g., NO_3 accumulation, discussed above) across the crop alley affected barley growth and development (see Thevathasan and Gordon, 1997). Barley biomass in 1994 and 1995 was influenced by the absence of poplar leaves on the experimental plots in fall 1993 and fall 1994, respectively. However, barley biomass yield in 1993 was influenced by the presence of poplar leaves from the previous year (fall 1992). The treatments were fixed distances from the poplar tree row. Mean barley above-ground biomass (AGB) declined from 517 g m^{-2} in the zone closest to tree row to 490 g m^{-2}, in the middle of the alley in 1993 but was relatively constant across the alley in 1995, most likely due to the removal of leaves during the 1993 and 1994 fall seasons and the resulting low NO_3 production as described above (Figure 2).

In 1993, total N concentration of barley grain in the 3 zones was 2.52, 1.69, and 1.65%, respectively, and in 1995 the recorded values were 1.69, 1.44, and 1.44%. Removal of poplar leaves from the field for two years resulted in a significant reduction in NO_3 accumulation across the crop alley (results presented above–experiment 1) and this may have affected NO_3 uptake by barley, causing a reduction in barley AGB and total N concentration in grain. Barley yield (grain) was not measured directly but the yield is normally approximately 37% of the recorded barley AGB for this particular variety of barley (OAC Kippen). It is

FIGURE 2. Above-ground biomass of barley as a function of distance from the tree row during three growing seasons at the Agroforestry Research Station, University of Guelph, Guelph, ON, Canada (adapted from Thevathasan and Gordon, 1997).

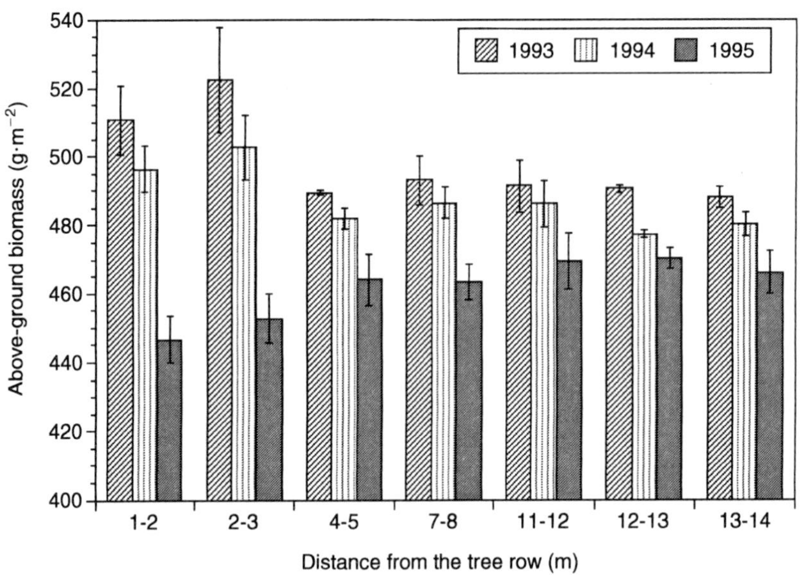

likely that a similar trend exists with respect to barley grain yield across the crop alley.

Soil organic carbon did not significantly ($p > 0.05$) change over the three-year period during which the above study was conducted with recorded SOC means of 3.25, 2.32, and 2.50%, in the 3 zones, respectively (Figure 3). This was to be expected as only 15 to 35% of added organic residue is actually incorporated into the permanent organic pool (humus) (Brady, 1990). Hence, it is unlikely that two years of addition or removal affected the total SOC pool close to the poplar tree row.

However, it should be emphasized that the high rate of poplar leaf biomass addition (1000 kg C $ha^{-1}year^{-1}$) over a period of 7 to 8 years has resulted in an increase in SOC of approximately 1% close to the tree row; this effect extends into the alley for approximately 4 m. This is approximately 30-35% increase in SOC close to the tree rows over the given period. The increase of SOC under tree canopies and the positive influence of agroforestry tree species in improving soil fertility has been well-reported (Kang, Reynolds, and Atta-Krah, 1990; Nair, 1993). For

FIGURE 3. Variation in soil organic carbon as a function of distance from the tree row in 1993, 1994, and 1995 at the Agroforestry Research Station, University of Guelph, Guelph, ON, Canada (adapted from Thevathasan and Gordon, 1997).

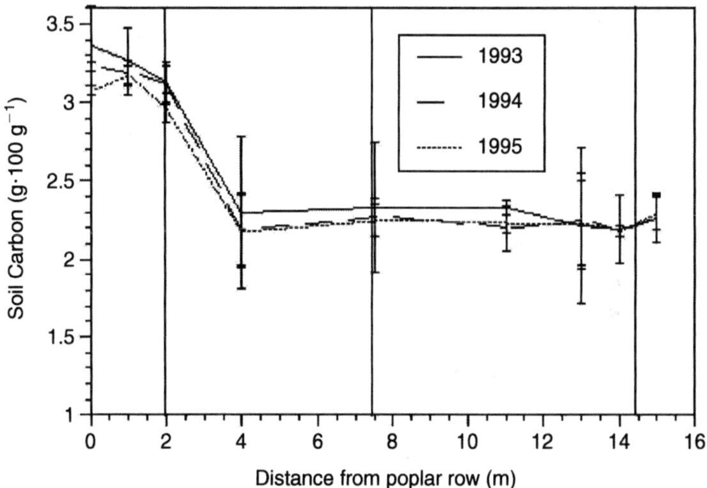

example, a gradual increase in SOC over 7 years under a black locust (*Robinia pseudoacacia* L.) agroforestry system was reported in the Azad Kashmir region of Pakistan (Ahmed, Mahler, and Ehrenreich, 1996). Increases in SOC in the cropping area may reduce soil erosion and could help to maintain soil fertility and stability, while at the same time contributing towards a positive C balance via soil sequestration.

NUTRIENT ADDITIONS THROUGH HYDROLOGIC PATHWAYS

Trees may intercept airborne particles by trapping them on leaf surfaces. These chemicals, along with chemical elements naturally occurring in tree components, especially in leaves, are then washed off into the soil by rainfall events. This process creates a sphere of influence beneath the tree canopy whereby soil nutrient levels may be slightly higher than in cropping systems that do not posses tree canopies. A study examined nutrient additions through two important hydrologic pathways: throughfall (rainwater falling through tree canopies) and stemflow (rainwater falling down the branches and stems). The annual nutrient inputs from tree leaf litterfall, net throughfall, and net stemflow for five intercropped tree species are presented in Table 3, adapted from Zhang (1999).

Hybrid poplar and silver maple (*Acer sacharrinum* L.) contributed the most N through these pathways, especially through leaf litterfall. The movement of nutrients through the hydrologic pathways varied depending on tree species and on the nutrient itself. For example, a higher percentage of potassium (K) was carried in the net throughfall than in the stemflow for all of the species observed. The nutrient contributions

TABLE 3. Annual nutrient inputs by litterfall, net throughfall and net stemflow at the University of Guelph Agroforestry Research Station, Guelph, ON (adapted from Zhang, 1999).

	Total-N	NO_3-N	NH_4-N	P	K	Ca	Mg
				$Kg \cdot ha^{-1} \cdot y^{-1}$			
Black walnut (*Juglans nigra* L.)							
Litterfall	1.74 a			0.15 a	0.50 a	3.08 a	0.77 a
Net throughfall		0.06 a	0.09 a	0.175 a	2.99 a	2.21 a	0.92 a
Net stemflow		0.01 a	0.004 a	0.32 a	0.08 a	0.06 a	0.02 a
Total		0.07	0.10	0.63	3.57	5.35	1.71
Red oak (*Quercus rubra* L.)							
Litterfall	4.6 b			0.35 a,c	1.87 b	11.09 b	3.29 b
Net throughfall		0.24 b	0.34 b	0.14 a	1.29 a	1.39 a	0.42 a
Net stemflow		0.01 a	0.01 a	0.13 a	0.08 a	0.05 a	0.01 a
Total		0.25	0.35	0.62	3.24	12.53	3.73
Silver maple							
Litterfall	15.22 c			0.99 b	3.13 c	15.51b	2.50 b
Net throughfall		2.73 c	3.05 c	0.12 a	7.67 b	8.93 b	2.77 b
Net stemflow		0.05 b	1.20 b	4.10 b	0.24 b	0.12 b	0.04 b
Total		2.78	4.25	5.21	11.04	24.56	5.31
Hybrid poplar							
Litterfall	10.99 d			1.06 b	5.49 d	28.51 c	7.27 c
Net throughfall		1.75 d	0.73 b	0.38 b	15.44 c	8.99 b	2.71 b
Net stemflow		0.03 b	0.01 a	3.56 b	0.29 b	0.11 b	0.05 b
Total		1.78	0.74	4.99	21.22	37.61	10.03
White ash							
Litterfall	4.86 b			0.51 c	1.84 b	13.38 b	2.37 b
Net throughfall		0.32 b	0.10 a	0.048 c	1.62 a	2.02 a	0.62 a
Net stemflow		0.01a	0.002 a	0.24 a	0.08 a	0.04 a	0.01 a
Total		0.33	0.11	0.79	3.54	15.44	3.01

By parameters and nutrient, values followed by the same letter across species are not significantly different (LSD, $p > 0.05$).

(available N, Ca, K, and P) through these hydrologic pathways and through leaf litter can enhance soil structure by increasing organic matter (OM) content and by producing a nutrient rich substrate for microbial processes. This, in conjunction with the well-known ability of trees to modify microclimate (e.g., reduced wind speed, low evapotranspiration demands; Brandle, Hodges, and Wight, 2000) likely contributed towards increased winter wheat yields close to the intercropped tree rows. Winter wheat grain yields in the intercropping system and in a monocropping system were 2450.3 kg ha^{-1} and 2244.4 kg ha^{-1}, respectively. This translates to an 8.4% wheat grain yield increase in the intercropping system compared to the monocropping system.

DYNAMICS OF EARTHWORM POPULATIONS UNDER TREE-BASED INTERCROPPING SYSTEMS

Increasing soil OM and reducing soil disturbance can increase the presence and activity of soil fauna, particularly invertebrates such as earthworms (Edwards and Lofty, 1977). Alternative land use systems, such as agroforestry, which encourage greater accumulation of OM and reduce soil disturbance could positively affect earthworm presence and activity.

A study of earthworm population dynamics in a temperate intercropping system was undertaken at the University of Guelph Agroforestry Research Station in 1997 and 1998 (Price and Gordon, 1999). Tree species played an important role in determining the spatial and temporal distribution of earthworms within the intercropping system. Significant differences ($p < 0.05$) in earthworm density and biomass were observed between sampling periods and tree species (Table 4). For example, earthworm densities in poplar and ash (*Fraxinus americanus* L.) tree rows were greater than those in silver maple or corn, possibly due to greater litter contributions or more rapid decomposition of leaf litter, or both. Earthworm numbers decreased during the summer period but densities close to tree rows were still significantly greater ($p < 0.05$) than those from a comparable conventionally cropped corn field. Differences in earthworm numbers between spring 1997 and spring 1998 likely resulted from differences in temperature and precipitation between the two sampling periods. The average monthly temperature in the spring of 1998 was almost double than that in the spring of 1997 (15.6°C versus 7.9°C); in addition, total precipitation was 34.2 mm in 1998 and 82.2 mm in 1997. It is likely that these differences contributed

TABLE 4. Comparison of total mean earthworm numbers (No. m^{-2}) and biomass ($g \cdot m^{-2}$) under an intercropped system and a conventionally cropped corn system at the University of Guelph Agroforestry Research Station, Guelph, ON (adapted from Price, 1999).

	Numbers (No. m^{-2})				Biomass ($g \cdot m^{-2}$)			
	Poplar	Maple	Ash	Corn	Poplar	Maple	Ash	Corn
1997								
Spring	394a(a)	57a(b)	379a(a)	11a(c)	457a(a)	440a(a)	735a(b)	6.07a(c)
Summer	119b(a)	42b(b)	61b(b)	4a(c)	245b(a)	89b(b)	153b(b)	4.54a(c)
Fall	257c(a)	196c(a)	268c(a)	30b(b)	345c(ab)	263a(b)	437c(a)	45.96 b(c)
1998								
Spring	90b(a)	63b(a)	46b(a)	3a(b)	181b(a)	144b(a)	161b(a)	3.12a(b)

Values followed by the same letter within a column are not significantly different (LSD, $p > 0.05$).
Values followed by the same letter across a row, within brackets, are not significantly different (LSD, $p > 0.05$).

greatly to the lower earthworm numbers in the spring of 1998, due to the environmental sensitivities of earthworms (Edwards and Bohlen, 1997; Lee, 1985).

The increased nutrient deposition due to stemflow, throughfall and litterfall under trees (Rhoades, 1997; Zhang, 1999) as compared to under corn may be responsible for the higher numbers and biomass of earthworms observed in this study. These OM and nutrient additions can significantly influence earthworm populations in the soil (Brown, Love, and Handley, 1963; Zwart et al., 1994; Willems, Marinissen, and Blair, 1996). Microclimatic differences attributed to the presence of trees in the intercropped situation may also have been responsible for greater earthworm numbers and biomass, even though further study is required to confirm this. We have observed distinct earthworm dispersal patterns within our tree-based intercropping system that might be linked to the microclimate gradient between tree rows and the middle of the crop alley (Price, 1999; Williams and Gordon, 1995). Similar effects have been noted under alley cropping systems in Africa (Tian, Brussaard, and Kang, 1993; Hauser, Asawalam, and Vanlauwe, 1998).

AVIAN DIVERSITY

The practice of tree-based intercropping adds a third vertical dimension, absent in conventional agricultural fields, to the land base, which helps to modify microclimatic parameters (discussed below) while at the same time providing habitat for small animals and birds.

A study was conducted on-site by Williams, Koblents, and Gordon (1995) to investigate the bird use of an intercropped cornfield, a conventional cornfield and an old-field site. The old-field site was comprised of various tall grasses and weeds including goldenrods (*Solidago* spp.), asters (*Aster* spp.), and milkweeds (*Asclepias* spp.). Only one species of bird nested in the conventional cornfield and the avian diversity in the intercropped area (7 species) was similar to that found in old-field (8 species). In addition, more bird species foraged in the intercropped plots (10 species) compared to the conventional cornfield (2 species) and old-field site (8 species).

INTERACTIONS RELATED TO MICROCLIMATIC MODIFICATIONS

Microclimatic parameters such as soil and air temperature, relative humidity, windspeed, surface horizon soil water, and light (photosynthetic photon flux density–PPFD) measured in any given agricultural field, may be modified by the introduction of trees. The magnitude of these microclimatic modifications in a tree-based intercropped system will vary between conditions found in an open agricultural field (no tree-influenced microclimatic modifications) to those found associated with dense woodlots (maximum tree-influenced microclimatic modifications). These tree-influenced microclimatic modifications may act in such a way as to increase the overall productivity of the associated agricultural crop. However, in some tree-crop combinations, the effect of the presence of trees on soil water and PPFD has negatively affected crop yields. The latter scenario will be discussed below in detail under 'Interactions related to competition.'

During the early stages of the establishment of tree-based intercropping systems, associated agricultural crops may cause microclimatic modifications along the tree rows. An experiment was established in 1992 to quantify the impact of these modifications on tree growth (height), as influenced by three agricultural crops: wheat, soybeans, and corn. Mean daily profiles for soil temperature, air temperature, windspeed, and PPFD were measured in the tree rows (see Williams and Gordon, 1995). Among the measured parameters, windspeed and soil moisture were significantly modified by the agricultural crops, with distinct treatment differences found in mean daily windspeed throughout the growing season. The above crop canopy (2.5 m) windspeed was measured over the entire year as a control and, as expected, the control windspeed was the highest recorded compared to that within tree rows (trees were less than 2.5 m in height). Early in the growing season, the largest reduction in the windspeed was found in tree rows growing adjacent to winter wheat, an early developing crop. However, windspeeds in corn and soybeans decreased gradually as crops matured, with the average windspeed in corn dropping below that of wheat in July and staying below 1 m s^{-1} for the reminder of the year (Williams and Gordon, 1995).

Under winter wheat, soil moisture content declined rapidly to a depth of 60 cm until harvest in early August. This corresponds to the early and rapid growth of this particular crop, as well as to the critical phase of growth and water demand of young trees. At this stage young trees are exploiting a relatively small and shallow soil zone with their shallow root system. The early demand for water by the winter wheat crop detrimentally affected the growth and development of young trees. Where corn was grown as the annual crop, a temporal change in water demand was observed. Water demand for corn early in the season was low and hence young trees experienced little competition for water. This positively influenced tree height growth in the corn treatment plots (Williams and Gordon, 1995).

The above study was undertaken while intercropped trees were being established, and indicates the importance of understanding these types of interactions at all stages of system development. Tree-influenced microclimatic interactions have also been studied in mature tree-based intercropping systems. A detailed account of these types of interactions can be found in Ong and Huxley (1996); Rao, Nair, and Ong (1998); and Garrett et al. (2000).

INTERACTIONS RELATED TO COMPETITION

The effects of shading on the productivity of corn (a C_4 plant) and soybean (a C_3 plant) in the intercropped system were studied during the 1997 and 1998 growing seasons (Simpson, 1999). Corn and soybeans were intercropped with hybrid poplar (clone-DN-177) and silver maple at a within-row spacing of 6 m and between-row spacing of either 12.5 or 15 m. Tree rows were oriented approximately north-south and trees were absent from all 'control' plots. Twelve locations were sampled around each tree at 2 and 6 m east and west of the tree. Generally, tree competition significantly reduced the growth of individual plants growing nearer to tree rows (2 m) (Table 5), and often reduced the size of plants furthest from competing trees (6 m) in comparison with those in the control treatment (Figure 4). Daily rates of C assimilation were generally lower near the trees where competition for photosynthetically-active radiation (PAR) was the greatest, resulting in lower crop yields.

Growth characteristics (height, leaf area, weight) of individual plants were significantly correlated with available PAR (r = 0.75 to 0.80–soybeans; r = 0.87 to 0.95–corn) and net assimilation (r = 0.73 to 0.83–soybeans; r = 0.92 to 0.96–corn), but poorly and non-significantly correlated (r = 0.02 to 0.16) with midday water potential. It was concluded that competition for light, and not water, within 6 m of the tree rows was the main factor that detrimentally affected corn and soybean yields (Simpson, 1999). However, several alternative management strategies can be recommended to reduce shading impacts from tree rows and to ensure an acceptable economic yield in the zone closest to the tree row. These could include, but are not limited to, pruning and crown thinning. Although growth reductions were not significantly correlated with plant water stress, the lowest plant water potentials observed were for crops grown in conjunction with silver maple. Concurrently, much higher intrinsic water use efficiencies (iWUE) were observed for corn plants compared to soybeans, and were highest in this tree treatment (Simpson, 1999). Collectively, these findings suggest that silver maple and corn may compete more vigorously for available soil moisture than poplar and soybean. It was concluded that, further study was needed to determine the extent to which tree roots were competing directly with crops for moisture.

An extensive survey of the rooting habits of potential intercropping tree species was undertaken by Gray (2000) at the study site. He found that soybean yield and tree root quantity were negatively correlated, and that numbers of roots diminished with distance from tree row. In conjunction with other results (Gray and Gordon, 1997), this information will be useful when considering lateral disc pruning of tree roots, especially in the first 15 cm of the soil layer where most of the crop roots are generally found. This management strategy may help reduce competition for water and nutrients close to the tree rows.

General observations made on crop yields at this research site have revealed that the presence of trees detrimentally affected C_4 plant yields to a greater extent than C_3 plant yields. We have documented (on average) a 20 to 25% yield reduction per unit land-base for corn when compared to conventional yield recorded from an adjacent field (control).

TABLE 5. Effects of tree competition on crop morphology and/or growth for individual soybean or maize plants harvested July 31, 1997.

Crop	Parameter	Control		Poplar		Maple	
		2 m	6 m	2 m	6 m	2 m	6 m
Soybean							
	Height (cm)	75.6a	82.7a	45.6b	67.5a	44.4b	69.4a
	Leaf area Single[1] trifoliate leaf (cm^2)	120.1a	111.5a	55.7b	104.2a	50.2 b	123.4 a
	Leaf weight single Trifoliate leaf (g)	0.52 a	0.46a	0.24b	0.42a	0.21 b	0.53 a
	Whole plant leaf area (cm^2)	796.2 b	1070.1a	317.4b	630.8a	247.1b	766.3a
	Whole plant leaf weight (g)	3.2b	4.1a	1.3b	2.3a	1.0b	2.9a
	Stem Weight (g)	3.7b	5.2a	1.4b	2.6a	1.0b	3.6a
	Number of pods (No.)	13.7a	20.0a	9.8b	15.0a	7.8b	15.5a
	Cumulative pod length (cm)	35.7b	61.1a	27.7b	39.8a	22.1b	44.4a
	Pod weight (g)	0.79b	1.38a	0.64b	0.84a	0.56b	1.00a
	Total aboveground weight (g)	7.6b	10.7a	3.4b	5.8 a	2.6b	7.4a
Corn	Height (cm)	196.0 b	209.3a	103.8 b	177.2a	126.2b	198.8a
	Whole plant leaf area (cm^2)	5386.9a	5389.3a	3769.2b	5026.5a	3758.5b	5302.0a
	Whole plant leaf weight (g)	30.2a	32.0a	17.3b	26.0a	17.6b	28.2a

[1]Single trifoliate leaf from top of plant.

By parameter and similar treatment, values in each row followed by the same letter are not significantly different (Tukey's HSD, $p > 0.05$).

FIGURE 4. Effects of tree competition on the height of individual corn plants as a function of distance from the tree row. The tree row represents (0) and the positive values are to the east side of the tree row and the negative values are to the west of the tree row.

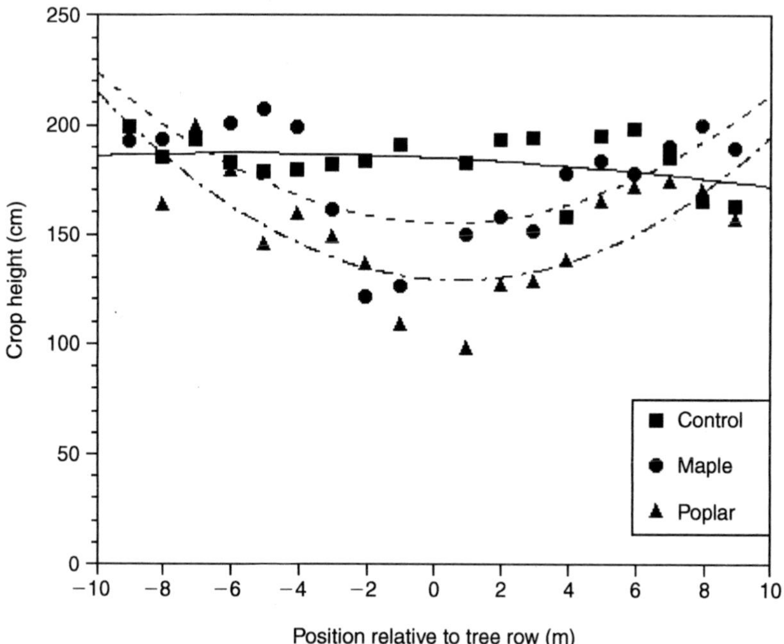

However, C_3 plant yields (soybean, wheat) are generally on par with conventional yields or slightly higher in tree-based intercropping especially under dry weather conditions. In the latter scenario, microclimatic modification effects by trees might have helped to enhance yields in the tree-based intercropping system when compared to a conventional agricultural system (Zhang, 1999).

Several other options are also available that may make the system more acceptable in the current economic environment. One of the primary goals of the research conducted at the University of Guelph Agroforestry Research Station is to grow trees with a high economic value. This is expected to offset the immediate economic loss of crop yield reduction due to shading and reduction of the land base for crop production. However, the long-term economics of such scenarios should be investigated in detail (e.g., Dyack, Rollins, and Gordon, 1999).

INTERACTIONS RELATED TO NUTRIENT AND SOIL CONSERVATION

Recent research conducted at the intercropping site has shown that N leaving the intercropping site as NO_3 in soil percolate can be potentially reduced by more than 50% when compared to losses from a monocropped barley field (Thevathasan, 1998). Actual leaching losses have been estimated to be approximately 9 kg N ha^{-1} y^{-1} at the intercropping site whereas leaching losses in a monocropped field adjacent to the above intercropped field were 20 kg N ha^{-1} y^{-1}. Intercropping appears to have reduced leaching losses by 11 kg N ha^{-1} y^{-1} (Thevathasan, 1998). Understanding N flow in these systems may lead to reduced NO_3 loadings to nearby waterways, and may also be useful for future fertilizer management recommendations. Furthermore, with respect to greenhouse gas emissions, less NO_3 leaching will reduce potential N_2O emissions into the atmosphere from agricultural fields. The latter scenario is discussed below under the sub-heading 'Interactions related to climate change mitigation.'

No research has been conducted on this site with respect to soil erosion. However, since the establishment of trees in 1987, from casual observation, soil erosion has been significantly reduced.

INTERACTIONS RELATED TO CLIMATE CHANGE MITIGATION

Many recent studies have identified tree-based land-use practices as a significant global opportunity to reduce the accumulation of CO_2 in the atmosphere (Brandle, Wardle, and Bratton, 1992; Kort and Turnock, 1999; Schroeder, 1993; Unruh, Houghton, and Lefebvre,1993). The United Nations has also estimated that agroforestry based land-use practices on marginal or degraded lands could sequester 0.82 to 2.2 Pg C y^{-1}, globally, over a 50-year period (Dixon et al., 1994). Apart from C sequestration, agroforestry practices, such as tree-based intercropping might also significantly reduce other greenhouse gas (GHG) emissions (e.g., N_2O).

Tree-based intercropping system may have a significant impact on climate change mitigation as a result of the following reasons: (1) Tree-based intercropping systems can be adopted in agricultural land classes from 1 through 4 (Canada) (Note: Land classes 1 to 3 = suitable for annual cultivation; all or many common field crops can be grown; Land class 4 = marginal for annual cultivation; choice of field crops that can be grown is limited). Therefore, the land base in Canada that could potentially be brought under tree-based intercropping is substantial (20-25 million ha) which in turn, can have a significant effect on C sequestration and GHG emission reduction; (2) The tree component occupies a part of the land base, reducing the land for agriculture and subsequently reducing the need for inputs such as supplemental N for crop production. The reduction in N_2O emissions will be directly proportional to the land base occupied by trees; (3) The presence of trees in a portion of the landscape will result in a variety of environmental interactions as previously discussed. Furthermore, the decrease in N moving out of the rooting zone will lead to reduced N_2O emissions now occurring as a result of denitrification in N-enriched surface water resources. The literature suggests that about 2.5% of the leached N is lost as N_2O; (4) The annual leaf fall will cycle some N back to the soil reserve, especially if the tree species are deciduous. While this N will be localized in the area close to the tree, it does constitute a quantifiable contribution of N to the subsequent agricultural crop. Thus, lower application rates of inorganic fertilizer (especially N) close to the tree row can result in reduced environmental losses and proportional reduction in N_2O emissions can be expected. Apart from reduction in GHG emissions, C sequestration in agricultural fields can be augmented through this type of land-use practice. Trees are a natural long-term sink for atmospheric CO_2 and, depending upon end-use, C sequestered in the wood could remain immobilized for an ecologically significant period of time. Furthermore, annual leaf litter input and fine root turnover can also significantly influence long-term soil organic C storage in agricultural soils.

Quantitative measurements on potential N_2O emission reduction and C sequestration were taken at the University of Guelph Agroforestry Research Station in 1999. N_2O estimates were derived from previously collected data at this site. Carbon sequestration potentials were calculated by destructively sampling a 12-year-old fast-growing fibre tree species (hybrid poplar). Estimates indicate that the quantity of N recycled to the agricultural zone in the leaf litter from fast-growing deciduous species can contribute up to 5 kg N ha^{-1} yr^{-1} (Thevathasan, 1998). This implies that the amount of inorganic fertilizer addition can potentially be reduced by this amount. Cole et al. (1996) suggest that N_2O emissions from agricultural land is directly related to the rate of N application and that 1.25% of the N applied is emitted directly from the land as N_2O. As previously indicated, N leaching losses were only 9 kg N ha^{-1} yr^{-1} at this site compared to 20 kg N ha^{-1} yr^{-1} calculated for an adjacent monocropped field, indicating that intercropping has reduced leaching losses by 11 kg N ha^{-1} yr^{-1} (Thevathasan, 1998). These results suggest that lower additions of inorganic fertilizer in concert with less NO_3 leaching losses could lead to a significant reduction in N_2O emissions as a result of adopting tree-based intercropping.

Quantitative measurements of C sequestered in both above- and below-ground woody components of a fast growing fibre tree species (hybrid poplar) indicate that over a period of 12 years, 9 Mg of C ha^{-1} have been sequestered (stems ha^{-1} = 111, at a spacing of 15 m between rows and 6 m within rows). Theoretically, therefore, trees alone have immobilized 36 Mg of CO_2 ha^{-1} over this 12-year period. In addition to the C sequestered in woody components, C contributions to the soil from annual litterfall alone were 1332 kg C ha^{-1}. Previous research indicates that the latter annual addition over a period of 10 years in concert with fine root turnover has increased soil organic C by 1% (absolute value) close to the tree rows (Thevathasan and Gordon, 1997). It is important to recognize that trees can significantly impact the C balance when introduced into agricultural fields. In a monocropped agricultural field, annual C input to the soil is in the range of 500 to 700 kg C ha^{-1} yr^{-1} (P. Voroney, 1999 pers. comm.) whereas with tree-based intercropping annual C input can be as high as 2500 kg C ha^{-1} y^{-1}, or about 5 times more than that found in monocropped agricultural fields.

CONCLUSIONS

In this new millennium, one of our fundamental priorities as policy makers, scientists and resource managers should be to ensure that management recommendations result in sustainable and economical production capabilities. Greater focus must be given to enhancing the attitude that we are essentially stewards of both soil and water resources. The adoption of agroforestry systems, especially intercropping as presented in this paper, shows much promise in this regard (Thevathasan, 1998).

Among the ameliorative agricultural practices currently under consideration in southern Ontario, tree-based intercropping systems remains a viable option. The beneficial effects of trees in relation to soil fertility, productivity and nutrient cycling, and microclimate can be positively exploited, especially in the context of developing systems for both marginal and prime agricultural lands (Gordon, and Newman, 1997). The success of intercropping hinges on the ability of the system components to maximize resource utilization while maintaining 'complementary' interactions between them (Rao, Nair, and Ong, 1998). When this occurs, productivity per unit land area is often enhanced resulting in higher economic returns. When components of an intercropping system vary dramatically (e.g., woody and non-woody plants), the demand for limited resources is generally staggered in space and time, and resource capture and productivity per unit land area may be maximized.

On a biological level, intercropping increases micro- and macro-faunal diversity and activity, both above- and below-ground. The increased range of faunal activity gives a clear indication of ecosystem 'health' within an intercropping system relative to that associated with conventional agricultural practices. From an ecological perspective, intercropping systems trap larger amounts of energy at different trophic levels, demonstrating higher energy utilization efficiency. In relation to CO_2 sequestration and other greenhouse gas (e.g., N_2O) emission reductions, tree-based intercropping systems have the potential to greatly contribute to climate change mitigation. The tangible benefits that are derived from the above described eco-biological processes, along with combined yields obtained from both trees and crops, place this land-use practice above conventional agricultural systems in terms of long term overall productivity.

However, the economics of tree-based intercropping systems need to be examined in more detail. Initial establishment costs, changes in labor requirements and the initial loss of revenue due to removing cropland from production often deter farmers from adopting these types of systems even though there are no additional costs required for specialized machinery. Policy measures, and/or tax incentives and cost-share programs should be investigated in order to enhance adoption rates of intercropping practices in southern Ontario and other appropriate geographical regions.

REFERENCES

Ahmed, I., R.L. Mahler, and J.H. Ehrenreich. (1996). Black locust (*Robinia pseudoacacia* L.) in agroforestry systems: Effect on soil properties. In *Proceedings of the Fourth North American Agroforestry Conference*, July, 1995, eds. J. H. Ehrenreich, and D.L. Ehrenreich, Boise, ID, pp. 35-39.

Ball, D.H. (1991). *Agroforestry in southern Ontario: A potential diversification strategy for tobacco farmers.* MSc Thesis, Dept. of Environmental Biology, Guelph, ON: University of Guelph.

Bezkorowajnyj, P.G., A.M. Gordon, and R.A. McBride. (1993). The effect of cattle foot traffic on soil compaction in a silvo-pastoral system. *Agroforestry Systems* 21:1-10.

Brady, N.C. (1990). *The Nature and Properties of Soils*. New York, NY: Macmillan Publishing Company.

Brandle, J.R., L. Hodges, and B. Wight. (2000). Wind break practices. In *North American Agroforestry: An Integrated Science and Practices*, eds. H.E. Garrett, W.J. Rietveld, R.F. Fisher, D.M. Kral, and M.K. Viney, Madison, WI: American Society of Agronomy. pp. 79-118.

Brandle, J.R., T.D. Wardle, and G.F. Bratton. (1992). Opportunities to increase tree planting in shelterbelts and the potential in impacts on carbon storage and conservation. In *Forest and Global Change*, eds. R.N. Sampson and D. Hair, Washington, DC: American Forests. pp. 157-176.

Brown, B.R., C.W. Love, and W.R.C. Handley. (1963). *Protein-Fixing Constituents of Plants*. London, UK: Report from Forest Resource, Part III, pp. 90-93.

Christrup, J. (1993). *Potentials of edible tree nuts in Ontario*. MScF Thesis, Faculty of Forestry, Toronto, ON: University of Toronto.

Cole, C.V., C. Cerri, K. Minami, A. Mosier, N. Rosengerg, and D. Sauerbeck. (1996). Agricultural options for mitigation of greenhouse gas emissions. In *Climate Change 1995. Impacts, Adaptations and Mitigation of Climate Change: Scientific Technical Analysis. Published for the Intergovernmental Panel on Climate Change*, eds. R.T. Watson, M.C. Zinyowera and R.H. Moss, Cambridge, UK: Cambridge University Press, pp. 745-771.

Dixon, R.K., J.K. Winjum, K.J. Andrasko, J.J. Lee, and P.E. Schroeder. (1994). Integrated land-use systems: Assessment of promising agroforest and alternative land-use practices to enhance carbon conservation and sequestration. *Climate Change* 27:71-92.

Dyack, B., K. Rollins, and A.M. Gordon. (1999). An economic analysis of market and non-market benefits of a temperate intercropping system in southern Ontario, Canada. *Agroforestry Systems* 44:197-214.

Edwards, C.A. and P.J. Bohlen. (1997). *Biology and Ecology of Earthworms*. London, UK: Chapman and Hall Press.

Edwards, C.A. and J.R. Lofty. (1977). *Biology of Earthworms*. London, UK: Chapman and Hall Press.

Garrett, H.E., W.J. Rietveld, R.F. Fisher, D.M. Kral, and M.K. Viney. (eds.). (2000). *North American Agroforestry: An Integrated Science and Practices*. Madison, WI: American Society of Agronomy.

Gordon, A.M. and S.M. Newman. (1997). *Temperate Agroforestry Systems*. Wallingford, U.K: CAB International Press.

Gordon, A.M., and P.A. Williams. (1991). Intercropping of valuable hardwood tree species and agricultural crops in southern Ontario. *Forestry Chronicle* 67: 200-208.

Gordon, A.M., S.M. Newman, and P.A. Williams (1997). Temperate agroforestry: An overview. In *Temperate Agroforestry Systems*, eds. A.M. Gordon, and S.M. Newman, Wallingford, UK: CAB International Press, pp. 1-8.

Gray, G.R.A. (2000). *Root distribution of hybrid polar in a temperate agroforestry intercropping system.* MSc Thesis, Dept. of Environmental Biology, Guelph, ON: University of Guelph.

Gray, R. and A.M. Gordon (1997). Application of a GIS for data management in a temperate intercropping system. In *Proceedings of the Fifth North American Agroforestry Conference, August, 1997, Ithaca, NY*, eds. L.E. Buck and J.P. Lassoie, pp. 136-139.

Hauser, S., D.O. Asawalam, and B. Vanlauwe. (1998). Spatial and temporal gradients of earthworm casting activity in alley cropping systems. *Agroforestry Systems* 41:127-137.

Howell, H. (2001). *Comparison of arthropod abundance and diversity in intercropping agroforestry and corn monoculture in southern Ontario.* MSc Thesis, Faculty of Forestry, Toronto, ON: University of Toronto.

Kang, B.T., L. Reynolds, and A.N. Atta-Krah. (1990). Alley farming. *Advances in Agronomy* 43:315-359.

Kenney, W.A. (1987). A method for estimating windbreak porosity using digitized photographic silhouettes. *Agricultural and Forest Meteorology* 39: 91-94.

Kort, J. and R. Turnock. (1999). Carbon reservoir and biomass in Canadian prairie shelterbelts. *Agroforestry Systems* 44:175-186.

Kotey, E. (1996). *Effects of tree and crop residue mulches and herbicides on weed populations in a temperate agroforestry system.* MSc Thesis, Dept. of Environmental Biology, Guelph, ON: University of Guelph.

Lee, K.E. (1985). *Earthworms: Their Ecology and Relationships with Solids and Land-Use.* NY: Academic Press.

Matthews, S., S.M. Pease, A.M. Gordon, and P.A. Williams. (1993). Landowner perceptions and adoption of agroforestry in southern Ontario, Canada. *Agroforestry Systems* 21:159-168.

Loeffler, A.E., A.M. Gordon, and T.J. Gillespie. (1992). Optical porosity and windspeed reduction by coniferous windbreaks in southern Ontario. *Agroforestry Systems* 17:119-133.

McLean, H.D.J. (1990). *The effect of corn row width and orientation on the growth of interplanted hardwood seedlings.* MSc Thesis, Dept. of Environmental Biology, Guelph, ON: University of Guelph.

Nair, P.K.R. (1993). *An Introduction to Agroforestry.* Dordrecht, The Netherlands: Kluwer Academic Publishers.

Ntayombya, P. (1993). *Effects of Robinia pseudoacacia on productivity and nitrogen nutrition of intercropped Hordeum vulgare in a agrosilvicultural system: Enhancing agroforestry's role in developing low input sustainable farming systems.* PhD Thesis, Dept. of Environmental Biology, Guelph, ON: University of Guelph.

Ntayombya, P. and A.M. Gordon. (1995). Effects of black locust on productivity and nitrogen nutrition of intercropped barley. *Agroforestry Systems* 29:239-254.

Oelbermann, M. and A. M. Gordon. (2001). Retention of leaf litter in streams from riparian plantings in southern Ontario, Canada. *Agroforestry Systems* 53:1-9.

Oelbermann, M. and A.M. Gordon. (2000). Quantity and quality of autumnal litterfall into a rehabilitated agricultural stream. *Journal of Environmental Quality* 29:603-611.

O'Neill, G.J., and A.M. Gordon. (1994). The nitrogen filtering capability of carolina poplar in an artificial riparian zone. *Journal of Environmental Quality* 23:1218-1223.

Ong, C.K. and P. Huxley. (1996). *Tree-Crop Interactions: A Physiological Approach.* Wallingford, UK: CAB International Press.

Price, G.W. (1999). *Spatial and temporal distribution of earthworms in a temperate intercropping system in southern Ontario.* MSc Thesis, Dept. of Environmental Biology, Guelph, ON: University of Guelph.

Price, G.W. and A.M. Gordon. (1999). Spatial and temporal distribution of earthworms in a temperate intercropping system in southern Ontario, Canada. *Agroforestry Systems* 44:141-149.

Rao, M.R., P.K.R. Nair, and C.K. Ong. (1998). Biophysical interactions in tropical agroforestry systems. *Agroforestry Systems* 38:3-50.

Rhoades, C.C. (1997). Single-tree influences on soil properties in agroforestry: Lessons from natural forest and savanna ecosystems. *Agroforestry Systems* 35:71-94.

Schroeder, P. (1993). Agroforestry systems: Integrated land use to store and conserve carbon. *Climate Research* 3:59-60.

Simpson, J.A. (1999). *Effects of shade on corn and soybean productivity in a tree based intercrop system.* MSc thesis, Guelph, ON: University of Guelph.

Thevathasan, N.V. (1998). *Nitrogen dynamics and other interactions in a tree-cereal intercropping systems in southern Ontario.* PhD Thesis. Guelph, ON: University of Guelph.

Thevathasan, N.V. and A.M. Gordon. (1995). Moisture and fertility interactions in a potted poplar-barley intercropping. *Agroforestry Systems* 29:275-283.

Thevathasan, N.V. and A.M. Gordon. (1997). Poplar leaf biomass distribution and nitrogen dynamics in a poplar-barley intercropped system in southern Ontario, Canada. *Agroforestry Systems* 37:79-90.

Tian, G., L. Brussaard, and B.T. Kang. (1993). Biological effects of plant residues with contrasting chemical compositions on plant and soil under humid tropical conditions: effects on soil fauna. *Soil Biology and Biochemistry* 25:731-737.

Unruh, J.D., R.A. Houghton, and P.A. Lefebvre. (1993). Carbon storage in agroforestry: An estimate for sub-Saharan Africa. *Climate Research* 3:39-52.

Willems, J.J.G.M., J.C.Y. Marinissen, and J. Blair. (1996). Effects of earthworm on nitrogen mineralization. *Biology and Fertility of Soils* 23:57-63.

Williams, P.A. and A.M. Gordon. (1992). The potential of intercropping as an alternative land use system in temperate North America. *Agroforestry Systems* 19:253-263.

Williams, P.A. and A.M. Gordon. (1994). Agroforestry applications in forestry. *Forestry Chronicle* 70:143-145.

Williams, P.A. and A.M. Gordon. (1995). Microclimate and soil moisture effects of three intercrops on the tree rows of a newly-planted intercropped plantation. *Agroforestry Systems* 29:285-302.

Williams, P.A., H. Koblents, and A.M. Gordon. (1995). Bird use of an intercropped corn and old fields in southern Ontario. In *Proceedings of the Fourth North American Agroforestry Conference*, July, 1995, Boise, ID, eds. J.H. Ehrenreich, and D.L. Ehrenreich, pp. 158-162.

Williams, P.A., A.M. Gordon, H.E. Garrett, and L. Buck. (1997). Agroforestry in North America and its role in farming systems. In *Temperate Agroforestry Systems*, eds. A.M. Gordon and S.M. Newman, Wallingford, UK: CAB International Press, pp. 9-84.

Zhang, P. (1999). *The impact of nutrient inputs from stemflow, throughfall, and litterfall in a tree-based temperate intercropping system, southern Ontario, Canada.* MSc Thesis, Dept. of Environmental Biology, Guelph, ON: University of Guelph.

Zwart, K.B., S.L.G.E. Burgers, J. Bloem, L.A. Bouwman, L. Brussard, G. Lebbink, W.A.M. Didden, J.C.Y. Marinissen, M.J. Vreeken-Buijs, and P.C. de Ruiter. (1994). Population dynamics in the below-ground food webs in two different agricultural systems. *Agriculture, Ecosystems and Environment* 51:187-198.

Agricultural Landscapes: Field Margin Habitats and Their Interaction with Crop Production

E. J. P. Marshall

SUMMARY. Most agricultural landscapes are a mosaic of farmers' fields, semi-natural habitats, human infrastructures (e.g., roads) and occasional natural habitats. Within such landscapes, linear semi-natural habitats often define the edges of agricultural fields. This paper reviews the role and interactions within and between the flora and fauna of these elements. In temperate, intensive agriculture, such field margin habitats, which historically had true agricultural functions, now are important refugia for biodiversity. As man-made habitats, field margins may also have important cultural roles as part of landscape heritage, e.g., hedges in Britain. Although field margins are not usually specific or characteristic habitat types, they contain a variety of plant communities in a variety of structures. These may range from aquatic elements, to ruderal and

E. J. P. Marshall is Director, Marshall Agroecology Ltd., 2 Nut Tree Cottages, Barton, Winscombe, North Somerset, BS25 1DU, UK (E-mail: jon.marshall@agroecol.co.uk).

The author thanks David Clements and Anil Shrestha for their invitation to contribute to this issue and to two referees who made very helpful comments on the paper. The author also thanks his erstwhile colleagues of the Agroecology Group, IACR–Long Ashton Research Station, University of Bristol.

Aspects of the work described were supported by the UK Department for the Environment, Food and Rural Affairs and by the Commission of the European Union.

[Haworth co-indexing entry note]: "Agricultural Landscapes: Field Margin Habitats and Their Interaction with Crop Production." Marshall, E. J. P. Co-published simultaneously in *Journal of Crop Improvement* (Food Products Press, an imprint of The Haworth Press, Inc.) Vol. 12, No. 1/2 (#23/24), 2004, pp. 365-404; and: *New Dimensions in Agroecology* (ed: David Clements, and Anil Shrestha) Food Products Press, an imprint of The Haworth Press, Inc., 2004, pp. 365-404. Single or multiple copies of this article are available for a fee from The Haworth Document Delivery Service [1-800-HAWORTH, 9:00 a.m. - 5:00 p.m. (EST). E-mail address: docdelivery@haworthpress.com].

http://www.haworthpress.com/web/JCRIP

Digital Object Identifer: 10.1300/J411v12n01_05

woodland communities, with combinations of them over small spatial scales. Studies demonstrate a variety of interactions between fields and their margins. Some margin flora may spread into crops, becoming field weeds. Margins also have a range of associated fauna, some of which may be pest species, while many are beneficial, either as crop pollinators or as pest predators. Agricultural operations, such as fertilizer and pesticide application, can have adverse effects on the margins. The biodiversity of the margin may be of particular importance for the maintenance of species at higher trophic levels, notably farmland birds, at the landscape scale. Margins can also contribute to the sustainability of production, by enhancing beneficial species within crops and reducing pesticide use, though further research on the predictability of these effects is needed. In northwestern Europe, a variety of methods to enhance diversity at field edges have been introduced, including sown grass and flower strips. While minor conflicts exist, notably for the conservation of rare arable weed species and the spread of some pests and weeds, the impacts of well-managed field margins on weed flora and arthropods indicate mostly beneficial effects. Thus field margin strips offer a practical means of providing on-farm biodiversity and enhancing more environmental and sustainable production.

KEYWORDS. Biodiversity, boundary, buffer strip, drift, field margin, headland, landscape, natural enemies, weeds

INTRODUCTION

Agricultural landscapes are extremely variable across the globe, varying with cropping system, topography and intensity of management. The interactions between land use and landform are profound, leading to landscape mosaics. Going back to the beginnings of agriculture, man has sought to define the areas of cropping and land ownership. As graziers, humans have also often sought to contain stock and protect them from wild animals. In extensive production systems, a balance between domestic and wild grazers is struck based on an understanding of carrying capacity. In intensive systems, land is typically enclosed and delineated with field boundaries. Most production areas are enclosed; that is, they are delimited or fenced into discrete areas. The advent of fencing had a profound impact on field margins and enclosure (Chapman and Sheail, 1994), allowing stock impoundment and field delineation to be done more cheaply and easily than before. Prior to post and wire fencing, other structures were used. The so-called "dead hedge," made of cut and woven shrub material is still used to some extent today, e.g., sheep hurdles made from hazel (*Corylus avellana*). In many areas of the world, hedges comprised of living trees, shrubs or other herbaceous flora, fulfill this role for stock impoundment and field delimitation. In England and northern France, the hedgerow has become a part of landscape heritage. Originally, these structures had important agricultural and production functions, though this is changing over time. Agricultural development has been aided by enclosure but at the same time the structure of the landscape has changed. The landscape is never given over totally to agriculture or any single land use. Depending on climate, terrain and population, there are usually areas of watercourse, woodland, road verge or other semi-natural habitats that are less exploited. The juxtaposition of land uses, particularly farmed areas and natural habitat, form mosaics in the landscape (Burel and Baudry, 1999; Forman, 1995).

Landscape mosaics are typically characterized by the matrix, patch and network model. Under this view, the matrix is formed of the agricultural fields and the network and patches are natural or semi-natural habitats. The linear elements of the landscape network include watercourses, road verges, hedges and field margins, the subject of this review. These may be particularly important for the survival of species and communities typical of the habitats present before the expansion of production systems.

In the past, the process of agricultural production has regarded the non-crop areas within the landscape as hostile and the source of limitations to production, such as weeds, pests and diseases. With a greater understanding of ecology and the interactions between fauna and flora, a more critical appraisal of interactions between cropped and non-cropped areas is being made. This is perhaps more advanced in Europe than elsewhere, reflecting population pressure on land. Approaches to nature conservation in Europe have, as elsewhere, relied on land designations for nature reserves and parks. However, fragmentation of habitat and the small size of many such areas may expose them to stochastic extinction events and threaten the long-term survival of these protected communities. A new paradigm is the conservation of species and communities within the farmed landscape as a whole (Mineau and McLaughlin, 1996). In the UK for example, there are almost no true wilderness areas and more than 75% of the land surface is farmed in one way or another.

With such extreme fragmentation of natural habitats (they are all semi-natural) and predominance of agriculture, agricultural ecologists have been examining in more detail the interactions between the crop and non-crop habitat. It is apparent that there are many species of fauna and flora that are adapted to farmland and many species dependent on these habitats. A worrying decline in the numbers and population ranges of species that were previously common and typical of farmland is now becoming apparent. In the UK, a number of common farmland birds have shown significant declines over the past 40 years, many in the 1980s (Chamberlain and Fuller, 2000; Fuller et al., 1995; Gillings and Fuller, 1998). While not ubiquitous, field edges usually have a boundary structure and associated habitats within them. The hedged areas of northwestern Europe and parts of Canada and the northeastern USA are perhaps the most striking examples of enclosed mosaic landscapes. In France, the hedge landscape of Brittany and Normandy has a specific name, the "bocage," reflecting its cultural importance (Bazin and Schmutz, 1994; INRA, 1976).

Clearly, from both agricultural and wildlife conservation viewpoints, the role of semi-natural habitat in affecting the crop and the effects of cropping on the wider landscape need to be understood. How does semi-natural habitat affect adjacent crops? How does cropping affect semi-natural habitat? Can semi-natural habitats in farmland be created and managed to enhance both agricultural production and farmland wildlife? This contribution reviews these questions for field boundaries, the commonest adjacent semi-natural features in agricultural areas.

FIELD MARGINS

Structure

Field margins in both arable and grassland farming are typified by having some form of boundary structure, usually with associated herbaceous vegetation, adjacent to the crop. In arable land, there may not be any physical border between blocks of different crops. However, there is almost invariably some form of margin between holdings and along roads, tracks and watercourses. The boundary may be a fence, a shelterbelt of trees, a hedge, a wall, a terrace, a ditch or drainage channel, a grass strip or a combination of these structures. In some situations, the boundary is simply another habitat type, such as woodland.

Increasingly, there are structures introduced at arable field edges that either enhance access, weed control or conservation of farmland wildlife. These can include simple farm tracks, or vegetation-free strips. In Europe, there are also set-aside strips, borders of sown perennial vegetation or so-called conservation headlands, where the cereal crop edge receives reduced pesticide and herbicide inputs (Rands, 1985; Rands and Sotherton, 1987). Greaves and Marshall (1987) defined the structure of the field margin in Britain as comprising the field boundary, the boundary strip (not always present) and the crop edge (where crop management may differ from the main field area) (Figure 1). Whilst a variety of structures are found in different parts of the world, this definition is applicable to most situations. The boundary is usually a structure or barrier, while the boundary strip provides an opportunity to modify field edge management for a variety of objectives.

The term "headland" is widely used to mean the crop edge, typically the area outside the outermost tractor wheeling or tramline through the

FIGURE 1. Components of an arable field margin, comprising a boundary structure, sometimes a field margin strip and the crop edge. Adapted from Greaves and Marshall (1987).

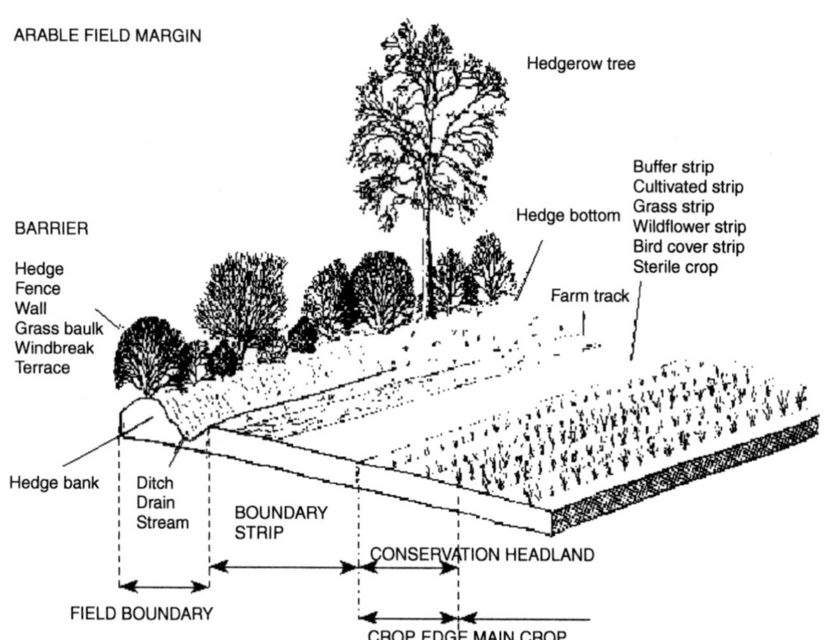

crop. Strictly, the headland is the area where machinery turns at the ends of the field, but the term is commonly used more loosely to mean the crop edge.

Habitat Types and Plant Communities in Field Boundaries

As field margins often comprise a variety of structures, including watercourses, banks, woodland edges, etc., a variety of plant communities can occur within them. There are almost no characteristic field margin plant species in Britain (Hooper, 1970). However, the flora can include segetal (associated with crop fields) and ruderal plants, typical of arable land or disturbed ground. In addition, grassland communities, tall herb communities, shrub, woodland and aquatic communities may be present. Field margins may therefore support a very wide range of plant species and their associated fauna.

Role of Field Margins

Field margins exist in enclosed landscape as they have, or had in the past, some agricultural functions. In stock farming areas, hedges, walls and fences were maintained to keep stock in or out. In arable land, field margins delineate the field edge and land ownership. More recently, a series of subsidiary roles have been identified, reflecting agricultural, environmental, conservation and cultural or historical interests (Table 1).

Udo de Haes (1995) identified four different interests involved in field margin management, as defined by de Snoo (1995):

- agronomic
- environmental
- nature conservation
- recreational use

From an agronomic viewpoint, field edge structures should not impact adversely on the adjacent crop. In the case of windbreaks and erosion control, margins have a positive role. Nevertheless, a traditional view has been that these structures are sources of weeds, pests and disease (Marshall and Smith, 1987). This has led to their removal or severe management in some cases. More positively, recent research is examining the possibilities of enhancing positive interactions on cropping (Marshall and Moonen, 2002).

TABLE 1. Functions of semi-natural field margins in good agricultural practice. Adapted from Marshall, 1993.

Interest	Function
Agronomic	Defining the field edge
	To be stock- or trespasser-proof, to keep animals in or out
	Providing shelter for stock
	Providing shelter for crops, particularly as windbreaks
	Promotion of ecological stability in crops
	Reducing pesticide use: exploiting pest predators and parasitoids
	Enhancing crop pollinator populations
	Providing a source of fruits, forage and wood
Environmental	Reducing soil erosion by wind or water
	Buffering agrochemical drift
	Reducing fertilizer and other pollutant movement, especially in surface run-off
Biodiversity	To act as a refuge or corridor for wildlife
	Promotion of biodiversity and farm wildlife conservation
Social and Recreational	Maintaining landscape diversity
	Promotion of game species
	Encouragement of non-farming enterprises
	Maintaining historical features, heritage and "sense of place"

Margins may have positive environmental roles. Some agronomic operations, notably pesticide and fertilizer application, can have adverse effects on adjacent habitats following drift events, erosion or leaching. Field margins can play a role as buffers (Daniels and Gilliam, 1996).

Whilst field margins seem to be sub-optimal habitats, notably for many bird species in Europe (Vickery, Carter, and Fuller, 2002) and Canada (Jobin, Choiniere, and Belanger, 2001), there is good evidence that these features are refugia for botanical diversity in intensively managed lowland landscapes. Thus, these features can have an important role in conserving species and communities representative of the original habitats before enclosure. At first view, this might appear irrelevant to large countries, particularly when the original habitats may be conserved sufficiently in large parks, designated wilderness areas or public lands administered by provincial or central government. However, even in these situations much original habitat remains only as small remnant fragments, e.g., tall-grass prairie (Boren et al., 1997), and field boundaries and small habitats may be important for their conservation in intensive agricultural areas, e.g., Boutin et al. (2002).

Margins can provide recreational opportunities. In the past, they have provided fencing, firewood and fruit. Grass strips can allow access, horse riding and nesting cover for game birds. The latter can be further enhanced with strips of game cover crops, such as kale (*Brassica oleracea*). At the landscape scale, hedges provide a characteristic network, particularly evocative of southern England and Brittany. This contributes to local tourism. In Britain, the oldest hedges were established in Saxon times over 1000 years ago (Rackham, 1976). Many are species rich and mark boundaries between parishes (Rackham, 1985), thus having cultural and heritage importance.

BIOLOGICAL DIVERSITY AND FIELD MARGINS

The reasons for the conservation of biodiversity are moral, aesthetic, social and economic. We steward other organisms for their intrinsic value and because these species may be of benefit to human society and have economic value. A culture that encourages respect for wildlife is preferable to one that does not (Anonymous, 1994). Biodiversity is easily lost but is difficult to regain, particularly if species are driven to extinction. Biodiversity, including genetic diversity, may provide economic benefits. Even at the level of landscape, biodiversity may influence tourism and sense of place. Perhaps of more concern is that biodiversity has a role in the function of ecosystems (Naeem et al., 1994; Tilman, Wedin, and Knops, 1996; Chapin et al., 1997). Erosion of diversity may thus result in damage to ecosystem function.

Traditionally, field margins had agricultural functions, notably impoundment of animals and field delineation. Under intensive arable production, such functions are less important to landowners and in the UK, for example, many margins and hedges have been removed since the Second World War (Pollard, Hooper, and Moore, 1974). The rates of hedge removal declined in the UK in the 1970s and it was not until the Countryside Survey 1990 (Barr et al., 1991; 1993), that more recent data has become available. These indicate that many hedges, particularly in livestock areas, are losing their structure and as a result, their value to agriculture, wildlife, and landscape. Changes within arable farming have also resulted in reduced diversity of arable weeds, such that many cornfield flowers are rare and even threatened by extinction (Wilson, 1993). Long-running census and atlas programs of the British Trust for Ornithology (BTO) have identified major declines in population size and range of farmland birds (Fuller et al., 1995). Many species, notably gray partridge (*Perdix perdix*), song thrush (*Turdus philomelos*), tree sparrow (*Passer montanus*), linnet (*Acanthis cannabia*), bullfinch (*Pyrrhula pyrrhula*), reed bunting (*Emberiza schoeniculus*), and corn bunting (*Emberiza calandra*), utilize cereal field margins to a large extent, but show marked declines, probably reflecting major changes in arable farming practice.

The flora of margins can be diverse. Phytosociological surveys of British field margins identify four scrub or underscrub communities, each associated with hedge species such as *Crataegus monogyna*, *Prunus spinosa*, and *Rubus fruticosus* (Rodwell, 1995). The plant species present may represent sub-optimal communities of a variety of other habitat structures. However, extensive surveys across Britain as part of Countryside Survey, an inventory of land use and its impact (Barr et al., 1993; Haines-Young et al., 2000), identify linear semi-natural features in lowland agricultural landscapes as containing the most diversity of plant species. Thus field margins, stream sides and verges are refugia for plant species in intensively managed land in the UK and a key to the conservation of plant diversity in the wider landscape (Bunce et al., 1994). This pattern is repeated across Europe (Burel and Baudry, 1995; Burel et al., 1998; Melman and Van Strien, 1993; Van Strien, 1991) and North America (Boutin and Jobin, 1998; Forman and Baudry, 1984; Freemark and Boutin, 1995; Jobin, Boutin, and DesGranges, 1997).

Birds also utilize margins and adjacent crops and are affected by structure and cropping patterns (Green, Osborne, and Sears, 1994; Parish, Lakhani, and Sparks, 1995). The network of hedgerows also supports invertebrates (Morris and Webb, 1987), such as beetles (Burel, 1989), some of which migrate into cereal crops in spring and feed on cereal aphids (Wratten, 1988). New approaches to understanding the ecology of these features include an integration of disciplines across agriculture and ecology and integration across spatial scales (Le Coeur et al., 2002). If biological diversity within agricultural systems also has a functional role, particularly in enhancing sustainability, then maintaining semi-natural habitat may be important. Field margins are clearly important for biodiversity and may provide a simple means of achieving habitat diversity at the farm scale.

INTERACTIONS: EFFECTS OF THE MARGIN ON THE CROP

Weed Ingress

There is a common but misguided perception that weed species are present in adjacent semi-natural areas and these invade crops (Marshall and Smith, 1987). Under conditions of land abandonment, this view seems borne out, with patterns of invading shrub species advancing into old-fields (Bard, 1952; McDonnell and Stiles, 1983). Nevertheless, the margin flora studied on a farm in Essex, UK (Marshall and Arnold, 1995) indicated no relationship with the arable weed flora at 5 m and 50 m into fields. Detailed examination of the patterns of occurrence of non-crop species at the edges of arable fields in the UK indicates a range of plant strategies (Marshall, 1989). Four distribution patterns have been identified (Table 2), including the absence of a significant proportion of the field boundary flora in the adjacent arable crop. Non-crop plants seem to segregate into those that are most abundant within the field, typically reducing in density towards the field edge, and those that are most abundant in the boundary and with decreasing density towards the field center. These patterns reflect a complex of environmental and management factors operating on the population dynamics of the species present. Similar patterns were reported by Wilson and Aebischer (1995). In temperate European conditions, the weed flora of arable fields is dominated by annual species, dependent on seed return and often having dormant seed (Chancellor and Froud-Williams, 1984; Whitehead and Wright, 1989), with a small proportion of perennial weeds, often with adaptations for vegetative propagation.

Investigations on a number of English arable farms have shown that the number of species that exhibit the pattern of spread from the boundary into the crop is limited. Typically, the proportion of herbaceous plant species of the boundary that is present more than 2.5 m into the crop is less than 25% of the flora of the boundary (Marshall and Smith, 1987). The numbers of important weed species that originate in margins are also limited, as illustrated in Table 3, where the commonest weed species in winter cereals in the UK are listed in decreasing order of occurrence. Only four of the 23 most important arable weed species spread from field boundaries. Of these, *Galium aparine* is particularly competitive (Froud-Williams, 1985) and *Bromus sterilis* and *Elytrigia repens* may also reduce crop yield and affect harvesting. Nevertheless, *B. sterilis* appears to be largely limited to crop headlands (Cussans et al.,

TABLE 2. Patterns of weed distribution at the field edge. Adapted from Marshall, 1985; 1989.

Distribution	Pattern	Example species
Limited to the boundary	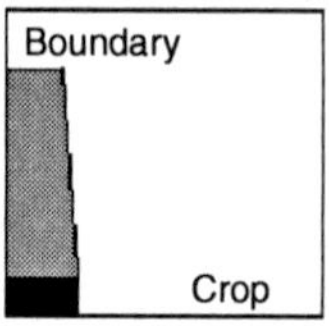 Boundary / Crop	• *Arum maculatum* • *Dactylis glomerata*
Limited to the crop	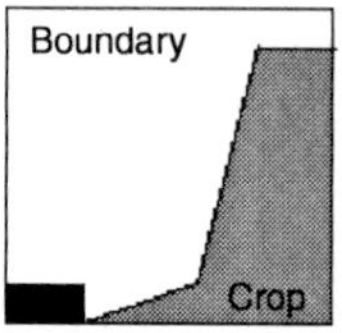 Boundary / Crop	• *Polygonum aviculare* • *Veronica persica*
Spreading from the boundary	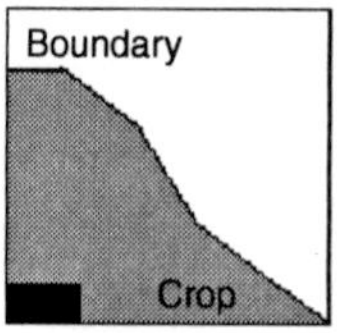 Boundary / Crop	• *Galium aparine* • *Bromus sterilis*
Headland	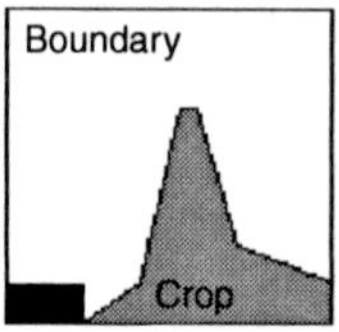 Boundary / Crop	• *Poa annua* • *Alopecurus myosuroides*

1994), while *E. repens* is well controlled with the herbicide, glyphosate (Smith and Oehme, 1992). Three of these four species are annuals (though *Poa trivialis* can persist) spreading by seed, while *E. repens* spreads mainly by rhizomes. Where winter annuals dominate the herbaceous flora, as is often the case in Mediterranean conditions with summer droughting, the boundary might be more significant in influencing weed flora. This requires further investigation. Overall, the importance

TABLE 3. Common weeds of winter cereals in decreasing order of percent (%) occurrence in 1476 fields in 1988 before herbicide treatment (adapted from Whitehead and Wright, 1989), with a note of their margin distribution. S = spreading from the boundary; F = field distribution.

Species	Total % occurrence	Distribution
Chickweed (*Stellaria media*)	94	F
Annual meadow grass (*Poa annua*)	79	F
Common speedwell (*Veronica persica*)	72	F
Mayweeds (*Matricaria* spp.)	67	F
Cleavers (*Galium aparine*)	58	S
Red deadnettle (*Lamium purpureum*)	47	F
Field pansy (*Viola arvensis*)	45	F
Wild-oats (*Avena* spp.)	42	F/S
Blackgrass (*Alopecurus myosuroides*)	38	F
Charlock (*Sinapis arvensis*)	36	F
Ivy-leaved speedwell (*Veronica hederifolia*)	30	F
Shepherd's purse (*Capsella bursa-pastoris*)	23	F
Volunteer rape	23	F
Couch grass (*Elytrigia repens*)	21	S
Common poppy (*Papaver rhoeas*)	18	F
Fumitory (*Fumaria officinalis*)	17	F
Ryegrass (*Lolium* spp.)	14	F
Fathen (*Chenopodium album*)	13	F
Barren brome (*Bromus sterilis*)	13	S
Parsley piert (*Aphanes arvensis*)	12	F
Cranesbills (*Geranium* spp.)	11	F
Rough-stalk meadow grass (*Poa trivialis*)	7	S
Volunteer cereals	7	F

of the field edge as a source of arable weeds is low in northern European systems.

Pests and Beneficial Insects

Both pest and beneficial insect species are found in field margins, but are they of economic significance? A review of the role of uncultivated land in the biology of crop pests by van Emden (1965) identified many records of pest species using uncultivated areas. The occurrence of invertebrate pest species in the surroundings of agricultural fields has often been cited as the source of crop losses. In a number of cases this is justified, for example with mollusk pests, e.g., Frank (1998); Marshall, Grant, and Fairbairn (1999). Often, polyphagous herbivores that are crop pests utilize a number of closely related plant species. For example, the weevil *Ceuthorrynchus apicalis* attacks the crop celeriac (*Apium graveolens* var. *rapaceum*), when the source of infestation is often naturally occurring *Heracleum sphondylium* in field margins and hedgerows. Certain pest species have alternative hosts for different parts of their life cycle, which involves movement from uncultivated to cultivated areas at particular times of year. For example, the winter host for the black bean aphid, *Aphis fabae*, includes *Euonymus europaeus*, the spindle shrub, a common component of hedgerows. The aphid is a serious pest of bean (*Vicia* sp.) crops. Certain insects are vectors of plant diseases. For example, cereal aphids can transmit barley yellow dwarf virus (Henry et al., 1993; Masterman, Holmes, and Foster, 1994). In Britain, aphids move from grasses to winter cereals during the autumn and virus can be transmitted to adjacent or distant crop fields. Uncultivated land can influence the distribution of insects in crops through the physical presence of hedges and windbreaks. Numbers of aphids can be higher in the crop edge adjacent to hedges, caused by wind vortices and settling (Lewis, 1969).

van Emden (1965) also reviewed the available literature on beneficial insects associated with uncultivated land. These insects can be defined as those that are either of direct benefit to the crop, typically as pollinators, or indirectly by controlling populations of crop pests or vectors of crop disease. These are often referred to as natural enemies. van Emden (1965) indicated that at that time there was insufficient information available "to establish whether, on balance, uncultivated land is beneficial or harmful to pest control." Nevertheless, he intimated that uncultivated land might have an overall beneficial effect. He also suggested that there was scope for managing field margins to enhance beneficial effects for pest control. Since such lines were written more than 35 years ago, there have been increasing amounts of research on the importance of beneficial insects and on the means to enhance them (see section on Managing Field Margins).

A number of studies have examined the occurrence of predatory arthropods in fields and their margins. Sotherton (1984; 1985) noted that a number of important predators of crop pests were found in field margins, particularly during the winter. Spiders (Araneae) and ground beetles (Carabidae, Coleoptera) have been shown to consume significant numbers of cereal aphids (Sunderland et al., 1987). Ground beetles are a diverse group, some of which are seed feeders; others are polyphagous predators that may be important natural enemies (Kromp, 1999). Several species overwinter in uncultivated areas and colonize adjacent crops in spring (Wratten, 1988). As well as feeding on aphids, a number of the larger ground beetle species can feed on slugs (Mair and Port, 2001). A number of spiders, notably the Linyphiidae, are also predators of aphids and other crop pests (Toft, 1995). These species can disperse over wide areas and contribute to pest predation (Halley, Thomas, and Jepson, 1996; Thomas and Jepson, 1997). Spiders can contribute to pest control in rice (*Oryza sativa* L.) and can be enhanced by the creation of suitable structures at field edges (Afun, Johnson, and Russell-Smith, 1999a; 1999b). Another group of generalist predators that are associated with field margins are the Syrphidae (hover flies). These can be important aphid predators (Ten Humberg and Poehling, 1995), but they require pollen and nectar resources that can be provided in field margins (Cowgill, Wratten, and Sotherton, 1993; Hickman and Wratten, 1996; MacLeod, 1999).

A novel field technique of marking individual ground beetles was developed, in order to study the dispersal characteristics of these invertebrates (Thomas, 1995), which are implicated in the control of aphid and slug pest populations. Markedly different spatial behavior was apparent (Thomas and Marshall, 1999; Thomas et al., 2001) (Figure 2). Certain species were associated with the hedge, notably *Harpalus rufipes*. Others, such as *Pterostichus cupreus*, are found within the fields, rather than the margins. *Nebria brevicollis* has an aestivation period, when it is limited to the field margin and hedgerow. In September in the UK, the adults of this species move out into the adjacent fields, where they may be involved in predation of aphids and other crop pests active at this time of year (Thomas et al., 2001). This species requires different habitats at different times of year. Many species were particularly mobile, moving from the field margin into the crop over short time intervals. Introduced field margin strips were rapidly colonized, so that over-wintering populations were similar to pre-existing margin habitat within 12 to 14 months of establishment (Thomas et al., 1994). Mark-recapture studies provided quantitative results on dispersal for use in subsequent modeling, for example, maximum (c. 100 m per day) and average daily displacement. Other significant findings were that patterns of occurrence were not uniform. Some species were found in consistently high

FIGURE 2. Cumulative trapping densities of different species of Carabidae (ground beetles) in a hedge (central shaded band) and in two adjacent arable fields over summer and autumn. Adapted from Thomas et al. (2001).

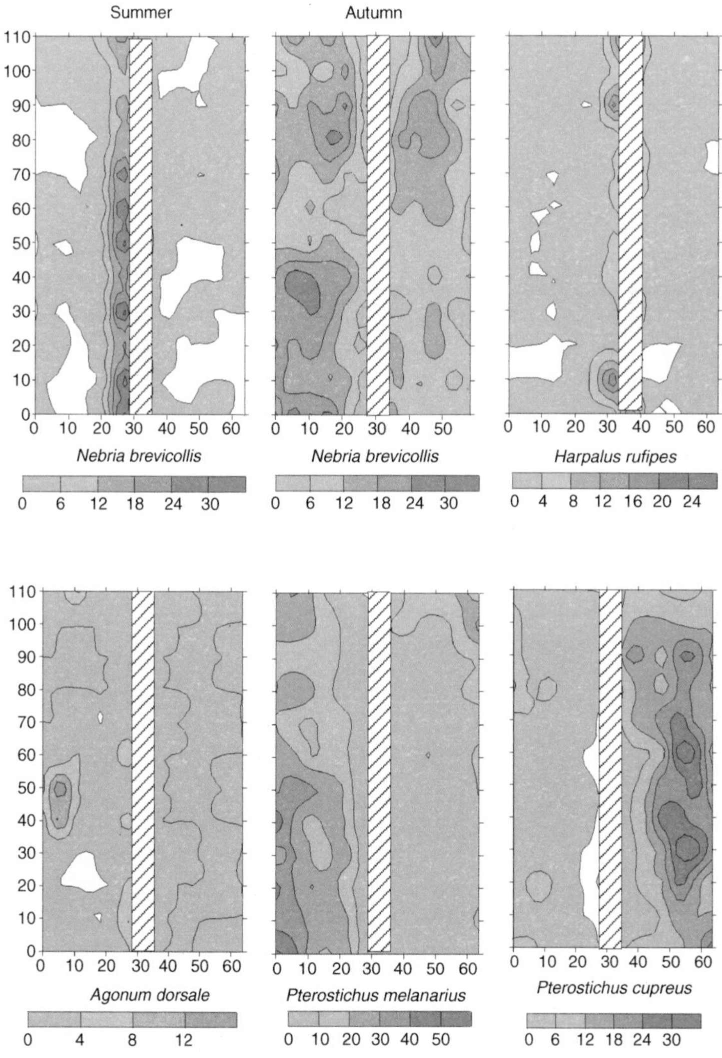

densities in certain locations, the reasons for which are unknown, but which might be manipulated to reduce pest populations (Thomas et al., 2001).

Quantitative measures of the impact of beneficial insects on pests within crops are scarce. Studies in Germany have demonstrated a significant local influence of the field margin on populations of aphids. The reduction of aphids up to 5 m or more from the margin is likely to have resulted from a complex of predator species associated with the margin (Marshall, 1997). Many studies nevertheless show that predator populations can be enhanced by suitable management of the field margin (Thomas, Wratten, and Sotherton, 1991; Wyss, 1995; Collins et al., 1997). It can be concluded that field margins and uncultivated land can support a diverse invertebrate fauna, some of which are pest species and some are natural enemies that can contribute to the agricultural control of pests. Further research on the economic impact and predictability of exploiting beneficial insects is needed, but there is convincing data to show that numbers of pest predators can be enhanced by modifying field margin management.

Crop Diseases

Uncultivated land can act as reservoirs for some crop diseases. The movement of disease to crop is often enhanced by insect vectors (Ahohuendo and Sarkar, 1995; Chancellor, Cook, and Heong, 1996; Dusi, Peters, and Van der Werf, 2000). Nevertheless, there is also evidence that disease spread can be contained by crop rotation or intercropping, e.g., Ahohuendo and Sarkar (1995). This might indicate that field margins may also contribute to disease control. In the case of the cereal virus, barley yellow dwarf virus (BYDV), grasses in field margins, cereal volunteers and grass weeds within crops can harbor the disease. There is some evidence that plant hosts within the crop are the most significant sources of disease, rather than those in field margins (Marshall and Smith, 1987; Coutts and Jones, 2000). Clearly, further research on the role of margins in crop disease transmission is required.

Crop Yield Effects

Studies on the yields of cereals at arable field edges have been made by several organizations in Europe (Boatman, 1992; Cook and Ingle, 1997; de Snoo, 1997; Marshall, 1967). Typically, yield at the field edge is reduced in comparison to the field center, as a result of a variety of factors, including shading, soil compaction and weed pressure. While boundaries may reduce yield immediately adjacent to them, the protective windbreak effect of some boundaries may also prevent yield depression further into the field (Marshall, 1967), though not under all conditions (Cleugh, 1998). There is nevertheless considerable site-to-site variation, reflecting different soil types and field histories. Some data indicate that headlands yield between 3 and 18% less than field centers (Boatman, 1992; Sparkes et al., 1998). Work on eight farm sites in the UK with predominantly southwest winds and with no modification to headland husbandry has shown that boundary height and aspect have significant effects on yield of winter wheat (*Triticum aestivum*) at the field edge (Cook and Ingle, 1997). Greatest yield reductions in the headland were given by tall hedges that were north or east facing (15.6% mean reduction at 2.5 m into the crop, compared with yield at 20 m). Adjacent to north or east-facing low hedges (< 2 m high), yields were reduced by an average of 9%, while south or west-facing low hedges had yields reduced by 3.6%. Recent work by Perry (1998) from a range of fields in Shropshire and Leicestershire has indicated a highly significant relationship between grain yield and distance from the field edge (Figure 3). Yield losses at the field edge can be estimated from these data, to match yields and profits foregone against incentives and costs to modify field edge management.

FIGURE 3. Relationship between grain yield (tonnes/ha) and distance into the crop, derived from a series of field edges in Shropshire and Leicestershire, UK. Adapted from Perry (1998) *The implications of improving the conservation value of field margins on crop production.* PhD Thesis. Harper Adams College.

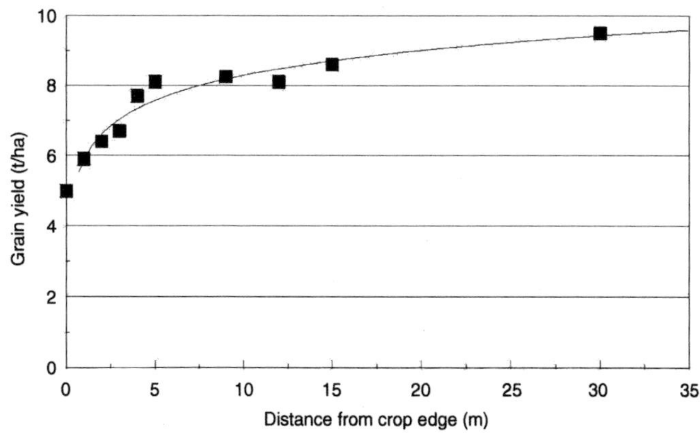

INTERACTIONS: EFFECTS OF CROP MANAGEMENT ON THE MARGIN

Fertilizer Movement

One of the important environmental impacts of production agriculture is eutrophication of soils and water (Addiscott, Whitmore, and Powlson, 1991). Nutrient additions are a major factor in plant biodiversity loss and a challenge to conservation (Marrs, 1993). Addition of nitrogen (N) fertilizer to natural ecosystems can often result in reduced species richness, e.g., Willis (1963). The relationship between increasing fertility (productivity) and species richness is typified by the "hump-back" model, in which diversity rises to an asymptote and then declines. Most ecologists agree that this is a good representation, though scale can influence observed results (Waide et al., 1999). The addition of fertilizers, most notably N, in agroecosystems has a major influence on plant species composition. Species diversity declines as adapted species become dominant. At high productivity, tall-growing, competitive species out-compete shorter subordinate species (Marrs, 1993). More fertilizer has been used within arable systems over the past century, though recent economic pressures have encouraged more targeted use (Jordan, 1998). It is likely that increased fertility within crops has encouraged more nitrophilous species. A good example is *Galium aparine* (cleavers), a weed that has increased markedly in frequency from 21% in the 1960s to 88% occurrence in fields in 1997 in central England (Sutcliffe and Kay, 2000). This species is particularly responsive to N (Froud-Williams, 1985). Thus fertilizer use may have been an important driver in changing weed assemblages.

Repeated surveys of land use and vegetation composition of fields and linear landscape elements in the UK (Countryside Survey) over the past 25 years indicate that there is a decline in botanical diversity in lowland landscapes (Barr et al., 1993; Haines-Young et al., 2000). Both eutrophication and disturbance are implicated in this continuing decline in plant diversity in the wider countryside.

Fertilizer misplacement from field applications into field margins commonly occurs (Rew, Theaker, and Froud-Williams, 1992). Studies by Tsiouris and Marshall (1998) demonstrated that pneumatic fertilizer applicators were accurate and did not contaminate adjacent habitat. However, simple spinning disk applicators could spread fertilizer considerable distances from the target area. A hedge at the field edge can act as a buffer, preventing fertilizer reaching adjacent habitat, but at the same time can concentrate fertilizer drift at the field edge at doses similar to the field (Figure 4). The impacts of fertilizer additions to field margins have been studied in several locations in the UK (Boatman et al., 1994; Rew, Froud-Williams, and Boatman, 1995; Theaker, Boatman, and Froud-Williams, 1995). Although clear adverse impacts in terms of increased weed productivity and species declines were not always observed, there is a tendency for these effects (Tsiouris and Marshall, 1998).

Nutrient movement by overland flow and leaching through the soil profile are other sources of off-field eutrophication (Addiscott, Whitmore, and Powlson, 1991). Overland movement of soil and water can be influenced by field margins and the creation of buffer strips (Pinay, Roques, and Fabre, 1993; Cooper, Smith, and Smith, 1995; Tim, Jolly, and Liao, 1995; Lowrance et al., 1997; Mander et al., 1997). Vegetated filter strips in North Carolina greatly reduced run-off of chemical loads into the watercourses, with up to 50% of P and NH_4 filtered out (Daniels and Gilliam, 1996). A multi-species riparian buffer strip system with four tree species, shrubs and a switchgrass strip placed along a stream in Iowa reduced NO_3-N concentrations considerably. In the buffer strip, levels never exceeded 2 mg l^{-1} whereas they were 12 mg l^{-1} in the adjacent arable fields. The system helped to control soil erosion, trapped and transformed N and P pollution, stabilized the streambank, provided

FIGURE 4. Patterns of fertilizer deposition from a spinning disk applicator fitted with a headland deflector plate set to operate over 6 m in an open field (open diamonds) and adjacent to a hedgerow (closed circles). Adapted from Tsiouris and Marshall (1998).

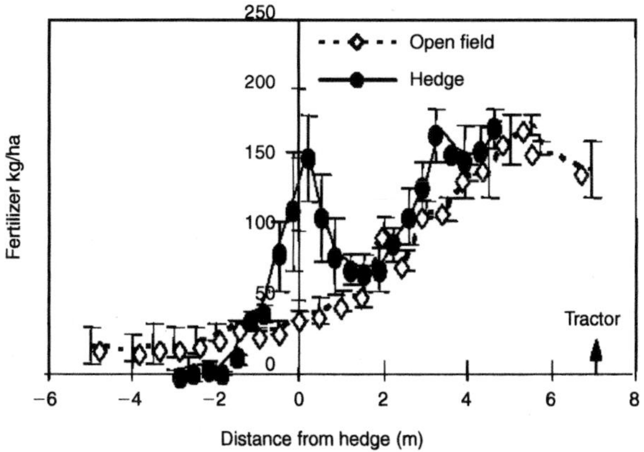

wildlife habitat, produced biomass for on-farm use, contains quality hardwood for the future and enhanced the aesthetics of the landscape (Schultz et al., 1995).

Pesticide Drift

Pesticide drift, as droplets and vapor, can occur from field applications (Breeze, Thomas, and Butler, 1992; Breeze et al., 1999) and measures to limit non-target effects are in place (Hewitt, 2000). A number of field measurements of drift have been made and are the basis for statutory risk assessment in Europe (Ganzelmeier et al., 1995). Following increased interest in the impacts of pesticide use on non-target habitats, e.g., Falconer (1998), a number of studies have shown that significant amounts of pesticide can reach field margins adjacent to sprayed fields (Davis et al., 1994; Longley et al., 1997; Longley and Sotherton, 1997). Studies of the impact of insecticide drift indicate the potential for adverse effects on fauna in field boundaries (Davis et al., 1991; Longley and Sotherton, 1997).

Although there are few clear reports of adverse effects of pesticides within field edges, recent work indicates significant impacts of agricultural operations on margin flora (Boutin and Jobin, 1998; Freemark and Boutin, 1995; Jobin, Boutin, and DesGranges, 1997). Determining the effects of herbicides on plant communities is not straight forward (Cousens, Marshall, and Arnold, 1988). Susceptibility of plants to pesticides is not a constant characteristic, as it is affected by many variables. Application variables, which include dose, timing, spray volume and spray deposition, interact with plant variables, such as growth stage, size and location within the plant canopy. Early work evaluated the direct effects of herbicides on field margin flora (Marshall and Birnie, 1985; Marrs et al., 1989; 1993). Low rates of some herbicides affected flowering and seed production, which may affect regeneration. Kleijn and Snoeijing (1997) made detailed studies of the effects of low levels of herbicide and fertilizer on field margin communities. Field experiments on a natural and a sown community were treated with a range of doses of fluroxypyr (0-50% of field rate) and fertilizer. Fertilizer contamination is likely to be a more important and more predictable factor in reducing botanical diversity in adjacent non-target areas, than herbicide drift. However, herbicide drift also resulted in reduced species richness, enhancing grass biomass and reducing biomass of flower species, notably the subordinate, lower-growing ones. Most significant effects were noted with the 50% rate, but 5% and 10% herbicide doses reduced biomass of colonizing herbs and increased extinctions. The herbicide had different effects on different species.

Overall, most agrochemical applications within fields with current machinery are not sufficiently accurate to prevent contamination of field boundary habitat within 1 or 2 m of the crop edge. Pesticide drift is unlikely to be greater than 1% of field application rates under most conditions, but fertilizer rates may be considerably greater. These disturbance levels will tend to reduce plant species diversity.

MANAGING FIELD MARGINS FOR WILDLIFE AND AGRICULTURE

Although there are some conflicts between cultivated and uncultivated land within farms, the challenge of managing the landscape mosaic for a variety of land uses is beginning to be met, e.g., Mineau and McLaughlin (1996). This can be regarded as part of the emerging new paradigm of matching crop production with conservation of biological resources (Paoletti et al., 1992) and the development of more sustainable systems. Within the field, this may require diversity by the maintenance of some weeds within crops (Marshall, 2001). At the field, farm and landscape scales, uncultivated areas may more simply provide diversity that can have positive impacts on agriculture, while providing habitat for fauna and flora (Dennis and Fry, 1992). For example, weed seed predation may be more effective in situations with small fields and hedgerows (Marino, Gross, and Landis, 1997). Likewise parasitoid diversity may be influenced by landscape structure (Marino and Landis, 1996).

A number of field margin initiatives have been developed in the UK and Europe, designed to conserve aspects of the farmland biota and provide agricultural benefits (Marshall and Moonen, 2002). These initiatives are listed and briefly described in Table 4. These initiatives can be divided into (a) those where the crop edge is managed in a different way to the main crop, such as conservation headlands, (b) the creation of new habitat features as a boundary strip (Figure 1), such as grass and wild flower strips, and (c) new features across fields.

Conservation Headlands

The loss of many annual cornfield weed species in Germany has been countered by measures to prevent the use of pesticides in the outside

TABLE 4. Types of modified field edge management aimed at enhancing both wildlife and agricultural production.

Type	Description	Benefits
Crop edge modification		
Conservation headland	Reduced pesticide inputs to the outside 6 to 12 m of cereal fields	Increased partridge survival (Rands, 1985); rare weed conservation
Uncropped wildlife strip	Cultivation, but no crop sown	Rare weed conservation
New boundary strip		
Grass strip	Sown perennial grass strip	Prevention of weed ingress; habitat for natural enemies; riparian buffer strips; nesting cover
Grass and wild flower strip	Sown perennial grassland vegetation strip	Prevention of weed ingress; habitat for wide range of natural enemies; wildlife conservation; riparian buffer strips; nesting cover
Flower strips	Sown flower mixtures	Enhancement of pollinators and some crop pest predators
Sterile strip	Elimination of all herbaceous vegetation by herbicide or cultivation	Weed and harvesting break
Set-aside margin	Natural regeneration of perennial vegetation	(Benefits depend on the vegetation structure and composition)
Sown wildlife mixtures (strips or blocks)	Sown mixtures for birds or bees	Resources for a range of wildlife, including gamebirds
New habitat across fields		
Beetle banks	Sown tussocky grasses on ridge across large fields	Overwintering habitat for ground beetles, encouraging field colonisation in spring
Weed/flower strips	Sown strips of flowers	Enhancement of pollinators and some crop pest predators

few meters of cereal fields. These "*ackerrandstriefen*" or weed strips were designed by Schumacher (1987) and have been supported by regional governments in a number of the German states (Jörg, 1995). The technique is known more widely as conservation headlands, in which herbicide and insecticide use in the first 6 to 12 m of the crop is restricted. Conservation headlands have been developed in the UK by the Game Conservancy Trust as a practical means of encouraging populations of gray partridge (*Perdix perdix*) (Rands, 1985; Rands and Sotherton, 1987). These gamebirds require suitable grassy nesting habitat and a supply of insects for chick feeding. Adult birds forage in cereal crops, where many chick food items are associated with dicotyledonous weeds. Improved weed control and losses of these items may be important factors in the decline of the partridge (Potts and Aebischer, 1995). Initially designed solely for gamebirds, the technique has been shown to provide many other conservation benefits (Chiverton, 1993; de Snoo et al., 1998; Dover, 1996; Sotherton, Rands, and Moreby, 1985). Although this management results in a weedier field edge that may require separate harvesting, the most pernicious weeds may still be controlled with selective herbicides.

Uncropped Wildlife Strips

Following observations that rare weed species can continue to survive in field edges, a management prescription is in place in certain areas as part of the UK Environmentally Sensitive Areas scheme. Under intensive arable conditions, the seed bank is not uniform across fields. Much lower seed densities and species richness have been found within the field than in the field boundary (Marshall, 1989). Where the conservation of rare cornfield weed species is encouraged, it is sensible to target the field edge. These species, which are often not competitive with modern crop cultivars, are more likely to persist in the seed bank at field edges (Wilson, 1994). Support is available to farmers who cultivate a field edge strip annually but do not plant a crop, allowing an annual weed flora to develop (Critchley, 1994).

Sown Grass Margins

A popular prescription that has financial support under the UK Countryside Stewardship Scheme is the creation of 2 m-wide or 6 m-wide margin grass strips. This proposal was first made by Marshall and Smith (1987) as a means of preventing weed ingress and as a response to the common practice of removing all perennial vegetation in field boundaries with broad-spectrum herbicides. Disturbance from agricultural operations tends to reduce plant species diversity in the field margin, favoring competitive ruderal species, typically annual species that are encouraged by the removal of perennials. Recreation of a perennial grass flora provides a barrier to weed ingress, a buffer against drift and habitat for some over-wintering natural enemies (Barker and Reynolds, 1999).

Beetle Banks

A variation on the grass margin strip is known as the "beetle bank" in the UK. This technique was developed by the Game Conservancy Trust as a means of encouraging the colonization of very large fields by ground beetle species that overwinter in tussock-forming grasses in field edges (Thomas, Wratten, and Sotherton, 1991; 1992). Simple grass mixtures are sown on a slightly raised ridge, created by ploughing, across the center of large fields. The beetle bank can be isolated from existing field boundaries, to facilitate machinery movement, or can be connected to become a new margin dividing a field. The tussocky grasses provide suitable overwintering habitat for some beetle species, as well as cover for ground-nesting birds. The impact on pest control is not clear (Collins et al., 1996; Collins et al., 1997), but populations of beetles do use these features.

Sown Grass and Wild Flower Margins

Sown grass and flower margins at the edges of arable fields have been investigated and developed over a number of years in the UK (Smith and McDonald, 1989; Marshall and Nowakowski, 1991; 1992; Smith et al., 1993; Marshall, Holmes, and Foster, 1994; Marshall and Nowakowski, 1995). The creation of diverse vegetation should support a diverse fauna (Thomas and Marshall, 1999). The use of seed mixtures containing perennial native grasses and flowers can be established successfully, if attention is paid to initial establishment conditions and weed control (Marshall and Nowakowski, 1991; West and Marshall, 1996). Interestingly, the retention of meadow strips at arable field edges is not new. William Cobbett noted these "veritable pleasure grounds" in Hertfordshire in his nineteenth century "Rural Rides" (Cobbett, 1853).

These perennial seed mixes can reduce weed problems at the field edge (Smith, Firbank, and Macdonald, 1999; West, Marshall, and Arnold,1997) by space pre-emption and competition. These sown strips can provide overwintering habitat for arthropod natural enemies (Pfiffner and Luka, 2000) that can be colonized within 12 to 14 months of establishment (Thomas et al., 1994). During the summer, flower margins support a range of invertebrates (Thomas and Marshall, 1999). Floral resources provided by sown grass and flower strips are important for

Syrphidae, which are effective aphid predators (Frank, 1999; Harwood, Wratten, and Nowakowski, 1992; MacLeod, 1999; Sutherland, Sullivan, and Poppy, 2001).

Flower Margins and Strips Across Fields

The potential of floral resources to encourage beneficial insects has led to work on the creation of flower strips, both at field edges and across fields. For example, the prolifically flowering *Phacelia tanacetifolia* can be sown in strips at field edges to encourage pollinators and pest predators (Gathmann, Grieler, and Tscharntke, 1994; Hickman and Wratten, 1996). Seed mixtures have also been used to promote bees, e.g., Carreck and Williams (1997) and ground beetles (Lys and Nentwig, 1992). Nevertheless, these weed strips may encourage certain pest species (Lethmayer, Nentwig, and Frank, 1997), such as mollusks (Frank, 1998), though crop damage is usually of little significance and limited to the first meter of field crop.

Set-Aside Margins

The creation of new features may not always require sowing, and on some soils and under some conditions natural regeneration of the local flora can be successful. This is usually where soil nutrients are low and there are other limitations to plant growth, such as high or low soil pH (West et al., 1999). Nevertheless, where weed species are already present, natural regeneration tends to result in encouraging these species (Marshall and Moonen, 1997; West, Marshall, and Arnold, 1997).

Wildlife Mixtures

The creation of blocks or strips of sown vegetation for the enhancement of particular fauna on farms has been practiced for many years. Initially, these have been established as cover for gamebirds (Game Conservancy, 1994). More recently, there has been interest in creating features for wild birds, invertebrates, bees, and other taxa (Marshall, 1998). The creation of riparian woodland strips or shelterbelts has also been undertaken in the past. The growth of novel crops, including biomass willows, as margin strips is currently being researched.

Sterile Strips

The creation of a "*cordon sanitaire*" at arable field edges offers some opportunity to limit the movement of weeds from field boundaries into the crop (Bond, 1987). Elimination of a strip of herbaceous vegetation at the field edge, usually with a herbicide, may also facilitate harvesting by providing a clean crop edge and preventing the combine harvester catching climbing weeds from the margin. This technique has been used by a number of farmers in the UK. However, the variability of location of the strip, using hand-held herbicide applicators, may exacerbate annual weed populations, such as *Bromus sterilis* (Marshall and Nowakowski, unpubl.) at the field edge.

Combined Approaches to Field Edge Management

As well as the main approaches outlined above, variations and combinations are used. For example, within a grass margin, the area nearest the hedge may be managed for tussocky grasses to provide nesting cover for gamebirds and over-wintering habitat for beetles and other invertebrates. The area closest to the crop may be managed more regularly to create a low sward, perhaps for access. These can be combined with an adjacent conservation headland within the crop.

Benefits of Modified Field Edge Management

The use of field margin strips provides indisputable benefits for conservation and environmental concerns. Conservation headlands can enhance invertebrates and gamebirds (de Snoo, Van der Poll, and Bertels, 1998; Dover, 1996; Rands and Sotherton, 1987; Sotherton, Rands, and Moreby, 1985). Sown margin strips can also enhance a range of farmland biota (Barker and Reynolds, 1999; de Snoo and de Wit, 1998; Huusela-Veistola, 1998; Moonen and Marshall, 2001; Thomas and Marshall, 1999). The creation of new margin strips on farmland may also protect existing features from drift and eutrophication (de Snoo and de Wit, 1998; Patty, Real, and Gril, 1997).

The evidence for agricultural benefits of modified field edge management is less clear-cut. There is good evidence that populations of pest predators and pollinating insects can be enhanced by field edge management (Lagerlöf, Stark, and Svensson, 1992; Thomas and Marshall, 1999; Sutherland, Sullivan, and Poppy, 2001). However, demonstrations of concomitant reductions in pest populations are limited to aphid reductions up to 10 m into a cereal crop from the edge (Marshall, 1997). Modifying habitats to enhance beneficial insects is also under investigation in orchards (Rieux, 1999; Wyss, 1995) and open glasshouses (Alomar, Goula, and Albajes, 2002). There is, however, good evidence that the creation of grass margins can reduce weed populations within the boundary (Moonen and Marshall, 2001). In a comparison of adjacent farms with similar crop rotations, but different margin management, Moonen and Marshall (2001) found differences in plant diversity and abundance of weed species. Where cropping was right up to hedges and sterile strips were used, the abundances of some weeds were significantly greater than where grass margins 4 m to 20 m wide were present. The abundances of *Bromus sterilis* on the two farms are represented on an ordination diagram of the herbaceous boundary floras in Figure 5. These data indicate that reduced weed pressure can result from modifying the management of the field edge.

FIGURE 5. PCA ordination showing the differences in abundance of *Bromus sterilis* within field boundary sites with or without sown grass margins. Sites without grass margins also had sterile strips created with herbicide. Adapted from Moonen and Marshall (2001).

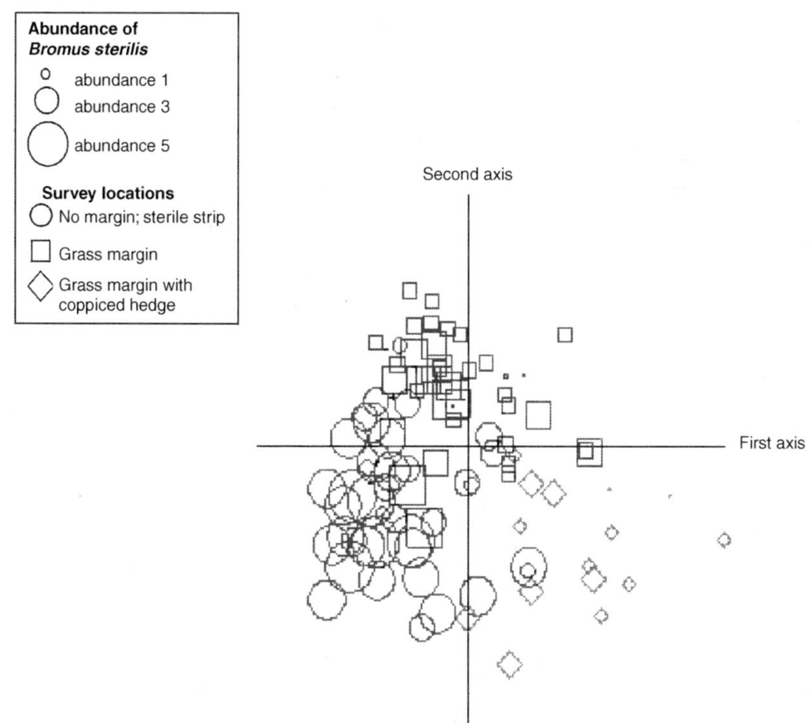

CONCLUSIONS

The interactions between cultivated and uncultivated land are complex. Both advantageous and disadvantageous effects occur in both directions. Nevertheless, margins are important for local biodiversity. The emerging evidence indicates that appropriate management of field margins can benefit both adjacent cropping and wildlife and provide some environmental protection by limiting agrochemical movement.

Where they exist, the diversity of structure that boundaries may have, including walls, hedges and ditches, can promote the diversity of plant communities that may occur there. The addition of conservation management in the form of permanent field margin strips or conservation headlands can further add to this diversity and protect existing habitat from some adverse effects of adjacent farm operations. Boundaries may have a diversity of communities, including woodland, shrub, tall herb, grassland, wetland, and arable plant species. However, often the diversity of the margin community is low, reflecting reduced structural diversity and disturbance from fertilizer, herbicide drift and cultivation. The approaches to management supported by agri-environmental schemes in several countries can promote diversity, partly by reducing disturbance and by encouraging an increase in the size of semi-natural habitat on farms.

The plant communities of the boundary may represent important refugia for species of habitats under threat in modern intensively managed landscapes. In the past, the perception that weed species spread from margins into the crop has colored the management applied by farmers. Many margins and hedges have been removed and broad-spectrum translocated herbicides were widely used in the UK. This resulted in the elimination of much of the perennial herbaceous flora and promotion of a weedy ruderal flora, often dominated by annuals adapted to germinating in shade under hedges and windbreaks. Studies now indicate that few perennial plant species spread successfully into adjacent regularly cultivated habitat. Perennial margin strips have a number of roles, including the reduction of spread of these few annual weeds into the crop. Grass strips also reduce fertilizer and pesticide drift reaching pre-existing boundary habitats, including watercourses, by moving tractor operations further into the field. Vegetated strips can also reduce surface movement of water into watercourses, buffering fertilizer and silt burdens, though subsurface flows will not be significantly affected. Habitat restoration with margin strips is supported in some agri-environmental schemes and is a management prescription often selected by farmers.

Studies have also shown that the seed bank of arable fields is often impoverished, but is larger and more diverse at the field edge. Within the crop habitat, there are annual plant species adapted to regular disturbance that are now rare. Prescriptions for conservation headlands and uncropped wildlife strips have been introduced to enhance the populations of these species. Clearly, permanent field margin strips and prescriptions for rare arable weeds are incompatible, as the disturbance regimes for one preclude the other. Nevertheless, in many situations it is possible to maintain a perennial margin with modified crop management alongside. Although there are yield penalties in removing land from production and by reducing inputs, as in conservation headlands, these can be mitigated by targeted financial support, and by the benefits associated with improved crop edge weed management. Further research on the augmentation of natural pest control and crop disease dynamics of field margins is needed.

REFERENCES

Addiscott, T. M., A. P. Whitmore, and D. S. Powlson. (1991). *Farming, Fertilizers and the Nitrate Problem.* Wallingford, UK: CAB International.

Afun, J. V. K., D. E. Johnson, and A. Russell-Smith. (1999a). The effects of weed residue management on pests, pest damage, predators and crop yield in upland rice in Cote d'Ivoire. *Biological Agriculture and Horticulture* 17:47-58.

Afun, J. V. K., D. E. Johnson, and A. Russell-Smith. (1999b). Weeds and natural enemy regulation of insect pests in upland rice; a case study from West Africa. *Bulletin of Entomological Research* 89:391-402.

Ahohuendo, B. C. and S. Sarkar. (1995). Partial control of the spread of African Cassava Mosaic-Virus in Benin by intercropping. *Zeitschrift Fur Pflanzenkrankheiten Und Pflanzenschutz-Journal of Plant Diseases and Protection* 102:249-256.

Alomar, O., M. Goula, and R. Albajes. (2002). Colonisation of tomato fields by predatory mirid bugs (Hemiptera: Heteroptera) in northern Spain. *Agriculture Ecosystems and Environment* 89:105-115.

Anonymous. (1994). *Biodiversity. The UK Action Plan.* London: HMSO.

Bard, G. E. (1952). Secondary succession on the Piedmont of New Jersey. *Ecological Monographs* 22:195-215.

Barker, A. M. and C. J. M. Reynolds. (1999). The value of planted grass field margins as habitat for sawflies and other chick-food insects. In *Aspects of Applied Biology 54*, Association of Applied Biologists, Wellesbourne, pp. 109-116.

Barr, C., D. Howard, R. Bunce, M. Gillespie, and C. Hallam (1991). *Changes in Hedgerows in Britain Between 1984 and 1990.* Institute of Terrestrial Ecology, UK.

Barr, C. J., R. G. H. Bunce, R. T. Clarke, R. M. Fuller, M. T. Furse, M. K. Gillespie, G. B. Groom, C. J. Hallam, M. Hornung, D. C. Howard, and M. J. Ness (1993). *Countryside Survey 1990.* Main Report. Department of the Environment, UK.

Bazin, P. and T. Schmutz. (1994). La mise en place de nos bocages en Europe et leur déclin. *Revue Forestiere Francaise* 46:115-118.

Boatman, N. D. (1992). Effects of herbicide use, fungicide use and position in the field on the yield and yield components of spring barley. *Journal of Agricultural Science* 118:17-28.

Boatman, N. D., L. J. Rew, A. J. Theaker, and R. J. Froud-Williams. (1994). The impact of nitrogen fertilisers on field margin flora. In *Field Margins: Integrating Agriculture and Conservation. Monograph No. 58*, British Crop Protection Council, 209-214.

Bond, S. D. (1987). Field margins, a farmer's view on management. In *Field Margins. British Crop Protection Council Monograph No. 35*, ed. J. M. Way and P. W. Greig-Smith, British Crop Protection Council, 79-83.

Boren, J.C., D. M. Engle, M. S. Gregory, R. E, Masters, T. G. Bidwell, and V. A. Mast. (1997). Landscape structure and change in a hardwood forest-tall-grass prairie ecotone. *Journal of Range Management* 50:244-249.

Boutin, C. and B. Jobin. (1998). Intensity of agricultural practices and effects on adjacent habitats. *Ecological Applications* 8:544-557.

Boutin, C., B. Jobin, L. Belanger, and L. Choiniere. (2002). Plant diversity in three types of hedgerows adjacent to cropfields. *Biodiversity and Conservation* 11:1-25.

Breeze, V., G. Thomas, and R. Butler. (1992). Use of a model and toxicity data to predict the risks to some wild plant species from drift of four herbicides. *Annals of Applied Biology* 121:669-677.

Breeze, V. G., E. J. P. Marshall, A. Hart, J. A. Vickery, J. Crocker, K. Walters, J. Packer, D. Kendall, J. Fowbert, and D. Hodkinson (1999). Assessing pesticide risks to non-target terrestrial plants. A desk study. Commission PN0923. MAFF Pesticides Safety Directorate.

Bunce, R. G. H., C. J. Barr, D. C. Howard, and C. J. Hallam (1994). The current status of field margins in the UK. In *Field Margins: Integrating Agriculture and Conservation. Monograph No. 58*, ed. N. D. Boatman, British Crop Protection Council, 13-20.

Burel, F. (1989). Landscape structure effects on carabid beetles spatial patterns in western France. *Landscape Ecology* 2:215-226.

Burel, F. and J. Baudry. (1995). Species biodiversity in changing agricultural landscapes: A case study in the Pays d'Auge, France. *Agriculture, Ecosystems and Environment* 55:193-200.

Burel, F. and J. Baudry. (1999). *Écologie du paysage: concepts, méthodes et applications.* Paris: Librairie Lavoisier, Editions TEC & DOC.

Burel, F., J. Baudry, A. Butet, P. Clergeau, Y. Delettre, D. Le Coeur, F. Dubs, N. Morvan, G. Paillet, S. Petit, C. Thenail, E. Brunel, and J.-C. Lefeuvre. (1998). Comparative biodiversity along a gradient of agricultural landscapes. *Acta Oecologia* 19:47-60.

Carreck, N. L. and I. H. Williams. (1997). Observations on two commercial flower mixtures as food sources for beneficial insects in the UK. *Journal of Agricultural Science, Cambridge* 128:397-403.

Chamberlain, D. E. and R. J. Fuller. (2000). Local extinctions and changes in species richness of lowland farmland birds in England and Wales in relation to recent changes in agricultural land-use. *Agriculture, Ecosystems and Environment* 78:1-17.

Chancellor, R. J. and R. J. Froud-Williams. (1984). A second survey of cereal weeds in central southern England. *Weed Research* 24:29-36.

Chancellor, T. C. B., A. G. Cook, and K. L. Heong. (1996). The within-field dynamics of rice tungro disease in relation to the abundance of its major leafhopper vectors. *Crop Protection* 15:439-449.

Chapin, F. S., B. H. Walker, R. J. Hobbs, D. U. Hooper, J. H. Lawton, O. E. Sala, and D. Tilman. (1997). Biotic control over the functioning of ecosystems. *Science* 277: 500-504.

Chapman, J. and J. Sheail. (1994). Field margins–an historical perspective. In *Field Margins: Integrating Agriculture and Conservation. Monograph No. 58*, ed. N. Boatman, British Crop Protection Council, pp. 3-12.

Chiverton, P. A. (1993). Large-scale field trials with conservation headlands in Sweden. In *Proceedings Crop Protection in Northern Britain 1993*, pp. 207-215.

Cleugh, H.A. (1998). Effects of windbreaks on airflow, microclimates and crop yields. *Agroforestry Systems* 41:55-84.

Cobbett, W. (1853). *Rural Rides*. London: Cobbett, A.

Collins, K. L., A. Wilcox, K. Chaney, and N. D. Boatman. (1996). Relationship between polyphagous predator density and overwintering habitat within arable field margins and beetle banks. In *1996 Brighton Crop Protection Conference–Pests and Diseases*, Farnham, 635-640.

Collins, K. L., A. Wilcox, K. Chaney, N. D. Boatman, and J. M. Holland. (1997). The influence of beetle banks on aphid population predation in winter wheat. In *Optimising Cereal Inputs: Its Scientific Basis. Part 2: Crop Protection and Systems. Aspects of Applied Biology 50*, eds. R. J. Froud-Williams, M. J. Gooding, W. P. Davis, and B. Hart, Association of Applied Biologists, Wellesbourne, UK, pp. 341-346.

Cook, S. K. and S. Ingle. (1997). The effect of boundary features at the field margins on yields of winter wheat. In *Aspects of Applied Biology 50. Optimising Cereal Inputs: Its Scientific Basis. Part 2: Crop Protection and Systems*, eds. R. J. Froud-Williams, M. J. Gooding, W. P. Davies, and B. Hart, Association of Applied Biologists, Wellesbourne, pp. 459-466.

Cooper, A. B., C. M. Smith, and M. J. Smith. (1995). Effects of riparian set-aside on soil characteristics in an agricultural landscape: Implications for nutrient transport and retention. *Agriculture, Ecosystems and Environment* 55:61-67.

Cousens, R., E. J. P. Marshall, and G. M. Arnold. (1988). Problems in the interpretation of effects of herbicides on plant communities. In *British Crop Protection Council Monograph No.40. Field Methods for the Study of Environmental Effects of Pesticides*, eds. M. P. Greaves, B. D. Smith and P. W. Greig-Smith, British Crop Protection Council, pp. 275-282.

Coutts, B. A. and R. A. C. Jones. (2000). Viruses infecting canola (*Brassica napus*) in south-west Australia: Incidence, distribution, spread, and infection reservoir in wild radish (*Raphanus raphanistrum*). *Australian Journal of Agricultural Research* 51: 925-936.

Cowgill, S. E., S. D. Wratten, and N. W. Sotherton. (1993). The selective use of floral resources by the hoverfly *Episyrphus balteatus* (Diptera, Syrphidae) on farmland. *Annals of Applied Biology* 122:223-231.

Critchley, C. N. R. (1994). Relationship between vegetation and site factors in uncropped wildlife strips in Breckland Environmentally Sensitive Area. In *Field Margins: Integrating Agriculture and Conservation. Monograph No. 58*, ed. N. D. Boatman, British Crop Protection Council, pp. 283-288.

Cussans, G. W., F. B. Cooper, D. H. K. Davies, and M. R. Thomas. (1994). A survey of the incidence of the *Bromus* species as weeds of winter cereals in England, Wales and parts of Scotland. *Weed Research* 34:361-368.

Daniels, R. B. and J. W. Gilliam. (1996). Sediment and chemical load reduction by grass and riparian filters. *Soil Science Society of America Journal* 60:246-251.

Davis, B. N. K., M. J. Brown, A. J. Frost, T. J. Yates, and R. A. Plant. (1994). The effects of hedges on spray deposition and on the biological impact of pesticide spray drift. *Ecotoxicology and Environmental Safety* 27:281-293.

Davis, B. N. K., K. H. Lakhani, T. J. Yates, and A. J. Frost. (1991). Bioassays of insecticide spray drift–The effects of wind speed on the mortality of *Pieris brassicae* Larvae (Lepidoptera) Caused by Diflubenzuron. *Agriculture Ecosystems and Environment* 36:141-149.

de Snoo, G. R. (1995). Unsprayed field margins: Implications for environment, biodiversity and agricultural practice. The Dutch Field Margin Project in the Haarlemmermeerpolder. PhD Thesis Rijksuniversiteit Leiden.

de Snoo, G. R. (1997). Arable flora in sprayed and unsprayed crop edges. *Agriculture, Ecosystems and Environment* 66:223-230.

de Snoo, G. R. and P. J. de Wit. (1998). Buffer zones for reducing pesticide drift to ditches and risks to aquatic organisms. *Ecotoxicology and Environmental Safety* 41:112-118.

de Snoo, G. R., R. J. Van der Poll, and J. Bertels. (1998). Butterflies in sprayed and unsprayed field margins. *Zeitschrift fur Angewandte Entomologie* 122:157-161.

Dennis, P. and G. L. A. Fry. (1992). Field margins–Can they enhance natural enemy population densities and general arthropod diversity on farmland. *Agriculture, Ecosystems and Environment* 40:95-115.

Dover, J. W. (1996). Conservation headlands: effects on butterfly distribution and behaviour. *Agriculture, Ecosystems and Environment* 63:31-49.

Dusi, A. N., D. Peters, and W. Van der Werf. (2000). Measuring and modelling the effects of inoculation date and aphid flights on the secondary spread of Beet mosaic virus in sugar beet. *Annals of Applied Biology* 136:131-146.

Falconer, K. E. (1998). Managing diffuse environmental contamination from agricultural pesticides. An economic perspective on issues and policy options, with particular reference to Europe. *Agriculture, Ecosystems and Environment* 69:37-54.

Forman, R. T. (1995). *Land Mosaics: The Ecology of Landscapes and Regions*. Cambridge, UK: Cambridge University Press.

Forman, R. T. T. and J. Baudry. (1984). Hedgerows and hedgerow networks in landscape ecology. *Environmental Management* 8:495-510.

Frank, T. (1998). Slug damage and numbers of the slug pests, *Arion lusitanicus* and *Deroceras reticulatum*, in oilseed rape grown beside sown wildflower strips. *Agriculture, Ecosystems and Environment* 67:67-78.

Frank, T. (1999). Density of adult hoverflies (Dipt. Syrphidae) in sown weed strips and adjacent fields. *Journal of Applied Entomology* 123:351-355.

Freemark, K. and C. Boutin. (1995). Impacts of agricultural herbicide use on terrestrial wildlife in temperate landscapes: A review with special reference to North America. *Agriculture, Ecosystems and Environment* 52:67-91.

Froud-Williams, R. J. (1985). The biology of cleavers (*Galium aparine*). In *Aspects of Applied Biology 9, The Biology and Control of Weeds in Cereals*. Association of Applied Biologists, Wellesbourne, pp. 189-195.

Fuller, R. J., R. D. Gregory, D. W. Gibbons, J. H. Marchant, J. D. Wilson, S. R. Baillie, and N. Carter. (1995). Population declines and range contractions among farmland birds in Britain. *Conservation Biology* 9:1425-1441.

Game Conservancy. (1994). *Game and Shooting Crops*. Fordingbridge, Hampshire, UK: Game Conservancy.

Ganzelmeier, H., D. Rautmann, R. Spangenberg, M. Streloke, M. Herrmann, H.-J. Wenzelburger, and H.-F. Walter. (1995). *Studies on the spray drift of plant protection products. Mitteilungen aus der Biologischen Bundesanstals fur Land- und Forstwirtschaft*. Berlin: Blackwell.

Gathmann, A., H. J. Greiler, and T. Tscharntke. (1994). Trap-nesting bees and wasps colonizing set-aside fields: Succession and body size, management by cutting and sowing. *Oecologia* 98:8-14.

Gillings, S. and R. J. Fuller. (1998). Changes in bird populations in sample lowland English farms in relation to loss of hedgerow and other non-crop habitats. *Oecologia* 116:120-127.

Greaves, M. P. and E. J. P. Marshall. (1987). Field margins: Definitions and statistics. In *Field Margins. Monograph No. 35*, ed. J. M. Way and P. J. Greig-Smith, British Crop Protection Council, Thornton Heath, Surrey, pp. 3-10.

Green, R. E., P. E. Osborne, and E. J. Sears. (1994). The distribution of passerine birds in hedgerows during the breeding season in relation to characteristics of the hedgerow and adjacent farmland. *Journal of Ecology* 31:677-692.

Haines-Young, R. H., C. J. Barr, H. I. J. Black, D. J. Briggs, R. G. H. Bunce, R. T. Clarke, A. Cooper, F. H. Dawson, L. G. Firbank, R. M. Fuller, M. T. Furse, M. K. Gillespie, R. Hill, M. Hornung, D. C. Howard, T. McCann, M. D. Morecroft, S. Petit, A. R. J. Sier, S. M. Smart, G. M. Smith, A. P. Stott, R. C. Stuart, and J. W. Watkins. (2000). *Accounting for Nature: Assessing Habitats in the UK Countryside*. London: DETR.

Halley, J. M., C. F. G. Thomas, and P. C. Jepson. (1996). A model for the spatial dynamics of linyphiid spiders in farmland. *Journal of Applied Ecology* 33:471-492.

Harwood, R. W. J., S. D. Wratten, and M. Nowakowski. (1992). The effect of managed field margins on hoverfly (Diptera: Syrphidae) distribution and within-field abundance. In *Brighton Crop Protection Conference–Pests and Diseases*, British Crop Protection Council, pp. 1033-1037.

Henry, M., S. George, G. M. Arnold, C. A. Dedryver, D. A. Kendall, Y. Robert, and B. D. Smith. (1993). Occurrence of Barley Yellow Dwarf Virus (BYDV) isolates in different farmland habitats in Western France and South-West England. *Annals of Applied Biology* 123:315-329.

Hewitt, A. J. (2000). Spray drift: Impact of requirements to protect the environment. *Crop Protection* 19:623-627.

Hickman, J. M. and S. D. Wratten. (1996). Use of *Phacelia tanacetifolia* strips to enhance biological-control of aphids by hoverfly larvae in cereal fields. *Journal of Economic Entomology* 89:832-840.

Hooper, M. D. (1970). The botanical importance of our hedgerows. In *The Flora of a Changing Britain*, ed. F. H. Perring, Botanical Society of the British Isles. Classey, Middlesex, 58-62.

Huusela-Veistola, E. (1998). Effects of perennial grass strips on spiders (Araneae) in cereal fields and impacts on pesticide side-effects. *Journal of Applied Entomology* 122:575-583.

INRA. (1976). Les Bocages: histoire, écologie, economie. In *Compte rendu de la table ronde CNRS "Aspects physiques, biologiques et humaines des écosystemes bocagers des régions tempérées humides" 5-7 Juillet 1976*, ed. INRA, CNRS-ENSA et Université de Rennes, 586.

Jobin, B., C. Boutin, and J. L. DesGranges. (1997). Effects of agricultural practices on the flora of hedgerows and woodland edges in southern Quebec. *Canadian Journal of Plant Science* 77:293-299.

Jobin, B., L. Choiniere, and L. Belanger. (2001). Bird use of three types of field margins in relation to intensive agriculture in Quebec, Canada. *Agriculture, Ecosystems and Environment* 84:131-143.

Joenje, W. and D. Kleijn. (1994). Plant distribution across arable field ecotones in the Netherlands. In *Field Margins: Integrating Agriculture and Conservation. Monograph No. 58*, ed. N.D. Boatman, British Crop Protection Council, Thornton Heath, Surrey, UK, pp. 323-328.

Jordan, V. W. L. (1998). The development of integrated arable production systems to meet potential economic and environmental requirements. *Outlook on Agriculture* 27:145-151.

Jörg, E., ed. (1995). *Field Margin-Strip Programmes. Proceedings of a Technical Seminar.* Mainz, Germany. Landesanstalt für Pflanzenbau und Pflanzenschutz.

Kleijn, D. and G. I. J. Snoeijing. (1997). Field boundary vegetation and the effects of agrochemical drift: Botanical change caused by low levels of herbicide and fertilizer. *Journal of Applied Ecology* 34:1413-1425.

Kromp, B. (1999). Carabid beetles in sustainable agriculture: A review on pest control efficacy, cultivation impacts and enhancement. *Agriculture, Ecosystems and Environment* 74:187-228.

Lagerlöf, J., J. Stark, and B. Svensson. (1992). Margins of agricultural fields as habitats for pollinating insects. *Agriculture, Ecosystems and Environment* 40:117-124.

Le Coeur, D., J. Baudry, F. Burel, and C. Thenail. (2002). Why and how we should study field boundary biodiversity in an agrarian landscape context. *Agriculture, Ecosystems and Environment* 89:23-40.

Lethmayer, C., W. Nentwig, and T. Frank. (1997). Effects of weed strips on the occurrence of noxious coleopteran species (Nitidulidae, Chysomelidae, Curculionidae). *Zeitschrift fur Pflanzenkrankheiten und Pflanzenschutz* 104:75-92.

Lewis, T. (1969). The distribution of insects near a low hedgerow. *Journal of Applied Ecology* 6:443-452.

Lewis, T. and D. E. Navas. (1962). Thysanopteran populations overwintering in hedge bottoms, grass litter and bark. *Annals of Applied Biology* 50:299-311.

Longley, M., T. Çilgi, P. C. Jepson, and N. W. Sotherton. (1997). Measurements of pesticide spray drift deposition into field boundaries and hedgerows: 1. Summer applications. *Environmental Toxicology and Chemistry* 16:165-172.

Longley, M. and N. W. Sotherton. (1997). Measurements of pesticide spray drift deposition into field boundaries and hedgerows: 1. Autumn applications. *Environmental Toxicology and Chemistry* 16:173-178.

Lowrance, R., L. S. Altier, J. D. Newbold, R. R. Schnabel, P.M. Groffman, J.M. Denver, D.L. Correll, J.W. Gilliam, J.L. Robinson, R.B. Brinsfield, K.W. Staver, W. Lucas, and A.H. Todd. (1997). Water quality functions of riparian forest buffers in Chesapeake Bay watersheds. *Environmental Management* 21:687-712.

Lys, J. A. and W. Nentwig. (1992). Augmentation of beneficial arthropods by strip-management . 4. Surface activity, movements and activity density of abundant carabid beetles in a cereal field. *Oecologia* 92:373-382.

MacLeod, A. (1999). Attraction and retention of *Episyrphus balteatus* DeGeer (Diptera: Syrphidae) at an arable field margin with rich and poor floral resources. *Agriculture, Ecosystems and Environment* 73:237-244.

Mair, J. and G. R. Port. (2001). Predation by the carabid beetles *Pterostichus madidus* and *Nebria brevicollis* is affected by size and condition of the prey slug *Deroceras reticulatum*. *Agricultural and Forest Entomology* 3:99-106.

Mander, U., V. Kuusemets, K. Lohmus, and T. Mauring. (1997). Efficiency and dimensioning of riparian buffer zones in agricultural catchments. *Ecological Engineering* 8:299-324.

Marino, P. C., K. L. Gross, and D. A. Landis. (1997). Weed seed loss due to predation in Michigan maize fields. *Agriculture, Ecosystems and Environment* 66:189-196.

Marino, P. C. and D. A. Landis. (1996). Effect of landscape structure on parasitoid diversity and parasitism in agroecosystems. *Ecological Applications* 6:276-284.

Marrs, R. H. (1993). Soil fertility and nature conservation in Europe: Theoretical considerations and practical management solutions. *Advances in Ecological Research* 24:241-300.

Marrs, R. H., A. J. Frost, R. A. Plant, and P. Lunnis. (1993). Determination of buffer zones to protect seedlings of nontarget plants from the effects of glyphosate spray drift. *Agriculture, Ecosystems and Environment* 45:283-293.

Marrs, R. H., C. T. Williams, A. J. Frost, and R. A. Plant. (1989). Assessment of effects of herbicide spray drift on a range of plant species of conservation interest. *Environmental Pollution* 59:71-86.

Marshall, E. J. P. (1985). Weed distributions associated with cereal field edges–some preliminary observations. In *Aspects of Applied Biology 9, The Biology and Control of Weeds in Cereals.*, Association of Applied Biologists, Wellesbourne, UK, pp. 49-58.

Marshall, E. J. P. (1989). Distribution patterns of plants associated with arable field edges. *Journal of Applied Ecology* 26:247-257.

Marshall, E. J. P. (1993). Exploiting semi-natural habitats as part of good agricultural practice. In *Scientific Basis for Codes of Good Agricultural Practice. EUR 14957,* ed. V. W. L. Jordan, Luxembourg: Commission for the European Communities, pp. 95-100.

Marshall, E. J. P., ed. (1997). *Field Boundary Habitats for Wildlife, Crop and Environmental Protection. Final Project Report of Contract AIR3-CT920476. Commission of the European Communities. 313 pp. + appendices.* Bristol: IACR-Long Ashton Research Station.

Marshall, E. J. P. (1998). *Guidelines for the Siting, Establishment and Management of Arable Field Margins, Beetle Banks, Cereal Conservation Headlands and Wildlife Seed Mixtures.* A report for the UK Ministry of Agriculture, Fisheries and Food. 90 pp. IACR–Long Ashton Research Station.

Marshall, E. J. P. (2001). Biodiversity, herbicides and non-target plants. In *2001 Brighton Crop Protection Conference–Weeds*, Brighton, UK: British Crop Protection Council, Thornton Heath, Surrey, pp. 855-862.

Marshall, E. J. P. and G. M. Arnold. (1995). Factors affecting field weed and field margin flora on a farm in Essex, UK. *Landscape and Urban Planning* 31:205-216.

Marshall, E. J. P. and J. E. Birnie. (1985). Herbicide effects on field margin flora. In *1985 British Crop Protection Conference–Weeds*, Brighton, UK: British Crop Protection Council, Thornton Heath, Surrey, pp. 1021-1028.

Marshall, E. J. P., A. J. Grant, and S. Fairbairn. (1999). Spatial patterns of gastropod occurrence and herbivory in a field-scale habitat mosaic in winter. In *Heterogeneity in Landscape Ecology: Pattern and Scale*, ed. M. J. Maudsley and E. J. P. Marshall, IALE (UK), pp. 239-244.

Marshall, E. J. P. and A. C. Moonen. (2002). Field margins in northern Europe: Their functions and interactions with agriculture. *Agriculture, Ecosystems and Environment* 89:5-21.

Marshall, E. J. P. and C. Moonen. (1997). Patterns of plant colonisation in extended field margin strips and their implications for nature conservation. In *Species Dispersal and Land Use Processes*, eds. A. Cooper and J. Power, Coleraine: IALE (UK), pp. 221-228.

Marshall, E. J. P. and M. Nowakowski. (1991). The use of herbicides in the creation of a herb-rich field margin. In *1991 Brighton Crop Protection Conference–Weeds*, British Crop Protection Council, Thornton Heath, Surrey, UK, pp. 655-660.

Marshall, E. J. P. and M. Nowakowski. (1992). Herbicide and cutting treatments for establishment and management of diverse field margin strips. In *Aspects of Applied Biology 29–Vegetation Management in Forestry, Amenity and Conservation Areas*, Association of Applied Biologists, Wellesbourne, UK, pp. 425-430.

Marshall, E. J. P. and M. Nowakowski. (1995). Successional changes in the flora of a sown field margin strip managed by cutting and herbicide application. In *Brighton Crop Protection Conference–Weeds*, British Crop Protection Council, Farnham, Surrey, pp. 973-978.

Marshall, E. J. P. and B. D. Smith. (1987). Field margin flora and fauna: Interaction with agriculture. In *Monograph No. 35. Field Margins*, ed. J. M. Way and P. J. Greig-Smith, British Crop Protection Council, Thornton Heath, Surrey, pp. 23-33.

Marshall, E. J. P., C. F. G. Thomas, W. Joenje, D. Kleijn, F. Burel, and D. Le Coeur. (1994). Establishing vegetation strips in contrasted European farm situations. In

Field Margins: Integrating Agriculture and Conservation. Monograph No. 58, ed. N.D. Boatman, British Crop Protection Council, Farnham, Surrey, UK, pp. 335-340.

Marshall, J. K. (1967). The effect of shelter on the productivity of grasslands and field crops. *Field Crop Abstracts* 20:1-14.

Masterman, A. J., S. J. Holmes, and G. N. Foster. (1994). Transmission of barley yellow dwarf virus by cereal aphids collected from different habitats on cereal farms. *Plant Pathology* 43:612-620.

McDonnell, M. J. and E. W. Stiles. (1983). The structural complexity of old field vegetation and the recruitment of bird-dispersed plant species. *Oecologia* 56:109-116.

Melman, C. P. and A. J. Van Strien. (1993). Ditch banks as a conservation focus in intensively exploited peat farmland. In *Landscape Ecology of a Stressed Environment*, eds. C. C. Vos and P. Opdam, London, pp. 122-142.

Mineau, P. and A. McLaughlin. (1996). Conservation of biodiversity within Canadian agricultural landscapes–integrating habitat for wildlife. *Journal of Agricultural and Environmental Ethics* 9:93-113.

Moonen, A. C. and E. J. P. Marshall. (2001). The influence of sown margin strips, management and boundary structure on herbaceous field margin vegetation in two neighbouring farms in southern England. *Agriculture, Ecosystems and Environment* 86:187-202.

Morris, M. G. and N. R. Webb. (1987). The importance of field margins for the conservation of insects. In *Field Margins. British Crop Protection Council Monograph No. 35*, eds. J. M. Way and P. W. Greig-Smith, British Crop Protection Council, Thornton Heath, UK, pp. 53-65.

Naeem, S., L. J. Thompson, S. P. Lawler, J. H. Lawton, and R. M. Woodfin. (1994). Declining biodiversity can alter the performance of ecosystems. *Nature* 368:734-736.

Paoletti, M. G., D. Pimentel, B. R. Stinner, and D. Stinner. (1992). Agroecosystem biodiversity–matching production and conservation biology. *Agriculture, Ecosystems and Environment* 40:3-23.

Parish, T., K. H. Lakhani, and T. H. Sparks. (1995). Modelling the relationship between bird population variables and hedgerow, and other field margin attributes. II. Abundance of individual species and of groups of species. *Journal of Applied Ecology* 32:362-371.

Patty, L., B. Real, and J. J. Gril. (1997). The use of grassed buffer strips to remove pesticides, nitrate and soluble phosphorus compounds from runoff water. *Pesticide Science* 49:243-251.

Perry, N. (1998). *The implications of improving the conservation value of field margins on crop production.* PhD Thesis, Harper Adams College, UK.

Pfiffner, L. and H. Luka. (2000). Overwintering arthropods in soils of arable fields and adjacent semi-natural habitats. *Agriculture, Ecosystems and Environment* 78:215-222.

Pinay, G., L. Roques, and A. Fabre. (1993). Spatial and temporal patterns of denitrification in a riparian forest. *Journal of Applied Ecology* 30:581-591.

Pollard, E., M. D. Hooper, and N. W. Moore. (1974). *Hedges*. London: Collins.

Potts, G. R. and N. J. Aebischer. (1995). Population-dynamics of the grey partridge *Perdix perdix* 1793-1993–monitoring, modelling and management. *Ibis* 137:S 29-S 37.

Rackham, O. (1976). *Trees and Woodland in the British Landscape*. London: Dent.

Rackham, O. (1985). Ancient woodland and hedges in England. In *The English Landscape: Past, Present and Future*, ed. S. R. J. Woodell, Oxford: OUP, pp. 68-105.

Rands, M. R. W. (1985). Pesticide use on cereals and the survival of grey partridge chicks: A field experiment. *Journal of Applied Ecology* 22:49-54.

Rands, M. R. W. and N. W. Sotherton. (1987). The management of field margins for the conservation of gamebirds. In *Field Margins. British Crop Protection Council Monograph No. 35*, eds. J. M. Way and P. W. Greig-Smith, British Crop Protection Council, Thornton Heath, Surrey, UK, pp. 95-104.

Rew, L. J., R. J. Froud-Williams, and N. D. Boatman. (1995). The effect of nitrogen, plant density and competition between *Bromus sterilis* and three perennial grasses: The implications for boundary strip management. *Weed Research* 35:363-368.

Rew, L. J., A. J. Theaker, and R. J. Froud-Williams. (1992). Nitrogen fertilizer misplacement and field boundaries. In *Aspects of Applied Biology No. 30, Nitrate and Farming Systems*, Association of Applied Biologists, Wellesbourne, UK, pp. 203-206.

Rieux, R. (1999). Role of hedgerows and ground cover management on arthropod populations in pear orchards. *Agriculture, Ecosystems and Environment* 73:119-127.

Rodwell, J. S., ed. (1995). *British Plant Communities. Volume 5. Maritime and Vegetation of Open Habitats*. Cambridge: Cambridge University Press.

Schultz, R. C., J. P. Colletti, T. M. Isenhart, W. W. Simpkins, C. W. Mize, and M. L. Thompson. (1995). Design and placement of a multispecies riparian buffer strip system. *Agroforestry Systems* 29:201-226.

Schumacher, W. (1987). Measures taken to preserve arable weeds and their associated communities. In *Field Margins. British Crop Protection Council Monograph No.35*, eds. J. M. Way and P. W. Greig-Smith, British Crop Protection Council, Thornton Heath, UK, pp. 109-112.

Smith, E. A. and F. W. Oehme. (1992). The biological activity of glyphosate to plants and animals–a literature review. *Veterinary and Human Toxicology* 34:531-543.

Smith, H., R. E. Feber, P. J. Johnson, K. McCallum, S. Plesner Jensen, M. Younes, and D. W. Macdonald (1993). *The Conservation Management of Arable Field Margins*. Peterborough, UK, English Nature.

Smith, H., L. G. Firbank, and D. W. Macdonald. (1999). Uncropped edges of arable fields managed for biodiversity do not increase weed occurrence in adjacent crops. *Biological Conservation* 89:107-111.

Smith, H. and D. W. McDonald. (1989). Secondary succession on extended arable field margins: Its manipulation for wildlife benefit and weed control. In *Brighton Crop Protection Conference–Weeds*, British Crop Protection Council, pp. 1063-1068.

Sotherton, N. W. (1984). The distribution and abundance of predatory arthropods overwintering on farmland. *Annals of Applied Biology* 105:423-429.

Sotherton, N. W. (1985). The distribution and abundance of predatory Coleoptera overwintering in field boundaries. *Annals of Applied Biology* 106:17-21.

Sotherton, N. W., M. R. W. Rands, and S. J. Moreby. (1985). Comparison of herbicide treated and untreated headlands on the survival of game and wildlife. In *1985 British Crop Protection Conference–Weeds*, British Crop Protection Council, Thornton Heath, Surrey, UK, pp. 991-998.

Sparkes, D. L., K. W. Jaggard, S. J. Ramsden, and R. K. Scott. (1998). The effect of field margins on the yield of sugar beet and cereal crops. *Annals of Applied Biology* 132:129-142.

Sunderland, K. D., N. E. Crook, D. L. Stacey, and B. J. Fuller. (1987). A study of feeding by polyphagous predators on cereal aphids using elisa and gut dissection. *Journal of Applied Ecology* 24:907-933.

Sutcliffe, O. L. and Q. O. N. Kay. (2000). Changes in the arable flora of central southern England since the 1960s. *Biological Conservation* 93:1-8.

Sutherland, J. P., M. S. Sullivan, and G. M. Poppy. (2001). Distribution and abundance of aphidophagous hoverflies (Diptera: Syrphidae) in wildflower patches and field margin habitats. *Agricultural and Forest Entomology* 3:57-64.

Ten Humberg, G. and H.-M. Poehling. (1995). Syrphids as natural enemies of cereal aphids in Germany: Aspects of their biology and efficacy in different years and regions. *Agriculture, Ecosystems and Environment* 52:39-43.

Theaker, A. J., N. D. Boatman, and R. J. Froud-Williams. (1995). The effect of nitrogen fertiliser on the growth of *Bromus sterilis* in field boundary vegetation. *Agriculture, Ecosystems and Environment* 53:185-192.

Thomas, C. F. G. (1995). A rapid method for handling and marking carabids in the field. *Acta Jutlandica* 70:57-59.

Thomas, C. F. G., H. Cooke, J. Bauly, and E. J. P. Marshall. (1994). Invertebrate colonisation of overwintering sites in different field boundary habitats. In *Arable Farming under CAP Reform. Aspects of Applied Biology No. 40*, eds. J. Clarke, A. Lane, A. Mitchell, M. Ramans and P. Ryan, Association of Applied Biologists, Wellesbourne, UK, pp. 229-232.

Thomas, C. F. G. and P. C. Jepson. (1997). Field-scale effects of farming practices on linyphiid spider populations in grass and cereals. *Entomologia Experimentalis Et Applicata* 84:59-69.

Thomas, C. F. G. and E. J. P. Marshall. (1999). Arthropod abundance and diversity in differently vegetated margins of arable fields. *Agriculture, Ecosystems and Environment* 72:131-144.

Thomas, C. F. G., L. Parkinson, G. J. K. Griffiths, A. Fernandez Garcia, and E. J. P. Marshall. (2001). Aggregation and temporal stability of carabid beetle distributions in field and hedgerow habitats. *Journal of Applied Ecology* 38:100-109.

Thomas, M. B., S. D. Wratten, and N. W. Sotherton. (1991). Creation of 'island' habitats in farmland to manipulate populations of beneficial arthropods: Predator densities and emigration. *Journal of Applied Ecology* 28:906-917.

Thomas, M. B., S. D. Wratten, and N. W. Sotherton. (1992). Creation of 'island' habitats in farmland to manipulate populations of beneficial arthropods: Predator densities and species composition. *Journal of Applied Ecology* 29:524-531.

Tilman, D., D. Wedin, and J. Knops. (1996). Productivity and sustainability influenced by biodiversity in grassland ecosystems. *Nature* 379:718-720.

Tim, U. S., R. Jolly, and H. H. Liao. (1995). Impact of landscape feature and feature placement on agricultural non-point-source-pollution control. *Journal of Water Resources Planning and Management-ASCE* 121:463-470.

Toft, S. (1995). Value of the aphid *Rhopalosiphum padi* as food for cereal spiders. *Journal of Applied Ecology* 32:552-560.

Tsiouris, S. and E. J. P. Marshall. (1998). Observations on patterns of granular fertiliser deposition beside hedges and its likely effects on the botanical composition of field margins. *Annals of Applied Biology* 132:115-127.

Udo de Haes, H. A. (1995). Akkerranden in perspectief. In *Akkerranden in Nederland*, eds. G. R. de Snoo, Rottevee, A.J.W. and Heemsbergen, H., Wageningen, pp. 7-14.

van Emden, H. F. (1965). The role of uncultivated land in the biology of crop pests and beneficial insects. *Scientific Horticulture* 17:121-136.

Van Strien, A. J. (1991). *Maintenance of plant species diversity on dairy farms*. PhD Thesis, Rijksuniversiteit te Leiden 143 pp.

Vickery, J., N. Carter and R. J. Fuller (2002). The potential value of managed cereal field margins as foraging habitats for farmland birds in the UK. *Agriculture Ecosystems & Environment* 89: 41-52.

Waide, R. B., M. R. Willig, C. F. Steiner, G. Mittelbach, L. Gough, S. I. Dodson, G. P. Juday, and R. Parmenter. (1999). The relationship between productivity and species richness. *Annual Review of Ecology and Systematics* 30:257-300.

West, T. M. and E. J. P. Marshall. (1996). Managing sown field margin strips on contrasted soil types in three Environmentally Sensitive Areas. In *Aspects of Applied Biology 44. Vegetation Management in Forestry, Amenity and Conservation Areas*, Association of Applied Biologists, Wellesbourne, UK., pp. 269-276.

West, T. M., E. J. P. Marshall, and G. M. Arnold. (1997). Can sown field boundary strips reduce the ingress of aggressive field margin weeds? In *1997 Brighton Crop Protection Conference–Weeds*, Brighton, UK: BCPC, 985-990.

West, T. M., E. J. P. Marshall, D. B. Westbury, and G. M. Arnold. (1999). Vegetation development on sown and unsown field boundary strips established in three environmentally sensitive areas. In *Aspects of Applied Biology 54. Field Margins and Buffer Zones: Ecology, Management and Policy*, Association of Applied Biologists, Wellesbourne, pp. 257-262.

Whitehead, R. and H. C. Wright. (1989). The incidence of weeds in winter cereals in Great Britain. In *1989 Brighton Crop Protection Conference–Weeds*, British Crop Protection Council, pp. 107-112.

Willis, A. J. (1963). Braunton Burrows: The effects on the vegetation of the addition of mineral nutrients to the dune soils. *Journal of Ecology* 51:353-374.

Wilson, P. J. (1993). Conserving Britain's cornfield flowers. In *1993 Brighton Crop Protection Conference–Weeds*, Brighton: British Crop Protection Council, 411-416.

Wilson, P. J. (1994). Managing field margins for the conservation of the arable flora. In *Field Margins: Integrating Agriculture and Conservation*, ed. N. D. Boatman, British Crop Protection Council, pp. 253-258.

Wilson, P. J. and N. J. Aebischer. (1995). The distribution of dicotyledonous arable weeds in relation to distance from the field edge. *Journal of Applied Ecology* 32:295-310.

Wratten, S. D. (1988). The role of field boundaries as reservoirs of beneficial insects. In *Environmental Management in Agriculture: European Perspectives*, ed. J. R. Park, London: Belhaven Press, pp. 144-150.

Wyss, E. (1995). The effects of weed strips on aphids and aphidophagous predators in an apple orchard. *Entomologia Experimentalis Et Applicata* 75:43-49.

Benefits of Re-Integrating Livestock and Forages in Crop Production Systems

E. Ann Clark

SUMMARY. Specialization is the foundation of contemporary agriculture, yet until the early decades of the 20th century, enterprise diversity was the norm. As agriculture has become more specialized in the intervening years, yields have increased dramatically, but so too have a range of economic, environmental, and health problems. This contribution presents evidence that specialization is an ecologically dysfunctional design for food production, a thesis which is supported by the predominance of problem-solving research in the contemporary literature. As demonstrated by Louis Bromfield in the 1940s and by organic and sustainable farmers today, the strategic integration of crops and livestock/forage enterprises avoids many of the problems of specialization, while also capturing economic and ecological synergies denied to specialist producers. In addition to providing human foodstuffs, forage-based livestock production also offers a range of novel opportunities to channel natural processes to the service of humanity, from carbon sequestration and site remediation to non-chemical vegetation management. *[Article copies available for a fee from The Haworth Document Delivery Service: 1-800-HAWORTH. E-mail address: <docdelivery@haworthpress.com> Website: <http://www.HaworthPress.com> © 2004 by The Haworth Press, Inc. All rights reserved.]*

E. Ann Clark is Associate Professor, Department of Plant Agriculture, University of Guelph, Guelph, ON, Canada N1G 2W1 (E-mail: eaclark@uoguelph.ca).

[Haworth co-indexing entry note]: "Benefits of Re-Integrating Livestock and Forages in Crop Production Systems." Clark, E. Ann. Co-published simultaneously in *Journal of Crop Improvement* (Food Products Press, an imprint of The Haworth Press, Inc.) Vol. 12, No. 1/2 (#23/24), 2004, pp. 405-436; and: *New Dimensions in Agroecology* (ed: David Clements, and Anil Shrestha) Food Products Press, an imprint of The Haworth Press, Inc., 2004, pp. 405-436. Single or multiple copies of this article are available for a fee from The Haworth Document Delivery Service [1-800-HAWORTH, 9:00 a.m. - 5:00 p.m. (EST). E-mail address: docdelivery@haworthpress.com].

http://www.haworthpress.com/web/JCRIP

Digital Object Identifer: 10.1300/J411v12n01_06

KEYWORDS. Specialization, soil management, pasture, grazing, environment, organic agriculture

INTRODUCTION

Contemporary agriculture is characterized by a high degree of specialization or '*dis*-integration' of crop and livestock production, both on individual farms and among regions. Grain and hay exported from specialized crop operations are utilized in confinement feeding operations which are ecologically disconnected with the crop operations, whether within the same region or thousands of kilometers away.

Specialization has afforded us a bountiful and seemingly cheap food supply partly by externalizing costs in the form of adverse impacts on human, livestock, and environmental health. Scientific literature relating to production agriculture focuses increasingly on rectifying such problems as manure nutrient and pathogen loading around confinement facilities (Davis, Young, and Ahnstedt, 1997; Entry et al., 2000a, b; Kovacic et al., 2000; Taylor and Rickerl, 1998; Whalen and Chang, 2001), soil degradation under continuous row crop agriculture (Gordon, Fjell, and Whitney, 1997; Nelson, 1997; Wheeler et al., 1997); surface- and groundwater contamination by biocides, fertilizers, and manure (Barbash et al., 2001; Karr et al., 2001; Russelle et al., 2001; Sharpley and Rekolainen, 1997; Troiano et al., 2001), and animal health and welfare in confinement (Mellon, Benbrook, and Benbrook, 2001).

This contribution explores the thesis that specialized agriculture is an ecologically dysfunctional design for food production because it widens opportunities for pests to proliferate, pollution to occur, and health to be compromised. It will be argued that re-integrating crop and livestock agriculture has the capacity to avoid these increasingly intractable problems, while also capturing useful synergies among enterprises.

The historic roots of specialization will first be reviewed, followed by consideration of contemporary problems created by the polarity of crop and livestock enterprises. Focusing primarily on soils, it will be proposed that livestock are, in fact, an essential prerequisite to ecologically sustainable farming practice. I will conclude with an overview of the largely untapped potential for using forages and livestock for environmental enhancement.

HISTORICAL CONTEXT

It is said that those who refuse to heed the lessons of history are condemned to repeat them. Nowhere is this better demonstrated than in the question of integrating crop and livestock enterprises to sustain agriculture.

As far back as 1947, Louis Bromfield argued in *Malabar Farm* that the 'general' highly diversified farm of his grandparents had 'outlived its usefulness and its economic justification' in favor of 'reasonably' specialized farming. By 'general' farm, he meant a self-sufficient farm, producing a few of each class of livestock as well as all the necessary crops, fruits, and vegetables to sustain the farm family. He reviewed the lessons derived from his 9 years of experience at Malabar Farm to conclude that specialized grass and small grain farming to support meat and milk production was the most efficient and profitable use of his rolling Ohio farmland.

What he called 'reasonably specialized' is archetypical mixed farming, with livestock and especially the grass consumed by the livestock fully integrated into the farming system. Perennial grass, clover (*Trifolium* sp.), and alfalfa (*Medicago sativa* L.) featured prominently in the prescription developed by Bromfield to return to health land which had been degraded by years of specialized, arable cropping. In a series of highly influential books (*Pleasant Valley*, 1945; *Malabar Farm*, 1948; *Out of the Earth*, 1950; *From My Experience*, 1955), along with farm tours attended by thousands, Bromfield provided dramatic evidence that grass–and hence, the livestock to convert the grass into profit–was essential to arable agriculture in the Midwest.

Yet, he then went on to conclude in the same text that the traditional three-or-four-year crop rotation was no longer necessary on productive farms. At the dawn of the modern synthetic chemical and fertilizer era, Louis Bromfield argued that the grass and livestock manure which he himself had shown to be indispensable could in fact be replaced with "lime, green manures, and humus," or more broadly, fertilizers and plowdown crops. He was impressed by Faulkner's *Plowman's Folly* (1943), which advocated conservation tillage coupled with heavy use of green manure crops. Despite his own extensive experience with integrated farming, Bromfield predicted that plowdown crops and chemical fertilizers would become the tools of preference for the next generation of specialized grain farmers.

That a visionary of the stature of Bromfield would have made such a prediction is testament to the allure of new technology–in this case, relying on fertilizer and conservation tillage to displace integrated agriculture and enable specialization. His prediction is all the more remarkable because these same forces of specialization starting earlier in the century had actually created the barren, rundown land which became Bromfield's canvas at Malabar Farm.

Although synthetic biocides did not become commercially available on a large scale until after World War II, farm enterprise specialization began much earlier in the 20th century. Gregson (1996) traced the forces driving enterprise specialization in the US from the mid-1800s to the present day. She cited evidence that diversified agriculture was the norm in the midwestern US (and in Ontario; see below) through the first quarter of the 20th century. From then onwards, however, developments in everything from farm equipment to fertilizer formulations supported increased specialization. Gregson (1996) showed, for example, that rate of synthetic fertilizer application accounted for two-thirds of the variation in specialization in Illinois between 1925 and 1969. She theorized that fertilizer served to homogenize growing conditions, allowing farmers to respond uniformly to market signals, regardless of the inherent suitability of their land for specific crops. More recently, Duvick (1989) emphasized the homogeneity of corn (*Zea mays* L.) genetics, stating:

> . . . there has been a change over the past 20 years toward widespread plantings of a relatively small number of single-cross hybrids . . . This change has probably increased Corn Belt-wise uniformity of reaction to climatic variables.

Increasing reliance on production inputs, including genetics, had the effect of standardizing both growing conditions and response to management and climate, removing the buffering effect of individual producer diversity in cropping method and cultivar/hybrid choice and increasing vulnerability to the market.

Ready access to effective chemical fertilizers also encouraged the perception that farmers were no longer dependent on biological nitrogen (N) fixation and nutrient cycling through livestock to maintain soil health. It took only a few decades for such thinking to produce the impoverished farmland which Bromfield purchased in 1939 as Malabar Farm. Paradoxically, the program of *revitalization through integration* which he chronicled so persuasively in his popular books was prematurely curtailed when government and academia as well as industry became proponents of specialization (Nelson, 1997).

The integral role of livestock in sustainable agriculture was widely known prior to the advent of chemical fertilizers and biocides. Leitch (1920) reported on a detailed economic survey of 385 beef farms in western Ontario focussing on the northern half of Middlesex County, which at that time was heavily devoted to grazing beef cattle. Although the primary product was beef, farms were 'mixed,' growing both clover *(Trifolium pratense* L.) and timothy (*Phleum pratense* L.) hay, wheat (*Triticum aestivum* L.), oats (*Avena sativa* L.), barley *(Hordeum vulgare* L.), corn, and potato (*Solanum tuberosum* L.). Most of the coarse grains were fed on the farm, while some grain crops, and specifically wheat, as well as alsike clover (*Trifolium hybridum* L.) seed were also sold as cash crops.

One of the practical questions addressed by the survey was "should crops be sold or fed?" Farms were placed in 6 categories, according to the fraction of total income that came from crop sales, ranging from 0-10% (#1) to >50% (#6). Over this range, total income increased with the fraction of income from crops (Figure 1).

The author's interpretation of this relationship reflected the perceived importance of grass in that era. He noted that while the evidence might suggest that the best thing to do would be to sell off the stock and just go cash crop entirely:

a. the line was curvilinear, not linear, with each additional increase in gross crop income producing disproportionately smaller increases in net farm income, and
b. about one-third of crop sales for the highest crop income group (#6) were from alsike clover, and 1918 was an unusually favorable year in that both yield and price were high.

He concluded that recognizing year-to-year variations in the weather and crop prices, the most profitable operations were those with 30-40% of their gross income from crops, with the balance from livestock, because "maintaining of this live stock insures the keeping up of soil fertility, which is a factor of no small consideration" (Leitch, 1920).

Within the same dataset, he asked "how much tillable land should be in pasture" for highest profit? Surveyed farms demonstrated that the answer varied with farm size (Figure 2). Small farms [< 100 acre (ac)] were most profitable with no more than 20% of tillable land in pasture. For medium to large farms, however, the most profitable fraction ranged from one-third to one-half of tillable land in pasture.

FIGURE 1. Net farm income as a function of gross crop income on Middlesex County beef operations in 1918 (adapted from Leitch, 1920).

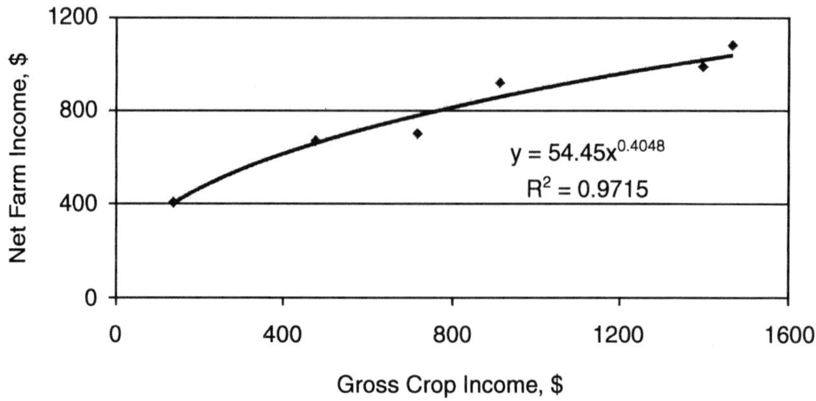

FIGURE 2. Net farm income as a function of the fraction of tillable land that was under pasture, as affected by farm size (< 100, 100-200, and > 200 ac) (adapted from Leitch, 1920).

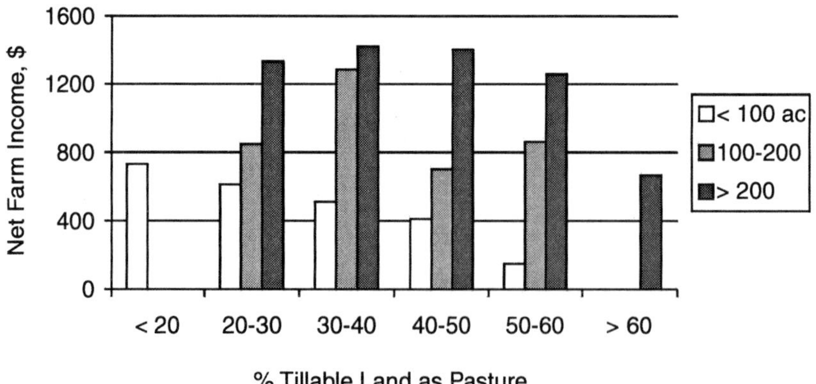

Clearly, prior to the arrival of synthetic fertilizers and other production inputs, both sustainability and profitability depended on the integration of livestock with crops. Neither crop nor livestock alone maximized profit in the long term. Too much of either crops or livestock reduced net income. Thus, in that era, farmers realized the benefits of integrating crops and livestock. With input costs rising much faster than the value of the commodities, this may be a lesson for contemporary farmers as well.

THE PROBLEMS OF POLARITY IN LIVESTOCK-DOMINATED AGRICULTURE

Livestock and the feedstuffs they consume dominate the agricultural landscape of North America and much of the developed world. Worldwide, the proportion of grain that is grown specifically for livestock feed is 40%, increasing with per capita income to as high as 90% in North America (Cox and Atkins, 1979). Baker and Raun (1989) reported that half of all agricultural receipts in the US came from the sale of livestock. Two-thirds of the feed grains and half of the soybeans [*Glycine max* (L.) Merr.] produced in the US were also absorbed in feeding livestock. Much of the remainder is destined for livestock feed in importing nations. Thus, the fortunes of the livestock sector are already intertwined with those of the grain crop sector. The strengths and weaknesses one sector affect the viability of the other sector.

Despite their interdependence, crops and livestock have become increasingly disconnected–spatially and functionally–in contemporary agriculture. Farmers and ranchers have been urged to specialize, to become experts at growing corn and soybeans or weaned calves and nothing else. Economies of scale are the cited argument, yet dis-economies of scale are becoming increasingly evident, particularly as society becomes increasingly unwilling to absorb the externalized costs of large scale agriculture. Thus, one argument for re-integrating livestock and forages into crop production systems is to avoid the economic, environmental, and health problems that have been incurred by specialization.

Economic

As reflected in declining farm populations and plummeting enrollments in agricultural majors, agriculture is no longer gainful, full time employment for most North American farmers. Bird, Bultena, and Gardner (1995) documented the inverse relationship between farm population and farm size in the upper Midwest and plains states. In Iowa, Minnesota, North Dakota, and Montana, for example, farm population declined by 72, 77, 82, and 74% between 1940 and 1990, while farm size increased by 88, 89, 123, and 121%, respectively.

John Ikerd, an agricultural economist, has written extensively about the broader economic implications of large scale agriculture, using swine in Missouri as a case example. One justification for large scale production is lower per unit cost of production, and hence, cheaper food for consumers. Ikerd argues that while cost of production is indeed slightly lower in large scale than in conventional hog operations, the impact on consumer costs is almost undetectable, because the cost of live hogs is less than 35% of the price paid by consumers (USDA Market Price Statistics). The remainder goes to processors, transportation, retailers, and other intermediaries.

Large scale hog production also employs a fraction of the people required to produce hogs in more conventional systems. In Missouri, for example, a composite farrow to finish contract operation would employ just 4.25 people to produce $1.3 million in sales of hogs (Ikerd, 1998). Producing roughly the same dollar value on independently operated hog farms would employ 12.6 people. Assuming a multiplier of 2.22, which reflects related jobs created in the feed business, construction, pharmaceutical, veterinary, and other suppliers, as well as jobs created through retail purchases of goods and services by new employees, then producing $1.3 million in hog sales under contract would 'employ' 9.4 people, compared to almost 28 people on conventional, independent hog farms (Ikerd, 1998). In this example, contract growing would displace 18.5 people, with ripple effects throughout the farm economy and rural community.

Specialization has generated a bountiful food supply, while driving farmers out of business. In an era of specialization, a farmer who can 'feed 128 of his fellows' has to take off-farm employment to put food on his own table. It is difficult to conclude that specialization has benefited either farmers or farm communities.

Environmental

The lack of integration of crop and livestock production amplifies the ecological footprint of agriculture. The unidirectional movement of feedstuffs from specialized cropland to specialized livestock feeding facilities, as is currently occurring with the installation of very large scale dairies based on imported feed in Kings and Tulare counties in California, depletes soil fertility at one end while concentrating nutrients to excess at the other. Decades of experience with manure and disease management on beef feedlots would predict future problems with large scale dairying.

Taylor and Rickerl (1998) compared statewide feedlot averages on the density of fed livestock per unit of available cropland to receive manure. Feedlots in the big four cattle states–Texas, Nebraska, Kansas, and Colorado–were 3.3 to 5.6 times as concentrated as those in South Dakota. Yet even so, the manure returned to cropland in the vicinity of 78 South Dakota feedlots exceeded recommended levels of N 41 and 51% of the time, and recommended levels of phosphorus (P) 52 and 71% of the time, for corn and wheat, respectively. Furthermore, due to economic disincentives for long distance hauling of manure, per hectare application of N and P in manure varied directly with feedlot size. The fraction of sites with cropland receiving in excess of 100 lb N ac^{-1} was 16% in the < 1000 head category, but increased to 33 and 75% in the 1001-2000 and > 2000 head categories, respectively. In effect, livestock capacity in feedlots has increased faster than the capacity of the landbase to receive the manure. As a result of agricultural practice, 98% of South Dakota lakes smaller than 5000 ac no longer meet standards for fishing and swimming (Taylor and Rickerl, 1998).

Colorado supports a cattle population of approximately 1 million head on feed at any point in time. Davis, Young, and Ahnstedt (1997) assessed nutrient status of 41 Colorado corn fields, each of which was within 8 km of a feedlot and had been receiving manure for several years. Nitrogen balance in excess of crop needs averaged 521 and 292 kg N ha^{-1} for sandy and clayey soils, respectively, with the difference due primarily to greater use of N-contaminated groundwater on the sandy soils. With one exception, all samples from both soil types were in the 'Very High' range of soil P. Repeated application of feedlot manure to corn land in Colorado has produced land at risk of both nitrate leaching and P runoff adsorbed to soil particles.

Whalen and Chang (2001) reported that Alberta has a 1.2 million head feedlot capacity, distributed among 4800 feedlots ranging in size from 1000 to 50,000 head. They examined P dynamics on experimental unirrigated and irrigated plots which had received a predetermined range of manure rates for 16 years. Solid manure was applied in the fall of each year after barley harvest, and immediately incorporated. They found that even using the provincially recommended rates of 30 and 60 Mg ha^{-1} provided 5-6 times as much P as is recommended for unirrigated and irrigated barley, respectively. They also documented that total and available P concentrations were higher, particularly at the higher rates of application, at every measured depth in the soil (down to 150 cm), when compared to unmanured controls. Whalen and Chang (2001) found that repeated applications of manure in excess of demand can result in risk of P leaching to groundwater even in calcareous soils. Evidence that soil P has reached the level of an environmental risk within confinement feeding zones in the Netherlands, Belgium, the northeastern USA, and Florida was reviewed by Sharpley and Rekolainen (1997).

Farming systems impacts on the environment may also include greenhouse gas production, and effects on global warming, eutrophication, and acidification (Haas, Wetterich, and Kopke, 2001). For example, in on-farm comparisons in Germany, up to half of the total energy cost per unit milk produced on intensively managed grass dairies came from purchased N fertilizer. Although not yet commonplace, these types of alternative indices of environmental impact appear likely to assume greater prominence in the future.

In sum, production systems which adopt cyclical rather than linear nutrient flows, and which rely more on biologically fixed N and less on fossil fuels impose a smaller 'ecological footprint.'

Health

Direct health risks associated with specialized agriculture may accrue from the fertilizers and biocides used in crop production and from the antibiotics, growth stimulants, and other products routinely used in confinement housing of livestock. Pathological risks associated with livestock manure, including cryptosporidia and *E. coli*, have assumed increasing prominence in recent years. Additional indirect risks can occur when biocides, manure, or fertilizer nutrients contaminate surface or groundwater. For example, a 5-year study by Porter, Jaeger, and Carlson (1999) compared endocrine, immune, and nervous system responses to aldicarb, atrazine, and nitrate–alone or in combinations–at the maximum allowable groundwater concentrations for these chemicals. Binary combinations of these chemicals elicited responses not seen with individual chemicals. Indices of aggression and thyroid activity as well as immune system function, measured as antibody production when challenged with a foreign protein, responded to combinations of either or both herbicides with nitrate.

CROPS

Perennial forages keep the soil covered year-around, enhancing infiltration and reducing surface runoff. In addition, a perennial sward establishes and maintains net nutrient uptake capacity earlier and later in the year, better accommodating natural cycles in mineralization and reducing leaching risk. In contrast, row crops are particularly prone to runoff and off-site contamination, both because the land is poorly covered for much of the year, and because timing of nutrient uptake is out-of-phase with soil mineralization (Staver and Brinfield, 1990). In Iowa, Schilling and Libra (2000) showed a positive linear relationship between the fraction of row crop land in a watershed and the concentration of nitrate in the watercourses draining the watershed, for both large and small watersheds:

Large watersheds: $y = 0.108x - 0.812$ ($r^2 = 0.65$)

Small watersheds: $y = 0.11x + 0.217$ ($r^2 = 0.94$)

In both types of watersheds, increasing the fraction of the land occupied by row crops increased nitrate in the watercourses at a rate of 0.11 units of nitrate (mg L^{-1}) per unit land area (%). The life cycle of the annual grain crops grown for confinement feeding, and particularly those which are heavily fertilized with N, predetermines risk of both leaching and off-site contamination.

Many of the annual crops used for livestock feed in North America are dependent upon biocides to control weeds, insects, and diseases. At present, almost a billion pounds (lbs) of biocide active ingredient (a.i.) are applied annually to US crop and rangeland (Benbrook, 1996), with the rates of application varying among crops. Kegley, Orme, and Neumeister (2000) reported that in 1998, California corn for grain and for silage received 2.7 and 4.9 lb a.i. ac^{-1}, respectively, in comparison with pasture and range, which received just 0.67 and 0.16 lb a.i. ac^{-1}, respectively.

According to Troiano et al. (2001), it was originally assumed that biocides were unlikely to leach to groundwater due to "dilution effects, low water solubility, high vapor pressure, rapid degradation, and binding to the soil." This belief has now been challenged by the ubiquitous finding of biocide contamination in both ground and surface water in large scale surveys, as reviewed by Barbash et al. (2001). The US National Water Quality Assessment program, sampled 2227 sites (surface and groundwater) distributed in 20 major hydrologic basins, between 1993 and 1995 (Barbash et al., 2001). Frequency of detection correlated positively with adjoining land usages for urban non-agricultural use (atrazine, cyanazine, simazine, alachlor, and metolachlor) and for rural agricultural use (atrazine, cyanazine, alachlor, and metolachlor).

In California, 16 biocide active ingredients and breakdown products have now been detected in groundwater as a result of legal agricultural use (Troiano et al., 2001). More than 50 biocide active ingredients have been classified as potential leachers. Well water surveys have identified Pesticide Management Zones (PMZs) which are vulnerable to leaching. When no mitigation measures are available and sampled concentrations are near the maximum contaminant level allowed in the state, use of specific biocides can be prohibited, as occurred with bentazon on rice (*Oryza sativa* L.).

Because California began testing wells in the mid to late 1980s, a considerable database has now evolved. Using the section (2.6 km^2) as the sampling unit, testing was designed to sample wells in 10% of the sections surrounding each PMZ. Detection of a previously detected a.i. in a section adjacent to a PMZ resulted in the sampled section being submitted for addition to the PMZ list. Thus, the number of adjacent sections to be monitored increased over time. For example, in an area covering parts of Fresno and Tulare Counties, biocides were detected in 74 sections in 1988, 163 in 1990, 253 in 1992, and 409 in 1995 (excluding DBCP; 1,2-dibromo-3-chloropropane).

Increasing detection reflected increased biocide dependence. According to Kegley, Orme, and Neumeister (2000), usage of biocide active ingredient increased from 22.1 to 39.9 million lb in Fresno County and from 11.0 to 18.3 million lb in Tulare County between 1991 and 1998. On an area basis, active ingredient averaged 29 and 26 lb ac^{-1} for these two counties in 1998.

The trend is clear. Increasing biocide dependence manifests itself in increasing detection in groundwater. The health implications remain uncertain, however, due to limited, independently validated testing, single residue testing, and lack of testing of newly identified risks.

California and other jurisdictions have documented biocide exposure due to concerns about human health. For example, some 70 biocide active ingredients are known to cause cancer in animals (Benbrook, 1996). Among the best studied is 2,4-D and other chlorophenoxy herbicides, which have been linked with non-Hodgkin's lymphoma (Zahm and Blair, 1992; various in Steingraber, 1997). Schreinemachers (2000) associated the incidence of cancer with use of chlorophenoxy herbicides in selected agricultural counties in Minnesota, North Dakota, South Dakota, and Montana. Most US wheat is grown in these states, with 90% of spring and durum wheat and 30% of winter wheat being treated with chlorophenoxy herbicides. Age-standardized cancer mortality (1980-1989) increased with acres sown to wheat for various types of cancer.

Thus, whether due to nitrate or biocide contamination of groundwater or risk of other biocide exposure, contemporary methods of growing crops on specialized grain farms, for feeding on specialized livestock operations, increases potential health risks.

Livestock

Large-scale confinement feeding systems foster potential human health risks, both from the manure they generate and from the pharmaceuticals that are required to keep stock healthy in high density confinement. In North Carolina, hog numbers grew from 3.7 to 9.8 million between 1991 and 1997, largely in a 5 county region around the world's largest slaughterhouse–which processes 24,000 hogs a day, 365 days a year (Nowlin, 1997). Agriculture–and primarily swine operations–accounted for 56 to 76% of the nutrient pollution of two major surface water systems, causing serious problems with eutrophication and with *Pfiesteria piscicida*, a dinoflagellate which is lethal to fish (Nowlin, 1997). More than 50% of the manure lagoons tested in North Carolina exhibited severe seepage losses, exposing groundwater to contamination. Karr et al. (2001) cited evidence that manure N produced in some North Carolina counties exceeds 500% of crop need for N, yet commercial fertilizers are widely used.

Antibiotic use has increased in confinement feeding systems, particularly for hogs and chickens (Mellon, Benbrook, and Benbrook, 2001). Fully 80% of the antibiotics used in contemporary farming are prophylactic in nature or growth promotants–not to cure animal illness. The US livestock industry annually consumes 24.6 million lbs of antimicrobials, such as tetracycline, penicillin, and erythromycin for non-therapeutic uses (in the absence of disease). Of these, hogs and poultry are the most dependent, consuming in excess of 10 million lbs each, compared to cattle at 3.7 million lbs. Pretty et al. (2000) cited evidence that resistant strains of *Salmonella*, *Campylobacter*, *Enterococci*, and *E. coli* resulted from use of these antimicrobials in farm animals.

There is diminishing societal tolerance for farm practices which externalize costs of production involuntarily to society and environment. Pretty et al. (2000) estimated the externalized costs of UK agriculture, many of which are a direct result of the functional isolation of crop from livestock enterprises. Just those which relate in one way or another to nutrient management, including N and P in drinking water, emissions of ammonia, and emissions of nitrous oxide from N fertilizer, account for over one-third of the externalized costs. Total value to 1996 UK agriculture of costs currently excluded from farm budgets was £2343 million or £208 ha^{-1} of arable land and permanent pasture. This conservative estimate of externalized costs amounts to 89% of average net farm income, reducing the benefit:cost ratio of UK agriculture to virtually 1:1.

Examples of diminishing societal tolerance of externalized costs are numerous. Neeteson (2000) described current regulations on nutrient management in Dutch agriculture, including a ban on winter-spreading of manure, obligatory cover for manure storage facilities, compulsory injection or immediate incorporation to reduce emissions during spreading, and fines for exceeding permissible levels of N and P surpluses. Troiano et al. (2001) reviewed the California legislation which mandated systematic water quality testing and constrained biocide use in 'PMZ.'

Whether economic, environmental, and health-related, specialized agriculture has engendered problems of increasing societal concern. In response, society appears to be moving toward a 'full cost accounting' approach, which would oblige farmers and others to more fully internalize their costs of production. Such a development may help to profile some of the potential positive synergies of integrated crop:livestock agriculture.

CAPTURING THE SYNERGIES OF LIVESTOCK-BASED AGRICULTURE

Historically, livestock were viewed as tools of production. Livestock were integrated into farm operations specifically to capture positive synergies among enterprises–to perform tasks and supply services to other enterprises–not just as a marketable commodity. As shown in the survey by Leitch (1920), livestock only made ecological and economic sense as part of a larger whole. And of course, the same could be said for grain or any other enterprise. In the traditional view, the value of a ewe or a corn field was more than just the saleable product. The economic value of their contributions to other enterprises, whether for weed control or nutrient cycling or animal health, was recognized.

Contemporary agriculture has been operating under the premise that farmers no longer *need* to make enterprises 'fit' together to accomplish tasks–they can just *buy* the services formerly achieved through integration. So long as costs of inputs were low, relative to the value of the commodities, this strategy appeared to work well. Now the cost of inputs is rising one or even two orders of magnitude faster than the value of the commodities produced, contributing to the economic crisis facing farmers today. For example, one of the newest inputs to production is genetically modified (GM) crops, such as Roundup Ready (RR) soybeans. Benbrook (1999; Table 7) documented that compared to growing non-GM soybeans, the cost of growing RR soybeans consumed an additional 2.3 to 12% of gross income per acre across 8 midwestern states.

Indeed, valuing crops and livestock simply as marketable commodities, as in specialized agriculture, not only creates costly and intractable problems (see above), but furthermore, blinds farmers to the potential for economic, agronomic, and environmental *synergies* between livestock and crop enterprises. Thus, enterprise integration has merit not simply to avoid generating problems, but also, to harness natural processes to human ends.

For example, crop rotation and associated management are known to influence perennial weed communities, the soil weed seed bank, and the soil reservoir of pests, pathogens, and nutrients as well as deleterious and beneficial soil microbes (Bezdicek and Granatstein, 1989; Dick, 1993; Reganold, 1995; Sturz et al., 2001). Entz, Bullied, and Katepa-Mupondwa (1995) surveyed 253 Manitoba and Saskatchewan farmers who were known to include forages in their crop rotations. Over 80% reported a reduction in weed pressure for 1, 2, or more years (11, 50, and 33%, respectively) following forages. Inclusion of forages in the rotation afforded good control of several of the most problematic weeds in western Canada, wild oat (*Avena fatua* L.), Canada thistle (*Cirsium arvensis* L.), wild mustard (*Sinapis arvensis* L.), and green foxtail (*Setaria viridis* (L.) Beauv.). The utility of forages in controlling wild oat and green foxtail is noteworthy, because both are already resistant to several common herbicide families. Over two-thirds of the farmers surveyed by Entz, Bullied, and Katepa-Mupondwa (1995) also reported higher grain yields following forages, particularly in the zones with higher moisture. Thus, this survey data corroborated the research findings that forages benefit the performance of subsequent crops in the rotation.

Likewise, Drinkwater et al. (1995) reviewed literature showing that soils managed for high soil organic matter (SOM) content in the absence of biocides showed higher microbial activity and lower levels of instantaneous mineral N pools, despite having higher levels of potentially mineralizable N. In a 2 year trial involving 20 California organic and conventional tomato (*Lycopersicon esculentum* Mill.) farms, low available soil N and high microbial activity were associated with lesser incidence and severity of corky root (*Pyrenochaeta lycopersici*). Thus, integrating crops and management practices conducive to maintaining high levels of SOM, such as perennial forages (see below), may enhance soil health for other crops in the rotation.

The contribution of livestock to ecologically sustainable farming is still recognized by organic and sustainable agriculturalists. The position statements of organic farming organizations such as the *International Federation of Organic Agriculture Movements (IFOAM)* include principles such as "to encourage and enhance biological cycles within the farming system, involving micro-organisms, soil flora and fauna, plants and animals" and "to create a harmonious balance between crop production and animal husbandry."

A survey of 2450 'sustainable' (based on a calculated sustainability index) and conventional producers in four states of the US found that 76 to 93% of the sustainable farms had livestock, compared with 37-58% for conventional (Bird, Bultena, and Gardner, 1995). The rationale for keeping livestock varied among regions, with intensively managed grazing preferred in the higher rainfall areas, compared with extensive grazing of range land in Montana. Some producers integrated grazed or hayed forages in their crop rotation, in part for weed control. Others valued livestock as a buffer, to convert weather damaged grain into income that would otherwise have been lost. On sustainably managed farms, livestock perform functions apart from simply production of a marketable commodity.

Although agroecosystems differ from natural ecosystems, integrating forages and livestock into a cohesive whole most closely mimics nature. By affording access to the synergies of a natural ecosystem, integrated systems minimize economic, environmental, and health problems, and hence, the need to purchase problem-solving inputs. In other words, integrated systems help to internalize costs of production. For example:

- Forage swards typically consist of multiple rather than a single species, utilizing biotic diversity to exploit diverse ecological niches (Clark, 2001), retain nutrients, maintain soil cover and productivity throughout the year, and discourage weed and pest proliferation.

- Withholding land from cultivation avoids the periodic disturbance characteristic of arable systems, which promotes SOM breakdown and exposes the land to erosion (see below).
- Vigorous perennial swards resist weed encroachment. Simple swards sown to 1 or 2 species can be vulnerable to weed encroachment, as shown in UK surveys by Morrison and Idle (1972) and Hopkins et al. (1985). However, weed contribution to sward yield was just 10 to 20% in a 9 year old pasture sown to a complex mixture in Ontario (Clark, 2001). In contrast, tillage and/or herbicides applied to annual crops set back succession, ensuring that pioneer invaders–weeds–will be an ongoing problem.
- Perennials establish and maintain a nutrient sink throughout a greater portion of the year, better diluting leaching risk in spring and fall. Studies cited by Pearson and Ison (1987) suggest that N leaching can be considerable under highly productive grazed legume pastures in regions of high rainfall. They conclude, however, that losses under permanent grassland are typically minor, perhaps 3% of what is taken up by the herbage, in contrast to losses of 10-30% of uptake under annual forages.

  In contrast, annual crops introduce a periodicity in nutrient uptake capacity which is out-of-phase with soil mineralization cycles, creating vulnerability to nutrient loss. In Maryland, Staver and Brinfield (1990) documented the ability of fall rye (*Secale cereale* L.) following corn to immobilize an additional 118 kg N ha^{-1} which would otherwise have been vulnerable to loss from corn land fertilized at the recommended rate. Aside from off-site contamination, nutrients lost due to the asynchrony between crop demand and nutrient supply must be replaced.
- Nutrient export in livestock products (meat, milk, or wool) is low, relative to that of grain or whole plant crops (hay or silage), as most ingested nutrients are excreted. Spedding, Walsingham, and Hoxey (1981) reported implied nutrient exports from grain and potato crops to range from 80 to 360 kg N ha^{-1}, from 12 to 52 kg P ha^{-1}, and from 17 to 230 kg K ha^{-1}, compared with 12 to 35 kg N ha^{-1}, 0.7 to 6 kg P ha^{-1}, and 0.1 to 11 kg K ha^{-1} in livestock products. Marketing feedstuffs as livestock in an integrated operation, rather than as feed in a specialized operation, promotes cyclical rather than linear nutrient flows and reduces dependence on purchased nutrients.
- Perennial swards can also serve as a reservoir for natural pest control agents. Langer (2001) cited evidence that clover:grass leys fa-

cilitated parasitoid control of cereal aphids (*Sitobion avenae* (F.)) in adjoining arable croplands. In Denmark, Langer (2001) found that an undersown grass/clover ley supported a population of aphids as well as parasitoids, thus maintaining a population of aphid-controlling parasitoids year-around.

- Livestock manure has been shown to be superior to chemical fertilizer, as shown in long-term studies by Baldock and Musgrave (1980). The beneficial effect appears to exceed what could be attributed to nutrients alone, suggesting that manure affects other yield enhancing processes, perhaps through soil structure or soil biotic diversity. For example, activity of the soil enzymes urease and amidase was shown to be much higher in soils amended with livestock manure than chemical N fertilizer (Dick, 1993). Both of these soil enzymes facilitate N cycling, and hence, the ability of soils to supply N to crops.

Thus, inclusion of forages and/or livestock within an integrated production system permits synergies which diminish dependence on both processes and purchased products which have become problematic in specialized agriculture.

Conserving/Improving Soil Health

Forages and livestock promote soil health, which is considered to be "... the capacity of soil to function within ecosystem boundaries to sustain biological productivity, maintain environmental quality, and promote plant and animal health" (Deng, Moore, and Tabatabai, 2000). Central to soil health is SOM status, which is an integrative index of the balance between organic matter (OM) input and breakdown. Soil organic matter is the energy which fuels the biotic processes which sustain soil health, as manifested in tilth, structure, water holding capacity, and resistance to both compaction and erosion. However, since the start of cultivation, SOM has declined exponentially to a new, lower plateau which is characteristic of specific management environments (Beauchamp and Voroney, 1994).

Soil organic matter can be replenished from the roots and stubble of sown crops or from green manure crops, or from pastured or applied livestock manure. The magnitude of root biomass that accumulates in perennial leys was reported by Eriksen and Jensen (2001). They assessed the effect of spring cultivation on 3-year-old swards of perennial ryegrass (*Lolium perenne* L.) (PRG) or PRG-clover. Under Danish conditions, root residues to a depth of 20 cm were 11.5 and 8.8 t ha^{-1} in the PRG vs. PRG-clover swards, respectively. Root tissues accounted for 95% of total plant residual dry matter. From early April to late June, cultivated swards evolved 2.6 t C ha^{-1} compared to 1.4 t C ha^{-1} in the uncultivated swards. Much of the N released through mineralization occurred within the first 4 weeks after cultivation. Conclusions which may be drawn from this study are: (a) that root tissues are a prominent source of the OM which enriches the land in grassland leys, (b) cultivation promotes C loss, and (c) capturing the N mineralized from incorporated swards requires rapid establishment of an effective nutrient sink, directly after cultivation.

In Norway, Breland and Eltun (1999) compared indices of soil microbial activity in 8-year arable vs. forage-based rotations. The forage-based rotations integrated a 3-year grass ley into the original arable crop rotation. After 5 years, total soil C in the arable vs. forage-based systems was 2.4 vs. 2.7%, while microbial biomass C was 242 vs. 303 mg C kg^{-1} soil and microbial biomass N was 41.1 vs. 51.4 mg N kg^{-1} soil, respectively. The metabolic quotient was also lower in the forage than in the arable rotation, 0.472 vs. 0.553 μg CO_2-C h^{-1} mg^{-1} biomass C. The metabolic quotient is inversely related to the maturity and species-diversity of soil microbial communities, and hence, to the efficiency with which OM is cycled. Greater root biomass and the addition of animal manure accounted for improved soil health in the forage-based vs. arable rotations (Breland and Eltun, 1999).

Deng, Moore, and Tabatabai (2000) compared soil health parameters in several 4-year crop rotations at long-term sites established in Iowa. When measured in 1996, soils with 1 or 2 years of alfalfa in the rotation were higher in microbial biomass C (380 and 429 mg C kg^{-1} soil, respectively) than soils under continuous corn or corn-soybean rotations (326 and 335 mg C kg^{-1} soil, respectively). Parallel trends were evident in microbial biomass N, with 60 and 66 mg N kg^{-1} soil in rotations with 1 or 2 years of alfalfa, respectively, compared to 57 and 58 mg N kg^{-1} soil in continuous corn or corn-soybean rotations, respectively. Microbial biomass is considered a sensitive indicator of the capacity for OM breakdown and nutrient cycling. Similar trends were shown for dehyrodrogenase activity, which is an index of microbial activity, and amidase activity, which relates to N mineralization. They concluded that including alfalfa in a crop rotation served to increase soil bioactivity compared to continuous arable cropping.

At the same sites in Iowa, Deng, and Tabatabai (2000) demonstrated that N mineralized during 24 weeks of aerobic incubation was higher on soils from rotations including 1 or 2 years of alfalfa (175 and 208 mg N kg^{-1} soil, respectively) than in continuous corn or corn-soybean rotations (132 and 140 mg N kg^{-1} soil, respectively). Parallel trends were observed when N mineralization was expressed relative to organic N content, namely, 8.25 and 9.48% for rotations with 1 or 2 years of alfalfa, respectively, vs. 6.97 and 7.57% for continuous corn or corn-soybean, respectively. Including alfalfa in crop rotations not only conserved soil C and N, but also served to strengthen the active N pools in soils and increase N availability (Deng and Tabatabai, 2000) for subsequent crops.

Land which had been under continuous arable cropping for 11 years provided the site for a New Zealand (NZ) comparison of microbial responses to grass vs. arable cropping (Haynes, 1999). After 5 years of treatment, responses to various grass and arable treatments were compared with those from a nearby long-term (LT) pasture. Throughout the 15 cm profile, soil organic C was highest under LT pasture and lowest under LT arable with the other treatments intermediate. The range between these two extremes was greatest in the 0-2.5 cm layer–65 vs. 30 g C kg^{-1} soil–declining to 40 vs. 30 g C kg^{-1} soil in the 10-15 cm layer. Just 5 years under grass, however, improved soil organic C in the 0-2.5 cm layer by 50%, compared to LT arable usage.

As noted by Brelund and Eltun (1999), Haynes (1999) found the metabolic quotient of LT arable land (stable at more than 80 μg CO_2-C mg^{-1} biomass day^{-1}), was clearly distinguishable from that of other treatments. The greatest contrast was that in the LT pasture, which increased from 45 to 55 μg CO_2-C mg^{-1} biomass day^{-1} with depth in the profile. Evidence from this and other studies led Haynes (1999) to infer an adverse effect of cultivation on the efficiency of use of C substrate. He concluded that the beneficial effect of including a grass-clover ley in an arable rotation was due not simply to the addition of C and N to the soil, but primarily to withholding the land from cultivation.

Weil, Lowell, and Shade (1993) compared 5 cropping systems in Maryland, of which one was continuous grass, while the others were rotations involving corn, soybean, and wheat, with varying intensities of both tillage and chemical use. Over 5 years, SOM in the surface 15 cm was significantly higher in the grass than the other treatments (2.28 vs. 1.68-1.95%), with parallel effects on total C (1.11 vs. 0.7-0.85%) and bulk density (1.15 vs. 1.29-1.38 g cm^{-3}). Consistent with effects on bulk density, infiltration rate was 10 times faster in the grass than in the continuous corn, with the other treatments intermediate. Total N was also significantly higher in grass than in the other rotations (0.1 vs.

0.07-0.08%), with the exception of an organic treatment (0.09%). The authors concluded that "Most dramatic improvement in porosity, OM accumulation and N mineralization ability came from 5 years continuous growth under grass sod, underscoring the potential role of grass in sustainable cropping systems . . ."

The effectiveness of plowdown crops as an OM source in stockless crop rotations was evaluated by Bulson et al. (1996). They compared the effect of 3, 4-year stockless rotations on agronomic performance and edaphic variables in the UK. In each rotation, 1 of 4 years was devoted to red clover for plowdown, to add N and OM to the system, with the remaining 3 years in combinations of cereals, potatoes, and beans. Over an 8 year period, SOM declined linearly from over 3% to about 2.5% in all three rotations. The authors observed that the decline in SOM occurred despite allocating 1 year in 4 to a plowdown crop, as well as incorporating all crop residue. They concluded that imported OM may be needed just to sustain existing levels of SOM. However, they also noted that initial SOM may have been unusually high, owing to having been in a 5-year grass ley prior to the trial.

Saviozzi et al. (1999) contrasted the utility of imported farmyard manure (FYM) and sewage sludge (SS) as C and nutrient sources in Italy. Both amendments were applied at a rate of 5 t ha^{-1} $year^{-1}$ to land in a long-term corn-wheat rotation. The control was undisturbed native grass. Total organic C was more than twice as high in the control as in the FYM and SS treatments (5.13 vs. 2.33 and 2.00 g 100 cm^{-3}, respectively). Biomass C was reduced in both the FYM and SS treatments, relative to the control (2340 and 2282 vs. 3541 mg 100 cm^{-3}, respectively). Similar trends were evident for other chemical and biochemical parameters, suggesting an overall deleterious effect of cultivation which was not compensated for by annual addition of FYM or SS at this rate.

Posner, Casler, and Baldock (1995) used 8 system descriptors to develop an integrative Agroecological Index to compare 6 crop rotations. Descriptors included: (a) aboveground productivity; (b) %DM recycled; (c) energy out:in ratio; (d) N fixation; (e) biodiversity index (# species planted); (f) energy subsidy; (g) cover crop factor (from USLE); and (h) frequency of disturbance. The rotations compared were R1 continuous corn, R2 OM-corn, R3 soybean-wheat/red clover-corn, R4 3 year alfalfa-corn, R5 oat/alfalfa-alfalfa-corn, and R6 grazing. Over these six rotations, the Agroecological Index ranged from 1 to 7.69, with the biggest jump between R3 and R4. The R3 rotation exemplifies continuous arable cropping including a plowdown crop, while R4, 5, and 6 integrated perennial swards into the production system. Thus, forage-based systems and especially the grazed system (R6) were clearly advantageous in most descriptor categories.

In effect, livestock provide the economic justification for inclusion of perennial forages in crop rotations, and perennial forages in turn confer multiple benefits to the system. Thus, in many cases, livestock provide the missing element needed to develop sustainable systems, particularly in terms of soil health.

FORAGES AND LIVESTOCK FOR ENVIRONMENTAL ENHANCEMENT

Apart from food production, forages and/or livestock have the potential to supply a range of additional services, both to farmers and to society at large.

Carbon Sequestration

Withholding land from cultivation under perennial crops, including forages, is the only known way to produce a net increase in SOM, effectively reducing the levels of carbon dioxide, a greenhouse gas, in the atmosphere. Beauchamp and Voroney (1994) developed a simple model to quantify OM addition from crop rotations typical of swine and dairy operations in Ontario. Crop C returned to the soil varied with crop type, depending on the harvest index, the C fraction contained in the roots and exudates, and the C fraction voided by the fed livestock. Estimates of crop C returned to the soil in a corn-soybean-wheat rotation for swine varied from 2.3 to 4.6 t C ha^{-1} for soybean and grain corn, respectively. For a dairy rotation of corn-corn silage-cereal-3 years of alfalfa-grass, the range in crop C returned to the soil was 3.5 to 4.6 t C ha^{-1} for silage and grain corn, respectively, compared with 4.7 to 6.8 (mean of 5.8) t C ha^{-1} $year^{-1}$ among the 3 alfalfa-grass years. Therefore, as a source of SOM, perennial forage swards averaged more than twice as much C per year as soybean, and 25% more than grain corn (Beauchamp and Voroney, 1994).

The proportion of fed C recovered in manure varies among classes of livestock, from 10% in broilers to 40% for non-lactating cattle, and from 20% for grain to 40% for perennial forages (Beauchamp and Voroney, 1994). In addition to contributing more C as roots and stubble, perennial swards also supported classes of livestock that void a larger fraction of fed C, and hence, return a larger fraction of C in the manure.

Higher levels of C input, together with simply withholding land from cultivation (Haynes, 1999), are the cause of the net increase in SOM which has been routinely observed under sod in LT research plots.

Soil Erosion

Perennial forages directly advantage sustainability by providing feedstuffs without exposing the soil to erosion. Rayburn (1993) compared pasture vs. recombinant bovine somatotropin (rBST) for increasing milk production in the Northeast. Using New York State as the example, he contrasted the acreage needed for pasture-based and confinement-based feeding systems at two levels of milk production–16,402 lb milk/cow 3 × 687,000 producing cows–versus 18,370 lb milk/cow 3 × 613,000 producing cows–with each scenario designed to produce 11.3 billion lb of milk projected for 1998. Each ration, whether pasture or confinement-based, was balanced using NRC nutrient requirements.

He concluded that more acreage would be needed to support the pasture option, particularly at the higher level of per-cow output, but that the type of crop grown would also change toward a larger proportion of soil conserving crops. The "C" value–an index of the crop rotation effect on soil erosion from the Universal Soil Loss Equation–for the pasture and confinement options was 0.108 and 0.178, respectively. Thus, while pasture-based systems required more total acreage, total potential soil loss was 33 and 27% less under the moderate and high production scenarios. Grass-based livestock production systems reduce soil erosion, as compared to confinement feeding dependent on annual grains.

Remediation

Contamination of aquifers is an increasingly prominent byproduct of specialized agriculture. Deep-rooting perennials, such as alfalfa, have the potential to intercept and capture labile nutrients which have leached below the rooting depth of annual crops. The perennial growth habit confers an additional advantage, in that it broadens the duration of effective 'sinkness' or nutrient extraction capacity to better correspond to the spring and fall leaching risk intervals which pertain in the humid temperate zone (Dinnes et al., 2002). Russelle et al. (2001) demonstrated the capacity of alfalfa to cleanse groundwater contaminated by a 1989 fertilizer spill in Minnesota. Over a 3 year period, Ineffective (non-N-fixing) Agate alfalfa removed 972 kg N ha^{-1}, compared to 287 kg N ha^{-1} in grain from corn (424 kg N ha^{-1} if all aboveground matter had been harvested from the annuals).

Potential for removal of fecal coliforms from surface applied hog manure was tested on several combinations of grass and forest vegetation (Entry et al., 2000a, b). Depletion of coliforms from the soil was unaffected by vegetation type, and responded primarily to soil moisture and temperature. Persistence of pathogens, including pathogenic *E. coli*, in surface applied manure may be prolonged in cool, humid climates.

Kovacic et al. (2000) tested the effectiveness of constructed wetlands based on native mesic species in removing nutrients from agricultural tile drainage. Subsurface tiles rapidly convey drainage water, including solutes and effluent from agricultural land, into surface water systems. Almost 50% of the arable land in southern Ontario is tile-drained (Barton, 1996), compared to 37% of Cornbelt and Great Lakes cropland (Kovacic et al., 2000). Over a 3-year period, artificial wetlands and adjoining 15 m buffer strips consisting of reed canary grass (*Phalaris arundinacea* L.), hop sedge (*Carex lupulina* Muhl. Ex Willd.), barnyard grass (*Echinochloa crus-galli* L. Beauv.), prairie cordgrass (*Spartina pectinata* Bosc ex Link), lady's thumb (*Polygonum amphibium* L.), and other native mesic species effectively removed 46% of the N and 2% of the P from effluent draining from tiled land growing corn and soybeans. Thus, herbage species were effective in cleansing N but not P from tiled land effluent.

Anthelmintic Properties

Forage species may have medicinal or behavioral effects which can be utilized to reduce dependence on synthetic worming agents. Increasing resistance to anthelmintics has encouraged a global search for alternatives (Niezen et al., 1996). Recent studies in NZ , Sweden, and the UK have shown that a range of forage and herb species, including chicory (*Cichorium intybus* L.), birdsfoot trefoil (*Lotus corniculatus* L.), lotus (*L. pedunculatus* Cav.), sulla (*Hedysarum coronarium* L.), meadow brome (*Bromus biebersteinii* Roem. & Schult.), dock (*Rumex obtusifolius* L.), and plantain (*Plantago lanceolata* L.) may act to reduce the degree of parasitic infestation in sheep (Moss and Vlassoff, 1993; Niezen et al., 1996, 1998; Robertson et al. 1995; Scales, Knight, and Saville, 1994;). In a 2-year NZ study, Fraser, Rowarth, and Knight (1997) reported that ram lambs grazing chicory, white clover (*Trifolium repens* L.), or birdsfoot trefoil grew faster and had a significantly lower adult nematode burden than lambs grazing perennial ryegrass or plantain. The anthelmintic properties of these species may involve: (a) secondary compounds (e.g., condensed tannins); (b) improved health of the livestock; or (c) a pasture canopy which morphologically limits larval development, survival, and mobility.

Windbreaks

Windbreaks can enhance water use efficiency and reduce physical damage from scouring of eroded soils. Bilbro and Fryrear (1997) compared the performance of rows of annual and perennial plants vs. slatted-fence in controlling wind erosion in Texas. When measured 10 m downwind, a single row of switchgrass (*Panicum virgatum* L.) reduced wind velocity to 39% of that upwind. In contrast, wind velocity downwind was 43% for two rows of slatted-fence, 48% for one row of slatted-fence, and 60% for 3 rows of grain sorghum (*Sorghum bicolor* L.). Thus, switchgrass may be used in place of slatted-fence for some wind abatement applications.

Steppuhn and Waddington (1996) monitored alfalfa yield and water conservation behind windbreaks created by double rows of tall wheatgrass (*Thinopyrum ponticum* L.) at 50 feet (15 m) intervals. Over a 7 year interval, alfalfa yield was 40% greater within the windbreak area than in adjoining open fields, not adjusted for the 8% area occupied by the wheatgrass. Water storage in the upper 1.2 m of soil was 78% greater in the windbreak area than in open fields. Thus, a hardy perennial utility species, tall wheatgrass, sufficiently improved the growing environment to improve yield in a valued hay species, alfalfa.

Biomass Plantations

Switchgrass may also have application as a source of biomass fuel. Vogel (1996) reviewed the historical replacement of pasture/hay land with grain crops in the US, and the associated increase in soil erosion. He presented evidence that 1.5 ac of switchgrass producing 6 to 7 t ac^{-1} could produce as much ethanol as 1.8 ac of 150 bu/ac corn. He estimated that 20 to 40 million ac of land could be sown to switchgrass biofuel plantations, essentially retiring grain crops from marginal land analogous to the Conservation Reserve Program but without large federal subsidies.

Vegetation Management

Grazing livestock can also enhance the environment by displacing herbicides for weed control. Baker and Raun (1989) noted that prior to WWII, sheep were used to graze weeds in corn. Wurtz (1995) reviewed the use of domestic geese as biological control agents in agriculture, including for Christmas tree plantations and tree nurseries. Reid (2002) reported on the use of free-range laying hens to weed commercial-scale raspberry (*Rubus idaeus* L.) and vegetable fields. The chickens rotated among 14 separately fenced paddocks. Raspberry fields were grazed virtually year-around, apart from in spring, to allow development of primal canes, and during the picking season from late June to mid August. Vegetable fields were grazed over the winter months, from November to April. The scratching and pecking action of chickens was particularly effective for both within-row weeding of the raspberries, eliminating a difficult and costly manual operation, and preparing the land for spring vegetables.

Cattle, sheep, and goats can also be used to graze out undesirable species during reforestation and on rangelands. Foster (1998) reviewed practical experience with employing grazing livestock to manage competing vegetation in forests. He referenced research starting in the 1950s, which proved commercially feasible in the western US starting in the early 1960s, and in B.C. since 1984. Experience in Ontario is more limited, starting in 1991.

The preference of most grazing livestock for herbaceous rather than woody vegetation is the key to releasing young trees from herbaceous weed pressure. Foster (1998) compiled a table of sheep grazing preferences in northeastern Ontario, identifying species considered to be high, medium, and low in palatability, as well as those unpalatable or poisonous. Species which are good candidates for removal by sheep include aster (*Aster* spp.), black bindweed (*Polygonum convolvulus* L.), honeysuckle (*Lonicera* spp.), and pin cherry (*Prunus pensylvanica* L.). Species of low palatability include balsam fir (*Abies balsamea* L. Mill.), blueberry (*Vaccinium* spp.), Labrador tea (*Ledum groenlandicum* Oeder), and sweetfern (*Comptonia peregrine* L. Coult.). Foster (1998) thoroughly discussed the managerial factors influencing the utility of grazers for vegetation management, including compliance with regulations, predator risks, and effective grazing management.

In sum, forages and the livestock that consume them have the potential to perform a range of services to society, in addition to food production. Some functions are remedial in nature, to rectify problems created by enterprise specialization and byproducts of modern society, while other applications channel natural processes to the service of humanity.

CONCLUSIONS

William McDonough stated that "regulations are an indication of design failure," and that the more we need to develop regulations, monitor for compliance, and test "to keep from killing each other too quickly," the more we should reconsider the fundamental design itself. By that definition, the design of specialized agriculture is fundamentally flawed and needs reconsideration. Ample evidence has been presented to support the thesis that specialization is ill-designed to meet the multiple needs of society for abundant, healthy foodstuffs, produced in accord with ecologically sound practices, combined in economically remunerative systems.

We are faced with two choices. We could continue to devote scarce scientific resources and government regulatory infrastructure to combat the diminished ecosystem integrity and collateral human losses caused by pursuing ecologically dysfunctional production systems. Alternatively, we could recognize that the root cause of these increasingly widespread problems is a fundamentally flawed system design and embrace a proven solution–the reintegration of crop and livestock agriculture.

REFERENCES

Baker, F.H. and N.S. Raun. (1989). The role and contributions of animals in alternative agricultural systems. *American Journal of Alternative Agriculture* 4(3/4):121-127.

Baldock, J.O. and R.B. Musgrave. (1980). Manure and mineral fertilizer effects in continuous and rotational crop sequences. *Agronomy Journal* 72:511-518.

Barbash, J.E., G.P. Thelin, D.W. Kolpin, and R.J. Gilliom. (2001). Major herbicides in ground water: Results from the National Water-Quality Assessment. *Journal of Environmental Quality* 30:831-845.

Barton, D.R. (1996). The use of Percent Model Affinity to assess the effects of agriculture on benthic invertebrate communities in headwater streams of southern Ontario, Canada. *Freshwater Biology* 36:397-410.

Beauchamp, E.G. and R.P. Voroney. (1994). Crop carbon contribution to the soil with different cropping and livestock systems. *Journal of Soil and Water Conservation* 49:205-209.

Benbrook, C.M. (1996). Pest *Management at the Crossroads*. Yonkers, NY: Consumers Union. 272 pp.

Benbrook, C. (1999). Evidence of the magnitude and consequences of the Roundup Ready™ yield drag from University-based varietal trials in 1998. AgBiotech InfoNet Technical Paper No. 1. (www.biotech-info.net/RR_yield_drag_98.pdf).

Bezdicek, D.F. and D. Granatstein. (1989). Crop rotation efficiencies and biological diversity in cropping systems. *American Journal of Alternative Agriculture*. 4:111-116.

Bilbro, J.D. and D.W. Fryrear. (1997). Comparative performance of forage sorghum, grain sorghum, kenaf, switchgrass, and slat-fence windbarriers in reducing wind velocity. *Journal of Soil and Water Conservation* 52:447-452.

Bird, E.A.R., G.L. Bultena, and J.C. Gardner. (1995). *Planting the Future: Developing an Agriculture that Sustains Land and Community*. Ames, IA: Iowa State University Press.

Breland, T.A. and R. Eltun. (1999). Soil microbial biomass and mineralization of carbon and nitrogen in ecological, integrated and conventional forage and arable cropping systems. *Biology and Fertility of Soils* 30:193-201.

Bromfield, L. (1947). *Malabar Farm*. Mattituck, NY: Aeonian Press.

Bulson, H.A.J., J.P. Welsh, C.E. Stopes, and L. Woodward. (1996). Agronomic viability and potential economic performance of three organic four year rotations without livestock, 1988-1995. *Aspects of Applied Biology* 47:277-286.

Clark, E. A. (2001). Diversity and stability in humid temperate pastures. In *Competition and Succession in Pastures*, eds. G. Tow and A. Lazenby, New York, NY: CAB International Publishing. pp. 103-118.

Cox, G.W. and M.D. Atkins. (1979). *Agricultural Ecology*. San Francisco, CA: W.H. Freeman and Co. 721 pp.

Davis, J.G., M. Young, and B. Ahnstedt. (1997). Soil characteristics of cropland fertilized with feedlot manure in the South Platte river basin of Colorado. *Journal of Soil and Water Conservation* 52:327-331.

Deng, S.P. and M.A. Tabatabai. (2000). Effect of cropping systems on nitrogen mineralization in soils. *Biology and Fertility of Soils* 31:211-218.

Deng, S.P., J.M. Moore, and M.A.Tabatabai. (2000). Characterization of active nitrogen pools in soils under different cropping systems. *Biology and Fertility of Soils* 32:302-309.

Dick, R.P. (1993). A review: Long-term effects of agricultural systems on soil biochemical and microbial parameters. In *Biotic Diversity in Agroecosystems*, eds. M.G. Paoletti and D. Pimentel, Amsterdam, The Netherlands: Elsevier Press. 356 pp.

Dinnes, D.L., D.L. Karlen, D.B. Jaynes, T.C. Kaspar, J.L. Hatfield, T.S. Colvin, and C.A. Cambardella. (2002). Nitrogen management strategies to reduce nitrate leaching in tile-drained midwestern soils. *Agronomy Journal* 94:153-171.

Drinkwater, L.E., D.K. Letourneau, F. Workneh, A.H.C. van Bruggen, and C. Shennan. (1995). Fundamental differences between conventional and organic tomato agroecosystems in California. *Ecological Applications* 5(4):1098-1112.

Duvick, D.N. (1989). Possible genetic causes of increased variability in U.S. maize yields. In *Variability in Grain Yields: Implications for Agricultural Research and Policy in Developing Countries*, eds. J.R. Anderson and P.B.R. Hazell, Baltimore, MD: Johns Hopkins University Press. pp. 147-156.

Eriksen, J. and L.S. Jensen. (2001). Soil respiration, nitrogen mineralization and uptake in barley following cultivation of grazed grasslands. *Biology and Fertility of Soils* 33:139-145.

Entry, J.A., R.K. Hubbard, J.E. Thies, and J.J. Fuhrmann. (2000a). The influence of vegetation in riparian filterstrips on coliform bacteria: I. Movement and survival in water. *Journal of Environmental Quality* 29:1206-1214.

Entry, J.A., R.K. Hubbard, J.E. Thies, and J.J. Fuhrmann. (2000b). The influence of vegetation in riparian filterstrips on coliform bacteria: II. Survival in soils. *Journal of Environmental Quality* 29:1215-1224.

Entz, M.H., W.J. Bullied, and F. Katepa-Mupondwa. (1995). Rotational benefits of forage crops in Canadian prairie cropping systems. *Journal of Production Agriculture* 8:521-529.

Foster, R.F. (1998). *Grazing Animals for Forest Vegetation Management*. NWST Technical Note TN-40. Vegetation Management Alternatives Program. Government of Ontario. 10 pp.

Fraser, T.J., J.S. Rowarth, and T.L. Knight. (1997). Pasture species effects on animal performance. ID No. 274. Proceedings of the XVIII International Grassland Congress, 8-19 June 1997, Winnipeg, MN.

Gordon, W.B., D.L. Fjell, and D.A. Whitney. (1997). Corn hybrid response to starter fertilizer in a no-tillage, dryland environment. *Journal of Production Agriculture* 10:401-404.

Gregson, M.E. (1996). Long-term trends in agricultural specialization in the United States: Some preliminary results. *Agricultural History* 70:90-101.

Haas, G., F. Wetterich, and U. Kopke. (2001). Comparing intensive, extensified, and organic grassland farming in southern Germany by process life cycle assessment. *Agricultural Ecosystems and Environment* 83:43-53.

Haynes, R.J. (1999). Size and activity of the soil microbial biomass under grass and arable management. *Biology and Fertility of Soils* 30:210-216.

Hopkins, A., E.A. Matkin, J.A. Ellis, and S. Peel. (1985). South-west England grassland survey 1983. 1. Age structure and sward composition of permanent and arable grassland and their relation to manageability, fertilizer nitrogen and other management features. *Grass and Forage Science* 40:349-359.

Ikerd, J.E. (1998). Sustainable agriculture, community development, and large-scale swine production. In *Pigs, Profits, and Rural Communities*, eds. K.M. Thu and E.P. Durrenberger, Albany, NY: State University of New York Press. pp. 157-169.

Karr, J.D., W.J. Showers, J. Wendell Gilliam, and A. Scott Andres. (2001). Tracing nitrate transport and environmental impact from intensive swine farming using Delta Nitrogen-15. *Journal of Environmental Quality* 30:1163-1175.

Kegley, S., S. Orme, and L. Neumeister. (2000). Hooked on Poison. Pesticide Use in California, 1991-1998. (www.panna.org/resources/documents/hookedAvail.dv.html).

Kovacic, D.A., M.B. David, L.E. Gentry, K.M. Starks, and R.A. Cooke. (2000). Effectiveness of constructed wetlands in reducing nitrogen and phosphorus export from agricultural tile drainage. *Journal of Environmental Quality* 29:1262-1274.

Langer, V. (2001). The potential of leys and short rotation coppice hedges as reservoirs for parasitoids of cereal aphids in organic agriculture. *Agriculture, Ecosystems, and Environment* 87:81-92.

Leitch, A. (1920). *Farm Management 1919. Part II. The Beef Raising Business in Western Ontario*. Ontario Department of Agriculture. Ontario Agricultural College. Bulletin 278. Toronto, ON.

Mellon, M., C. Benbrook, and K. Lutz Benbrook. (2001). *Hogging It: Estimates of Antimicrobial Abuse in Livestock*. Washington, DC: Union of Concerned Scientists.

Morrison, J. and A.A. Idle. (1972). A pilot survey of grassland in S.E. England, Grassland Research Institute. Technical Report No. 10; cited in J. Morrison (1979). Botanical change in agricultural grassland in Britain. In *Changes in Sward Composition and Productivity*, eds. A.H. Charles and R.J. Haggar, Proceedings of Occasional Symposium No. 10, Hurley, UK: British Grassland Society.

Moss, R.A., and A. Vlassoff. (1993). Effect of herbage species on gastro-intestinal roundworm populations and their distribution. *New Zealand Journal of Agricultural Research* 36:371-375.

Neeteson, J.J. (2000). Nitrogen and phosphorus management on Dutch dairy farms: Legislation and strategies employed to meet the regulations. *Biology and Fertility of Soils* 30:566-572.

Nelson, P.J. (1997). To hold the land: Soil erosion, agricultural scientists, and the development of conservation tillage techniques. *Agricultural History* 71:71-90.

Niezen, J.H., W.A.G. Charleston, J. Hodgson, A.D. Mackay, and D.M. Leathwick. (1996). Controlling internal parasites in grazing ruminants without recourse to anthelmintics: Approaches, experiences and prospects. *International Journal for Parasitology* 26:983-992.

Niezen, J.H., W.A.G. Charleston, J. Hodgson, C.M. Miller, T.S. Waghorn, and H.A. Robertson. (1998). Effect of plant species on the larvae of gastrointestinal nematodes which parasitise sheep. *International Journal for Parasitology* 28:791-803.

Nowlin, M. (1997). Point counterpoint: Regulating large hog lots. *Journal of Soil and Water Conservation* 52:314-37.

OMAFRA (Ontario Ministry of Agriculture, Food, and Rural Affairs). (1999) Farm Operators' Income from Farm and Off-Farm Sources, Ontario by Region, 1999. (www.gov.on.ca/OMAFRA/english/stats/finance/reginc99.html).

Pearson, C.J. and R.L. Ison (1987). *Agronomy of Grassland Systems*. New York, NY: Cambridge University Press. 169 pp.

Porter, W.P., J.W. Jaeger, and I.H. Carlson. (1999). Endocrine, immune, and behavioral effects of aldicarb (carbamate), atrazine (triazine) and nitrate (fertilizer) mixtures at groundwater concentrations. *Toxicology and Industrial Health* 15:133-150.

Posner, J.L., M.D. Casler, and J.O. Baldock. (1995). The Wisconsin integrated cropping systems trial: Combining agroecology with production agronomy. *American Journal of Alternative Agriculture* 10:99-107.

Pretty, J.N., C. Brett, D. Gee, R.E. Hine, C.F. Mason, J.I.L. Morison, H. Raven, M.D. Rayment, and G. van der Bijl. (2000). An assessment of the total external costs of UK agriculture. *Agricultural Systems* 65:113-136.

Rayburn, E.B. (1993). Potential ecological and environmental effects of pasture and BGH technology. In *The Dairy Debate*. ed. W.C. Liebhardt. Davis, CA: University of California SAREP. pp. 247-276.

Reganold, J.P. (1995). Soil quality and profitability of biodynamic and conventional farming systems: A review. *American Journal of Alternative Agriculture* 10:36-45.

Reid, F. (2002). Integrating layer chickens into a certified organic raspberry and vegetable farm. p. 80. In *Proceedings of the 14th IFOAM Organic World Congress*, 21-24 Aug 2002, Victoria, BC.

Robertson, H.A., J.H. Niezen, G.C. Waghorn, W.A.G. Charleston, and M. Jinlong (1995). The effect of six herbages on liveweight gain, wool growth and faecal egg count of parasitised ewe lambs. *Proceedings of the New Zealand Society of Animal Production* 55:199-210.

Russelle, M.P., J.F.S. Lamb, B.R. Montgomery, D.W. Elsenheimer, B.S. Miller, and C.P. Vance. (2001). Alfalfa rapidly remediates excess inorganic nitrogen at a fertilizer spill site. *Journal of Environmental Quality* 30:30-36.

Saviozzi, A., A. Biasci, R. Riffaldi, and R. Levi-Minzi. (1999). Long-term effects of farmyard manure and sewage sludge on some soil biochemical characteristics. *Biology and Fertility of Soils* 30:100-106.

Scales, G.H., T.L. Knight, and D.J. Saville. (1994). Effect of herbage species and feeding level on internal parasites and production performance of grazing lambs. *New Zealand Journal of Agricultural Research* 38:237-247.

Schilling, K.E. and R.D. Libra. (2000). The relationship of nitrate concentrations in streams to row crop land use in Iowa. *Journal of Environmental Quality* 29:1846-1851.

Schreinemachers, D.M. (2000). Cancer mortality in four northern wheat-producing states. *Environmental Health Perspectives* 108:873-881.

Sharpley, A.N. and S. Rekolainen. (1997). Phosphorus in agriculture and its environmental implications. In *Phosphorus Loss from Soil to Water*, eds. H. Tunney, O.T. Carton, P.C. Brookes, and A.E. Johnston, Wallingford, UK: CAB International. pp. 1-53.

Spedding, C.R.W., J.M. Walsingham, and A.M. Hoxey. (1981). *Biological Efficiency in Agriculture*, London, UK: Academic Press. 383 pp.

Staver, K.W. and R.B. Brinfield. (1990). Patterns of soil nitrate availability in corn production systems: Implications for reducing groundwater contamination. *Journal of Soil and Water Conservation* 45:318-322.

Steingraber, S. (1997). *Living Downstream: An Ecologist Looks at Cancer and the Environment*. Reading, MA: Addison Wesley Publishing. 357 pp.

Steppuhn, H. and J. Waddington. (1996). Conserving water and increasing alfalfa production using a tall wheatgrass windbreak system. *Journal of Soil and Water Conservation* 51:439-445.

Sturz, A.V., B.G. Matheson, W. Arsenault, J. Kimpinski, and B.R. Christie. (2001). Weeds as a source of plant growth promoting rhizobacteria in agricultural soils. *Canadian Journal of Microbiology* 47:1013-1024.

Taylor, D.C. and D.H. Rickerl. (1998). Feedlot manure nutrient loadings on South Dakota farmland. *American Journal of Alternative Agriculture* 13:61-68.

Troiano, J., D. Weaver, J. Marade, F. Spurlock, M. Pepple, C. Nordmark, and D. Bartkowiak. (2001). Summary of well water sampling in California to detect pesticide residues resulting from nonpoint-source applications. *Journal of Environmental Quality* 30:448-459.

Vogel, K.P. (1996). Energy production from forages (or American agriculture–back to the future). *Journal of Soil and Water Conservation* 51:137-139.

Weil, R.R., K.A. Lowell, and H.M. Shade. (1993). Effects of intensity of agronomic practices on a soil ecosystem. *American Journal of Alternative Agriculture* 8:5-14.

Whalen, J.K. and C. Chang. (2001). Phosphorus accumulation incultivated soils from long-term annual applications of cattle feedlot manure. *Journal of Environmental Quality* 30:229-237.

Wheeler, T.A., J.R. Gannaway, H.W. Kaufman, J.K. Dever, J.C. Mertley, and J.W. Keeling. (1997). Influence of tillage, seed quality, and fungicide seed treatments on cotton emergence and yield. *Journal of Production Agriculture* 10:394-400.

Wurtz, T.L. (1995). Domestic geese: Biological weed control in an agricultural setting. *Ecological Applications* 5:570-578.

Zahm, S.H. and A. Blair. (1992). Pesticides and non-Hodgkins' lymphoma. *Cancer Research* 52 (suppl):5485s-5488s.

Nitrogen Efficiency in Mixed Farming Systems

Egbert A. Lantinga
Gerard J. M. Oomen
Johannes B. Schiere

SUMMARY. The problem of nitrogen surpluses in Northwestern Europe is related in part to recent segregation of animal and crop production. A long-term solution can be found by re-integration of the main agricultural production components into mixed farming systems. In a new classification scheme of farming systems, high-input systems are placed in a sequence of modes in agriculture that each address sustainability problems in different ways. In this classification scheme, New-Conservation Agriculture (NCA) is considered to be a new mode of farming that aims to replace losses from the system, whilst not overloading it through critical use of non-renewable resources. Mixed farming systems that integrate crops and livestock are a typical example of NCA. The advantageous environmental features of mixed farming systems are illustrated by the favorable nitrogen balance of two experimental prototypes (a conventional and an organic one) where arable, dairy and sheep farming are integrated to a high degree. However, particularly

Egbert A. Lantinga (E-mail: Egbert.Lantinga@wur.nl) and Gerard J. M. Oomen are Associate Professors, Group of Biological Farming Systems, Wageningen University, Marijkeweg 22, 6709 PG Wageningen, The Netherlands.

Johannes B. Schiere is Research Officer, International Agricultural Centre, P.O. Box 88, 6700 AB Wageningen, The Netherlands.

[Haworth co-indexing entry note]: "Nitrogen Efficiency in Mixed Farming Systems." Lantinga, Egbert A., Gerard J. M. Oomen, and Johannes B. Schiere. Co-published simultaneously in *Journal of Crop Improvement* (Food Products Press, an imprint of The Haworth Press, Inc.) Vol. 12, No. 1/2 (#23/24), 2004, pp. 437-455; and: *New Dimensions in Agroecology* (ed: David Clements, and Anil Shrestha) Food Products Press, an imprint of The Haworth Press, Inc., 2004, pp. 437-455. Single or multiple copies of this article are available for a fee from The Haworth Document Delivery Service [1-800-HAWORTH, 9:00 a.m. - 5:00 p.m. (EST). E-mail address: docdelivery@haworthpress.com].

http://www.haworthpress.com/web/JCRIP

Digital Object Identifer: 10.1300/J411v12n01_07

the plowing of grass/clover swards caused serious problems on both farms regarding seedling survival and product quality in sugarbeet, maize, onion, and potato crops. This was due to the occurrence of large pest populations of leatherjackets (*Tipula paludosa*) and wireworms (Coleoptera: Elateridae).

KEYWORDS. Integrated, organic, conventional, new-conservation agriculture, nitrogen balance, nitrate leaching, grass/clover leys, leatherjackets, *Tipula paludosa*, wireworms, Elateridae

INTRODUCTION

Mixed farming systems occur in several forms (Schiere and Kater, 2001). While some consist of different components that function rather independently, others consist of components that function together by exchanging resources. These are called integrated mixed farming systems, the main topic of this paper. In it, the production of crops and livestock is integrated on the same farm or between farms at a regional level. Mixed farming systems cover world-wide about 2.5 billion ha of land, of which 1.1 billion ha are arable rain-fed crop land, 0.2 billion ha are irrigated cropland and 1.2 billion ha are grassland (FAO, 1996). Mixed farming systems produce 92% of the world's milk supply, all buffalo meat and approximately 70% of the sheep and goat meat. About half of the meat and milk produced in this system is produced in the OECD, Eastern Europe and the CIS, and the remainder comes from the developing world.

Mixed farming is probably the most benign agricultural production system from an environmental viewpoint because it is at least partially closed to external inputs except energy. It provides opportunities for recycling of nutrients and organic farming. It also provides opportunities for a varied landscape; i.e., mixed farming is the favorite system of many agriculturists and environmentalists.

From an agro-ecological viewpoint (Oomen et al., 1998), integrated mixed farming systems can:

- maintain soil fertility by recycling some nutrients;
- allow broad rotations with various crops, thus decreasing the risks of soil-borne pests and diseases, and weed invasions;

- maintain or improve soil biodiversity and soil organic matter content, and minimize soil erosion; and
- make the best use of crop residues.

A possible disadvantage of integrated mixed farming systems is the occurrence of soil-borne pests like wireworms (Coleoptera: Elateridae), leatherjackets (*Tipula paludosa* Meigen) and, especially on light soils, the nematodes *Meloidogyne fallax*, *M. chitwoodi*, and *Paratrichodorus teres* after plowing the leys (Halley and Soffe, 1988; Korthals, Brommer, and Molendijk, 1990).

Renaissance of Mixed Farming Systems in Developed Countries

Considering the developed countries, mixed farms have disappeared to a large extent within West-European agriculture during the last decades, but still dominate in regions such as the Canterbury Plains in New Zealand and North East Scotland in the UK. In The Netherlands, mixed farms were dominating until the early 1960s, but more recently they are generally considered an out-dated way of farming. However, over the last decades agricultural production systems have developed in Northwestern Europe which are suboptimal in agronomic and environmental terms (Lantinga and Rabbinge, 1997). A high degree of specialization, narrow crop rotations, and high external inputs of chemical fertilizers, biocides, and feedstuffs characterize them. This development has led to unacceptable environmental and ecological impacts, but also to negative economical and social effects (Rabbinge, 1992). Therefore, there is a need to develop and test new systems that decrease the undesirable effects of specialization and intensification and that utilize the symbiosis between several objectives. Therefore, a 'renaissance' of mixed farming systems on a farm or regional level is desirable (Oomen et al., 1998).

According to Nguyen, Haynes, and Goh (1995), mixed farming systems incorporate most aspects of sustainable agricultural systems as defined by Keeney (1990). An appropriate blend of arable fields and grass/clover leys can be the key to sustainable management of soils as well as to minimising nutrient surpluses. The grass/clover ley serves for fertility accumulation and improves soil structure. When the grass/clover sward is plowed for arable cropping, mineralization of soil organic matter occurs to provide nutrients for the subsequent crops.

The rate of change to such systems will, however, be restricted by the absence of structure and management skills for livestock in specialized arable areas and the large capital requirements for change (Wilkins, 2000). However, this can be overcome by establishing mixed farming between farms, e.g., with systems at a regional level (Bos and Van de Ven, 1999). They are characterized by intensive co-operation between two or more specialized farms, each producing crop or animal products.

This review first covers the historical background of the nitrogen (N) problem in Northwestern Europe. Secondly, it discusses a new classification scheme of farming systems and the place of mixed farming systems in it. Finally, the desirable perspectives of modern, mixed farming systems are compared to the less sustainable, specialized farms in Northwestern Europe by discussing the favorable N balance of two prototypes in The Netherlands where arable, dairy and sheep farming are integrated to a high degree.

HISTORICAL BACKGROUND OF THE NITROGEN PROBLEM IN THE NETHERLANDS

Nitrogen has always been one of the keys to achieve higher yields over the past centuries. In medieval Europe soil fertility could be maintained at a relatively low level by integrating livestock, arable cropping and fallow periods, e.g., the dehesa system and the Flemish three-course system (Lord Ernle, 1962; Slicher van Bath, 1963). In the Flemish system, the animals were kept in a system of zero grazing on deep litter. They were fed on crop residues and on crops that could fix atmospheric N (legumes) or mobilize soil phosphorus (Cruciferae). Another system was the infield/outfield system that uses animals to graze "outfields" distant from the village in order to concentrate part of the excreted nutrients on relatively small cropping areas around the villages, the "infields" (Thirsk, 1997; Schiere and Kater, 2001). Thus, in the infield/outfield system, a large area was mined to keep a small part more fertile. The system has existed in many forms throughout the world for many centuries, in pre-medieval Europe and in Russia until the early 20th century, and is still found today in many countries in the tropics. It can employ animals that are only stocked on the crop fields during the night while being grazed on the outfields during the daytime, it can be intensified by constructing special stables in which straw from the crop land, litter from the forest or even topsoil from the outfield are used to conserve the nutrients of the dung and urine from the animals. Moreover, the use of human manure ("nightsoil") on the crop lands was an additional way to maintain soil fertility by further recycling the nutrients.

Until the end of the 17th century, the main arable crops belonged to the so-called Neolithic Near East crop assembly consisting of barley (*Hordeum vulgare* L.), wheat (*Triticum aestivum* L.), oats (*Avena sativa* L.), and rye (*Secale cereale* L.) together with N-fixing legume crops like peas (*Pisum sativum* L.), lentils, and vetches (*Vicia* spp.) (Oomen et al., 1998). Animal manure, N deposition and biological fixation of N provided a moderate input of about 20 kg N ha^{-1}, just enough to maintain grain production at a level of about 1000 kg ha^{-1}. An ecological crisis due to diminishing amounts of available N developed after the Middle Ages due to a decrease in the area of fallow land (Kjaergaard, 1994). During the agricultural revolution in the 17th century, production was increased substantially with the introduction of the legume clover, e.g., in the mixed crop-livestock system of the Norfolk four-course system (Slicher van Bath, 1963). Clover was brought from Moorish Andalusia, which maintained the tradition from antiquity, via Christian Spain and regions under Spanish rule to the more northern parts of Europe. "The fertile Romantic 19th century landscape with red, white and green fields, with humming bees and endless herds of cattle was created by clover and the measures connected with clover" (Kjaergaard, 1995). Consequently, the areas of more demanding and more productive crops such as potato (*Solanum tuberosum* L.) and beet (*Beta vulgaris* L.) could be expanded.

In the beginning of the 20th century, many farms in Europe were still mixed farms. Animals were raised on farms that produced both food and fodder crops. After the Second World War, the cheap artificial fertilizers and concentrates from other continents induced specialization of farms and regions in Europe. During the 1950s, mixed farming systems also became more and more expensive to maintain in The Netherlands with the need for mechanization. Together with an increased use of external inputs this resulted in a fast increase in field and farm scale (Van der Ploeg and Van Dijk, 1995). This trend was amplified and continues due to increasing international trade. Entire infrastructures of supply firms, processing firms, trade structures and education, extension and research institutes were established around the regionally most lucrative branches of agriculture.

Nitrogen Use (In)efficiency

Intensification of Dutch agriculture following the introduction of the Common Agricultural Policy by the European Community in the 1950s was based on the use of large amounts of external inputs, especially artificial fertilizers and concentrates. It was an easy and profitable, and therefore common way to increase yields, albeit inefficient in terms of N use. For example, during the period 1950-1985, the use of external N inputs in the Dutch dairy farming sector increased by a factor of 8.5, but the output increased only by a factor of 2.5 (Van Keulen, Van der Meer, and De Boer, 1996). Input of artificial fertilizers increased drastically from about 50 to 290 kg N ha^{-1} yr^{-1}, while Lantinga and Groot (1996) showed that maximum production levels of predominantly grazed grassland is already achieved from fertilization with 200 kg N ha^{-1} yr^{-1}. This is supported by the data of Van Keulen, Van der Meer, and De Boer (1996), from which it can be calculated that the marginal N use efficiency, i.e., the fraction of the extra applied fertilizer N that is recovered in animal products, was only 0.05 over the period 1950-1985. Besides, Lantinga (2000) showed that grass/clover mixtures receiving only slurry can be just as productive as monocultures of heavily fertilized perennial ryegrass (*Lolium perenne* L.). High mineral N contents of slurry, improper manure handling after excretion and during storage (no rapid removal from the floor; uncovered storage units), inappropriate allocation techniques (surface spreading) and high soil mineral N contents due to high fertilizer rates resulted in heavy losses and a low efficiency of N use from the slurry, especially on sandy soils. Consequently, the apparent N recovery by forage crops from artificial fertilizer and manure together was below 50% in the mid 1980s (Van Bruchem, Bosch, and Oosting, 1996). Low N efficiency caused pollution of deep groundwater, eutrophication of lakes and surface waters, acidification, eutrophication of forests and emission of nitrous oxide.

Consequently, since the early 1990s, farmers have not been permitted to spread slurry on grassland in The Netherlands. Instead, slurry must be injected into the grass sward. On arable land, slurry must also be injected or plowed-in immediately after application. Hence, the level of ammonia (NH_3) volatilization from animal manure has been reduced by 30% between 1980 and 2000 (RIVM, 2001). However, since the N content in animal manure decreased by only 10% between 1984 and 2000 (RIVM, 2001), the potential for nitrate (NO_3) leaching to groundwater and surface waters has increased, as well as the possible emission of nitrous oxide.

To remain viable in the 21st century, the agricultural sector has been required to improve the efficiency of N use in a situation of greater environmental consciousness. This will be achieved by reducing N losses within separate production systems, as well as by using N output from one production system as input in another such as in integrated mixed farming systems. However, the N problem should be solved in such a way that social interests will not be jeopardized either. These interests include food supply for the European population, basic income and profit for farmers, employment in rural areas, an unpolluted environment, biodiversity in flora and fauna, an appropriate and attractive landscape, and welfare for humans and animals (Vereijken, 1997a).

CLASSIFICATION OF FARMING SYSTEMS

Mixed farming systems occur in different forms and the discussion about them can be based on a classification scheme of farming systems (FSs) as a framework to rethink of, e.g., the role of livestock in high-input agriculture. The classification scheme presented in Table 1 is elaborated by Schiere and De Wit (1995) and it explains what is meant by high-input systems by placing them in a sequence of modes in agriculture that address sustainability problems in different ways. It assumes that differences among FSs can be explained on the basis of relative ac-

TABLE 1. A classification of farming systems based on the production factors land, labor, and capital (based on Schiere and De Wit, 1995). EXPAGR = expansion agriculture; LEIA = low-external-input agriculture; HEIA = high-external-input agriculture; NCA = new-conservation agriculture.

Mode of farming	EXPAGR	LEIA	HEIA	NCA
Relative access to:				
– land	+	–	–	–
– labor	–	+	–	+/–
– capital	–	–	+	+/–
Characteristic issues related with soil fertility	Emphasis on throughput of nutrients, ultimately mining the outfields	Intensive circulation of nutrients within farms by relying on high labor inputs, ultimately exhausting the system	High throughput of nutrients causing local eutrophication and mining of phosphate and oil reserves	Emphasis on cycling of nutrients in the system and on 'precision farming'
Examples	• Slash and burn grain production • Infield/outfield cattle farming • Pastoralism/ ranching	• Rainfed grain farming • Stall feeding/use of crop residues/ collection of dung	• Green revolution grain farming • Greenhouse horticulture • Specialized dairy, industrial poultry, pigs, etc.	• Multiple cropping • Alley farming • Stall feeding • Legume-based farming • Mixed farming

cess to the resources land, labor, and capital. Land is considered as an aggregate of land quantity and quality, and labor is an aggregate of individual skills and numbers of persons. Access to capital is defined as access to inputs such as fertilizer and commercially compounded feeds.

Four modes of FSs are distinguished: Expansion Agriculture (EXPAGR), Low-External-Input Agriculture (LEIA), High-External-Input Agriculture (HEIA) and New-Conservation Agriculture (NCA). The classification in Table 1 shows that EXPAGR and HEIA systems are grounded on throughput, at the expense of reserves, also called biophysical capital (Giampietro, Cerretelli, and Pimentel, 1992). As a sharp contrast one can say that LEIA and NCA aim at circulation of nutrients. If EXPAGR exhausts local resources, the people can migrate to other areas or they can use nutrients from other areas by sending animals for grazing. Local and regional migration occurs as shifting cultivation and transhumance. LEIA typically exists where shortage of land cannot be overcome by migration and where there is no access to external inputs other than from solar radiation, deposition and soil weathering. This lack of access to external inputs implies that only increased use of labor and skills offers a way out. This in turn implies modified practices, where demand is adjusted to resource availability. If not managed properly, this can result in mining of soil and/or collapse of systems (Schiere and Grasman, 1997). HEIA has good access to the (external) resources, i.e., the goals can dictate the required resources. HEIA is based on the old German concept of conservation agriculturpe that aimed to replace nutrients lost from the soil. In due time, however, HEIA began to overcompensate the nutrient losses, resulting in the modern problems of waste disposal. Livestock production in HEIA is the equivalent of industrial manufacturing of pork, poultry and milk. Therefore, HEIA can lead to environmental problems, whereas LEIA leads to exhaustion of local outfields. This can result in the need for a new mode of farming that is called New-Conservation Agriculture (NCA). It applies the principle of Conservation Agriculture that aims to replace losses from the system but it stresses the need to avoid overloading of the system. Thus, it combines the positive elements of both LEIA and HEIA and is characterized by a critical use of non-renewable resources, recycling of resources, and reduction of losses. Mixed farming systems are a typical example of NCA (Table 1).

HEIA: The Reaction in The Netherlands

Government regulation and public opinion in The Netherlands since the mid-1980s have forced agriculture in general and animal production in particular into measures that were technically challenging and administratively difficult. The HEIA mode had become unsustainable and difficult choices had to be made that became even more difficult due to the output restrictions (quota) and market conditions that tend to pay low prices for the products. Dutch restrictions on nutrient use began by limiting phosphorus surpluses at farm level. Target levels for decreases in NH_3 emissions at a national scale and limits on N surpluses at farm level followed (RIVM, 2001). Eventually, the regulations led to complicated quota systems, farmers' protests, judicial problems and political haggling. Targets for permissible levels were difficult to enforce because of problems with measurements of NH_3 volatilization and NO_3 leaching at farm level (RIVM, 2001). The technical response by the livestock sector to these governmental measures included various options to improve performance of individual components of the system such as the use of other feeds containing less proteins, non-emission housing systems, remixing of crops and livestock, growing more feeds on farm, other ways of manure application and a ban on spreading of slurry outside the growing season. All of this served to re-introduce return of nutrients in the system, a shift to the NCA mode. Most of such technical measures were put successfully into practice on the experimental farms De Marke (Aarts et al., 1999) and De Minderhoudhoeve (Oomen et al., 1998).

These so-called 'linear' approaches resulted in drastic reductions in nutrient emissions but there were trade-offs due to system dynamics. Perhaps the most drastic non-linear response to the changed conditions was the emigration of farmers to countries such as Canada, Denmark, and Eastern Europe. Some farmers did shift to NCA, but even though the total number of ecological farms increased from less than 300 in 1986 to more than 1000 in 2001, this was still less than 2% of the total farmed area in The Netherlands. Other farmers have pursued precision agriculture, or perpetuated the intensive methods of HEIA by purchase of "manure production rights" or by arranging "manure delivery contracts" (Schiere and Van Keulen, 1999).

Institutional changes accompanied the search for other ways, including a shift from only reductionism to the inclusion of holistic approaches in which the behavior of the whole system was taken into account, and interactions among the sub-systems. Wageningen Agricultural University established the chairs of Animal Production Systems, Biological Farming Systems, and Plant Production Systems and began whole-farm research (Lantinga and Oomen, 1998; Oomen et al., 1998). This fits the tradition of, for example, De Marke (Aarts et al.,

1999) and earlier farms (Zadoks, 1989). Increasing use has been made of farmers' participation, where new ways are sought to solve problems by working with farmers (Vereijken, 1997b), and models are also employed (Van de Ven, 1996; Rossing et al., 1997). On the basis of these initiatives, many changes are now taking place in co-operation between legislators, scientists, extension agents and farmers, through iterative procedures between design of innovative farming systems and introduction in practice.

RESEARCH ON MIXED FARMING SYSTEMS: THE MINDERHOUDHOEVE PROJECT

The re-integration of animal and crop production in the NCA mode outlined above is studied at the Minderhoudhoeve, the experimental farm of Wageningen Agricultural University at Swifterbant in Oostelijk Flevoland, The Netherlands. The farm covers an area of 247 ha on a calcareous marine loam soil reclaimed from the sea about 45 years ago with a good water supply. A start was made in 1995 to set up on this farm two prototypes of mixed farming systems: a conventional and an organic one. The aim of this research project was not to compare the two farms but to study the development of sustainable farming systems through monitoring, analysis and adaptation. Within both the conventional and the organic system, three agricultural components were mixed and integrated to a high degree: dairy, sheep, and arable (including field-grown vegetables) farming. The expected advantages of such mixed farming systems were:

- reduction of external inputs and increased efficiency of their use through inclusion of home-grown, energy-rich feeds with a low N content (like corn (*Zea mays* L.) silage, whole-crop wheat silage, and fodder beets) and by-products like straw, potato waste, and beet pulp;
- more efficient nutrient use of animal manure through manipulation of composition (e.g., a high C:N ratio) resulting in a reduction of nutrient losses through volatilization, denitrification, and leaching;
- incorporation of short-term grasslands (up to 4 years) into the crop rotation which reduces the excessive accumulation of soil N typical for long-term grasslands resulting in a more efficient utilization of mineralized N after plowing (Whitmore, Bradbury, and Johnson, 1992);

- broadening the crop rotation which results in a decreased use of herbicides (suppression of weeds by grass/clover) and pesticides (fewer problems with some soil-borne pests and diseases);
- optimal use of legumes for biological N fixation; and
- a more even distribution of labor input and spreading of income risks.

Farm Profiles

The farming systems research at the Minderhoudhoeve has been used a total area of 225 ha of which 135 ha for the conventional farm and 90 ha for the organic farm. The most important objective of the conventional farm was to achieve an efficient and high production with minimal N surplus and biocide use per unit product, while maintaining high product quality. This aim is characteristic of the so-called globally-oriented agriculture where high productivity and efficiency are combined by making maximal use of the ecological processes that dictate the functioning of these agricultural systems (WRR, 1995; Rabbinge and Van Latesteijn, 1998). This type of agriculture allows the use of inputs such as fertilizers and biocides, but it aims to maximize their efficiency and effectiveness at the global scale. The organic farm has been developed in such a way that the farm acts as a model for a highly-productive, organic farm on a good soil with low emissions per unit area by closing all cycles as much as possible at the lowest level (locally-oriented agriculture). The organic farm disallows use of artificial N fertilizers and biocides and about 5% of the total area was used for natural habitat to establish an ecological infrastructure. The objective of the conventional farm (maximized efficiency and effectivity) and of the organic farm (maximized productivity without synthetic inputs) can produce complementary ideas. A good example of this has been the intensive use of legumes (which are indispensable in organic farming systems) on the conventional farm since 1996. The main characteristics of both mixed farms are given in Table 2.

Discussion of Minderhoudhoeve Results

Nitrogen Balance

The results from the period 1996-1998 show that on both mixed farms considerable reductions in N losses (about one-third; Table 3) have been realized compared to the period 1991-1993 when the Minderhoudhoeve consisted of two specialized farms (a dairy farm of 145 ha and an arable

TABLE 2. Main characteristics of the two integrated mixed farms at The Minderhoudhoeve.

Characteristic	Conventional farm	Organic farm
Total area (ha)	135	90
Crop rotation	1:6	1:7
Area fodder crops : food crops	45:55	55:45
Main fodder crops	grazed and ensiled grass/clover, whole-crop wheat silage	grazed and ensiled grass/clover, whole-crop oat silage
Main food crops	potatoes, wheat, barley, onions, peas, beans, oilseed rape, sugar beets	potatoes, wheat, onions, carrots, peas, sugar corn, white cabbage
Grazed grass/clover pastures in the crop rotation?	yes	No
Animals	90 dairy cows + young stock; 60 ewes + lambs	60 dairy cows + young stock; 40 ewes + lambs
Stall type for dairy cattle	cubicle houses	tie-up cowhouse
Fertilization	animal manure and artificial fertilizers	animal manure and compost
Animal manure	slurry	farmyard and liquid manure
Application of animal manure	slit-injection of all slurry in grass/clover swards (spring and summer)	slit-injection of liquid manure in grass/clover swards (spring) and plowing-in of farmyard manure in arable fields (autumn)
Milk production per cow (kg year^{-1})	8500	7500
Purchased concentrates (kg per kg of milk)	0.2	0.1
Sowing of grass/clover	after seed potatoes (late summer)	under cover crop (silage oat; spring)
Pest and disease management	crop rotation, pesticides	crop rotation
Weed control	crop rotation, first mechanical weeding, herbicides	crop rotation, mechanical weeding

farm of 90 ha). This was also a great improvement compared to the annual surplus of Dutch agriculture as a whole which amounted to about 350 kg N ha^{-1} in 1989 (Van Keulen, Van der Meer, and De Boer, 1996). It should be noted, however, that the farming activities on the Minderhoudhoeve were not representative for Dutch agriculture as a whole, since there is no pig, poultry, and veal production. From statistical data it can be derived that these animals produced an average of 100 kg N ha^{-1} yr^{-1} in the manure at the end of the 20th century, over 2 million ha of agricultural land (Van Keulen, Van der Meer, and De Boer,

TABLE 3. The average annual nitrogen balances (kg N ha^{-1} yr^{-1}) of the two mixed farms on the Minderhoudhoeve for 1996-98 compared with 1991-93 (two specialized farms on the same area, a dairy farm of 145 ha and an arable farm of 90 ha).

	Organic ('96-'98)	Conventional ('96-'98)	Minderhoudhoeve ('91-'93)
Input			
• Fertilizer	0	72	168
• Deposition	30	30	34
• Biological N fixation	78	38	8
• Compost	15	3	0
• Roughages and by-products	0	14	0
• Concentrates	27	34	33
	150	191	243
Output			
• Sold crops	35	57	63
• Sold milk	24	27	26
• Sold cattle	6	4	9
	65	88	98
Surplus (input-output)	85	103	148
Net accumulation in soil	−12	50	50
Environmental losses (surplus-accumulation)	97	53	98
Efficiency (output/input)	0.43	0.46	0.40
Surplus per unit output	1.31	1.17	1.51
Environmental losses per unit output	1.49	0.60	1.00

1996). On the conventional farm, only 45% of the total area was used for growing fodder crops, while on the organic farm this was 55% (Table 2), a percentage which is nearly representative of the Dutch average. Together with a higher crop yield per ha (~20%) on the conventional farm this explained the higher N output per ha by crops on this farm. The main constraint for high milk production per unit area on the organic farm was the restricted use of purchased concentrates in organic agriculture (Table 2).

On the conventional farm, the average N surplus including atmospheric deposition amounted to 103 kg N ha^{-1} yr^{-1} (Table 3). Between February 1996 and February 1999, the amount of total soil N in the upper 90 cm accumulated at an average rate of 50 kg N ha^{-1} yr^{-1}. This was despite the fact that nearly all of the old permanent grassland fields, which accounted for about 40% of the farm area, were plowed during the first three autumns. Consequently, N losses via NO_3 leaching, NH_3volatilization, and denitrification could be estimated as 53 kg N ha^{-1} yr^{-1}. Between February 1999 and February 2002, average soil N accumulation increased to 90 kg N ha^{-1} yr^{-1} on this farm. On the organic farm, the estimated losses to the environment were 97 kg N ha^{-1} yr^{-1} (Table 3). These higher losses can be associated with (i) the grassland history of the whole area on this farm leading to increased NO_3 leaching after plowing swards more than 20 years old, and (ii) the use of farmyard manure which resulted in higher levels of NH_3 volatilization than from slurry. The slurry produced on the conventional farm was slit-injected into the grass-clover swards and had a very high C:N-ratio of about 13 due to more fibrous and less-protein containing feed-stuffs than used previously, and the use of chopped straw in the cubicles. The basal feed of the dairy cows on this farm included grazed and ensiled grass-clover, whole crop wheat or corn silage, cereal straw, beet pulp, and brewer's grains. Wheat silage, corn silage, straw, and beet pulp were used to increase the C:N ratio of the slurry so as to reduce NH_3 emission, production of nitrous oxides and NO_3 leaching. On the organic farm, the input of N via biological fixation was very high in 1996 due to the large area of grass-clover. Between 1996 and 1998 the grass-clover area has been reduced gradually. On the conventional farm, the level of fertilizer N use has been reduced since 1996 due to a further replacement of grass monocultures by grass-clover mixtures, leading to an increased input via biological N-fixation.

N efficiency, defined as output divided by input, increased only slightly on both mixed farms (Table 3) since the reduced inputs were partly compensated by reduced outputs. When expressed per unit output, N losses to the environment decreased considerably on the conventional farm and increased considerably on the organic farm. On the conventional farm this was the result of severely reduced environmental losses per ha and only a slight reduction in output per ha. On the organic farm the environmental losses per ha remained unchanged, but N output per ha decreased by one-third. Nitrogen surplus per unit output decreased on both farms, especially on the conventional farm. Both N surplus and environmental losses of N per unit output are comparative indicators of N efficiency. Both values were lowest on the conventional farm (Table 3).

Nitrate Leaching

In two ditches within each of both farms the NO_3 concentration has been measured since the end of 1996. According to the Nitrate Directive of the European Union, surface freshwater or groundwater that is intended for drinking water may not contain more than 11.3 mg $NO_3\ l^{-1}$. It can be seen in Figure 1 that after some unallowable peaks during the winters 1996/97 and 1997/98, the NO_3 concentration in the ditch water on both farms was generally below this target from the autumn of 1998 onward. The initial peaks were in all probability caused by the plowing of old grassland paddocks, especially on the organic farm, which would have enabled a large increase in N mineralization and potential leaching of NO_3. From 1999, the average concentration of NO_3 in the ditch water of both farms remained below the target of the European Union, especially between autumn 2001 and spring 2002 (Table 4).

FIGURE 1. Concentration of NO_3 in the surface water of two internal ditches on the conventional mixed farm (C, K38/39 and K21/22) and the organic mixed farm (O, K24/25 and K25/26) between December 1996 (12/96) and March 1999. The EU-directive on NO_3 in water is a maximum concentration of 11.3 mg N l^{-1}.

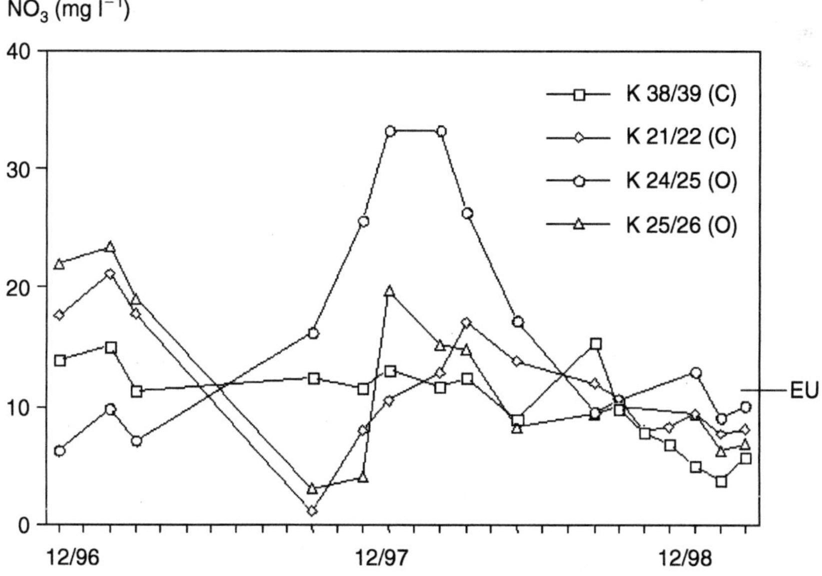

TABLE 4. Average concentration of NO_3 (mg l^{-1}) in the surface water of the internal ditches on the conventional mixed farm and the organic mixed farm.

	Conventional	Organic
Autumn 1999-spring 2000	9.2	8.6
Autumn 2000-spring 2001	8.0	10.0
Autumn 2001-spring 2002	4.5	3.5

Pests

Although the inclusion of short-term grass-clover swards in the rotation has shown to have many advantages (e.g., improvement of soil structure and fertility, suppression of weeds and some nematodes), a main disadvantage can be the occurrence of wireworms and leatherjackets in the first years after plowing. Relatively low populations of these larvae can already lead to poor establishment of seedlings, especially in crops like sugarbeets, corn and onion (*Allium cepa* L.). On potatoes, losses in yield due to wireworms are generally negligible but relatively low populations can cause serious economic loss due to the incidence of small round holes in the tubers. Although no concrete data were collected, these negative aspects were indeed observed on both farms. However, visual observations on the conventional farm revealed that damage due to leatherjackets and wireworms could be minimized when the grass/clover sward with all its insects and eggs was killed by spraying with glyphosate before plowing in autumn. In addition, since wireworm activity is highest in autumn this is unlikely to be a major problem on early harvested potato varieties.

CONCLUSIONS

Mixed farming offers an opportunity for environmentally friendly agriculture if it emphasizes the exchange of resources between its components, e.g., between crops and livestock. This paper discussed some different forms of mixing and the results of two prototypes at The Minderhoudhoeve in The Netherlands. The results demonstrated the perspectives of modern integrated mixed farming systems as a way towards sustainable agriculture in terms of reducing N surpluses and NO_3 leaching. Especially, the N losses to the environment were low on the conventional mixed farm, both per unit area and per unit output. These losses could be minimized due a restricted N surplus on the farm balance and the accumulation of N in the soil. This was the result of an adapted feeding strategy including more fibrous products with low protein content. Consequently, cattle slurry was produced with a very high C:N ratio of about 13. At the end of the 20th century, the average ratio on commercial dairy farms in The Netherlands was 6.5. On the organic farm, the environmental N losses were higher, certainly when expressed per unit output. This was due to a net release of N from the topsoil when most of the grassland swards of more than 20 years old were plowed, and to the production of farmyard manure instead of slurry (more NH_3 volatilization). However, at the end of the first three years of the experiment, the average NO_3 concentration in the surface of the ditches on both farms was on average below the Nitrate Directive of the European Union of 11.3 mg l^{-1} NO_3. During the second three years of the project, the concentrations remained below this target level.

Regarding pests, both farms had serious problems with wireworms and leatherjackets after plowing grass-clover swards. These larvae were especially detrimental to seedlings of sugarbeet, corn, and onion. Moreover, in some years the wireworms considerably reduced the market value of potatoes. Because insecticides to kill these larvae are no longer allowed in The Netherlands, an option for conventional mixed farms would be to kill the grass-clover mixture with all its insects and eggs by spraying glyphosate before plowing.

REFERENCES

Aarts, H.F.M., B. Habekotté, G.J. Hilhorst, G.J. Koskamp, G.J., F.C. van der Schans, and C.K. De Vries. (1999). Efficient resource management in dairy farming on sandy soil. *Netherlands Journal of Agricultural Science* 47:153-167.

Bos, J.F.F.P. and G.W.J. Van de Ven. (1999). Mixed specialized farming systems in Flevoland (The Netherlands): Agronomic, environmental and socio-economic effects. *Netherlands Journal of Agricultural Science* 47:185-200.

FAO (1996). Mixed farming systems and the environment: http://www.fao.org/docrep/x5303e/x5303e09.htm

Giampietro, M., G. Cerretelli, and D. Pimentel. (1992). Energy analysis of agricultural ecosystem environment: Human return and sustainability. *Agriculture, Ecosystems and Environment* 38:219-244.

Halley, R.J and R.J. Soffe, eds. (1988). *The Agricultural Notebook*, 18th Edition. Oxford, UK: Blackwell Scientific Publications, 689 pp.

Keeney, D. (1990). Sustainable agriculture: Definition and concepts. *Journal of Production Agriculture* 3:281-285.
Kjaergaard, T. (1994). *The Danish Revolution, 1500-1800; An Ecohistorical Interpretation.* Translated by David Hohnen. Cambridge, UK: Cambridge University Press.
Kjaergaard, T. (1995). Agricultural development and nitrogen supply from an historical point of view. *Biological Agriculture & Horticulture*, Vol. II:3-14.
Korthals, G.W., E. Brommer, and L.P.G. Molendijk. (1990). *Meloidogyne chitwoodi* and *Meloidogyne fallax* a threat to potato production? In *Proceedings of the Fourth World Potato Congress.* Wageningen, The Netherlands: Wageningen Pers, pp. 207-208.
Lantinga, E.A. (2000). Management and output of grass-clover swards in mixed farming systems. In *Grazing Management*, eds. A.J. Rook and P.D. Penning, Reading, UK: The British Grassland Society, pp. 241-242.
Lantinga, E.A. and J.C.J. Groot. (1996). Optimization of grassland production and herbage feed quality in an ecological context. In *Utilisation of Local Feed Resources in Dairy Cattle*, eds. A.F. Groen and J. Van Bruchem, EAAP Publication No. 84, Wageningen, The Netherlands: Wageningen Press, pp. 58-66.
Lantinga, E.A. and G.J.M. Oomen. (1998). The Minderhoudhoeve project: Development of an integrated and an ecological mixed farming system. In *Mixed Farming Systems in Europe*, eds. H. van Keulen, E.A. Lantinga, and H.H. van Laar, Wageningen, The Netherlands: APMinderhoudhoeve-reeks nr. 2, Wageningen Agricultural University, pp. 115-118.
Lantinga, E.A. and R. Rabbinge. (1997). The renaissance of mixed farming systems: A way towards sustainable agriculture. In *Gaseous Nitrogen Emissions from Grasslands*, eds. S.C. Jarvis and B.F. Pain, Wallingford, UK: CAB International, pp. 408-410.
Lord Ernle. (1961). *English Farming, Past and Present*, 6th Edition. London, UK: Frank Cass & Co., 559 pp.
Nguyen, M.L., R.J. Haynes, and K.M. Goh. (1995). Nutrient budgets and status in three pairs of conventional and alternative mixed cropping farms in Canterbury, New Zealand. *Agriculture, Ecosystems and Environment* 52:149-162.
Oomen, G.J.M., E.A. Lantinga, E.A. Goewie, and K.W. van der Hoek. (1998). Mixed farming systems as a way towards a more efficient use of nitrogen in European Union agriculture. *Environmental Pollution* 102, S1:697-704.
Rabbinge, R. (1992). Options for integrated agriculture in Europe. In *Proceedings of an International Conference Organized by the IOBC/WPRS*, Veldhoven, The Netherlands, eds. J.C. van Lenteren, A.K. Minks, and O.M.B. Ponti, Wageningen, The Netherlands: Pudoc Scientific Publishers, pp. 211-218.
Rabbinge, R. and H.C. Van Latesteijn. (1998). Sustainability, risk perception and the perspectives of mixed farming systems. In *Mixed Farming Systems in Europe*, eds. H. van Keulen, E.A. Lantinga, and H.H. van Laar, Wageningen, The Netherlands: APMinderhoudhoeve-reeks nr. 2, Wageningen Agricultural University, pp. 3-6.
RIVM (2001). http://www.rivm.nl/milieucompendium
Rossing, W.A.H., J.E. Jansma, F.J. de Ruijter, and J. Schans. (1997). Operationalizing sustainability: Exploring options for environmentally friendlier flower bulb production systems. *European Journal of Plant Pathology* 103:217-234.

Schiere, J.B. and J. de Wit. (1995). Livestock and farming systems research. II: Development and classifications. In J.B. Schiere, *Cattle, Straw and System Control, a Study of Straw Feeding Systems*. PhD Thesis, Wageningen, The Netherlands: Wageningen Agricultural University.

Schiere, J.B. and J. Grasman. (1997). Agro-ecosystem health: aggregation of systems in time and space. In *Proceedings of a Seminar on Agro-Ecosystem Health, Wageningen, 26 September 1996*, The Hague, The Netherlands: NRLO Report No. 97/31, pp. 22-36.

Schiere, J.B. and H. van Keulen. (1999). Harry Stobbs Memorial Lecture (1997). Rethinking high input systems of livestock production: A case study of nitrogen emissions in Dutch dairy farming. *Tropical Grasslands* 33:1-10.

Schiere, J.B. and L. Kater, (2001). *Mixed Crop-Livestock Farming: A Review of Traditional Technologies Based on Literature and Field Experiences*. Rome: FAO, 73 pp.

Slicher van Bath, B.H. (1963). *The Agrarian History of Western Europe, A.D. 500-1850*. Translated by O. Ordish. London, UK: Arnold, 364 pp.

Thirsk, J. (1997). *Alternative Agriculture, a History from the Black Death to the Present Day*. Oxford, UK: Oxford University Press, 365 pp.

Van Bruchem, J., M.W. Bosch, and S.J. Oosting. (1996). Nitrogen efficiency of grassland-based dairy farming–New perspectives using an integrated approach. In *Utilisation of Local Feed Resources in Dairy Cattle*, eds. A.F. Groen and J. Van Bruchem, EAAP Publication No. 84, Wageningen, The Netherlands: Wageningen Press, pp. 99-101.

Van der Ploeg, J.D. and G. van Dijk, G., eds. (1995). *Beyond Modernization*. Assen, The Netherlands: Van Gorcum.

Van de Ven, G.W.J. (1996). *A mathematical approach to comparing environmental and economic goals in dairy farming on sandy soils in The Netherlands*. PhD Thesis, Wageningen, The Netherlands: Wageningen Agricultural University.

Van Keulen, H., H.G. Van der Meer, and I.J.M. De Boer. (1996). Nutrient balances of livestock production systems in the Netherlands. In *Utilisation of Local Feed Resources in Dairy Cattle*, ed. A.F. Groen and J. Van Bruchem, EAAP Publication No. 84, Wageningen, The Netherlands: Wageningen Press, pp. 3-18.

Vereijken P., ed. (1997a). *Programming Study Multifunctional Agriculture*, Report 7-278hu, Vol. I, 90 pp. Wageningen, The Netherlands: AB-DLO (in Dutch).

Vereijken, P. (1997b). A methodical way of prototyping integrated and ecological farming systems (I/EAFS) in interaction with pilot farms. *European Journal of Agronomy* 7:37-43.

Whitmore, A.P., N.J. Bradbury, and P.A. Johnson, P.A. (1992). Potential contribution of ploughed grassland to nitrate leaching. *Agriculture, Ecosystems and Environment* 39:221-233.

Wilkins, R.J. (2000). Grassland in the twentieth century. In *IGER Innovations*, ed. T. Gordon, Aberystwyth, UK: IGER, pp. 26-33.

WRR (1995). *Sustained Risks: A Lasting Phenomenon*, Report No. 44, Netherlands Scientific Council for Government Policy, The Hague, The Netherlands: Sdu Uitgeverij.

Zadoks, J.C., ed. (1989). *Development of Farming Systems: Evaluation of the Five-Year Period 1980-1984*, Wageningen, The Netherlands: Pudoc.

Ecological Context for Examining the Effects of Transgenic Crops in Production Systems

Jennifer A. White
Jason P. Harmon
David A. Andow

SUMMARY. Ecological processes can have strong effects on production systems, many of which prevent crops from reaching their maximal yield. Past and present management strategies have been developed to mitigate the negative interactions, thereby improving crop production. However, when applied to complex and variable agroecosystems, some of these strategies have resulted in unintended ecological effects that ultimately hindered production. Transgenic crops are a widespread and powerful management option, and there is a tremendous need to understand their intended ecological effects as well as potential unintended direct and indirect effects. The purpose of this review is threefold: (1) to discuss four of the major ecological effects that limit crop production: competition, herbivory, disease, and abiotic stresses, (2) to describe how

Jennifer A. White (E-mail: whit0079@tc.umn.edu) is affiliated with the Department of Ecology, Evolution and Behavior, and Jason P. Harmon and David A. Andow are affiliated with the Department of Entomology and Center for Community Genetics, University of Minnesota, St. Paul, MN 55108.

The authors thank J. Hinton, E. Hladilek, M. Liebman, T. Stodola and two anonymous reviewers for constructive comments on the manuscript.

JW was funded by an EPA STAR fellowship.

[Haworth co-indexing entry note]: "Ecological Context for Examining the Effects of Transgenic Crops in Production Systems." White, Jennifer A., Jason P. Harmon, and David A. Andow. Co-published simultaneously in *Journal of Crop Improvement* (Food Products Press, an imprint of The Haworth Press, Inc.) Vol. 12, No. 1/2 (#23/24), 2004, pp. 457-489; and: *New Dimensions in Agroecology* (ed: David Clements, and Anil Shrestha) Food Products Press, an imprint of The Haworth Press, Inc., 2004, pp. 457-489. Single or multiple copies of this article are available for a fee from The Haworth Document Delivery Service [1-800-HAWORTH, 9:00 a.m. - 5:00 p.m. (EST). E-mail address: docdelivery@haworthpress.com].

http://www.haworthpress.com/web/JCRIP

Digital Object Identifer: 10.1300/J411v12n01_08

biotechnology is addressing these problems, and (3) to examine the ways these biotechnological solutions may cause ecological effects that unintentionally hinder crop production. We find that, to date, there has been little diversity in the types of transgenic crops available and the approaches they use to improve crop production. Transgenic crops, like previous agricultural technologies, are designed to enhance a singular plant trait in order to solve a specific production problem. When viewed in a simplified system, transgenic crops seem to provide effective means to mitigate negative ecological effects. However, when approached via a larger ecological context, it is clear that transgenic crops have already had and will continue to have unintended ecological effects that can ultimately affect crop production through mechanisms such as gene flow, resistance evolution, community interactions, and production practices. No management option, including transgenics, is universally beneficial or detrimental. The ecological context of specific agroecosystems may improve predictions of the benefits, limitations, and consequences of a given management tactic within that system. As we increase our understanding of the ecological context of crop production problems, we may be able to improve our control efforts to maximize production and minimize potential problems in the future.

KEYWORDS. Abiotic stress, Bt, disease, genetic engineering, herbicide-resistant crops, insect herbivores, salinity, transgenic crops, viruses, weeds

INTRODUCTION

Agricultural production and distribution processes must continue to be improved if we are to feed our global population in a sustainable manner. Advocates have argued that biotechnology will lead the next revolution in agricultural production, and substantial economic resources are being used to bring this vision about. Indeed, the biotechnology industry has used the issue of world hunger as a cornerstone of its public relations campaign. What is lost in all of this enthusiasm is an analysis of the cultural and ecological contexts into which these biotechnological innovations could fit. This problem is extremely complex, so here we focus specifically on the potential ecological effects of biotechnology on crop production.

The basis of any crop production system is the plant, which consumes light, water, and nutrients to produce a harvestable crop. An idealized, overly simplified production system might consist only of this crop and its resources, allowing yield to be maximized unfettered by any realistic constraints. As with all ecological systems, however, agroecosystems are not that simple. The crop and its resources are part of a complex web of interactions involving many species and an ever-changing abiotic environment. These ecological factors can have strong effects on the system, many of which prevent the crop from reaching its maximal yield. In particular, consumption by herbivores, competition with other plants, attack by pathogens, and abiotic stresses can hinder plant fitness or growth, reducing or even decimating crop production. While it is relatively easy to recognize these direct negative interactions, mitigating them is less straightforward.

The interconnectedness of an agroecosystem means that any perturbation may have many consequences beyond the intended effect. For example, suppose an exotic species of predator is introduced to kill an insect pest. Let us assume that the crop, pest, and exotic predator are the only species with strong ecological interactions relevant to the crop. We would expect that the predator would act directly to suppress the pest, thus indirectly helping the crop by being an enemy to the crop's enemy. However, if our assumption was incorrect and the exotic predator also consumed other natural enemies that had been important in controlling the pest, our introduction could actually harm the crop by disrupting pest control and allowing the pest to increase (Rosenheim, 1998). Alternatively, by killing one pest species, the exotic predator might release another pest from competition with the first species, potentially allowing the second pest to cause even greater damage to the crop (Sih et al., 1985). Moreover, if we consider evolutionary processes, an exotic predator is capable of exerting strong selection on a pest by killing many individuals in the pest population. By shifting the gene frequency in the surviving pests, the predator may select for pest individuals that are more resistant to the predator's attacks (Henter and Via, 1995). This would make the pest species more difficult to control and ultimately cause more damage to the crop. We can see from this example that managerial intervention that appears to work well under simplifying assumptions can fail under more complicated, realistic conditions, because they produce ecological chain reactions with unexpected indirect effects that ultimately harm, rather than help crop production.

The unexpectedness of these outcomes stems from our tendency to develop standardized technological solutions to crop production problems while ignoring their larger ecological and evolutionary context. Often this assumption proves inconsequential, and technologies work as planned. Occasionally, however, technologies fail to work solely as planned and unintended ecological effects occur. For example, extensive herbicide use has selected for herbicide-resistant weeds (Heap, 2001), pesticide application has caused outbreaks of other pest species (Pimentel et al., 1980), and widespread planting of homogeneous crops has encouraged susceptibility of the crop to disease (Ullstrup, 1972). When relatively simplistic technologies are overlaid on complex agroecosystems, many ecological interactions can be affected, and numerous outcomes become possible.

The advent of biotechnology, genetic engineering in particular, has given us new tools with which to try to improve crop production. Genetic engineering is the process of changing an organism's genetic makeup, usually by inserting specific genes from one species into another, resulting in novel traits for the recipient organism. The processes of genetic engineering to create crops with novel characteristics, often called transgenic crops, have the potential to improve agriculture as a whole. Certain transgenic crops are already quite common in commercial production. In 2000, transgenic crops [primarily corn (*Zea mays*) and soybean (*Glycine max*)] were grown on over 44 million ha, predominantly in the US (James, 2000; USDA NASS, 2001).

The purpose of current transgenic crops is fundamentally the same as that of their technological predecessors: to mitigate ecological factors limiting crop production, thereby increasing yield. There is no reason to expect that transgenic crops will be less likely to generate unexpected ecological consequences than their technological predecessors. Indeed, the very novelty of genetically engineered traits makes it even more difficult to predict their potential ecological effects (Marvier, 2001). Transgenic crops have been created to address complicated ecological problems in complex agroecosystems, but they are still based on simplifying assumptions about the ecological and evolutionary interactions of the system, leaving open the possibility of unintended consequences.

The purpose of this review is to: (1) discuss four of the major ecological effects that limit crop production: competition, herbivory, disease, and abiotic stresses; (2) describe how current biotechnology is addressing these problems; and (3) examine the ways biotechnology may unintentionally hinder crop production. In particular, we will look at how genetic variability and ecosystem complexity may interact with transgenic crops to produce unintended negative consequences for crop production. In doing so, we will elaborate on our theme that technological innovation in agriculture must be viewed from a broader ecological and evolutionary context. We will not attempt to estimate these ecological risks quantitatively, but will focus on identifying potential hazards to crop production. We will provide minimal attention to the effects of agricultural biotechnology on other ecosystems or organisms not directly related to crop production (see Snow and Moran Palma, 1997; Wolfenbarger and Phifer, 2000 for coverage). While these non-crop issues are important in their own right and will be crucial for determining how society uses biotechnology, we find in this review that additional analysis of transgenic crops is needed before we can be assured they will be beneficial for agriculture.

COMPETITION

Magnitude of the Problem

By creating an environment ideal for the growth of crop plants, agriculture also provides excellent habitat for other plant species. As long as there have been crops, there have been weeds, and agriculturalists have struggled to limit or eliminate the unwanted invaders. Uncontrolled, weeds can cause almost complete yield loss (Holm et al., 1977; Lacey, 1985). Despite control efforts, annual losses in the US have been estimated at over 4 billion US dollars (Bridges, 1992).

Competition is the primary ecological mechanism for crop losses. Weeds and crops compete for nutrients, water, space, and light. Hence, increases in weed biomass often directly translate into decreased crop biomass and subsequent yield losses (Ross and Lembi, 1999). For example, the perennial sedge *Cyperus rotundus* L. is one of the world's worst weeds, both for its widespread distribution (found in more than 90 countries, infesting more than 50 crops) and its tenacity (Holm et al., 1977). Because the cost of removal or control is often extremely high, entire fields have been abandoned to *C. rotundus*, rather than attempt its eradication (Holm et al., 1977). Other weeds are even more aggressive, either directly parasitizing crops (e.g., *Striga* spp.), or secreting allelopathic chemicals that are toxic to the crop (e.g., *Sorghum halepense* L. Pers. Friedman and Horowitz, 1970). In addition, weeds can act as reservoirs for disease and pest populations (Zimdahl, 1999). Moreover, weeds can interfere with harvest machinery and contaminate the harvested product, further hindering crop production (Ross and Lembi, 1999). It should be noted, however, that some weeds can play beneficial roles (Andow, 1988), such as reducing herbivore attack.

Biotechnology Applications

Plant breeding in general, and biotechnology in particular, offer the potential to improve crop competitive ability and diminish losses due to weeds. To date, development of transgenic crops has focused almost exclusively on herbicide resistant crops (HRCs) (Mazur and Falco, 1989; Dyer, 1994). As of 2001, more than 21 HRCs have been field tested in the US (USDA APHIS, 2001). While HRCs are easy to use and extremely effective, there is concern that HRCs will lead to even greater reliance on agrichemicals (Snow and Moran Palma, 1997) and detract from pursuit of more environmentally-friendly weed control solutions (Goldburg, 1992). A number of alternative weed control approaches using biotechnology have been suggested, such as improving crop competitive ability by improving morphological or nutrient acquisition characteristics (Gressel, 2000). These alternatives would diminish reliance on chemical treatments, but have only received minimal attention.

The majority of transgenic HRCs have been engineered to tolerate broad-spectrum herbicides, such as glyphosate (RoundUp™) or glufosinate (Liberty™), that are characterized by rapid degradation, low volatility, and low vertebrate toxicity (Baylis, 2000; Malik, Barry, and Kishore, 1989; Moorman and Keller, 1996). Glyphosate inhibits 5-enolpyruvyl-shikimate-3-phosphate synthase (EPSPS), an enzyme involved in the biosynthesis of essential amino acids. Most plants have a sensitive form of EPSPS, but a few microbes have EPSPS with decreased affinity for glyphosate. Transgenic "Roundup Ready™" (RR) crop varieties have been created by inserting a gene for EPSPS from one of these microbes (Padgette et al., 1996). Glufosinate resistance, in contrast, has been engineered by introducing an enzyme that inactivates the herbicide. The bar gene from *Streptomyces hygroscopius* codes for an enzyme that acetylates glufosinate and prevents its interference with glutamine sythatase, an essential enzyme used in plant nitrogen metabolism (Vasil, 1996). Herbicide resistant crops, particularly RR soybean and cotton (*Gossypium hirsutum* L.), have been widely adopted in the US, propelled largely by the low cost of glyphosate and simplified application relative to conventional herbicides (Gianessi and Carpenter, 2000). In 2001, 68% of US soybean (20.8 million ha) and 56% of upland cotton (3.6 million ha) were herbicide resistant (USDA NASS, 2001).

Potential Ecological Effects on Agricultural Production

Cultivation practices–A shift to HRCs and their associated broad spectrum herbicides has been predicted to reduce herbicide usage. This reduction may occur either by replacing mixtures of multiple herbicides for different weed species (Wilcut et al., 1996), or by allowing a shift to targeted post-emergence applications based on weed thresholds, rather than extensive prophylactic treatments (Burnside, 1992; Hess, 1996; Wilcut et al., 1996). By 1998, it appeared that increases in HR soybean acreage corresponded to a 9% decrease in application acres. That is, the average number of herbicide treatments per acre multiplied by the number of crop acres was lower in 1998 than 1995, despite a substantial increase in soybean acreage over this period (Gianessi and Carpenter, 2000). This reduction may have occurred because glyphosate replaced multiple herbicide mixtures, effectively decreasing the number of herbicide treatments. While the application acres decreased during this period, the average amount of active ingredient applied per acre increased (Gianessi and Carpenter, 2000), perhaps because glyphosate replaced herbicides with lower application rates. Thus, while it is clear that HRCs have shifted herbicide use in this system (potentially towards more environmentally benign chemicals), it remains to be seen whether total herbicide application rates will be substantially diminished by introduction of HRCs.

HRCs could indirectly benefit crop production by increasing the use of no-till agriculture thereby reducing erosion from row crops (Marshall, 1998; Wilcut et al., 1996). By relying on post-emergence herbicides, HRCs do not require tillage for weed control, so farmers could adopt no-till practices. It should be possible to determine if this has already happened with the widespread use of HR soybean and HR cotton, but definitive studies have not been conducted.

Gene flow–One of the greatest concerns with GM crops is the possibility that transgenes might introgress, or "escape," to a wild or cultivated relative via hybridization (Ellstrand, 1988; Hails, 2000; Snow and Moran Palma, 1997; Wolfenbarger and Phifer, 2000). Most major crops co-occur with wild relatives somewhere in the world (Keeler, Turner, and Bolick, 1996), and hybridization between them has frequently been documented (e.g., Mikkelsen, Andersen, and Jorgensen, 1996; reviewed in Ellstrand, Prentice, and Hancock, 1999; Jarvis and Hodgkin, 1999). Under the heavy selection pressure provided by cultivation, introgression can create wild varieties that mimic the crop, resulting in tenacious weeds that are difficult to eliminate (Barrett, 1983). For example, cultivated rice (*Oryza sativa* L.) fields are frequently infested with non-cultivated wild rice (also *Oryza sativa*) that is virtually indistinguishable from cultivated rice in the vegetative state (Dave, 1943; Holm, 1997). In India, a purple-leafed variety of cultivated rice was bred to allow easy discrimination from weeds, but within only a few years, purple-leafed races of the wild rice had developed (Dave, 1943; Harlan, 1975). Traditionally, such weedy relatives are often the most difficult to control chemically: their physiological similarity to the crop species makes it problematic to develop an effective herbicide that does not also destroy the crop. Transgenic HRCs can circumvent this problem, but introgression of the herbicide resistance genes to the weed would quickly diminish the effectiveness of the herbicide, thus negating the value of the transgenic crop (Dale, 1994). Given the large variety of environmental contexts into which transgenic crops will be introduced and the extreme selection pressure often exerted on weed populations, such transgene escape seems inevitable (Kareiva, Morris, and Jacobi, 1994).

HRCs themselves can also act as weeds, changing their intended role in the agroecosystem. Crops that reestablish themselves following harvest, known as volunteers, can be serious weeds in agricultural rotation. It was estimated that 22% of surveyed winter crops in the UK contained volunteers (Talbot, 1993). These volunteers can act as sources of inoculum for pathogens, and can carry soil borne disease through crop rotation in a field (Yarham and Gladders, 1993). Removal of HRC volunteers will be more difficult than their non-biotech counterparts. In particular, if subsequent crops in the rotation are resistant to the same herbicide, herbicide applications will be ineffective against the transgenic volunteers (Dyer, 1994; Marshall, 1998). Another concern is that cross-fertilization among different herbicide resistant varieties might create volunteers with multiple-resistance, a phenomenon already observed in canola (Hall et al., 2000). Other genetically engineered traits, such as insect resistance or increased stress tolerance, could also increase the competitive ability of volunteers, increasing their weediness within agricultural systems, or even allowing greater establishment of weedy feral populations.

Recent reports of transgenic DNA in native landraces of corn in Mexico (Quist and Chapela, 2001) highlight a further concern with gene flow: erosion of agricultural biodiversity. Genetic diversity within a crop species is essential for traditional breeding endeavors, and is considered to be vital for global food security (Hawkes, 1991; NRC, 1993). If transgenes confer a fitness advantage (or are neutral or even slightly deleterious), population genetics theory suggests that they and linked genetic elements should spread throughout the associated landraces. In the geographic centers of agricultural genetic diversity, such "genetic assimilation" would likely increase the homogeneity in parts of the genome within and among landraces, undercutting their adaptability and breeding potential. Genetic assimilation is less of a concern if the transgene has a substantial fitness cost because the transgene is less likely to spread through the entire population. However, it could spread into local landraces, reducing fitness and causing the extinction of the race. In some cases, any genetic contamination of non-GMO crops by the transgene can lead to crop losses. Presently crops produced organically can be down-graded if transgenes are detected. Some biotechnology proponents have countered that engineering crops that produce sterile seed (terminator technology) would prevent transgenic escape. Unfortunately, terminator technology would simultaneously prevent use of seed for subsequent plantings, forcing growers to purchase seed annually. While this would not be a large burden in developed areas where growers routinely purchase new hybrid varieties every year, the self-sufficiency of traditional farmers would be seriously threatened.

Resistance–Reliance on only a few herbicides may ultimately undercut HRC effectiveness because weeds can adapt or shift community composition towards more tolerant species (Baylis, 2000; Holt, 1994; Shaner, 2000). As of 2001, 252 resistant weed biotypes have been reported in 154 different species (Heap, 2001). Recently, there have been reports of weeds resistant to glyphosate (Powles et al., 1998; Lee and Ngim, 2000), even though it had previously been thought unlikely that weeds would be able to adapt to this herbicide (Bradshaw et al., 1997; Malik, Barry, and Kishore, 1989). It seems inevitable that other resistant and tolerant weed biotypes will become apparent. Once these biotypes become problematic, chemical mixtures and/or increased frequency and intensity of herbicide application will once again become prevalent, thus negating the potential advantages of HRCs and making it difficult to protect crops from weeds.

Community effects–Despite the negative effects of weed competition, weeds sometimes play a beneficial role in the agroecosystem. Weeds increase the vegetational diversity of agricultural fields, which in turn can decrease crop susceptibility to pest and disease outbreaks (Andow, 1988). The pronounced herbicide efficacy allowed by HRCs is capable of restricting weed densities well below economic thresholds (Radosevich, Ghersa, and Comstock, 1992), providing "near total weed control" (Malik, Barry, and Kishore, 1989). The near absence of weeds from agricultural landscapes could have important ecological consequences. Decreased plant diversity could potentially affect organisms dependent on these plants, such as beneficial generalist predators (Root, 1973), or seed-eating birds (Watkinson et al., 2000).

Case Study: Maize in Africa

Corn (*Zea mays* L.) is one of the most important crops in Africa. Three critical factors that limit its yield are lepidopteran stemborers, parasitic weeds in the genus *Striga*, and soil erosion. In Kenya, there are at least 4 stemborer species that can inflict significant damage, but the most widespread and important species are the spotted stemborer, *Chilo partellus* Swinhoe, and the maize stemborer, *Busseola fusca* Fuller (De Groote et al., 2001). Losses to stemborers probably average between 15 and 40% of the corn crop (Abate, van Huis, and Ampofo, 2000). *Striga* spp. are widespread in sub-Saharan Africa, but are particularly prevalent in infertile soils in drier regions (Ransom, 2000). The tiny seeds of this species can lay dormant in the soil for up to 20 years (Hosmani, 1995). Upon germination, the seedlings remain underground, attaching themselves to the root systems of nearby host plants, causing the hosts to funnel water, nutrients, and photosynthate to the parasite (Hosmani, 1995). *Striga* spp. can cause complete yield loss, and estimates of losses due to *Striga* range from 311 million to several billion US dollars annually (Abayo et al., 1995; Ransom, 2000). Corn is a particularly erosive crop in Africa, because it takes nearly 1 1/2 months before it can cover the ground, and most of the rain falls during this period. Moreover, it is often grown on hillsides where erosion rates can be extremely high, creating conditions favorable for *Striga*.

Transgenic varieties have been suggested as solutions to corn production problems. For the stemborers, varieties of *Bt* corn have been tested and found effective against most Kenyan species except *B. fusca* (Mugo and Poland, 2001). Additional Cry toxins have recently been commercialized that have not yet been tested, but it is expected that some combination of the commercially available Cry transgenes will provide adequate protection against the African stemborers. Two biotechnological solutions have been offered up to the *Striga* problem. First, crop plants could be engineered to be directly resistant to *Striga* attack (Verkleij and Kuiper, 2000). Many plant species are not susceptible to *Striga* (see below), so it is theoretically possible to transfer genes from a non-host species to the host that would render the host invulnerable to attack. The second solution is to use herbicide-resistant corn, so that herbicides can be used to kill the parasitic weeds (Abayo et al., 1995; Gressel, 1992). As long as the herbicide resistance mechanism does not involve metabolic degradation of the herbicide by the host, the host plant would shuttle undegraded herbicide to the parasite along with other metabolites, and the parasite would perish (Gressel, 1992). No transgenic solution to the erosion problem has been suggested. In each of the proposed solutions, transgenic varieties provide a way to cope with isolated environmental problems, but do not provide opportunities to improve the environment to eliminate the underlying causes of the problems.

An alternative solution is the "push-pull strategy," which relies on diverse plantings to reduce stemborers, *Striga*, and erosion (Khan et al., 1997a; Khan et al., 2000). The stemborers *B. fusca* and *C. partellus* are generalists that feed on many graminaceous host species (Khan et al., 1997b), but prefer Sudan grass (*Sorghum vulgare sudanense*), and Napier grass (*Pennisetum purpureum*) over corn (Khan et al., 1997b; 2000). Stemborer survival, however, is lower on these hosts, due to host plant physiology and/or increased attractiveness to parasitoids (Khan et al., 1997a). Furthermore, other plants, such as molasses grass (*Melinis minutiflora*) and *Desmodium* spp., repel ovipositing stemborers (Khan et al., 1997a; Khan et al., 1997b). Thus, intercropping molasses grass or *Desmodium* with corn would "push" the pests away from corn, and nearby Sudan grass could "pull" them away. The push-pull strategy can also confer protection against *Striga* when *Desmodium* is used as the intercrop with corn. *Desmodium* is not a host for *Striga*, and it also exerts an allelopathic effect to kill *Striga* (Khan et al., 2000). Corn fields intercropped with *Desmodium* spp. had vastly lower levels of *Striga* than corn monocultures, and corn yield was significantly increased relative to the monoculture (Khan et al., 2000). Some genetic engineers have proposed to isolate the genes from *Desmodium* and introduce them into corn to produce transgenic corn that suppresses *Striga* without herbicides. Finally, it must be noted that *Desmodium* are nitrogen-fixing legumes that improve soil fertility and provide early season cover, which should reduce erosion and eliminate conditions favorable for *Striga*. In contrast to transgenic solutions, the push-pull strategy offers a wide variety of benefits while providing opportunities to eliminate some of the underlying causes of the problems. Hence it is essential to consider the broader ecological context and alternative management strategies when assessing the merits of technological innovation in agriculture.

HERBIVORY

Magnitude of the Problem

Insect herbivores have always taken advantage of human efforts to promote the rapid growth of a handful of edible plant species. For example, plagues of locusts are described destroying crops in some of our earliest written documents (e.g., Exodus 10:11-15). Cotton-boll weevil (*Anthonomus grandis* Boheman) devastated the cotton industry in the Southern US in the early 20th century and "wreaked more destruction upon American agriculture than any other single insect" (Wagner, 1980). In many places weevil damage caused farmers to abandon their land in search for work in towns and cities (Fite, 1984) ultimately contributing to important social, economic, and agricultural changes. In the US, preharvest losses to insects have been estimated at ~13%, despite a billion dollar pesticide industry (Andow and Davis, 1989; Pimentel et al., 1980).

Direct consumption of plant biomass is the primary ecological mechanism of crop loss. Herbivores reduce the quantity or quality of crop yield by a variety of means. Herbivores such as corn earworm (*Helicoverpa zea* Boddie) infest the harvested portion of the plant, making it inedible or unmarketable. Other herbivores remove photosynthetic material, decreasing plant resources for allocation to the harvested product. Similarly, phloem feeders such as aphids can alter a plant's source-sink relationships, and divert resources away from harvested products. Other pests, such as corn rootworm (*Diabrotica* spp.) can cause structural damage to the host plant, making mechanical harvest difficult or impossible. Indirectly, herbivores such as the green peach aphid (*Myzus persicae* Sulzer) are also important vectors of disease, transmitting viruses from plant to plant contaminating vast acreage of crops.

Biotechnology Applications

Genetic engineering can help reduce herbivore pressure by enhancing defenses already existing within crops or by inserting transgenes with insecticidal traits. To date, however, only transgenes from *Bacillus thuringiensis* (*Bt*), a soil bacterium, have been used commercially in transgenic crops. Other transgenes, which are plant-derived, result in slower insect growth but little outright mortality and are not being commercialized (Gatehouse and Gatehouse, 1998; Hilder and Boulter,

1999). Consequently, commercial development has focused almost exclusively on acutely toxic *Bt* crops.

Strains of *Bt* produce a variety of different crystal (Cry) proteins that are toxic to different groups of insects. Usually an individual Cry protein is active against many insects within a single taxonomic order (e.g., beetles or flies). Cry proteins bind to receptors in the epithelium of an insect's midgut, causing lysis of the gut, cessation of feeding, digestive shutdown, and death from starvation within a few days (Peferoen, 1997). Commercial *Bt* crops (corn, cotton and potato) have a transgene with a shortened or activated form of a Cry toxin coupled with a constituitively expressing promoter, resulting in continuous production of transgenic protein in all plant tissues. The construct is inserted into the crop genome virtually at random, so different transformation events can vary in their expression of the transgenic product (Peferoen, 1997). For example, different varieties of commercially available transgenic *Bt* corn can originate from one of five different transformation events that differ substantially in quantity and tissue specificity of toxin production (Andow, 2001). Under field conditions, most transgenic *Bt* corn varieties give nearly complete control of the target pests, but some non-target pests can survive and damage the crop (Andow and Hutchison, 1998; Gould and Tabashnik, 1998). During 2000, *Bt* corn was grown on 5.8 million ha (19%) and *Bt* cotton was grown on 2.7 million ha (42%) in the US (USDA-NASS, 2001).

Potential Ecological Effects on Agricultural Production

Insecticide use–While the low survival rates of target pests are impressive, to evaluate the efficacy of a *Bt* crop, one must determine whether it is a viable alternative to other pest control options. Given the premium price growers pay for transgenic seed, transgenic crops need to provide substantially improved pest control, and/or substantially reduced costs (labor, chemical, environmental) relative to alternative strategies (Letourneau, Hagen, and Robinson, 2001). In Arizona, *Bt* cotton reduced insecticide and labor inputs for growers, providing equivalent yield for lower cost. In contrast, growers have received less benefit from *Bt* corn. Because the primary pests of corn are usually found deep within plant tissue, conventional insecticides are rarely used. Hence, *Bt* corn may prevent loss of yield, but has little effect on insecticide use (Letourneau, Hagen, and Robinson, 2001). Furthermore, transgenic crops are not effective against all insects attacking the crop. For example, in the southeastern US, herbivorous bugs can cause economically significant damage to cotton, but are not controlled by transgenic varieties (Turnipseed et al., 1995). Thus, insecticides can still be required, and use of *Bt* cotton has not resulted in clear reductions in insecticide use in this region.

Gene flow–If gene flow allows a weed to acquire an insecticidal trait, the prevalence and competitiveness of the weed might increase (Snow and Moran Palma, 1997). A number of studies have shown that herbivores reduce weed fitness (Letourneau, Hagen, and Robinson, 2001), thus insecticidal transgenes could release weeds from the negative effects of herbivory. It is even possible that these transgenes could enable an innocuous relative to undergo ecological release, increasing its weediness (Kareiva, Morris, and Jacobi, 1994; Schmitt and Linder, 1994). For example, the present distribution of St. John's wort (*Hypericum perforatum* L. Per.) is restricted to shady habitats where there is little herbivory (Shepherd, 1985), and ecological studies in these habitats would indicate that St. John's wort is controlled by something other than herbivory. Yet we know that this exotic, invasive plant was initially reduced by a herbivorous beetle, *Chrysolina quadrigemina* Suffr., in open habitats (Harris and Maw, 1981), allowing remnant St. John's wort populations to persist in the shade only because there were few beetles. Without knowing the biological control history of this system, it would not be evident that *C. quadrigemina* is the actual limiting factor. It is likely that many plant species are similarly limited in distribution by their herbivores, and it is possible that acquisition of an insecticidal trait may enable a previously innocuous plant to become a serious weed.

Resistance–One of the principle concerns with transgenic insecticidal crops is the potential for resistance to the insecticidal toxin to evolve in the targeted pest species. The history of resistance evolution to insecticides demonstrates that insects will evolve resistance to any toxin that is used sufficiently over a long enough period of time. Numerous insect species have evolved resistance to Cry toxins in *Bt* insecticidal sprays (Tabashnik, 1994). In all, more than 500 species of insects have become resistant to one or more conventional insecticides (Georghiou and Lagunes-Tejeda, 1991) often with disastrous consequences. The collapse of the cotton industry in northeastern Mexico caused by resistant tobacco budworm (*Heliothis virescens* Fabricius) is a dramatic example. During the 1950s and 1960s, more than 700,000 acres of cotton was grown in this area, but by 1970 the budworm was resistant to four classes of insecticide and less than 1,000 acres remained in production (Adkisson, 1969; 1972).

Transgenic insecticidal crops, with their constitutive expression and acute toxicity to targeted pests, will place pests under intense selection pressure and are probably more likely to encourage resistance evolution than conventional insecticides (Gould, 1998). Once resistance becomes widespread, *Bt* toxins will become ineffective, and transgenic *Bt* crops will have no production advantage. Furthermore, topical *Bt* sprays could also lose their efficacy against resistant pests, a disastrous outcome for farmers who rely on this as their primary pest control alternative (Mellon, 1998). Resistance evolution to transgenic insecticidal crops would ultimately limit pest management options, making it necessary to use less desirable options. Strategies to delay the evolution of resistance have been undertaken for many of these insecticidal crops to prolong the durability of this technology (Andow, 2001; Gould, 1998).

Community effects–Compared to broadly toxic conventional insecticides, *Bt* crops should have fewer adverse non-target effects. Conventional pesticides often negatively affect non-target arthropods (Croft, 1990). For example, insecticides used to control boll weevil from 1945 to 1970 destroyed many natural enemies in cotton that normally kept the tobacco budworm in check. The secondary pest outbreak that ensued elevated tobacco budworm from minor pest to destroyer of an entire industry (Pimentel et al., 1980). The specificity of *Bt* crops should reduce non-target effects relative to conventional pesticides.

Bt crops can cause non-target effects, however. The vast majority of non-target species remain untested, but lethal and sublethal responses to *Bt* crops have been reported for some natural enemy species (Hilbeck, 2001). Even subtle changes in life history parameters can profoundly affect predator/prey dynamics (Hassell, 2000), and it is certainly possible that negative effects on natural enemy populations could release some pests, negating some of the benefits of the insecticidal plant. In some instances *Bt* crops could even encourage secondary outbreaks. This could occur if a secondary pest is being controlled by direct competition with the primary pest (Denno, McClure, and Ott, 1995), a shared natural enemy of the primary pest (Abrams, 1987), or the previously used control tactic.

Bt crops might also affect the soil community in agricultural fields. Root exudates from *Bt* corn contain biologically active Cry toxins (Saxena, Flores, and Stotzky, 1999), and in some soil types these toxins may retain insecticidal activity for months (Stotzky, 2001). When tested against non-target species groups in the soil community (e.g., earthworms, bacteria, fungi), these Cry toxins had no measurable detrimental effect on overall density (Saxena and Stotzky, 2001; Stotzky, 2001), but shifts in community composition in response to Cry toxin may still occur under field conditions. Additionally, *Bt* corn decomposes more slowly than nontransgenic isolines (Stotzky, 2001). Slower decomposition may improve soil structure, reduce erosion and sequester carbon, but greater persistence of Cry toxin in the soil could increase the potential for adverse non-target effects (Stotzky, 2001).

Widespread adoption of *Bt* crops and their associated resistance management strategies will change the spatial arrangement and density of associated pest populations, which could have important consequences for natural enemy populations. For example, *Bt* corn probably will dramatically reduce corn borer populations. This in turn might make it difficult for specialist natural enemies to persist and many populations may be extirpated locally. Although growers using *Bt* corn may have little immediate need for these specialists, their loss could lead to increased corn borer infestations in non-*Bt* corn or other crops attacked by corn borers, and make control particularly difficult if corn borers evolve resistance to Cry toxin.

DISEASE

Magnitude of the Problem

There is tremendous variation in the types of plant pathogens, the ways in which they cause disease, and the types of damage they can inflict (Lucas, Campbell, and Lucas, 1985). Phytopathogenic viruses, bacteria, fungi, and nematodes inflict the vast majority of diseases in crop plants (Fuchs and Gonsalves, 1997; Kahl and Winter, 1995). Control of viral diseases has been problematic, because there are no chemical treatments that act directly against viruses in the field (Galun and Breiman, 1997) and natural sources of resistance are scarce (Galun and Breiman, 1997; Power, 2001). There are over 50,000 diseases of economic plants that cause billions of dollars in damage worldwide (Fuchs and Gonsalves, 1997; Lucas, Campbell, and Lucas, 1985). In the US alone, plant diseases cause annual losses that average around 15% of total agricultural production (Lucas, Campbell, and Lucas, 1985).

Disease epidemics, such as southern corn leaf blight, can cause massive yield loss, often with substantial social and economic consequences (see Strange, 1993; Scheffer, 1997 for reviews of plant disease epidemics). In 1970, approximately 85% of corn had been bred to exhibit cytoplasmic male sterility (CMS), a maternally inherited condition that restricts self-pollination in the production of hybrid seed (Ullstrup, 1972). CMS is caused by a membrane change in the mitochondria (Ward, 1995), which also leaves the mitochondria highly sensitive to pathotoxins produced by some strains of the fungus that causes Southern corn leaf blight (Leavings, 1990; Ward, 1995). A combination of weather conditions and widespread susceptibility of cultivars resulted in a major outbreak in 1970 that caused 1 billion US dollars in losses (Ullstrup, 1972). Thus, although CMS seemed to be unrelated to disease susceptibility, it triggered a major disease epidemic.

Biotechnology Applications

With the tremendous variation in pathogens and disease mechanisms, there are many potential targets for genetic engineering to address. Except for preliminary investigations into genetic engineering for resistance against bacteria (Lindow, Panopoulos, and McFarland, 1989), transgenic crops thus far have only been field tested for viruses. Two genetically engineered virus-resistant crops have been released for production (Fuchs and Gonsalves, 1997), and at least ten more have been developed and tested experimentally (Fuchs and Gonsalves, 1997; Kahl and Winter, 1995). Part of the reason for this focus on viruses may be the well developed techniques for engineering virus resistance, which are lacking for other pathogens (Fuchs and Gonsalves, 1997; Kahl and Winter, 1995).

Viral resistance is achieved using genes derived from the viruses themselves (Sanford and Johnston, 1985). This idea is derived from the traditional method of exposing plants to a mild strain of virus to provide the crop with cross-protection against other, more virulent, strains. In transgenic crops, a gene from the virus is inserted into the plant genome, thus increasing tolerance to subsequent viral exposure (Fuchs and Gonsalves, 1997; Galun and Breiman, 1997). Many viral sequences have been inserted in this fashion, including genes for coat proteins (Beachy, Loesch-Fries, and Tumer, 1990), replicase (Carr et al., 1994), and movement proteins (Lapidot et al., 1993), as well as non-coding regions of the viral genome (Zaccomer et al., 1993). The most commonly used viral sequences are coat protein genes. They are effective because a virus must shed its own coat protein before it can reproduce, and plants that produce viral coat protein can continually recoat the virus, inhibiting viral reproduction.

The first transgenic virus resistant crop to be released was a summer squash (*Cucurbita pepo* L.) expressing coat proteins of the zucchini yellow mosaic potyvirus and watermelon mosaic potyvirus 2 (Fuchs and Gonsalves, 1997). The only other commercially available transgenic virus-resistant crop is papaya (*Carica papaya* L.) that expresses coat protein to combat papaya ringspot virus (PRV) in Hawaii (Gonsalves, 1998). PRV is the most widespread and damaging virus infecting papaya and conventional methods of dealing with this pathogen have proven ineffective or extremely costly (Gonsalves, 1998). Transgenic papaya, in contrast, demonstrates an extremely high rate of resistance to some strains of PRV, good fruit production (Gonsalves, 1998), and has protected papaya in actual production systems in Hawaii.

Potential Ecological Effects on Agricultural Production

Gene flow–As with other transgenic traits, escape of disease resistance to related species could increase weed competition both within and outside of agriculture. In the case of transgenic squash, gene flow has already been documented between cultivated squash varieties and their wild relatives (Kirkpatrick and Wilson, 1988; Wilson, 1990). Therefore, it is reasonable to expect that genes for virus resistance can spread to wild squash plants. If viral diseases act as an ecological constraint for these species, then development of resistance could allow ecological release, change their population dynamics, and potentially increase their weediness or invasiveness.

There is also the possibility of transgene movement from the transgenic plant directly to other viruses via recombination, resulting in the creation of new viruses. Although many of these viral modifications will probably be inconsequential, significant changes in viral host range, virulence, or transmission characteristics could occur (Power, 2001). If viral fitness is increased, the new virus could act like an invading novel pathogen, ultimately harming crop production.

Resistance–As previously discussed, resistance can result in serious difficulties for pest control and crop production. Because viruses replicate rapidly and are not genetically uniform, they are capable of rapid adaptation (Elena et al., 2000). While the transgenic varieties may not harbor viral populations themselves, nearby crops or weeds could harbor large viral reservoirs. Transgenic virus-resistant crops could create strong selection pressure on these viral populations potentially causing production of viral strains that overcome plant resistance and render the transgenic construct useless (Salomon, 2000).

Community effects–Transgenic crops could alter plant-viral community interactions in a couple of ways. First, transgenic plants may increase the frequency of viral transcapsidation (de Zoeten, 1991), which in turn can enable a virus to infect a new host. Transcapsidation occurs in mixed infections of multiple viruses when one virus becomes encapsidated by the coat protein of a different virus. While this process does not alter the genetic makeup of the virus and causes no permanent change, the transcapsidated virus may have different transmission dynamics and vector specificity than the original virus (Rochow, 1977), and could potentially infect new hosts, including other crop species (Power, 2001). Second, viral transgenes could exacerbate viral problems in the original host plant. Viral synergism occurs when disease symptoms of mixed infections are worse than the additive effects of each virus on its own (Power, 2001). If transgenic products act synergistically with other viruses, the transgenic crop could be more susceptible to disease than its conventional counterparts.

ABIOTIC STRESSES

Magnitude of the Problem

Plant stresses result from a number of abiotic factors and ecological mechanisms that together play a significant role in limiting the development and production of crops worldwide. For example, excessive salt contamination has rendered about one-third of the world's irrigated land unsuitable for crops (Frommer, Ludewig, and Rentsch, 1999). High levels of contaminants, such as solubilized aluminum in acidic soils, are toxic to crops (de la Fuente et al., 1997). Cold stress limits the geographic areas where many crops can be successfully developed and at times causes extensive damage to crops in their normal production range, such as when unseasonably low temperatures caused 60 million US dollars in damage to cotton seedlings in 1980 (Christiansen and St. John, 1981; McKersie and Leshem, 1994). Excessive heat and drought can also have a tremendous effect on crop production, causing billions of dollars in losses during years of low rainfall (Anonymous, 1984).

The degradation of Mesopotamian agriculture is a stunning example of how abiotic stress can change and limit crop production. In the Tigris-Euphrates alluvial plain, Mesopotamians used irrigated agriculture as the foundation for at least eleven empires (Gelburd, 1993). The development of new irrigation technologies helped agriculture thrive and flourish in this era, but it also led to progressive increases in soil salinity and sedimentation that contributed to the demise of the civilization and left a problem that plagues the area to this day (Jacobsen and Adams, 1958). The opening of a canal from the Tigris River resulted in increased flooding, seepage, and over-irrigation, ultimately increasing soil salinity problems. The rise in soil salinity resulted in a shift from the less salt-tolerant wheat (*Triticum aestivum* L.) to the moderately tolerant barley (*Hordeum vulgare* L.) (Jacobsen and Adams, 1958). Increased salinity also led to decreased productivity. Despite an almost tripling of seeding rates (Gelburd, 1993), barley yields fell from about 2,500 L per ha in 2,400 B.C. to about 900 L per ha by 1700 B.C. (Jacobsen and Adams, 1958). While short-term practices and new technologies extended the productivity of the area for some time longer, they were unable to avoid the long-term consequences of the high salinity that unintentionally arose from their management practices and destroyed their agricultural productivity.

Biotechnology Applications

The development of transgenic anti-stress crops has been difficult (Kahl and Winter, 1995): none were commercially available in 2001, nor are any likely to become available in the near future. There are, however, numerous strategies being developed as attempts to produce crops tolerant of abiotic stresses (Kahl and Winter, 1995). Tolerance for high salinity has been observed in plants engineered with a bacterial gene to synthesize a sugar alcohol (Tarczynski, Jensen, and Bohnert, 1993), and by plants with an overexpression of a vacuolar antiport that compartmentalizes sodium ions (Apse et al., 1999; Zhang and Blumwald, 2001). Tobacco and papaya plants engineered to overexpress a citrate synthase gene were able to maintain root growth in solutions with high levels of aluminum (de la Fuente et al., 1997). Tobacco plants have also been engineered to survive exposure to high cadmium concentrations by producing a human protein (Chen and Gu, 1993), to resist drought by synthesizing trehalose (Holmstrom et al., 1996), and to resist heat by silencing the synthesis of a chloroplast membrane fatty acid (Murakami et al., 2000). Overexpression of cold regulated genes in transgenic *Arabidopsis* has resulted in increased resistance to cold tolerance (Jaglo-Ottosen et al., 1998).

Potential Ecological Effects on Agricultural Production

Gene flow–Because stress tolerant crops are only in the developmental stages, it is difficult to predict all of their unintended ecological consequences. However, many of the same ecological and evolutionary assumptions underlying other transgenic technologies hold for stress tolerant crops. In particular, gene flow to weeds could have important consequences if it removes a weed's limiting constraint (Dale, 1994; Schmitt and Linder, 1994). Environmental stresses limit the distributions of many weed species (Holm, 1997), suggesting that acquisition of transgenic stress tolerance traits could result in more vigorous weeds or allow them to invade new habitats.

Community effects–Improvement of stress tolerance could increase the range and/or extend the growing season of transgenic crops. This, in turn can have important effects on pests and pathogens of the crop. Extension of the growing season through irrigation, for example, has increased populations of the Asian stem borer (*Chilo agamemnon* Strand) in Israel, by allowing a greater number of pest generations per year, and allowing populations to build to higher numbers. The extended growing season also increased onion fly (*Delia antiqua* Meigen) populations, because more host plants were available at the insect's emergence from estivation (Rivnay, 1972). Likewise, dry season irrigation in Nicaragua has been credited with increasing incidence of maize stunt virus by providing continuous host availability for the disease vector, the corn leafhopper (*Dalbulus maidis* DeLong and Wolcott) (Hruska, Gladstone, and Obando, 1996). It seems likely that any trait that extends the growing season (e.g., water, heat or cold tolerance) has the potential to similarly affect the arthropod and pathogen communities. Furthermore, these traits might allow range expansions for crop species, resulting in many new associations between the transgenic plants and the surrounding biotic and abiotic communities. While many of these changes may be inconsequential, some of these associations could produce new direct and indirect interactions that ultimately harm crop production.

Case Study: Salt-Affected Lands in Australia

Salt accumulation in soil is one of the major causes of loss of agricultural land worldwide. Approximately 10% of the earth's land mass is salt-affected (Szabolcs, 1988), and it is estimated that more land is lost to salinization every year than is gained by clearing of forests (Frommer, Ludewig, and Rentsch, 1999). Most crop plants have low tolerance for salt, suffering osmotic stress, reduced leaf growth rate, inhibited cell expansion in growing tissues, and ultimately death under highly saline conditions (McKersie and Leshem, 1994; Wyn Jones, 1981). While many natural phenomena cause soil salinization (which are known collectively as primary salinization), secondary salinization due to human activity is also widespread. Irrigation, in particular, causes salinization by increasing the amount of water that percolates to subterranean aquifers, raising the water table in areas of poor drainage, and bringing salt to the surface (Hillel, 2000; Szabolcs, 1988).

The Murray-Darling Basin in Southeastern Australia is plagued by a multitude of salt accumulation problems (Ghassemi, Jakeman, and Nix, 1995). The area is naturally prone to salinization, but human settlement has exacerbated the problem. The water table has been rising at an estimated 10-20 cm per year, due to extensive irrigation and deforestation of the deep-rooted native vegetation. By 1990, 720,000 ha of land had a shallow water table, and an estimated 1,300,000 ha are estimated to be in this category by 2040 (Ghassemi, Jakeman, and Nix, 1995). Furthermore, deforestation and overgrazing have increased salinity problems in non-irrigated areas of the basin, allowing erosion to expose saline sub-soils and creating "scalds" incapable of supporting vegetation (Working Party on Dryland Salting in Australia, 1982). This sort of dryland salinity affects an estimated 100,000 ha in Victoria (Ghassemi, Jakeman, and Nix, 1995). The annual agricultural losses in the basin due to salinization have been estimated at 208 million US dollars (Ghassemi, Jakeman, and Nix, 1995).

Transgenic technology offers the potential to recoup some of these losses by creating plants more tolerant to salt, thus allowing agricultural practices to continue on salt-affected lands. Apse et al. (1999) engineered *Arabidopsis thaliana* (L.) Heynh. that overexpressed an Na^+/H^+ antiport protein, that transports and compartmentalizes sodium ions into vacuoles. These transgenic plants were capable of normal growth and development at 200 mM NaCl, a concentration that severely stunted growth of unmodified plants. This gene was subsequently transferred to tomato, producing plants that accumulate salt in their foliage but not in their fruit (Zhang and Blumwald, 2001), suggesting that feasible salt-tolerant transgenic crops could be developed to allow productive use of salt-affected lands. Furthermore, such salt-tolerant crops could potentially be irrigated with lower quality water that contains higher initial concentrations of salt, such as runoff from other irrigated fields (Rhoades, Kandiah, and Mashali, 1992). This sort of water re-use would increase irrigation efficiency, and conserve scarce water resources (Johnson, Revenga, and Echevarria, 2001; Rhoades, Kandiah, and Mashali, 1992).

It is clear, however, that transgenic salt tolerant crops would not provide a long-term solution to salinity problems. As illustrated by the ancient Mesopotamian example described in the text, switching to more tolerant crop varieties allows accommodation to increased salinity levels, but does not stop salt accumulation. Even highly tolerant transgenic crops have salt thresholds (Apse et al., 1999), and continued salt accumulation would eventually limit the growth of engineered species. Moreover, the same processes that salinize soil also tend to waterlog soil, produce sodic soils, and destroy soil structure by reducing porosity (Ghassemi, Jakeman, and Nix, 1995; Szabolcs, 1988). Salt-tolerance does not confer tolerance to these correlated inhospitable conditions. Thus, salt tolerant varieties might not lead to improved production on a large portion of salt-affected soils worldwide. Furthermore, salinization of agricultural lands has implications beyond the affected fields, due to contamination of water resources. In the Murray-Darling basin, approximately 30% of the salt load is estimated to be human-caused, and soil salinity is apparently increasing annually (Ghassemi, Jakeman, and Nix, 1995). Unless salt-accumulating species substantially reduced the salinity of effluent from agricultural lands, they will not improve water quality in this area, and brackish, non-potable water will continue to be a concern. Hence, we observe that transgenic varieties do not address the fundamental ecological causes for agricultural problems, but merely enable us to cope imperfectly with the present troublesome conditions.

CONCLUSIONS

To date, there has been little diversity in the types of transgenic crops available for the improvement of agricultural production. Traits for herbicide resistance and insect resistance for corn, soybean and cotton predominate. Compared to the promise of biotechnology, relatively little has come to fruition. While genetic engineering is in its infancy and additional innovations are expected in the coming years, herbicide resistance and insect resistance are expected to continue to dominate improvements in agricultural production in the near future (NRC, 2002).

Genetically engineered crops for the improvement of agricultural production have been developed to address complex ecological problems with simplified standardized solutions (e.g., glyphosate for weeds, *Bt* for insect pests, etc.). Standardization is necessary to ensure a sufficient market for the transgenic crop seed, and within the context of conventional plant breeding, such standardized solutions have led to substantial increases in agricultural production. Transgenic crops have a similar potential, but the very complexity of the ecological problems that transgenic crops are meant to address makes it increasingly unlikely that a standardized solution will be sufficient.

In the summary of a 1968 conference discussing the ecological costs of technological development, Farvar and Milton (1972) and Commoner (1972) concluded that the litany of ecological disasters could be attributed to one overriding factor: the reductionist bias of modern science. "... viewed through the blinders of reductionism, we see this inherently complex environment only in terms of separate parts. As a result, new technologies are designed, not to fit into the environment as a whole, but only to enhance a singular desired effect" (Commoner, 1972). Because these technologies are then implemented without full understanding of the system as an interacting whole, unforeseen ecological consequences inevitably accrue.

Genetically engineered transgenic crops are subject to the same shortcomings as the technologies described over 30 years ago. As we have shown, specific transgenic crops are designed only to enhance a singular desired effect to solve a specific crop production problem. When released into the reality of complex ecological systems, however, the narrow technological scope provided by these transgenic creations may result in unforeseen effects that ultimately hinder, rather than help crop production. This potential for harm has been recognized by governmental agencies, and attempts have been made to anticipate problems, regulate biotechnological releases, and monitor ecological interactions. However, transgenic crops are an unprecedented technological achievement, thus we lack a historical basis for predicting all of their potential effects. Furthermore, even when such a knowledge base exists, as with long-standing agricultural technologies, unanticipated ecological effects continue to occur.

For example, the recent increase in potato leafroll virus (PLRV) and potato virus Y infestation of potato (*Solanum tuberosum* L.) in the Red River Valley of Minnesota provides a striking example of surprising consequences to a technological fix. In the early 1990s, exotic strains of *Phytophthora infestans* (Mont.) de Bary, the causative agent of potato late blight, rapidly invaded the potato growing regions of North America (Goodwin et al., 1998). These strains are extremely virulent and resistant to metalaxyl, the only fungicide that had been available to treat infected plants (Fry and Goodwin, 1997; Goodwin, Sujkowski, and Fry, 1996). Outbreaks of disease caused by these new strains could be prevented only by extensive application of other fungicides before disease symptoms appeared. Such practices, however, greatly increased the total amount of fungicide applied (Kato et al., 1997). Unfortunately, these fungicides also kill certain entomopathogenic fungi that control populations of green peach aphid (GPA), *Myzus persicae* Sulzer. The decline in pathogenic fungi has likely contributed to the population outbreaks of GPA that started in the late 1990s (Lagnaoui and Radcliffe, 1997; Suranyi, 2000). Because GPA is an efficient vector of several potato diseases, increased GPA has resulted in increased incidence of PLRV and potato virus Y. Thus, a technological shift, such as a change in fungicide practices, can unintentionally contribute to a wide variety of consequences, including outbreaks of aphids, increased potato viral disease, and ultimately the decimation of the seed potato industry in the region.

As with their technological predecessors, transgenic crops have vast potential to provide simple solutions to the ecological limitations of crop production. However, numerous potential problems can develop as a result of these actions. It is impossible to predict the full magnitude or even the likelihood of these complicated ecological consequences across all current and potential agroecosystems. No management option, including transgenics, is universally beneficial or detrimental. However, we can use the ecological context of specific agroecosystems to make better predictions of the benefits, limitations, and consequences of a given management tactic within that system. As we increase our understanding of the ecological context of crop production problems, we may be able to improve our control efforts to maximize production and minimize potential problems in the future.

REFERENCES

Abate, T., A. van Huis, and J. K. O. Ampofo. (2000). Pest management strategies in traditional agriculture: An African perspective. *Annual Review of Entomology* 45: 631-659.

Abayo, G. O., T. English, R. E. Eplee, F. K. Kanampiu, J. K. Ransom, and J. Gressel. (1995). Control of parasitic witchweeds (*Striga* spp.) on corn (*Zea mays*) resistant to acetolactate synthase inhibitors. *Weed Science* 46:459-466.

Abrams, P. A. (1987). Indirect interactions between species that share a predator: Varieties of indirect effects. In *Predation: Direct and Indirect Impacts on Aquatic Communities*, eds. W. C. Kerfoot and A. Sih, Hanover, NH: University Press of New England, pp. 38-54.

Adkisson, P. L. (1969). How insects damage crops. In *How Crops Grow–A Century Later. Connecticut Agricultural Experiment Station Bulletin*, pp. 155-164.

Adkisson, P. L. (1972). The integrated control of the insect pests of cotton. *Proceedings of the Tall Timbers Conference of Ecological Animal Control by Habitat Management* 4:175-188.

Andow, D. and W. Hutchison. (1998). Bt-corn resistance management. In *Now or Never: Serious New Plans to Save a Natural Pest Control*, eds. M. Mellon and J. Rissler, Cambridge, MA: Union of Concerned Scientists.

Andow, D. A. (1988). Management of weeds for insect manipulation in agroecosystems. In *Weed Management in Agroecosystems: Ecological Approaches*, eds. M. A. Altieri and M. Liebman, Boca Raton, Florida: CRC Press, pp. 265-301.

Andow, D. A. (2001). Resisting resistance to Bt corn. In *Genetically Engineered Organisms: Assessing Environmental and Human Health Effects*, eds. D. K. Letourneau and B. E. Burrows, Boca Raton, FL: CRC Press, pp. 99-124.

Andow, D. A. and D. P. Dasvis. (1989). Agricultural chemicals: Food and environment. In *Food and Natural Resources*, eds. D. S. Pimentel and C. W. Hall: Academic Press, pp. 191-234.

Anonymous. (1984). Drought: Blight in fertilizer demand–Southern Africa, Peru, Australia. *Fertilizer International* 181:11-13.

Apse, M. P., G. S. Aharon, W. A. Snedden, and E. Blumwald. (1999). Salt tolerance by overexpression of a vacuolar Na+/H+ antiport in *Arabidopsis*. *Science* 285:1256-1258.

Barrett, C. H. (1983). Crop mimicry in weeds. *Economic Botany* 37:255-282.

Baylis, A. D. (2000). Why glyphosate is a global herbicide: Strengths, weaknesses and prospects. *Pest Management Science* 56:299-308.

Beachy, R. N., S. Loesch-Fries, and N. E. Tumer. (1990). Coat protein-mediated resistance against virus infection. *Annual Review of Phytopathology* 28:451-474.

Bradshaw, L. D., S. R. Padgette, S. L. Kimball, and B. H. Wells. (1997). Perspectives on glyphosate resistance. *Weed Technology* 11:189-198.

Bridges, D. C. (1992). *Crop Losses Due to Weeds in the United States*. Champaign, IL: Weed Science Society of America.

Burnside, O. C. (1992). Rationale for developing herbicide-resistant crops. *Weed Technology* 6:621-625.

Carr, J. P., G. O. Amit, P. Palukaits, and M. Zaitlin. (1994). Replicase mediated resistance to cucumber mosaic virus in transgenic plants involves suppression of virus replication in the inoculated leaves and long distance movement. *Virology* 199: 439-447.

Chen, Z. and H. Gu. (1993). Plant biotechnology in China. *Science* 262:377-378.

Christiansen, M. N. and J. B. St. John. (1981). Plant function and symptoms of chilling injury. In *Analysis and Improvement of Plant Cold Hardiness*, eds. C. R. Olien and M. N. Smith, Boca Raton, FL, USA: CRC Press.

Commoner, B. (1972). Summary of the conference. In *The Careless Technology: Ecology and International Development*, eds. M. T. Farvar and J. P. Milton, Garden City, NY: The Natural History Press, pp. xxi-xv.

Croft, B. A. (1990). *Arthropod Biological Control Agents and Pesticides*. New York, NY: John Wiley and Sons.

Dale, P. J. (1994). The impact of hybrids between genetically modified crop plants and their related species: General considerations. *Molecular Ecology* 3:31-36.

Dave, B. B. (1943). The wild rice problem in the central provinces and its solution. *Indian Journal of Agricultural Science* 13:46-53.

De Groote, H., W. A. Overholt, L. Macopiyo, and S. Mugo. (2001). Guiding technology development through a GIS-based ex ante impact assessment model: The case of insect resistant maize in Kenya. *IRMA Updates* 2:5.

de la Fuente, J. M., V. Ramirez-Rodriguez, J. L. Cabera-Ponce, and L. Herrera-Estrella. (1997). Aluminum tolerance in transgenic plants by alteration of citrate synthesis. *Science* 276:1566-1568.

de Zoeten, G. A. (1991). Risk assessment: Do we let history repeat itself? *Phytopathology* 81:585-586.

Denno, R. F., M. S. McClure, and J. R. Ott. (1995). Interspecific interactions in phytophagous insects: Competition reexamined and resurrected. *Annual Review of Entomology* 40:297-331.

Dyer, W. E. (1994). Herbicide-resistant crops: A weed scientist's perspective. *Phytoprotection* 75S:71-77.

Elena, S. F., R. Miralles, J. M. Cuevas, P. E. Turner, and A. Moya. (2000). The two faces of mutation: Extinction and adaptation in RNA viruses. *IUBMB Life* 49:5-9.

Ellstrand, N. C. (1988). Pollen as a vehicle for the escape of engineered genes? *Trends in Ecology and Evolution* 3:S30-S31.

Ellstrand, N. C., H. C. Prentice, and J. F. Hancock. (1999). Gene flow and introgression from domesticated plants into their wild relatives. *Annual Review of Ecology and Systematics* 30:539-563.

Farvar, M. T. and J. P. Milton, eds. (1972). *The Careless Technology: Ecology and International Development.* Garden City, NY: The Natural History Press.

Fite, G. C. (1984). *Cotton Fields No More: Southern Agriculture 1865-1980.* Lexington, KY USA: The University Press of Kentucky.

Friedman, T. and M. Horowitz. (1970). Phytotoxicity of subterranean residues of three perennial weeds. *Weed Research* 10:382-385.

Frommer, W. F., U. Ludewig, and D. Rentsch. (1999). Taking transgenic plants with a pinch of salt. *Science* 285:1222-1223.

Fry, W. E. and S. B. Goodwin. (1997). Resurgence of the Irish potato famine fungus. *BioScience* 47:363-371.

Fuchs, M. and D. Gonsalves. (1997). Genetic engineering. In *Environmentally Safe Approaches to Crop Disease Control*, eds. N. A. Rechcigl and J. E. Rechcigl, Boca Raton, FL: CRC Press, pp. 333-368.

Galun, E. and A. Breiman. (1997). *Transgenic Plants.* London, UK: Imperial College Press.

Gatehouse, A. M. R. and J. A. Gatehouse. (1998). Identifying proteins with insecticidal activity: Use of encoding genes to produce insect-resistant transgenic crops. *Pesticide Science* 52:165-175.

Gelburd, D. E. (1993). Is our future written in the past? A lesson from Mesopotamia. In *Culture and Environment: A Fragile Coexistence. Proceedings of the Archaeological Association of the University of Calgary*, eds. R. W. Jamieson, S. Anbonyi and N. Mirau, Calgary, AB, Canada: University of Calgary, pp. 95-98.

Georghiou, G. P. and A. Lagunes-Tejeda. (1991). *The Occurrence of Resistance to Pesticides in Arthropods.* Rome: Food and Agriculture Organization.

Ghassemi, F., A. J. Jakeman, and H. A. Nix. (1995). *Salinisation of Land and Water Resources.* Wallingford, UK: CAB International.

Gianessi, L. P., and J. E. Carpenter. (2000). Agricultural biotechnology: Benefits of transgenic soybeans. Available www.ncfap.org/reports/biotech/rrsoybeanbenefits.pdf

Goldburg, R. J. (1992). Environmental concerns with the development of herbicide-tolerant plants. *Weed Technology* 6:647-652.

Gonsalves, D. (1998). Control of papaya ringspot virus in papaya: A case study. *Annual Review of Phytopathology* 36:415-437.

Goodwin, S. B., C. D. Smart, R. W. Sandrock, K. L. Deahl, Z. K. Punja, and W. E. Fry. (1998). Genetic change within populations of *Phytophthora infestans* in the United States and Canada during 1994 to 1996: Role of migration and recombination. *Phytopathology* 88:939-949.

Goodwin, S. B., L. S. Sujkowski, and W. E. Fry. (1996). Widespread distribution and probable origin of resistance to metalaxyl in clonal genotypes of *Phytophthora infestans* in the United States and western Canada. *Phytopathology* 86:793-800.

Gould, F. (1998). Sustainability of transgenic insecticidal cultivars: Integrating pest genetics and ecology. *Annual Review of Entomology* 43:701-726.

Gould, F. and B. E. Tabashnik. (1998). Bt-cotton resistance management. In *Now or Never: Serious New Plans to Save a Natural Pest Control*, eds. M. Mellon and J. Rissler, Cambridge, MA: Union of Concerned Scientists.

Gressel, J. (1992). Addressing real weed science needs with innovations. *Weed Technology* 6:509-525.

Gressel, J. (2000). Molecular biology of weed control. *Transgenic Research* 9:355-382.

Hails, R. S. (2000). Genetically modified plants–the debate continues. *Trends in Ecology and Evolution* 15:14-18.

Hall, L., K. Topinka, J. Huffman, L. Davis, and A. Good. (2000). Pollen flow between herbicide-resistant *Brassica napus* is the cause of multiple-resistant *B. napus* volunteers. *Weed Science* 48:688-694.

Harlan, J. R. (1975). *Crops and Man*. Madison, WI: American Society of Agronomy and the Crop Science Society of America.

Harris, P. and M. Maw. (1981). *Hypericum perforatum* L., St. John's wort (Hypericaceae). In *Biological Control Programmes Against Insects and Weeds in Canada 1969-1980*, eds. J. S. Kelleher and M. A. Hulme, Slough, England: Commonwealth Agricultural Bureaux, pp. 171-178.

Hassell, M. P. (2000). *The Spatial and Temporal Dynamics of Host-Parasitoid Interactions*. Oxford Series in Ecology and Evolution. Oxford, UK: Oxford University Press.

Hawkes, J. G. (1991). The importance of genetic resources in plant breeding. *Biological Journal of the Linnean Society* 43:3-10.

Heap, I. (2001). The International survey of herbicide resistant weeds. Available www.weedscience.com

Henter, H. J. and S. Via. (1995). The potential for coevolution in a host-parasitoid system. I. Genetic variation within an aphid population in susceptibility to a parasitic wasp. *Evolution* 49:427-438.

Hess, F. D. (1996). Herbicide-resistant crops: Perspectives from a herbicide manufacturer. In *Herbicide-Resistant Crops: Agricultural, Environmental, Economic, Regulatory, and Technical Aspects*, ed. S. O. Duke, Boca Raton, FL: CRC Press, pp. 263-270.

Hilbeck, A. (2001). Transgenic host plant resistance and non-target effects. In *Genetically Engineered Organisms: Assessing Environmental and Human Health Effects*, eds. D. K. Letourneau and B. E. Burrows, Boca Raton, FL: CRC Press, pp. 167-186.

Hilder, V. A. and D. Boulter. (1999). Genetic engineering of crop plants for insect resistance–a critical review. *Crop Protection* 18:177-191.

Hillel, D. (2000). *Salinity Management for Sustainable Irrigation*. Washington, DC: The World Bank.

Holm, L. G. (1997). *World Weeds: Natural Histories and Distribution*. New York, NY: John Wiley & Sons, Inc.

Holm, L. G., D. L. Plucknett, J. V. Pancho, and J. P. Herberger. (1977). *The World's Worst Weeds*. Honolulu, HI: University Press of Hawaii.

Holmstrom, K.-O., E. Mantyla, B. Welin, A. Mandal, T. E. Palva, P. E. Tunnela, and J. Londesborough. (1996). Drought tolerance in tobacco. *Nature* 379:683-684.

Holt, J. S. (1994). History and identification of herbicide-resistant weeds. *Weed Technology* 6:615-620.

Hosmani, M. M. (1995). *Striga (A Noxious Root Parasitic Weed)*. Dharwad, India: Sarasjakshi Hosmani.

Hruska, A. J., S. M. Gladstone, and R. Obando. (1996). Epidemic roller coaster: Maize stunt disease in Nicaragua. *American Entomologist* 42:248-252.

Jacobsen, T. and R. M. Adams. (1958). Salt and silt in ancient Mesopotamian agriculture. *Science* 128:1251-1258.

Jaglo-Ottosen, K. R., S. J. Gilmour, D. G. Zarka, O. Schabenberger, and M. F. Thomashow. (1998). Arabidopsis CBF1 overexpression induces COR genes and enhances freezing tolerance. *Science* 280:104-106.

James, C. (2000). *Global Status of Commercialized Transgenic Crops: 2000*. ISAAA Briefs No. 21: Preview, Ithaca, NY: ISAAA.

Jarvis, D. I. and T. Hodgkin. (1999). Wild relatives and crop cultivars: Detecting natural introgression and farmer selection of new genetic combinations in agroecosystems. *Molecular Ecology* 8:S159-S173.

Johnson, N., C. Revenga, and J. Echevarria. (2001). Managing water for people and nature. *Science* 292:1071-1072.

Kahl, G. and P. Winter. (1995). Plant genetic engineering for crop improvement. *World Journal of Microbiology and Biotechnology* 11:449-460.

Kareiva, P., W. Morris, and C. M. Jacobi. (1994). Studying and managing the risk of cross-fertilization between transgenic crops and wild relatives. *Molecular Ecology* 3:15-21.

Kato, M., E. S. Mizubuti, S. B. Goodwin, and W. E. Fry. (1997). Sensitivity to protectant fungicides and pathogenic fitness of clonal lineages of *Phytophthora infestans* in the United States. *Phytopathology* 87:973-978.

Keeler, K. H., C. E. Turner, and M. R. Bolick. (1996). Movement of crop transgenes into wild plants. In *Herbicide-Resistant Crops: Agricultural, Environmental, Economic, Regulatory, and Technical Aspects*, ed. S. O. Duke, Boca Raton, Florida: CRC Press, pp. 303-330.

Khan, Z. R., K. Ampongo-Nyarko, P. Chiliswa, A. Hassanali, S. Kimani, W. Lwande, W. A. Overholt, J. A. Pickett, L. E. Smart, L. J. Wadhams, and C. M. Woodcock. (1997a). Intercropping increases parasitism of pests. *Nature* 388:631-632.

Khan, Z. R., P. Chiliswa, K. Ampong-Nyarko, L. E. Smart, A. Polaszek, J. Wandera, and M. A. Mulaa. (1997b). Utilisation of wild gramineous plants for management of cereal stemborers in Africa. *Insect Science Application* 17:143-150.

Khan, Z. R., J. A. Pickett, J. van den Berg, L. J. Wadhams, and C. M. Woodcock. (2000). Exploiting chemical ecology and species diversity: Stem borer and striga control for maize and sorghum in Africa. *Pest Management Science* 56:957-962.

Kirkpatrick, K. J. and H. D. Wilson. (1988). Interspecific gene flow in *Cucurbita*: *Cucurbita texana* vs. *Cucurbita pepo*. *American Journal of Botany* 75:519-527.

Lacey, A. J. (1985). Weed control. In *Pesticide Application: Principles and Practice*, ed. P. T. Haskell, Oxford, UK: Oxford University Press, pp. 456-485.

Lagnaoui, A. and E. B. Radcliffe. (1997). Interference of fungicides with entomopathogens: Effects of entomophthoran pathogens of green peach aphids. In *Ecological Interactions and Biological Control*, eds. D. A. Andow, D. W. Ragsdale and R. F. Nyvall, Boulder, CO: Westview Press, pp. 301-315.

Lapidot, M., R. Gafny, B. Ding, S. Wolf, W. J. Lucas, and R. N. Beachy. (1993). A dysfunctional movement protein of tobacco mosaic virus that partially modifies the plasmodesmata and limits virus spread in transgenic plants. *Plant Journal* 4:959-970.

Leavings, C. S. I. (1990). The Texas cytoplasm of maize: Cytoplasmic male sterility and disease susceptibility. *Science* 250:942-947.

Lee, L. J. and J. Ngim. (2000). A first report of glyphosate-resistant goosegrass (*Eleusine indica* (L) Gaertn.) in Malaysia. *Pest Management Science* 56:336-339.

Letourneau, D. K., J. A. Hagen, and G. S. Robinson. (2001). Bt-crops: Evaluating benefits under cultivation and risks from escaped transgenes in the wild. In *Genetically Engineered Organisms: Assessing Environmental and Human Health Effects*, eds. D. K. Letourneau and B. E. Burrows, Boca Raton, FL: CRC Press, pp. 33-98.

Lindow, S. E., N. J. Panopoulos, and B. L. McFarland. (1989). Genetic engineering of bacteria from managed and natural habitats. *Science* 244:1300-1307.

Lucas, G. B., C. L. Campbell, and L. T. Lucas. (1985). *Introduction to Plant Diseases: Identification and Management*. Westport, CT: The AVI Publishing Company, Inc.

Malik, J., G. Barry, and G. Kishore. (1989). The herbicide glyphosate. *BioFactors* 2:17-25.

Marshall, G. (1998). Herbicide-tolerant crops-real farmer opportunity or potential environmental problem? *Pesticide Science* 52:394-402.

Marvier, M. (2001). Ecology of transgenic crops. *American Scientist* 89:160-167.

Mazur, B. J. and S. C. Falco. (1989). The development of herbicide resistant crops. *Annual Review of Plant Physiology and Plant Molecular Biology* 40:441-470.

McKersie, B. D. and Y. Y. Leshem. (1994). *Stress and Stress Coping in Cultivated Plants*. Dordrecht: Kluwer Academic Press.

Mellon, M. (1998). UCS introduction. In *Now or Never: Serious New Plans to Save a Natural Pest Control*, eds. M. Mellon and J. Rissler, Cambridge, MA: Union of Concerned Scientists, pp. 1-11.

Mikkelsen, T. R., B. Andersen, and R. B. Jorgensen. (1996). The risk of crop transgene spread. *Nature* 380:31.

Moorman, T. B. and K. E. Keller. (1996). Crop resistance to herbicides: Effects on soil and water quality. In *Herbicide-Resistant Crops: Agricultural, Environmental, Economic, Regulatory, and Technical Aspects*, ed. S. O. Duke, Boca Raton, FL: CRC Press, pp. 283-302.

Mugo, S. and D. Poland. (2001). IRMA bioassay shows *Bt* effective against Kenyan borers. *IRMA Updates* 2:1-2.

Murakami, Y., M. Tsuyama, Y. Kobayashi, H. Kodama, and K. Iba. (2000). Trienoic fatty acids and plant tolerance of high temperature. *Science* 287:476-479.

NRC. (1993). *Managing Global Genetic Resources: Agricultural Crop Issues and Policies*. Washington, DC: National Academy Press.

NRC. (2002). *Environmental Effects of Transgenic Plants: The Scope and Adequacy of Regulation*. Washington, DC: National Academy Press.

Padgette, S. R., D. B. Re, G. F. Barry, D. E. Eichholtz, X. Delannay, R. L. Fuchs, G. M. Kishore, and R. T. Fraley. (1996). New weed control opportunities: Development of soybeans with a Roundup Ready™ gene. In *Herbicide-Resistant Crops: Agricultural, Environmental, Economic, Regulatory, and Technical Aspects*, ed. S. O. Duke, Boca Raton, FL: CRC Press, pp. 53-84.

Peferoen, M. (1997). Insect control with transgenic plants expressing *Bacillus thuringiensis* crystal proteins. In *Advances in Insect Control: The Role of Transgenic Plants*, eds. N. Carozzi and M. Koziel, London, UK: Taylor and Francis, pp. 21-48.

Pimentel, D., D. Andow, R. Dyson-Hudson, D. Gallahan, S. Jacobson, M. Irish, S. Kroop, A. Moss, I. Schreiner, M. Shepard, T. Thompson, and B. Vinzant. (1980). Environmental and social costs of pesticides: A preliminary assessment. *Oikos* 34:126-140.

Power, A. G. (2001). Ecological risks of transgenic virus-resistant crops. In *Genetically Engineered Organisms: Assessing Environmental and Human Health Effects*, eds. D. K. Letourneau and B. E. Burrows, Boca Raton, FL: CRC Press, pp. 125-142.

Powles, S. B., D. F. Lorraine-Colwill, J. J. Dellow, and C. Preston. (1998). Evolved resistance to glyphosate in rigid ryegrass (*Lolium rigidum*) in Australia. *Weed Science* 46:604-607.

Quist, D., and I. H. Chapela. (2001). Transgenic DNA introgressed into traditional maize landraces in Oaxaca, Mexico. *Nature* 414:541-543.

Radosevich, S. R., C. M. Ghersa, and G. Comstock. (1992). Concerns a weed scientist might have about herbicide-tolerant crops. *Weed Technology* 6:635-639.

Ransom, J. K. (2000). Long-term approaches for the control of *Striga* in cereals: Field management options. *Crop Protection* 19:759-763.

Rhoades, J. D., A. Kandiah, and A. M. Mashali. (1992). *The Use of Saline Water for Crop Production*. FAO Irrigation and Drainage Paper, Rome, Italy: Food and Agriculture Organization.

Rivnay, E. (1972). On irrigation-induced changes in insect populations in Israel. In *The Careless Technology: Ecology and International Development*, eds. M. T. Farvar and J. P. Milton, Garden City, NY: The Natural History Press, pp. 349-364.

Rochow, W. F. (1977). Dependent virus transmission from mixed infections. In *Aphids as Virus Vectors*, eds. K. Harris and K. Maramorosch, New York, NY: Academic Press, pp. 253-273.

Root, R. B. (1973). Organization of a plant-arthropod association in simple and diverse habitats: The fauna of collards. *Ecological Monographs* 43:95-120.

Rosenheim, J. A. (1998). Higher-order predators and the regulation of insect herbivore populations. *Annual Review of Entomology* 43:421-447.

Ross, M. A. and C. A. Lembi. (1999). *Applied Weed Science*. Upper Saddle River, NJ: Prentice Hall.

Salomon, R. (2000). The evolutionary advantage of breeding for tolerance over resistance against viral plant disease. *Israel Journal of Plant Sciences* 47:135-139.

Sanford, J. C. and S. A. Johnston. (1985). The concept of parasite-derived resistance–Deriving resistance genes from the parasite's own genome. *Journal of Theoretical Biology* 113:395-405.

Saxena, D., S. Flores, and G. Stotzky. (1999). Insecticidal toxin in root exudates from Bt corn. *Nature* 402:480.

Saxena, D. and G. Stotzky. (2001). *Bacillus thuringiensis* (*Bt*) toxin released from root exudates and biomass of *Bt* corn has no apparent effect on earthworms, nematodes, protozoa, bacteria, and fungi in soil. *Soil Biology and Biochemistry* 33:1225-1230.

Scheffer, R. P. (1997). *The Nature of Disease in Plants*. Cambridge, UK: Cambridge University Press.

Schmitt, J. and C. R. Linder. (1994). Will escaped transgenes lead to ecological release? *Molecular Ecology* 3:71-74.

Shaner, D. L. (2000). The impact of glyphosate-tolerant crops on the use of other herbicides and on resistance management. *Pest Management Science* 56:320-326.

Shepherd, R. C. H. (1985). The present status of St. John's Wort (*Hypericum perforatum* L.) and its biological control agents in Victoria, Australia. *Agriculture, Ecosystems and Environment* 12:141-149.

Sih, A., P. Crowley, M. McPeek, J. Petranka, and K. Strohmeier. (1985). Predation, competition and prey communities: a review of field experiments. *Annual Review of Ecology and Systematics* 16:269-311.

Snow, A. A. and P. Moran Palma. (1997). Commercialization of transgenic plants: Potential ecological risks. *BioScience* 47:86-96.

Stotzky, G. (2001). Release, persistence, and biological activity in soil of insecticidal proteins from *Bacillus thuringiensis*. In *Genetically Engineered Organisms: Assessing Environmental and Human Health Effects*, eds. D. K. Letourneau and B. E. Burrows, Boca Raton, FL: CRC Press, pp. 187-222.

Strange, R. M. (1993). *Plant Disease Control: Towards Environmentally Acceptable Methods*. London, UK: Chapman and Hall.

Suranyi, R. A. (2000). Management of potato leafroll virus in Minnesota. *Entomology*. St. Paul, MN, University of Minnesota.

Szabolcs, I. (1988). *Salt-Affected Soils*. Boca Raton, FL: CRC Press.

Tabashnik, B. E. (1994). Evolution of resistance to *Bacillus thuringiensis*. *Annual Review of Entomology* 39:47-79.

Talbot, M. N. (1993). The occurrence of volunteers as weeds of arable crops in Great Britain. In *Volunteer Crops as Weeds*, eds. R. J. Froud-Williams, C. M. Knott, and P. J. W. Lutman, Wellesbourne, Warwick, UK: The Association of Applied Biologists, Horticulture Research International, pp. 231-235.

Tarczynski, M. C., R. G. Jensen, and H. J. Bohnert. (1993). Stress protection of transgenic tobacco by production of the osmolyte mannitol. *Science* 259:508-510.

Turnipseed, S. G., M. J. Sullivan, J. E. Mann, and M. E. Roof. (1995). Secondary pests in transgenic Bt cotton in South Carolina. *Proceedings of the Beltwide Cotton Conferences*, Memphis, TN. pp. 768-769.

Ullstrup, A. J. (1972). The impact of the southern corn leaf blight epidemics of 1970-1971. *Annual Review of Phytopathology* 10:37-50.

USDA APHIS (2001). Biotechnology permits database. Available www.nbiap.vt.edu

USDA NASS (2001). Acreage. Available jan.mannlib.cornell.edu/reports/nassr/field/pcp-bba/acrg2001.pdf

Vasil, I. K. (1996). Phosphinothricin-resistant crops. In *Herbicide-Resistant Crops: Agricultural, Environmental, Economic, Regulatory, and Technical Aspects*, ed. S. O. Duke, Boca Raton, FL: CRC Press.

Verkleij, J. A. C. and E. Kuiper. (2000). Various approaches to controlling root parasitic weeds. *Biotechnology and Development Monitor* 41:16-19.

Wagner, F. (1980). *The Boll Weevil Comes to Texas*. Corpus Christi, TX, USA: The Friends of the Corpus Christi Museum.

Ward, G. C. (1995). The Texas male-sterile cytoplasm of maize. In *The Molecular Biology of Plant Mitochondria*, eds. C. S. Levings, III and I. K. Vasil, Dordrecht, the Netherlands: Kluwer Academic Publishers, pp. 433-460.

Watkinson, A. R., R. P. Freckleton, R. A. Robinson, and W. J. Sutherland. (2000). Predictions of biodiversity response to genetically modified herbicide-tolerant crops. *Science* 289:1554-1557.

Wilcut, J. W., H. D. Coble, A. C. York, and D. W. Monks. (1996). The niche for herbicide resistant crops in U.S. agriculture. In *Herbicide-Resistant Crops: Agricultural, Environmental, Economic, Regulatory, and Technical Aspects*, ed. S. O. Duke, Boca Raton, FL: CRC Lewis Publishers, pp. 213-230.

Wilson, H. D. (1990). Gene flow in squash species. *BioScience* 40:449-455.

Wolfenbarger, L. L. and P. R. Phifer. (2000). The ecological risks and benefits of genetically engineered plants. *Science* 290:2088-2092.

Working Party on Dryland Salting in Australia. (1982). *Salting of Non-Irrigated Land in Australia*. Victoria, Australia: Soil Conservation Authority.

Wyn Jones, R. G. (1981). Salt tolerance. In *Physiological Processes Limiting Plant Productivity*, ed. C. B. Johnson, London, UK: Butterworths, pp. 271-292.

Yarham, D. J. and P. Gladders. (1993). Effect of volunteer plants on crop diseases. In *Volunteer Crops as Weeds*, eds. R. J. Froud-Williams, C. M. Knott, and P. J. W. Lutman, Wellesbourne, Warwick, UK: The Association of Applied Biologists, Horticulture Research International, pp. 75-82.

Zaccomer, B., F. Cellier, J. C. Boyer, A. L. Haenni, and M. Tepfer. (1993). Transgenic plants that express genes including 3′ untranslated region of the turnip yellow mosaic virus (TYMV) genome are partially protected against TYMV infection. *Gene* 136:87-94.

Zhang, H.-X. and E. Blumwald. (2001). Transgenic salt-tolerant tomato plants accumulate salt in foliage but not in fruit. *Nature Biotechnology* 19:765-768.

Zimdahl, R. L. (1999). *Fundamentals of Weed Science*. San Diego, CA: Academic Press.

Redesigning Pest Management: A Social Ecology Approach

Stuart B. Hill

SUMMARY. Social ecology provides a broad framework for critically examining the theory and practice of pest management, and the institutional and psychosocial contexts within which all theories and practices are developed and implemented. Such an analysis finds most current theories and practices seriously wanting, but also reveals extensive opportunities for innovation and improvement. The greatest need is for a shift in emphasis from a responsive, symptom-based, linear approach based on biocides and their substitutes to a proactive, holistic approach based on knowledge-intensive, ecosystem design and redesign, and whole systems management (based particularly on the further development of cultural controls). Although this will require broader parallel changes within individuals and throughout society, significant improvement can be made by focusing on small achievable initiatives and their public celebration to facilitate their spread. The developments in pest management being proposed here are regarded as one expression of the ongoing psychosocial evolution of our species. *[Article copies available for a fee from The Haworth Document Delivery Service: 1-800-HAWORTH. E-mail address: <docdelivery@haworthpress.com> Website: <http://www.HaworthPress.com> © 2004 by The Haworth Press, Inc. All rights reserved.]*

Stuart B. Hill is Foundation Chair of Social Ecology, School of Social Ecology and Lifelong Learning, University of Western Sydney-Hawkesbury Campus, Locked Bag 1797, Penrith South Distribution Centre, NSW 1797, Australia.

[Haworth co-indexing entry note]: "Redesigning Pest Management: A Social Ecology Approach." Hill, Stuart B. Co-published simultaneously in *Journal of Crop Improvement* (Food Products Press, an imprint of The Haworth Press, Inc.) Vol. 12, No. 1/2 (#23/24), 2004, pp. 491-510; and: *New Dimensions in Agroecology* (ed: David Clements, and Anil Shrestha) Food Products Press, an imprint of The Haworth Press, Inc., 2004, pp. 491-510. Single or multiple copies of this article are available for a fee from The Haworth Document Delivery Service [1-800-HAWORTH, 9:00 a.m. - 5:00 p.m. (EST). E-mail address: docdelivery@haworthpress.com].

http://www.haworthpress.com/web/JCRIP

Digital Object Identifer: 10.1300/J411v12n01_09

KEYWORDS. Change, cultural controls, ecosystem, holistic, pest management, psychosocial, redesign, social ecology, sustainability, systems

INTRODUCTION

The main aim of this paper is to encourage all engaged in pest management, from producers to policy makers, to be more comprehensive and responsible in considering the consequences of choice of ecosystem design and patterns of management (including pest management), decision-making processes and specific actions (and inaction). Unfortunately, these areas are commonly characterized by unexamined assumptions, lack of critical thought, and associated narrow habitual (often defensive) patterns of thinking, deciding and acting (Hill, Vincent, and Chouinard, 1999). Consequently, in addition to the many negative consequences of our dominant pest control practices, there are also numerous opportunities for improving pest management. Failure to openly engage in this type of reflection and associated transformative change at this time will certainly disadvantage future generations (Clements and Shrestha, 2004).

As our species continues to evolve psychosocially (Huxley, 1952; 1964), we must be open to progressive changes in our values, understandings, designs and relationships. This includes the design of managed ecosystems, our perception of their functions, our attitude towards pests, and the actions taken and tools used to manage them. This evolutionary process is challenging us to move on from a situation in which ecosystem designs are highly simplified, and management decisions are primarily concerned with inputs and externally imposed controls. It is currently common for management decisions in agriculture to be based primarily on productivity, sales and profits. Pests are invariably regarded as enemies to be detected and managed or eliminated, primarily by means of biocides developed by, and under the advice of, technical experts (biocide is used here in preference to pesticide because the latter term incorrectly implies specificity, which is not possible in situations such as this where the target is defined primarily by its economic and not its biological properties). Within this context, most progress in pest control has focused on using biocides more efficiently, and on substituting less impacting interventions, such as biological controls and, more recently, genetically manipulated crops and biocontrols (which bring a new set of problems that will not be discussed here). Despite the progress that has been made, it is important to realize that such strategies tend to divert our attention from those causes of pest outbreaks that are related to the design of agroecosystems, and they condemn producers to the recurring costs of purchasing curative control products and services. In contrast, design strategies to pest management have the potential to remove the causes of outbreaks and, in some cases, to do this permanently without recurrent costs (Hill and MacRae, 1995; Hill, Vincent, and Chouinard, 1999; MacRae et al., 1989a).

SOCIAL ECOLOGY AS A FRAMEWORK FOR IMPROVING PEST MANAGEMENT

Social ecology (SE) can support this project by helping us to systematically examine our history, our dominant modes of thinking and acting, the contexts in which we are operating, and the range of future choices available to us and their possible consequences. Also, SE can provide a framework for effectively implementing desirable change. Although there is some overlap with the term 'human ecology,' SE's explicit link with values and situation improvement distinguishes it from that more neutral field of study. Indeed, some regard the holistic and progressive nature of SE approaches as expressions of emerging next steps in our ongoing psychosocial evolution (Bookchin, 1995; Hill, 1999).

Julian Huxley (1964) was one of the first people to use the term social ecology. SE has evolved considerably, however, since his limited perception of it as "man's [sic] social relations, both within and between human societies." Today there are as many perceptions of SE as there are academic programs (of which there are still very few), and among staff and students within them the emphasis usually varies quite widely. The name most associated with SE is Murray Bookchin, who established the first degrees in this meta-field at Goddard College in Vermont in the 1970s. Bookchin is an anarchist who has argued passionately in many books (e.g., 1980, 1982, 1986, 1996) and on many public platforms for local participatory control over local affairs, for a more caring relationship with nature and one another, and for the dismantling of most centralized bureaucracies. Due partly to his willingness to take a stand where others have vacillated and postponed, he has also attracted much opposition, even resulting in books with titles like "Beyond Bookchin" (Watson, 1996) and "Social Ecology After Bookchin" (Light, 1998; this book includes numerous references to the broad literature on SE). My own definition of SE as the study of the theory and practice of holistic and integrated approaches to personal, social and environmental health, sustainability and change, based on the critical application and integration of ecological, humanistic, community and 'spiritual' values (Hill, 1999) evolved through my association with the SE group at the University of Western Sydney-Hawkesbury in Australia. Here, I take 'sustainability' to mean the rehabilitation and maintenance of systems in a healthy condition; 'society' to embrace all of the institutional structures and processes that characterise cultures, including but not priviledging economics (any more than politics, education, technology, business or religion–all being, in a sense, 'tools' to enable us to act on our values); 'change' to be preferentially concerned with development and situation improvement within whole systems (with an emphasis on whole persons, equity across time, space and sector, and social justice); and 'spiritual' to refer to the vast unknowness, mystery and wonder that is always present and that must be taken into account and engaged with (by trying to impose structures on the unknown, most religions have paradoxically become barriers to effective engagement in this area). The importance of acknowledging the extent of this unknowness when designing resource management systems was well recognized by Andre Voisin (1959), one of the great pioneers of agroecosystem design, and the 'father' of rotational grazing. Thus, within a SE framework, all initiatives and interventions, including those in pest management, must be tested for their appropriateness against such a list of criteria. A preliminary attempt to formalize these is provided in Table 1. While acknowledging that acting on these 'testing questions' fully at this particular time in our history would present many challenges, few would argue that the world would not be improved by doing so. Hence the challenge is to recognize and act on the changes that can be made in this direction under whatever contextual conditions one is operating.

Clearly most present pest control practices would fail most of these testing questions. On the other hand, if we were to take a systems-design or redesign approach to solving our pest problems it would soon become clear that both the opportunities and benefits (within this broader framework of evaluation) would be extensive. Thus, by crossing boundaries, thinking 'outside of the box' and integrating the knowledge bases of the arts and sciences, especially philosophy, psychology, sociology and ecology, and being open to working across difference and also with the unknown as well as the known (always keeping in mind that the 'known' is always provisional and a very small fraction of what is), SE can help us to extend our thinking and provide a more critical framework for analysis, the development of more effective ap-

TABLE 1. Testing questions for evaluating social ecology pest management initiatives (all of these could be evaluated qualitatively and quantitatively).

Personal

1. Does it support: empowerment, awareness, creative visioning, values clarification, acquisition of essential literacies and competencies, responsibility, wellbeing and health maintenance, vitality and ***spontaneity*** (building & maintaining personal capital–personal sustainability)?
2. Does it support: caring, loving, responsible, mutualistic, ***negentropic relationships*** with diverse others (valuing equity & social justice), other species, place and planet (home & ecosystem maintenance)?
3. Does it support: positive total life-cycle ***personal development*** and 'progressive' change?

Socio-Political

4. Does it support: accessible, collaborative, responsible, creative, celebrational, ***life-promoting community and political structures and functions*** (building & maintaining social capital–cultural [including economic] sustainability)?
5. Does it support: the valuing of 'functional' high ***cultural diversity*** and mutualistic relationships?
6. Does it support: positive ***cultural development*** and co-evolutionary change?

Environmental

7. Does it support: effective ***ecosystems functioning*** (building & maintaining natural capital–ecological sustainability)?
8. Does it support: 'functional' high ***biodiversity***, and prioritised use and conservation of resources?
9. Does it support: positive ***ecosystem development*** and co-evolutionary change?

General

10. Does it support: proactive (*vs.* reactive), ***design/redesign*** (*vs.* efficiency & substitution) and ***small meaningful collaborative initiatives*** that together one can guarantee to carry through to completion (*vs.* heroic, Olympic-scale, exclusive, high risk ones) and their ***public celebration*** at each stage–to facilitate their spread–thereby making wellbeing and environmental caring 'contagious'?
11. Does it focus on: key opportunities and ***windows for change*** (pre-existing change locations & 'moments')?
12. Does it explain: how it will effectively monitor and evaluate its progress (broad, long-term, as well as specific & short-term) by identifying and using ***integrator indicators*** *and* ***testing questions***, and by being attentive to all ***feedback and outcomes*** (& redesigning future actions & initiatives accordingly)?

proaches, and wiser decision-making and action. SE is particularly concerned with deepening our understanding, and working with change that is life affirming, progressive, and personally, socially and ecologically sustainable (Hill, 1999).

SE's focus is simultaneously the big picture and small meaningful initiatives that can contribute to desirable change. Its perception and concern for the consequences of our actions is both global and local. All action is encouraged to be contextually sensitive and emergent from shared concerns, responsibilities and collaborations. As such it draws on both quantitative and qualitative methods of inquiry (Denzin and Lincoln, 2000), and particularly on the rich diversity of approaches to action research (Reason and Bradbury, 2001).

A preliminary SE-analysis of pest management, including personal, social, political, ecological and 'spiritual' comparisons between current dominant practices and alternative possibilities is provided in Table 2. The aim of providing such a comparison is to stimulate reflection and sow the seeds of alternative progressive thinking.

One of the most effective ways of visioning such more desirable futures is to engage a diverse group of people in 'storying' them (including re-storying the past and present), and then constructing composite stories. Commonly, a broad spectrum of stories emerge, ranging from the extreme technocentric to the extreme ecocentric. For this process to work, it is necessary to be open to learning as much as one can about all of these alternative perceptions and not to prematurely dismiss any of them. Rather, the approach is to aim to deepen one's understanding of their key features, implications,consequences and assumptions upon which they are based. It is important to realize that the frameworks and practices that each of us favor are all embedded within and are emergent from our own partially shared personal and cultural stories. Because of the pain and hurt that is often associated with such stories, (much of which exists only in the subconscious), our perceptions are invariably to some extent distorted (yet we usually do not think of them as being distorted and may energetically deny any such suggestion). Unless we are open to critically examining and mapping these stories and their origins, we are vulnerable to being unknowingly imprisoned by them, and our efforts to make progress are likely to be largely automatic extrapolations of the past (Hill, Vincent, and Chouinard, 1999; Hill, 2001; Miller, 1999).

My own preference is to emphasize the ecocentric, and personal and community well-being end of the values spectrum, as this is where I believe our greatest potential for improvement lies. A typical generalized

TABLE 2. Attitudinal differences between conventional and social ecology approaches to pests and the systems within which they occur.

	CONVENTIONAL	SOCIAL ECOLOGY
Personal	Pests regarded as **enemies** to be controlled or eliminated by means of chemical, physical or biological reactive, symptom-focused interventions. Acting out of **fear and separation** from ecosystems; diminishing the human spirit (loss of personal capital) and threatening wellbeing. Proactive strategies largely similar to **reactive** ones.	Pests regarded as **indicators** of maldesigned, mismanaged systems that need to be redesigned. Minimally disruptive curative controls only used in emergencies. Sense of **love, and oneness** with ecosystems (being part of a meta-organism/Gaia); supportive of wellbeing. **Proactive** strategies fundamentally different from reactive interventions and based largely on the **substitution** of **knowledge and skills for inputs,** and on system design/redesign.
Social	All produce regarded as **commodities** to dispose of anywhere for the highest price; associated loss of social capital not considered.	Produce regarded as valued '**gifts**' from nature for meaningful distribution, sale or exchange to meet basic needs of valued people; building social capital.
Political	Food and other resources used to generate **export earnings** and as **political tools** and weapons of **power** to manipulate opinion and compliance.	Equitably meeting **basic needs**, such as nourishment, **from ecologically sustainably managed renewable resources** a political priority (open to "rights-to-food" discussions). Farmers also rewarded for **landscape maintenance.**
Ecological	Highly **simplified**, controlled, input-dependent, managed ecosystems becoming increasingly degraded–losing species and other natural capital over time.	Complex, uniquely designed in space and time, co-evolving systems that are building and maintaining natural capital and biodiversity.
'Spiritual'/ Unknown	Outputs are commodities, **jobs** and, for the diminishing number of producers, a way of life. Frontiers of development: increased control through increasing dependence on purchased inputs and expertise (technologies, GM crops, etc.). Success linked to **control**, **productivity** and **profit** A sense of **panic** as systems degrade, old interventions fail, communities and relationships break down and uncertainty and dependence increase. Reaching for ever more powerful and potentially disruptive interventions.	For many–**a calling**–a connection with and a commitment to the land–building personal, rural community and landscape health.Ever expanding reverence and respect for the wonders of nature and experiences of deep discovery and synchronicity. Success linked to **meaning**, **wellbeing** and **sustainability**. A sense that we've hardly scratched the surface of our understanding of the wonders and intricacies of the processes and relationships within nature. Much **hope** regarding emergent understandings and opportunities.

story with this focus might envision a world in which ecosystems are designed and managed to achieve a range of compatible personal, social, ecological and 'spiritual' goals (rather than the very limited goals of productivity and profit). More specific goals might include the equitable meeting of basic needs, such as the nourishment of primarily local communities, the provision of high quality habitat for a great diversity of other species, particularly those concerned with the maintenance of ecosystem health and social and personal wellbeing, and the provision of meaningful and fulfilling work and play. Clearly, implementing such a vision would require fundamental changes throughout society, both locally and globally. It is important to be both aware of these potential barriers and to not use them as excuses for postponement and inaction. Rather, every initiative should be designed to integrate with all others that are compatible to achieve such much-needed social change. The argument that such thinking is impractical within our present frameworks is a formula for cultural paralysis and ongoing environmental, social and personal degradation (Hill, 1993; 1998, Table 1), and is often the product of unacknowledged fears (Hill, 1991; 2001). The challenge is to engage in the thinking and action that is possible, especially through collaboration with others who have overlapping concerns, and in this way bring about the changes than can be made.

Typical areas that require such transformative changes include the manipulated consumer demand for cosmetic perfection in fruits and vegetables, and the associated regulations concerned with false quality standards (MacRae et al., 1990; MacRae, Henning, and Hill, 1993). Similarly, we must change the institutional structures, regulations and processes that support research, education, services and access to economic advantages for those associated with curative, 'magic bullet' pest management approaches (Hill, 1985; 1990a; MacRae et al., 1989a; 1989b; MacRae, Henning, and Hill, 1993). These barriers to progress also make it almost impossible for both researchers and pest managers to work in a transdisciplinary way with whole systems, and to take an integrated institutional, personal development and ecosystem redesign approach to minimizing pest problems (Hill, 1998; Hill, Vincent, and Chouinard, 1999; Odum and Barrett, 2003).

Although the dominant institutional structures, processes and worldviews comprise enormous barriers to individual and small group efforts, these are only likely to be changed as a result of transformative grass roots efforts, as all of history tells us. This is why collaboration among interest groups across diverse domains of concern and action is so important. The greatest barriers to such transformational change are the persistent beliefs that small efforts in this direction will not make a difference, and that problems should be solved by experts. Paradoxically, most equitable, progressive and sustainable change can be traced back to such small-scale collaborative initiatives. One key is to generate a range of goals associated with such initiatives that are appropriate to the full spectrum of modes of action. These might include, at the personal level, new learning and relationships, the satisfaction from engaging in values-based action, the copying of such action by others (much harder to achieve for larger initiatives), and the capacity of such initiatives to act as stepping-stones for subsequent ones. At the social level, appropriate goals might include improved public knowledge and competence, access to needed resources, responsibility, political and community participation, equity, freedom, and support to make informed decisions, and a general growth in social capital would be likely goals. At the environmental level, key gains might include enhanced ecosystem resilience, biodiversity (Thrupp, 2003), carrying capacity and aesthetic qualities of the environment.

EXAMPLE OF A PRODUCER-LEVEL 'REDESIGN' INITIATIVE TO A PEST PROBLEM

A farmer friend in Canada, with a community spirit, used to maintain small plots down one side of a field to teach local children how to garden. He was passionate about carrots (*Daucus carota* L.), so he started them off growing carrots. In his own fields of carrots, he had problems with carrot rust fly (*Psila rosae* (Fabricius)), and so he was surprised when it came time to harvest the children's carrots that some plots were attacked and some not. He knew there had to be a reason for this difference, so he walked up and down the plots searching for the causes. It took him a while to notice the difference, but it was clear once he saw it.

In the attacked plots, the soil had been compacted. In these, the children had thinned and weeded by walking on the soil along the rows, whereas the pest-free plots had loose and open soil because the children had reached in from the edge to thin and weed, never walking on the soil. The carrot rust fly egg and first stage larva are both highly susceptible to desiccation, and so their survival would be much higher in the more protected environment of a compacted soil. Realizing this, he postulated that in his field the damage from carrot rust fly would be likely to be highest in the rows of carrots adjacent to where the tractor wheels would have compacted the soil. He took several transect samples across his field and sure enough he was correct in his assumption. His solution was to design and make a scratcher to go behind the tractor wheel. In his soil and climate conditions this, together with timing his seeding to avoid the main carrot rust fly flight, was sufficient to solve his problem with this pest indefinitely (in contrast to pesticide solutions that may only be temporary and that must be changed as resistance and secondary pests inevitably develop). It should be added that under different soil and climate conditions this problem may not have been so easily solved, and that many pest problems provide a much greater challenge. Also, it should be noted that many other potentially beneficial initiatives, such as insectary plant management, might have been integrated into this producer's response. Nevertheless, much may be learned from this example with respect to the largely untapped potential for designing knowledge-intensive, ecologically sustainable systems of pest management. These approaches are likely to be reliant on such local observational knowledge and fine-tuned for specific local conditions, and, in most situations, on the design and maintenance of more complex agroecosystems, as well as the presence of adjacent healthy non-managed ecosystems. This contrasts with the more usual context-insensitive, expert and technology-intensive, purchased product-based systems of pest control that keep producers dependent and fail to capitalize on their local knowledge and skills.

REDESIGN APPROACHES AT THE PEST MANAGER LEVEL

The starting point for such thinking is having a comprehensive map of the systems one is working with. At the 'field' level (Figure 1), this involves recognizing that pest damage is related to the particular characteristics of the pests involved, their dispersal abilities, the availability of suitable food and space and the presence and effectiveness of natural controls. These latter factors are influenced by our plant production practices, listed on the left in Figure 1 (selection of crop and site, planting design, maintenance of site, and harvesting and distribution of the produce). At every stage in this process there are numerous opportunities to make decisions that will favor both the crop and the pest's natural controls, and that will be unfavorable for the pest. What is clear from approaching pest management in this way is the vast largely untapped pool of opportunities for creative thinking, experimentation and especially paradoxical initiatives. Among the latter approaches, I have

FIGURE 1. Framework for understanding the 'field causes' of pest and disease outbreaks.

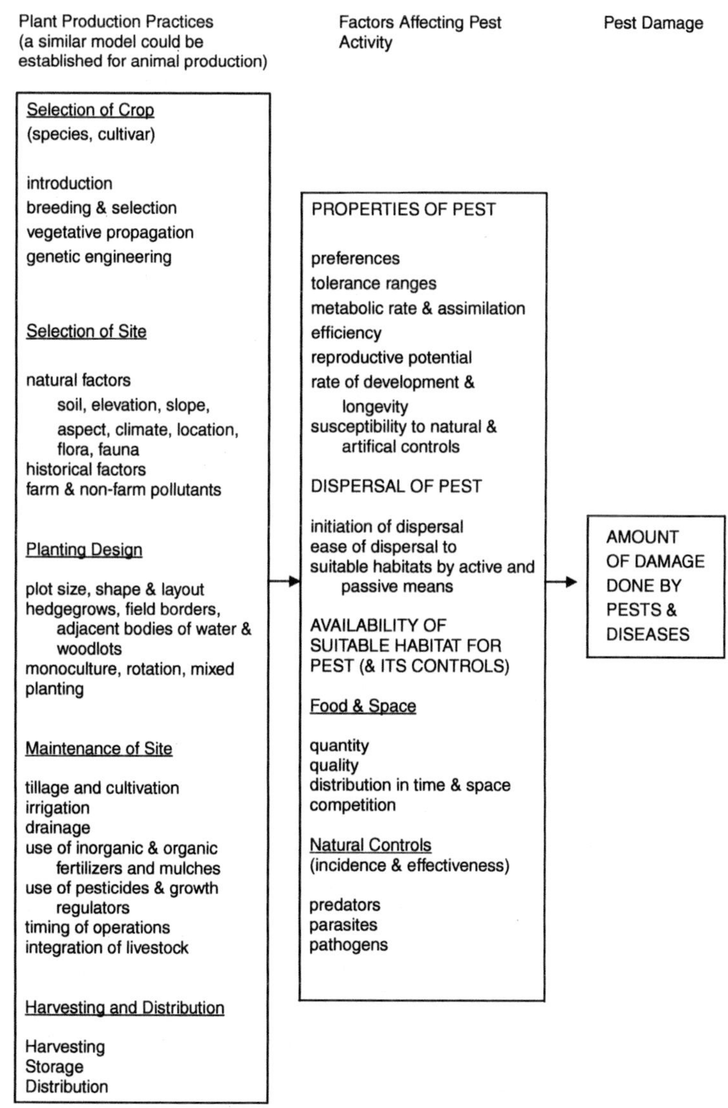

found it helpful to, in a sense, 'become the pest' in an effort to better define its needs, preferences and vulnerabilities. The meeting of these needs and preferences in prescribed locations may be used to concentrate pests and facilitate their effective control. Steps towards doing this with the plum curculio (*Conotrachelus nenuphar* (Herbst)) in apple orchards in south-western Quebec have achieved some success. Until a whole-systems approach can be taken, however, the full potential of 'redesign' initiatives for controlling this pest will not be realized (Lafleur and Hill; 1987; Lafleur, Hill, and Vincent, 1987; Hill, Vincent, and Chouinard, 1999; Vincent, Chouinard, and Hill, 1999). This will particularly require the further development of cultural methods of pest control. It should be noted that there is some overlap between my approach to the 'redesign' of agroecosystems (Hill, 1998) and Savory and Butterfield's (1999) Holistic Management approach.

CULTURAL METHODS OF PEST, PRIMARILY INSECT, CONTROL

Cultural controls are the oldest methods that have been used to manage pest populations. With the development of synthetic biocides, however, these controls were rapidly abandoned or de-emphasized and research on them was largely discontinued. Because most cultural controls are preventative rather than curative they are dependent on long-range planning. Also, because they are dependent on detailed knowledge of the ecology of crop-pests relationships, most of which in the past were poorly understood, the results were very variable, and it was often difficult to evaluate their effectiveness. It is understandable that most producers eventually adopted the chemical solutions to pest problems, which initially appeared more reliable and demanded less knowledge and skill.

Today, the situation is very different from those early days of pest control. We have a much better understanding of the ecological relationships within crop systems; predictive computer models are available for some pests; and the social climate is demanding reduced dependence on toxic chemicals to solve problems. Such concerns have developed as a result of numerous biocide accidents; detection of residues in recreational environments, drinking water, foods and human tissues; the increase in the incidence of biocide-related allergy and petrochemical sensitivity; and the growing interest in the relationships between food quality and human health (and the associated expanding organic food market (Hill and MacRae, 1992); the decreasing effectiveness of many biocides as more and more pests become resistant to them; their increasing costs; and effects on natural controls and other non-pest species. The literature supporting the above list of 'driving forces' is extensive and will not be reviewed here.

Pest control scientists have responded to this situation by promoting the philosophy and practice of integrated pest management (IPM). At first these approaches, encouraged by the 'chemical' companies, emphasized the efficient use of biocides, claiming that problems associated with these chemicals could usually be traced to their misuse. There is ample evidence, however, that problems arise even when biocides are used as recommended. Because of this, it seems likely that we will eventually have to restrict their use to socially important emergencies.

The next strategy in the development of IPM was to search for more benign substitutes for biocides: hormones, sex attractants, traps, biological control agents, including bio-engineered pathogens, release of sterile males, and now genetically manipulated crops and biocontrols. Such approaches, although useful as less intrusive curative methods, still avoid confronting the causes of the problem. They keep the producer dependent on experts and suppliers of products, and they perpetuate the image of pests strictly as enemies to be eliminated or controlled.

We are now on the threshold of a third phase in the development of IPM systems that recognizes pests not as enemies, but as indicators of problems in the design and management of systems (Hill 1985, 1998; Hill, Vincent, and Chouinard, 1999). It is also becoming clear that broad approaches to ecological thinking hold most of the keys to appropriate environmental designs and methods of management capable of keeping pest numbers below unacceptable levels (Mulligan and Hill, 2001). In this approach, potential pests are prevented from becoming problems by means of the integration of a range of cultural and bio-ecological controls. The more familiar intrusive controls, including certain biocides, are then reserved for emergencies.

Further advances in this approach are dependent on identifying and responding to the barriers to the development and implementation of cultural methods of pest control (and their integration with other agricultural goals and practices). Foremost among these are the lack of appropriate research, training, services, equipment, and crop species and cultivars. In addition to responding to these deficiencies, changes in human values and attitudes will also be required, such as a shift in emphasis from cosmetic to nutritional food quality; and from pest elimination to management below thresholds related to broader social values.

In their present forms, cultural controls do not offer a panacea for pest prevention and control. These approaches range from environmentally supportive, knowledge- and skill-intensive techniques, such as the optimal design and management of agroecosystems in time and space (e.g., the design of integrated polycultures, management of adjacent environments [Marshall, 2004], use of companion crops, rotations, and timing of seeding, harvesting and field operations; Odum and Barrett, 2004), to the more heavy-handed interventions that can result in soil erosion and environmental degradation (e.g., intensive cultivation, summer fallowing, burning of crop residues, flooding, and destruction of uncultivated areas containing alternative hosts of pests). Clearly the former group of approaches are preferred over the latter.

Cultural controls employ designs and system management practices that make the environment less attractive to pests and less favorable for their survival, dispersal, growth and reproduction, and that promote the pest's natural controls and the vitality of the crop (Figure 1). The objective is to achieve reduction in pest numbers, either below economic injury levels, or sufficiently to allow natural and/or biological controls to take effect. Cultural control strategies aim to:

- make the crop or habitat unacceptable to pests by interfering with their oviposition preferences, host plant discrimination or location by both adults and immatures;
- make the crop unavailable to the pest in space and time by utilizing knowledge of the pest's life history, especially its dispersal and overwintering habits; and
- reduce pest survival on the crop by enhancing its natural enemies, or by altering the crop's susceptibility to the pest.

To design and effectively implement cultural controls it is necessary to have accurate knowledge of crop, pest and natural control biology, ecology and phenology, and of the weak links in pest-crop interactions.

A key advantage of most cultural controls is that they are generally the cheapest of all control measures because they usually only require modifications to normal production practices. Sometimes they do not even require extra labor, only careful planning. Often they are the only control measures that are profitable for high acreage of low value crops.

Cultural controls are often highly dependable, and in many cases can be specific. Of major importance is the fact that they do not possess some of the detrimental side-effects of biocides, namely the creation of biocide-resistance, undesirable residues in food, feed crops and the environment, the poisoning of non-target organisms and the creation of secondary pests. Their disadvantages are that they require long-term planning for greatest effectiveness and they need careful timing. They are often based on the substitution of knowledge and skills for purchased inputs and, as such, demand higher levels of competence among producers and extension agents. They may be effective for one pest, but ineffective against a closely related species. Effectiveness of cultural controls is difficult to assess and they do not always, on their own, provide complete economic control of pests. Furthermore, some cultural controls have adverse effects on fish and wildlife and may also cause erosion problems. Further details of cultural controls are provided in Hill (1990b) and Bugg and Pickett (1998).

IMPLEMENTATION OF ECOLOGICALLY BASED 'REDESIGN' APPROACHES TO PEST MANAGEMENT

For the further development of cultural controls and for redesign approaches to achieve their potential, the following initiatives will need to be taken. Although this will require considerable courage, creativity and collaboration, only by integrating these approaches can the future of effective pest management be assured.

Policy

Governments will need to establish a clear food and agriculture policy that recognizes that the primary function of the food system is to nourish all members of the population in a sustainable way, without undermining the earth's capacity to meet the other basic needs of present and future generations. Similar policies will be required for other renewable resources. Such policies will necessarily imply the prioritization of more appropriate strategies for the management of agricultural pests. As a first step towards this, collaborative inquiry and action research groups (Reason and Bradbury, 2001) could be established (with representation from the appropriate departments, together with representation from the appropriate private and public sectors) to draw up such policies for public discussion and subsequent implementation.

Regulations

Legislation must be developed to permit the appropriate government departments to, if necessary, introduce regulations that will require producers and other renewable resource managers to cooperate in the application of certain cultural control programs (e.g., collection and destruction of June drop and fallen apples (*Malus* sp.) in orchards; destruction of certain types of prunings and crop residues, synchronized planting and seeding of certain crops; and destruction of alternative hosts of certain pests). Such regulations could build on those that already exist in certain countries.

Legislation is also needed to ensure that users of biocides have not only achieved competence in their 'safe' use, but are also competent in the use of safe alternatives, including the relevant cultural controls, and in design-approaches to pest management. Governments must also evaluate how their current programs, policies and regulations encourage producers to rely on biocides, e.g., programs that inadvertently encourage the use of biocides to meet top grading standards based on cosmetic appearance (MacRae et al., 1990; MacRae, Henning, and Hill, 1993).

Research

Adequate government support is required to establish research programs related to sustainable food production systems, including research on the use of cultural methods of pest control and redesign approaches (Cox, Picone and Jackson, 2003; Menalled, Landis and Dyer, 2003). A list of priorities need to be established, with emphasis on long-term participatory (on-farm), multidisciplinary team research. Given the trend to fund research through matching grants with industry, it is imperative that areas such as cultural controls and redesign approaches, which tend to focus on structures and processes rather than on products, be designated as priority areas for support by government. To stimulate interest in this area among researchers, governments should provide funds to hold conferences specifically on such approaches to pest management. Research in this area must be done both in-house and through the awarding of grants to researchers in universities, the private sector, and also to farm organizations and innovative farmers. There is already a growing perception that all or most of the costs of research on biocides, especially their testing and evaluation, should be borne by the companies that profit from their sales, and not by taxpayer's money. This would free-up government funding for research into alternative pest management strategies (MacRae et al., 1989a).

Services

At the present time, a widely reported complaint by producers wishing to avoid the use of biocides is that most extension agents are unable to help them solve their pest problems by alternative means. To correct this situation, short courses must be available for these agents to enable them to become familiar with alternative strategies, including cultural and redesign methods for managing pests. Support will also need to be available for those seeking to establish companies providing services and supplies relating to safe and effective alternative approaches.

Training and Education

The success of most of the above recommendations are dependent on widespread access to comprehensive educational programs covering design and integrated system-management approaches to pest control (Francis, 2003; Hill and MacRae, 1988). Options include support for university 'chairs' in this area, for the establishment of courses in redesign approaches and cultural controls, and for the preparation of appropriate teaching materials, for the establishment of demonstration plots, and for the holding of relevant conferences.

Public Education

Although public awareness is shifting, much work still needs to be done to help the public to realize that most insects and their relatives are beneficial, and that cosmetic quality of food is not a reliable indicator of nutritional quality. A greater appreciation of the essential role of farmers and other renewable resource managers in society is called for, and an expanded responsibility for considering our actions in relation to the needs of future generations. A creative and intensive public education campaign will need to be mounted to promote such understanding.

FINAL THOUGHTS FOR REFLECTION

The approach to further developments in pest control being promoted here is based on the following three assumptions (Hill et al., 2001).

1. There is a vast, largely untapped territory available to all of us for thinking and acting in new and effective ways with respect to any area of endeavor–in this case pest control.

2. The knowledge required to get started in transforming our thinking and action already exists–it includes general understandings in a range of areas–especially ecology, biology and the natural sciences–but also psychology (Hill, 2001), philosophy, sociology, and all of the areas dealing with social processes including politics, economics, business, education, the arts, and spirituality. It is particularly concerned with deep understandings of the structure (design) and functioning (management) of systems, their diversity with respect to time and place (contextual qualities and other contextual factors), maintenance (sustainability), and processes involved in change (co-evolution) over time.
3. The key to progressive change is to recognize, acknowledge, and celebrate past generically related initiatives and to support the taking of further small meaningful initiatives that are consistent with the emerging paradigm. The high visibility of such do-able initiatives enables others with similar interests to copy them, and those in other areas to be inspired by them. During this process, it is important to critically evaluate all actions and outcomes and so learn our way into more sustainable and meaningful futures.

REFERENCES

Bookchin, M. (1980). *Towards an Ecological Society.* Montreal, QC: Black Rose Books.

Bookchin, M. (1982). *The Ecology of Freedom.* Palo Alto, CA: Cheshire Books.

Bookchin, M. (1986) [1971]. *Post-Scarcity Anarchism*, 2nd edn. Montreal, QC: Black Rose Books.

Bookchin, M. (1995) [1990]. *The Philosophy of Social Ecology*, 2nd edition. Montreal, QC: Black Rose Books.

Bugg, R. L. and C. H. Pickett. (1998). *Enhancing Biological Control; Habitat Management to Promote Natural Enemies of Agricultural Pests.* Berkley, CA: University of California Press.

Clements, D. R. and A. Shrestha. (2004). New dimensions in agroecology for developing a biological approach to crop production. *Journal of Crop Improvement* 11(1/2): 1-20.

Cox, T.A., Picone, J.C. and W. Jackson. (2004). Research priorities in natural systems agriculture. *Journal of Crop Improvement* 12(1/2): 511-531.

Denzin, N. K. and Y. S. Lincoln, ed. (2000). *Handbook of Qualitative Research*, 2nd edition. Thousand Oaks, CA: Sage.

Francis, C. A. (2003). Education in agroecology and integrated systems. *Journal of Crop Improvement* 11(1/2): 21-43.

Hill, S. B. (1985). Redesigning the food system for sustainability. *Alternatives* 12 (3/4): 32-36.
Hill, S. B. (1990a). Pest control in sustainable agriculture. *Proceedings of the Entomological Society of Ontario* 121:5-12.
Hill, S. B. (1990b). Cultural methods of pest, primarily insect control. *Proceedings of the Annual Meeting of the Canadian Pest Management Society* 36:35-49.
Hill, S. B. (1991). Ecological and psychological pre-requisites for the establishment of sustainable prairie agricultural communities. In *Alternative Futures for Prairie Agricultural Communities*, ed. J. Martin, Edmonton, AB: Faculty of Extension, University of Alberta, pp. 197-229.
Hill, S. B. (1993). Environmental sustainability and the redesign of agroecosystems. In *Sustainable Agriculture in Egypt*, ed. M. A. Faris and M. H. Khan, Boulder, CO: Lynne Rienner Press, pp. 47-58.
Hill, S. B. (1998). Redesigning agroecosystems for environmental sustainability: A deep systems approach. *Systems Research and Behavioral Science* 15:391-402.
Hill, S. B. (1999). Social ecology as future stories. *A Social Ecology Journal* 1: 197-208.
Hill, S. B. (2001). Working with processes of change, particularly psychological processes, when implementing sustainable agriculture. In *The best of . . . Exploring Sustainable Alternatives: An Introduction to Sustainable Agriculture*, ed. H. Haidn, Saskatoon, SK: Canadian Centre for Sustainable Agriculture, pp. 125-134.
Hill, S. B. and R. MacRae. (1988). Developing sustainable agriculture education in Canada. *Agriculture and Human Values* 5(4): 92-95.
Hill, S. B. and R. MacRae. (1992). Organic farming in Canada. *Agriculture, Ecosystems and Environment* 39:71-84.
Hill, S. B. and R. MacRae. (1995). Conceptual frameworks for the transition from conventional to sustainable agriculture. *Journal of Sustainable Agriculture* 7:81-87.
Hill, S. B., C. Vincent, and G. Chouinard. (1999). Evolving ecosystems approaches to fruit insect pest management. *Agriculture, Ecosystems and Environment* 73:107-110.
Hill, S. B., S. Wilson, K. Watson, and K. Lambert. (2001). Learning ecology and transformative change in education. Toronto, ON: Fourth International Conference of Transformative Learning.
Huxley, J. (1952). *Evolution in Action*. London: Scientific Book Club.
Huxley, J. (1964). *Essays of a Humanist* (Education and Humanism p.132). Harmondsworth, UK: Penguin.
Lafleur, G. and S. B. Hill. (1987). Spring migration, within-orchard dispersal, and apple-tree preference of plum curculio (Coleoptera: Curculionidae) in southern Quebec. *Journal of Economic Entomology* 80:1173-1187.
Lafleur, G., S. B. Hill, and C. Vincent. (1987). Fall migration, hibernation site selection, and associated winter mortality of plum curculio (Coleoptera: Curculionidae) in a Quebec apple orchard. *Journal of Economic Entomology* 80:1151-1172.
Light, A., ed. (1998). *Social Ecology After Bookchin*. New York: Guilford Press.
MacRae, R. J., S. B. Hill, J. Henning, and G. R. Mehuys. (1989a). Agricultural science and sustainable agriculture: A review of the existing scientific barriers to sustainable food production and potential solutions. *Biological Agriculture and Horticulture* 6(3): 173-219.

MacRae, R. J., S. B. Hill, J. Henning, and G. R. Mehuys. (1989b). Farm-scale agronomic and economic conversion from conventional to sustainable agriculture. *Advances in Agronomy* 43:155-198.

MacRae, R. J., S. B. Hill, J. Henning, and A. J. Bentley. (1990). Policies, programs and regulations to support the transition to sustainable agriculture in Canada. *American Journal of Alternative Agriculture* 5(2):76-92.

MacRae, R. J., J. Henning, and S. B. Hill. (1993). Strategies to overcome barriers to the development of sustainable agriculture in Canada: the role of agribusiness. *Journal Agriculture and Environmental Ethics* 6(1):21-51.

Marshall, E. J. P. (2004). Agricultural landscapes: Field margin habitats and their interaction with crop production. *Journal of Crop Improvement* 12(1/2): 365-404.

Menalled, F. D., D. A. Landis, and L. E. Dyer (2004). Research and extension supporting ecologically-based IPM systems. *Journal of Crop Improvement* 11(1/2): 153-174.

Miller, A. (1999). *Environmental Problem Solving: Psychological Barriers to Adaptive Change.* New York: Springer.

Mulligan, M. and S. B. Hill. (2001). *Ecological Pioneers: A Social History of Australian Ecological Thought and Action.* Melbourne, VIC: Cambridge University Press.

Odum, E. P. and G. W. Barrett. (2004). Redesigning industrial agroecosystems: Incorporating more ecological processes and reducing pollution. *Journal of Crop Improvement* 11(1/2): 45-60.

Reason, P. and H. Bradbury, ed. (2001). *Handbook of Action Research.* London: Sage.

Savory, A. and J. Butterfield. (1999). *Holistic Management: A New Framework for Decision Making*, 2nd edition, Washington DC: Island Press.

Thrupp, L. A. (2004). The importance of biodiversity in agroecosystems. *Journal of Crop Improvement* 12(1/2): 315-337.

Vincent, C., G. Chouinard, and S. B. Hill. (1999). Progress in plum curculio management: A review. *Agriculture, Ecosystems and Environment* 73:167-175.

Voisin, A. (1959). *Soil, Grass and Cancer.* London: Crosby Lockwood.

Watson, D. (1996). *Beyond Bookchin: Preface to a Future Social Ecology.* Brooklyn, NY: Autonomedia.

Research Priorities in Natural Systems Agriculture

T. S. Cox
C. Picone
W. Jackson

SUMMARY. Over the course of millennia and most dramatically in recent years, agriculture has eroded natural capital as it has supplied human demands. These tendencies can be offset by infusions of fossil fuels for fertility, pest control and traction, but only partially and not over the long term. This could be called the "problem *of* agriculture," in contrast to problems *in* agriculture. Natural ecosystems are unmatched for efficient nutrient recycling, solar energy use, and biodiversity preservation, but they cannot feed dense human populations. An ecological agriculture that is a synthesis of natural and agricultural systems can save soils from erosion and reduce chemical and water use. Reducing chemical pollution of air, water and soils will conserve wildlife habitat, improve water quantity and quality and protect human health. By practicing an agriculture that has conservation as a direct result, farmers can become better stewards of the land. Lower input costs and inherent sustained soil fertility will mean more profit for farmers and communities. Principles of natural systems agriculture (NSA) are applicable to any food- or fiber-production system worldwide. In the central Great Plains of the

T. S. Cox is Senior Research Scientist, C. Picone is Research Scientist, and W. Jackson is President, The Land Institute, 2440 East Water Well Road, Salina, KS 67401 (E-mail: theland@landinstitute.org).

[Haworth co-indexing entry note]: "Research Priorities in Natural Systems Agriculture." Cox, T. S., C. Picone, and W. Jackson. Co-published simultaneously in *Journal of Crop Improvement* (Food Products Press, an imprint of The Haworth Press, Inc.) Vol. 12, No. 1/2 (#23/24), 2004, pp. 511-531; and: *New Dimensions in Agroecology* (ed: David Clements, and Anil Shrestha) Food Products Press, an imprint of The Haworth Press, Inc., 2004, pp. 511-531. Single or multiple copies of this article are available for a fee from The Haworth Document Delivery Service [1-800-HAWORTH, 9:00 a.m. - 5:00 p.m. (EST). E-mail address: docdelivery@haworthpress.com].

http://www.haworthpress.com/web/JCRIP

Digital Object Identifer: 10.1300/J411v12n01_10

United States, agriculture should mimic the native prairie, which is a polyculture of herbaceous perennial plants. The Land Institute has spent 25 years studying the structure and function of prairies. Now we and other research organizations are breeding perennial grain crops, including cool- and warm-season grasses, legumes, and composites. We are following two parallel strategies: selecting wild perennials for greater seed production and other traits of domestication and hybridizing wild perennials with annual crops to combine desired traits. Although the breeding process will take many years, the effort will be repaid many times over if we achieve our goal: a lasting solution to the 10,000-year-old problem of agriculture.

KEYWORDS. Perennial polyculture, prairie, agroecology, sustainable agriculture, plant breeding

INTRODUCTION

Since the beginning of agriculture 10,000 years ago, the human diet has depended primarily on the deficit spending of ecological capital. All of the important cereal seeds that sustain us–wheat, rice, corn, beans, and others–are annuals that require disturbance of the soil and sowing every growing season. That disturbed soil is vulnerable to the forces of wind and rain. Yet without the calories from annual grains, humanity would lose more than 70% of its diet and most of us would quickly starve. Wild plants, the ancestors of the high-yielding annual crops in our food inventory, made it possible for our ancestors to launch the agriculture revolution. Although the world's population was far smaller in ancient times, even then we were spending our ecological capital. From the beginning, soil erosion has been linked repeatedly to the collapse of ancient civilizations (Lowdermilk, 1953; Runnels, 1995).

Let us assume we do manage to feed 10 billion people by 2050, as foreseen by Evans (1998). If the feat is accomplished by current methods, every kilogram (kg) of grain produced between now and then represents food stolen from the tables of future generations. Tillage on sloping land always risks soil erosion (Lowdermilk, 1953; Hillel, 1991), and very few fields have zero slope. Due to industrial agricultural practices, the earth has lost a third of its arable land just in the past few decades, and natural soil fertility is declining worldwide (Pimentel et al., 1995). Therefore, wholly new agricultural systems will be necessary if we are to feed ourselves while reversing the millennia-old trend of wasting the ecological capital on which agriculture depends.

With such a large proportion of the earth's surface now needed to feed humanity, we can no longer afford to grow our food "here" while protecting nature "there." The two functions must be performed on the same landscape. The natural systems paradigm, designed to move agriculture from an extractive to a renewable economy, could help set the precedent for moving society in general toward a sustainable future.

Measures such as no-till farming, contour plowing, and crop rotation have slowed soil erosion , but not stopped it. As recently as 1997, over 4500 kg of soil eroded annually from an average hectare (ha) of US farmland (USDA 1999)–far more than can be replaced. We need systems that can be maintained for centuries, not decades. Research such as that we are doing at The Land Institute is needed to eliminate the deficit spending of ecological capital.

THE PRAIRIE AS A MODEL

As a model for a sustainable food-producing ecosystem, The Land Institute has looked to native prairies that once covered the central Great Plains of North America. We use prairies as a model for several reasons. First, they are very productive, yielding 400-570 gm^{-2} of aboveground biomass per year (Knapp and Smith, 2001; Knapp et al., 1999; Piper, 1994). Second, native prairies, in contrast to cropland, exhibit exceptionally high soil quality and low rates of erosion and leaching (e.g., Noll et al., 1995; Baer et al., 2000). Third, prairies feature efficient nutrient cycling through complementary plant groups and complex soil communities. Fourth, they avoid epidemics of insects and pathogens while adequately managing weeds (Weaver, 1954). Finally, like all native terrestrial ecosystems, they run on contemporary sunlight rather than on fossil fuels. In addition to these functions, our domestic prairie would be expected to supply grains for human consumption, something the native system does not do. We propose that by mimicking the structure of the prairie, a domestic, grain-producing analog can achieve similar functions.

In order to have a crop system that emulates the structure of prairie, we must assemble a diverse community of perennial crops, allowing for grazing or burning but little or no soil disturbance (Piper, 1999). Prairie perennials–in contrast to annual crops–have extensive root systems that reach over two meters down into the soil that they thoroughly permeate year-round. As a consequence, the prairie uses water and nutrients more efficiently than annual crops, and it is much less prone to environmental extremes such as drought (Weaver 1954). Moreover, the perennial habit of prairie plants is critical to soil health. Perennial grasses have been repeatedly shown to improve soil quality relative to tilled crops. For example, restored prairie soils under the U.S. government's Conservation Reserve Program have better aggregate structure, increased microbial biomass, greater amounts of fungus hyphae, and reduced labile nitrates, than do cropped soils (Karlen et al., 1999). Perennial plants also invest more carbon (C) in below-ground structures than do annuals. Even when compared to no-till crops of annuals, perennial plants sequester far more soil C (J. Brown, USDA-NRCS, pers. comm; Gilley et al., 1997).

Plant diversity is a critical component of prairie structure (e.g., Weaver 1954; Piper, 1999); a perennial crop in monoculture would likely be poor, unsustainable substitute. More specifically, wherever there is native prairie in North America, the plant community is dominated by four functional groups: warm and cool season grasses, legumes and sunflower family. In prairies studied by The Land Institute, grasses comprised 59 to 88% of the plant biomass, which is overwhelmingly dominated by warm-season grasses (Piper, 1994). Legumes and composites comprise 0 to 17% and 7 to 20%, respectively, depending on soil fertility. All other plant groups generally make up less than 10% of the biomass, so a crop system could conceivably function like a prairie with species from the four functional groups listed above. It may be fortuitous that the major crop providing most of the calories which feed humanity fall into those four groups. In the section on "The agro-ecological principles for building a domestic prairie," we will describe how we have attempted to predict the levels of plant diversity and optimal species combinations that are required to construct a mimic of the prairie.

A fully functioning natural systems agriculture, or NSA (Piper, 1999), in the form of a domestic prairie, must await development of perennial, grain-producing crops. But its experimental development can proceed by incorporating populations of perennial crops-in-the-making ("prototypes") as they are spun off by the breeding programs. The Land Institute has studied polycultures of prairie plants for many years (Piper, 1999; Jackson and Jackson, 1999), and the next step is to incorporate perennial, grain-producing plants into such experiments. As our whole-system research begins to reveal what a future NSA will look like and how it will work, those same experiments will provide an environment in which to test our breeding populations as they evolve. The two research streams–agroecology and plant breeding–will continue informing each other until, eventually, they flow together to form a full-scale NSA.

BREEDING GRAIN CROPS FOR A DOMESTIC PRAIRIE

Evolution has produced two broad categories of herbaceous plant species: perennials with low to moderate seed production in any single season, and annuals with higher seed production (Harper, 1977; Snaydon, 1984). Humans over the past 10,000 years have taken the path of least resistance, pushing the grain yields of annual grasses, legumes, and composites to higher and higher levels (Whyte, 1977; Wagoner, 1990; Cox et al., 2002).

The negative relationship often noted in wild plants between pereniality and seed yield (Wagoner, 1990) is consistent with life-history theory (e.g., Gadgil and Solbrig, 1972). Gardner (1991) has provided a physiological application of the theory to crop plants. But these theories are based on observation of existing species, which are products of divergent selection. Correlations based on phenotypes of extant species cannot predict response to selection in breeding populations derived from hybridizing different genotypes or species.

Selection by humans almost always pushes the phenotypes of crop plants well beyond the range of their wild ancestors. Therefore, perennial genotypes that produce acceptable grain yields are not inconceivable, given proper selection criteria. The pool of C to be shared by reproductive and vegetative structures is not necessarily fixed genetically (Jackson and Jackson, 1999). Grain production can be sink-limited (Slafer and Savin, 1994), so that more or larger reproductive structures will induce greater photosynthetic output. Basal or rhizome-derived tillers on a grass plant are largely self-sufficient (Nyahoza, Marshall, and Sagar, 1973; Jackson and Dewald, 1994), and much of the photosynthate for seed development can come from floral structures (Blum, 1985). Perennials also may maintain green tissue late in the growing season, after seed harvest, and use the additional photosynthate to build up reserves for the next year (Scheinost et al., 2001). For these and other reasons, there need not be a gram-for-gram tradeoff between grain and structures that hold reserves for regrowth.

Even when yield and persistence are negatively associated, selection should not be ruled out. Plant breeders routinely work against negative correlations, but simultaneous selection for negatively correlated traits can succeed if some compromise is made between traits (Cox et al., 2002).

In developing a domestic prairie, we must first breed perennial grain-producing representatives of the four functional groups that dominate North American prairies. Depending upon the species involved, we can either select perennial species for higher seed yield or combine the chromosomes of related perennial and annual species to develop wholly new crop species. For some species, both options are possible.

In our breeding programs, we are concentrating on the cool-season grasses wheat (*Triticum* spp.), rye (*Secale* spp.), triticale (× *Triticosecale*), and intermediate wheatgrass (*Thinopyrum intermedium*); the warm-season grasses sorghum (*Sorghum* spp.) and eastern gamagrass (*Tripsacum dactyloides*); the legumes Illinois bundleflower (*Desmanthus illinoensis*) and chickpea (*Cicer* spp.); and the composite sunflower (*Helianthus* spp.). These are the species that we can handle with our current research capacity, but Cox et al. (2002) listed at least nine additional species of perennial grains that could, conceivably, be developed by breeding. Many other perennial species could be considered candidates for domestication (Jackson and Jackson, 1999). To breed a full range of perennial grains will require a cooperative effort by many research institutions.

Past and current approaches to breeding perennial cool-season grasses provide a useful comparison between direct domestication and interspecific hybridization. Researchers at the Rodale Institute and USDA-ARS have carried out a 10-year program to select intermediate wheatgrass (a distant, perennial relative of wheat) for high multi-year grain yield, larger seeds, and other traits (Wagoner, 1995; Wagoner, van der Grinten, and Drinkwater, 1996). Resulting yield increases have been a dramatic 25% per cycle, but because the grain yield of the original wheatgrass population was very low, and cycles are long in a perennial, it would take, hypothetically, 50 years to achieve grain yields comparable to that of annual wheat. On the other hand, 50 years does not appear to be very long when compared with the 10,000 years it has taken to domesticate and breed today's small grain crops.

Wheat (the hexaploid species *Triticum aestivum* and the tetraploid *T. turgidum*) can be hybridized with more than 50 different perennial species (Jiang, Friebe, and Gill, 1994; Sharma, 1995). Perennial wheat breeding programs have concentrated on wheatgrasses of the genus

Thinopyrum, including *Th. intermedium*. Efforts in the then-USSR, the USA, and Canada to combine the perenniality of several wheatgrass species with the grain yield of wheat have achieved yields tending to fall between 50% and 65% of those of comparable annual wheats (Suneson, El Sharkawy, and Hall, 1963; Tsitsin, 1965; Fatih, 1983; Wagoner, 1990; Scheinost et al., 2001); however, because they are usually partial amphiploids, most such "perennial wheats" are often chromosomally unstable (Cai, Jones, and Murray, 1998, 2001; Jones, Zhang, and Wang, 1999; Cox et al., 2002). Furthermore, no accessions in a collection of wheat/perennial amphiploids survived or re-grew after harvest in 2001 in our initial observation plots at The Land Institute. Our own breeding program, initiated in 2001, is aimed at developing perennial wheat hybrids that are better adapted to conditions in the southern Great Plains. We have begun crossing Kansas-adapted hard winter wheats (*T. aestivum*), winter durums (*T. turgidum*), and the ancient wheat *T. turgidum* ssp. carthlicum with perennial species of *Thinopyrum* and *Secale*. We will backcross both to the annual and to the perennial parents.

Rye (*Secale cereale*) has a high degree of chromosome homology with its direct perennial ancestor, *S. montanum*, providing an opportunity for development of a perennial rye for grain production. However, breeders have been stymied by the strong association between perenniality and sterility in *S. cereale/S. montanum* populations (Reimann-Philipp, 1995). A chain of translocations involving three of rye's seven pairs of chromosomes separates the two species (Stutz, 1957; Devos et al., 1993), and gene(s) from *S. montanum* governing perenniality appear to be are located on one or more of the translocated chromosomes. Because of the resulting gametic duplications or deficiencies, most perennial plants in interspecific populations have low seed set. Induced tetraploidy has improved seed size (Reimann-Philipp, 1995), and backcrossing to *S. montanum* has produced a useful perennial forage rye (Oram, 1996), but sterility is still a problem.

The Land Institute's longest-running project in breeding a warm-season grass is aimed at a perennial grain sorghum. Annual grain sorghum (*Sorghum bicolor*), a diploid, is well-adapted in the central and southern Great Plains of the U.S., as is a related perennial weed, johnsongrass (*S. halapense*), a tetraploid. We have dozens of strains of perennial sorghum in the breeding program. To produce them, Piper and Kulakow (1994) crossed *S. halapense* with induced-tetraploid grain sorghum lines. In an interspecific F_4 population, approximately 40% of plants were rhizomatous, but rhizome production dropped to near zero when hybrid plants were backcrossed to tetraploid *S. bicolor* in an effort to increase grain yield.

We have selected for winter survival among rare rhizomatous BC_2 plants and their selfed progeny. The phenotypes of winterhardy selections remain dissimilar from those of the cultivated parent. Most selections are much taller and later-maturing than either parent, have open panicles and small seed, and produce many tillers, although they produce fewer tillers and much shorter rhizomes than johnsongrass. A moderately tillering sorghum might well be desirable, and Cox, House, and Frey (1985) found no significant correlation between compactness of the panicle and grain yield in crosses involving wild annual sorghums. However, selection for shortness, earliness, and larger seeds will be essential. Piper and Kulakow (1994) concluded that development of a winterhardy sorghum with acceptable yield is feasible, through selection for greater biomass and reallocation of photosynthate to seed production. We have pollinated diverse male-sterile diploid grain sorghums with tetraploid perennials to obtain tetraploid progenies via unreduced gametes (Hadley, 1958). Our goal is to improve and diversify the tetraploid gene pool as rapidly as possible.

The Land Institute has a long history of work with another warm-season grass, eastern gamagrass. Wagoner (1990) and Cox et al. (2002) described in detail its status as a potential grain crop, and the species' most discouraging characteristic: very low seed yield. At that time, interest had been stimulated by the discovery of a gynomonoecious, or pistillate, mutant (DeWald and Dayton, 1985) in which pistillate and perfect spikelets replaced the staminate spikelets of the normal inflorescence, providing a threefold yield increase (Jackson and Jackson, 1999).

Aside from the pistillate mutant, there is a huge pool of genetic variability available in the *T. dactyloides* (Newell and deWet, 1974; Wright, Taliaferro and Horn, 1983). The maximum seed yield of eastern gamagrass in plots at The Land Institute has been 240 kg ha^{-1}, in the third year after sowing (Piper, 1999). It remains to be seen how rapidly yield can be improved through selection within pistillate or other populations. The food quality of eastern gamagrass is excellent (Bargman, 1989). But the hard fruitcase of *Tripsacum* weighs almost three times as much as the seed and makes processing difficult.

It was an extremely rare mutation in annual teosinte that freed the kernel from the fruitcase and allowed its use as a grain and the development of maize (Iltis, 2000). That gene, and others, may have to be introgressed from maize (*Zea mays*) if the needed variation cannot be found in eastern gamagrass. The two species have been hybridized many times, beginning with the work of Manglesdorf and Reeves (1931). Plants of the diploid (2n = 36) or tetraploid (2n = 72) races may be crossed with maize. If *T. dactyloides* is used to pollinate maize, embryo rescue is necessary (deWet et al., 1973; James, 1979), but if maize is used as the male, some hybrid seed may be obtained without rescue (deWet et al, 1973). In addition, several strains of popcorn, when pollinated with tetraploid *T. dactyloides*, produce large amounts of hybrid seed that does not require embryo rescue (Kindiger and Beckett, 1992). However, all previous attempts to develop perennial maize from crosses with *T. dactyloides*, the tropical tetraploid *Zea perennis*, or the tropical diploid *Z. diploperennis* have failed (see Cox et al., 2002).

We are breeding perennial legumes for two purposes: as producers of high-protein grain and as sources of nitrogen (N) for the grass and composite crops. In the medium term, the perennial prairie legume Illinois bundleflower is our prime candidate as a grain legume for NSA. It has the capacity to produce seed yields similar to those of annual wheat, but with much higher protein content (Kulakow, Benson, and Vail, 1990; Kulakow, 1999). The seed has never been utilized for human food, but it contains no known toxins or digestive inhibitors, and our initial investigations indicate that it can be made palatable through simple processing. We are launching a breeding program to improve yield and other traits in bundleflower. Meanwhile, we have developed non-shattering strains of the species ready to use as NSA prototypes.

We plan to expand our grain-legume breeding program to include chickpea (*Cicer arietinum*), which has perennial relatives. The perennial species are little studied; small numbers of accessions of 12 species are maintained at the Western Region Plant Introduction Station in Pullman, Washington. Two species, *C. anatolicum* and *C. songaricum*, have survived for 10 years in the field and continue to produce seed. One accession of *C. songaricum* is white-flowered, has some degree of shattering resistance, and produces seed similar to the commercial 'desi' type. The presence of these traits suggests that *C. songaricum* was once cultivated in Asia (F.J. Muehlbauer, USDA-ARS, personal communication).

The prairie legume with the highest grain yield in our studies (Piper, 1992, 1993) is wild senna (*Cassia marilandica*). Its crop-like growth habit and high yield make it an attractive prospect. However, we have not yet pursued a breeding program in this species, because it does not form associations with N-fixing *Rhizobium* bacteria.

As non-grain-producing, N-fixing components of NSA, we have available a very wide range of prairie species, including false indigo

(*Amorpha fruticosa*), leadplant (*A. canescens*), purple prairie clover (*Dalia purpurea*), white prairie clover (*D. candida*), and roundhead lespedeza (*Lespedeza capitata*). We also are investigating non-native forage legumes. We will not have full-scale breeding programs in these species, but we are testing them to determine which will be most useful as non-harvested legumes in polycultures.

The prairie harbors many perennial species of *Helianthus*, the genus to which the domestic sunflower (*H. annuus*) belongs. The Land Institute has studied the common prairie species *H. maximiliani* as a candidate for direct domestication (Jackson and Jackson, 1999). It is a vigorous prairie plant with small but edible seed. Seed yield estimates have varied widely (Jackson and Jackson, 1999; Piper, 1999), and yields tend to decline after the first year of propagation in monoculture. We are now collecting and evaluating hundreds of accessions of Maximilian sunflower to identify potential domesticates as well as ones that should be crossed to cultivated sunflower to produce a new, perennial oilseed crop.

According to data compiled by Jan (1997), along with a recent study by Sukno et al. (1999), 20 perennial species have been hybridized with *H. annuus*. In approximately half of the crosses, F_1 plants were perennial; sometimes one accession of a species produced perennial F_1s, while another produced annuals. The majority of hybrids have enough fertility to be backcrossed. We began a perennial sunflower-breeding program in 2001, beginning with crosses between *H. annuus* and two perennial species, *H. maximiliani* and *H. salicifolius*.

Developing the first perennial grain genotype with acceptable yield will be a demanding process in itself, but that will be only the beginning; no new agricultural system can be based on a single genotype per species. Furthermore, breeding of perennial grains will quickly reach a dead end without sufficient genetic diversity. During the domestication process, most crop species suffered a similar "founder effect" that restricted the gene pool on which early cultivators drew (Ladizinsky, 1985). If new perennial crops are to be developed on a shorter time scale, the initial genetic bottleneck must be forced wider, sooner.

To this point, we have not mentioned a role for genetic engineering in NSA. That is because no role for transgenes has become apparent. No research to date suggests that perenniality is governed by a single gene, or even two or three genes, in any crop or crop relative. Although some wheat amphiploids with as little as 25 to 30% of their genome derived from the perennial parent are themselves perennial (Cai, Jones, and Murray, 2001), more typically (in rye, triticale, sorghum, maize, soybean, and sunflower), dilution of the perennial-derived genome to below approximately 50% often eliminates perenniality (Cox et al., 2002). Furthermore, breeders have yet to backcross any gene or chromosome conditioning perenniality into any annual genotype.

Transgenic technology may be useful, once perennial grain crops have been developed, in improving their pest resistance, food quality, or other more simply inherited traits; however, we do not intend to use such techniques at The Land Institute in the foreseeable future. Before we would consider utilizing transgenic technology, *all* of the following conditions would have to be met: (a) the gene(s) to be transferred would govern expression of a necessary trait that we could not introduce via any other practical method; (b) plant varieties carrying the gene(s) would be non-patented, unburdened by any other intellectual property agreements, and open to free public use; and (c) in our judgement, the gene(s) and methodologies would be thoroughly tested and represent no threat to gene pools, the environment, human health, or social and economic justice.

None of these conditions is fulfilled by any engineered gene today, and we do not expect all of them to be fulfilled for many years, if ever. But sexual hybrids and traditional breeding, augmented by other forms of genetic technology, will be sufficient to develop perennial grains.

THE AGROECOLOGICAL PRINCIPLES FOR BUILDING A DOMESTIC PRAIRIE

Development of perennial grains is only one step in bringing NSA to the field. Monocultures of perennial grains, lacking the opportunity for annual crop rotation, would probably experience greater problems with pests and nutrient deficiencies than do annual monocultures. To have a resilient agroecosystem, we must assemble a range of species into a polyculture that will (a) produce adequate yields of edible grain over several years without tillage or re-sowing; (b) yield as much as or more than an equivalent set of monocultures; (c) manage insects, pathogens, and weeds; and (d) compensate for N, phosphorus, and other nutrients removed in harvest (Piper, 1999; Jackson and Jackson, 1999).

Lacking perennial grain crops to grow as a mixture, we have so far depended on the native prairie to provide the basic information on how a perennial polyculture functions. Over our first 25 years, Land Institute staff and interns have studied the prairie and its component species, to answer questions about the relative productivity of mixtures versus monocultures, the effects of weeds, insects, and plant diseases, and the cycling of soil nutrients (Jackson and Jackson, 1999).

Not surprisingly, perennials in mixtures will typically out-yield the component species in monoculture. In the second year of growth in one study (Jackson and Jackson, 1999), bicultures of bundleflower and wild senna had yields 29% higher than the mean of their monocultures. Other studies showed that the grasses and legumes exploit different regions of the soil for water and nutrients (Jackson and Jackson, 1999). Bicultures of bundleflower and eastern gamagrass had a 19% advantage (Piper, 1999), and tricultures of bundleflower, gamagrass, and mammoth wild-rye had a 26% advantage (Barker and Piper, 1995).

Growing plants in polycultures also can reduce damage by pests. Populations of a bundleflower pest, the beetle *Anomoea flavokansiensis*, were significantly reduced from year to year in polycultures, relative to monocultures (Piper, 1996). Maize dwarf mosaic virus and sugarcane mosaic virus infections were reduced in eastern gamagrass when it was grown in mixtures with bundleflower (Piper, Handley, and Kulakow, 1996). A grass understory cushions raindrops and prevents infection of bundleflower by splashborne leaf-spot diseases (Jackson and Jackson, 1999). Mitchell, Tilman, and Groth (2002) recently demonstrated that fungal disease of perennials is significantly reduced when they are grown in diverse mixtures.

Future research on the role of diversity and spatial arrangements of crops will be even more necessary in NSA than in the current agricultural system. Annual plants can escape many pathogens and other pests by allowing all their vegetative structures to die back each year; they survive to the next year(s) as seed. Pests can then be further thwarted by crop rotation. In contrast, perennials must have vegetative structures and energy reserves that persist through the whole year, and then from year to year, thus making them more susceptible to disease and herbivores. Lack of tillage will also allow litter to accumulate, which can increase pressure from certain diseases. A solution to these problems does not lie simply in synthetic pesticides. Their use tends to create new problems (Soule and Piper, 1992), and our current, increasing reliance on these toxins is already exhibiting global impacts on human and ecological health (Steingraber, 1998; Tilman et al., 2001). The only long-term solution to the pressures on perennial crops from pathogens and insects is to build in the resilience to pest outbreaks that is found in native prairie.

In order to mimic the prairie's resilience to pathogens and insects, several fundamental questions will need to be answered for future perennial polycultures. First, how much species "diversity" is required to thwart such pests in a particular crop system? A polyculture would be nearly impossible to manage if it required the high diversity found in a native prairie, but such high species richness is probably excessive for this particular ecosystem function. Second, what level of functional-group diversity is required? Or as a corollary, which combinations of functional groups are more effective than others? Finally, what will be the optimal spatial arrangements of species mixtures? Perhaps resilience to pests can emerge from intercropped strips of different species and functional groups. This arrangement would be easiest to plant, weed, and harvest. Or it may be necessary to grow polycultures in complete mixtures to gain the prairie's resilience.

Annual weeds are likely to be best suppressed by crop communities that are both perennial *and* diverse. For example, the native perennials eastern gamagrass and Maximilian sunflower are very effective in suppressing annual weeds (Piper, 1993; Jackson and Jackson, 1999). In an untilled soil, perennials have a competitive advantage over annuals, especially weeds. In addition, higher diversity of perennial species is very effective in quickly eliminating annual weeds (Piper, 1999).

Another area of current research is the role of diversity in "community assembly rules" (Pimm, 1991, 1997). That is, how does diversity affect the rate at which the plant community reaches a relatively stable state? Our preliminary results show that a community founded with only four prairie species (one representative of each functional group) does not lead quickly enough to a stable, diverse polyculture, largely because annual weeds are initially so abundant. In contrast, a community founded with eight or more species reaches stability much faster, and is more effective at controlling weeds (Piper, 1999). However, much remains to be learned about how initial conditions affect the later species and genotype composition of the polyculture.

Recently, our research has reached well below the soil surface. In order to reduce dependence on synthetic fertilizers, NSA must treat soil not simply as a substrate for crops, but as an ecosystem. Specifically, a healthy, productive soil requires complex foodwebs that include symbiotic microbes, decomposers, detritivores, and macrofauna. Our research in this area has focused on the community ecology of one of these essential groups–arbuscular mycorrhizal (AM) fungi.

AM fungi are found in the roots of about 80% to 90% of plant species, including most crops (Smith and Read, 1997). Plants provide carbohydrates to the fungi in return for improved nutrient uptake, and thus the fungi act as "biofertilizers." In certain conditions, the fungi can also help resist root pathogens (Azcón-Aguilar and Barea, 1996), suppress non-host weeds (Jordan, 2000), reduce effects of toxic metals (Kaldorf et al. 1999), and–perhaps most importantly–bind soil particles together into aggregates that improve soil structure (Jastrow, Miller, and Lussenhop, 1998).

Our as-yet unpublished research in this area has shown that tillage decreases fungus diversity by about 23%, and that untilled crops can recover high fungus diversity after only a few years without soil disturbance. Of 20 common species, at least seven are significantly reduced by tillage, while one is more abundant in tilled soils. In contrast to soil disturbance, plant diversity has shown minimal effect on the community composition of AM fungi. The particular combination of species in a crop system has had a much larger effect on fungus community structure than has diversity *per se*. We are now assessing which AM fungus species associate best with different plant species and with different functional groups. We will soon begin assessing the effectiveness of different species in terms of the multiple roles of these fungi, from nutrient uptake to soil aggregation. Ultimately, we hope to understand the cropping practices that will help promote an optimal AM fungus community for particular types of perennial polycultures.

In addition to studies of the soil community, other research is addressing nutrient dynamics directly. Some of our studies have suggested that legumes in the prairie can supply at least some of the N needs of adjacent grasses. For example, eastern gamagrass produced higher yields over years in mixtures with bundleflower (Piper, 1998).

One of the fundamental questions of N fertility will be the management of a likely trade-off between high grass-seed yield and high N fixation by legumes (T. Crews, Prescott College, personal communication). Grasses typically cannot reach maximal seed yield unless soil N is abundant; but legumes will not fix much atmospheric N unless soil N is limited. If NSA is to reduce or eliminate inputs of synthetic N, it must address these contradictory requirements. We are currently comparing seed yields of a grass that is grown in different spatial and temporal arrangements with the perennial legume alfalfa (*Medicago sativa*). Perhaps maximal seed yield would be achieved by deviating somewhat from the prairie model and isolating the grass polycultures from legume polycultures in space or time. Soil N could be kept high in grass strips, and low in legume strips, by moving legume mulch to the grass strips. Alternatively, if grasses are more aggressive than legumes at foraging for soil N, the trade-off would only be apparent at a very small scale, and these functional groups would not need to be planted separately.

DESIGNING DOMESTIC PRAIRIES FOR GRAIN PRODUCTION

We can learn much from the prairie, but we will never be able to apply that knowledge or even fully answer questions about yield, pests, fertility, or other aspects of food production until we establish communities of plants more similar to the crops that will someday form the basis of NSA. Well before we have perennial-grain cultivars suitable for production, we can and should begin incorporating our breeding lines and populations into experimental polycultures, where they can serve as stand-ins or prototypes for future NSA crops. While not suitable for agricultural production, these prototypes will provide more useful data than will unselected prairie species. Just as important, data from the polyculture experiments will help the breeding program formulate parental germplasm, selection criteria, and testing methodology.

A simplified illustration of our long-term plan (Figure 1) shows how our breeding and agro-ecology research programs will interpenetrate each other and eventually merge into a new kind of agricultural research. The complexity of such a research scheme cannot be overestimated. To account simultaneously for a large number of species (crop and non-crop), an even larger number of genotypes within each species, and an almost limitless number of possible combinations of genotypes, species, and environmental factors will require research methods that go beyond traditional statistics. Weaving individually reductionist studies involving genetic analysis, nutrient cycling, etc., into a nonreductionist whole will be a challenge.

Simple experience may teach us how to assemble grain-producing polycultures that behave like a prairie. However, with the vast literature of ecology and evolutionary biology on which to draw, and the still-evolving sciences of complexity (Kauffman,1993) and dialectics (Levins, 1998) to provide an overall framework, we can certainly find a better methodology than trial and error. In addition, we will need to draw scientists from many other disciplines–molecular and population genetics, soil biology, plant physiology, agronomy, range management and others–into NSA research. Calls for interdisciplinary research have become automatic when addressing new research problems, but we will need to go beyond perfunctory relationships between co-investigators and each learn to view NSA through the lens of the others' disciplines.

FIGURE 1

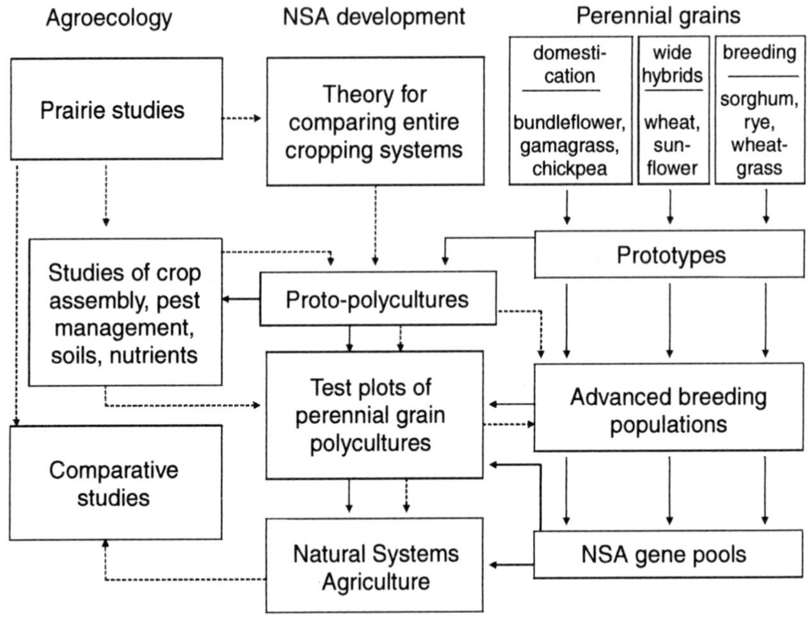

PROGNOSIS

The principles on which we are basing NSA can be applied to ecosystems worldwide, in developing cropping, forestry, or livestock systems that can take over the regenerative functions once performed by the local native flora and fauna. But while we are attracted by the great possibilities opened up by such research, the realities of our time are sobering. The last one percent of the history of agriculture, the twentieth century, gave humanity its largest increase in food production. That accomplishment is unlikely to recur. Most of the elasticity for yield increase has been absorbed. Moreover, it was a Faustian bargain: much of the gain in grain yield came at the cost of accelerated soil degradation by erosion, chemical contamination, and salinization. In addition, the spread of industrial agriculture's brittle economics has dislodged thousands of traditional farmers and torn much of the social and cultural fabric standing behind production.

Population growth will end one day, voluntarily or otherwise. The first cries of the newborns who arrive that day will likely be heard in a very crowded world. Not only they, but their descendants must have the means to feed themselves. No one can foretell the time when the population growth curve will flatten, or under what circumstances, but we can be certain that by then, liquid fossil fuel supplies will be severely reduced. Natural gas now serves as the feedstock for N fertilization, responsible for 40 percent of the current standing crop of humans (Smil, 2001). When fossil fuels begin to run short, the remaining natural soil fertility will be humanity's best friend.

But there is hope. The now-maturing disciplines of ecology and evolutionary biology are available to merge with agriculture and assist in making possible food production that does not erode soil or poison the environment. Soils are the key not only to agriculture but to civilization. Without them, there would have been no pyramids, no Parthenon, no temple of Solomon, no Teotihuacan, no Forbidden City, no Chartres–no New York City or San Francisco, no United States of America. And soil is as much a non-renewable resource as is oil.

The NSA we seek will function as an ecosystem, recycling materials and running on contemporary sunlight. By beginning to make agriculture sustainable we will have taken the first step toward someday declaring humanity's independence from the extractive economy.

REFERENCES

Azcón-Aguilar, C., and J.M. Barea. (1996). Arbuscular mycorrhizas and biological control of soil-borne plant pathogens–An overview of the mechanisms involved. *Mycorrhiza* 6:457-464.

Baer, S.G., C.W. Rice, J.M. Blair. (2000). Assessment of soil quality in fields with short and long term enrollment in CRP. *Journal of Soil and Water Conservation* 55:142-146.

Bargman, T.J., G.D. Hanners, R. Becker, R.M. Saunders, and J.H. Rupnow. (1989). Compositional and nutritional evaluation of eastern gamagrass (*Tripsacum dactyloides* (L.) L.), a perennial relative of maize (*Zea mays* L.). *Lebensmittel–Wissenschaft und Technologie* 22:208-212.

Barker, A.A. and J.K. Piper. (1995). Growth and seed yield of three grassland perennials in monocultures and mixtures. In *Prairie Biodiversity: Proceedings of the 14th North American Prairie Conference*, D.C. Hartnett, ed. Manhattan, KS: Kansas State University Press. pp. 193-197.

Blum, A. (1985). Photosynthesis and transpiration in leaves and ears of wheat and barley varieties. *Journal of Experimental Botany* 36:432-440.

Cai, X., S.S. Jones, and T.D. Murray. (1998). Molecular cytogenetic characterization of *Thinopyrum* and wheat-*Thinopyrum* translocations in a wheat-*Thinopyrum* amphiploid. *Chromosome Research* 6:183-189.

Cai, X., S.S. Jones, and T.D. Murray. (2001). Molecular cytogenetic characterization of *Thinopyrum* genomes conferring perennial growth habit in wheat-*Thinopyrum* amphiploids. *Plant Breeding* 120:21-26.

Cox, T.S., M. Bender, C. Picone, D.L. Van Tassel, J.B. Holland, E.C. Brummer, B.E. Zoeller, A.H. Paterson, and W. Jackson. (2002). Breeding perennial grain crops. *Critical Reviews in Plant Science* 21:59-91.

Cox, T.S., L.R. House, and K.J. Frey. 1985. Trait associations in intreogressed populations of sorghum. *Zeitschrift fur Pflanzenzuchtung* 94:265-277.

Devos, K.M., M.D. Atkinson, C.N. Chinoy, H.A. Francis, R.L. Harcourt, R.M.D. Koebner, C.J. Liu, P. Masojc, D.X. Xie, and M.D. Gale. (1993). Chromosomal rearrangements in the rye genome relative to that of wheat. *Theoretical and Applied Genetics* 85:673-680.

Dewald, C.L. and R.S. Dayton. (1985). Registration of gynomonoecious germplasm (GSF-I and GSF-II) of eastern gamagrass. *Crop Science* 25:715.

deWet, J.M.J., J.R. Harlan, L.M. Engle, and C.A. Grant. (1973). Breeding behavior of maize–*Tripsacum* hybrids. *Crop Science* 13:254-256.

Evans, L.T. (1998). *Feeding the Ten Billion*. Cambridge, UK: Cambridge Univ. Press.

Fatih, A.M.B. (1983). Analysis of the breeding potential of wheat-*Agropyron* and wheat-*Elymus* derivatives. *Hereditas* 98:287-295.

Gadgil, M. and O.T. Solbrig. (1972). The concept of *r*- and *K*-selection: evidence from wild flowers and some theoretical considerations. *American Naturalist* 106:14-31.

Gardner, J.C. (1991). The biology of annual and perennial grasses in the plains. In *Grass or Grain?: Intermediate Wheatgrass in a Perennial Cropping System for the Northern Plains*, eds. P. Wagoner, J.C. Gardner, B.G. Schatz, F. Sobolik, and D. Watt, Fargo, ND and Kutztown, PA: North Dakota State Univ. Press and Rodale Institute Research Center, pp. 4-7.

Gilley, J.E., J.W. Doran, D.L. Karlen, and T.C. Kaspar. (1997). Runoff, erosion, and soil quality characteristics of a former Conservation Reserve Program site. *Journal of Soil and Water Conservation* 52:189-193.

Hadley, H.H. (1958).Chromosome numbers, fertility, and rhizome expression of hybrids between grain sorghum and Johnson grass. *Agronomy Journal* 50:278-282.

Harper, J.L. (1977). *The Population Biology of Plants*. New York: Academic Press.

Hillel, D. (1991). *Out of the Earth: Civilization and the Life of the Soil*. Berkeley, CA: University of California Press.

Iltis, H.H. (2000). Homeotic sexual translocations and the origin of maize (*Zea mays*, Poaceae): A new look at an old problem. *Economic Botany* 54:7-42.

Jackson, L.L. and C.L. Dewald (1994). Predicting evolutionary consequences of greater reproductive effort in *Tripsacum dactyloides*, a perennial grass. *Ecology* 75:627-641.

Jackson, W. and L.L. Jackson. (1999). Developing high seed yielding perennial polycultures as a mimic of mid-grass prairie. In *Agriculture as a Mimic of Natural Systems*, eds. E.C. Lefroy, R.J. Hobbs, M.H. O'Connor, and J.S. Pate, Dordrecht, Netherlands: Kluwer Academic Publishers, pp. 1-37.

James, J. (1979). New maize × *Tripsacum* hybrids for maize improvement. *Euphytica* 28:239-247.

Jan, C.C. (1997). Cytology and interspecific hybridization. In *Sunflower Technology and Production*, ed. Schneiter, A.A., Madison, WI: ASA-CSSA-SSSA (Agronomy Monograph 35), pp. 497-558.

Jastrow, J.D., R.M. Miller, and J. Lussenhop. (1998). Contributions of interacting biological mechanisms to soil aggregate stabilization in restored prairie. *Soil Biology and Biochemistry* 30:905-916.

Jiang, J., B. Friebe, and B.S. Gill. (1994). Recent advances in alien gene transfer in wheat. *Euphytica* 73:199-212.

Jones, T.A., X.-Y. Zhang, and R.R.-C. Wang. (1999). Genome characterization of MT-2 perennial and OK-906 annual wheat × intermediate wheatgrass hybrids. *Crop Science* 39:1041-1043.

Jordan, N.R., J. Zhang, and S. Huerd. (2000). Arbuscular-mycorrhizal fungi: Potential roles in weed management. *Weed Research* 40:397-410.

Kaldorf, M., A.J. Kuhn, W.H. Schroder, U. Hildebrandt, and H. Bothe. (1999). Selective element deposits in maize colonized by a heavy metal tolerance conferring arbuscular mycorrhizal fungus. *Journal of Plant Physiology* 154:718-728.

Karlen, D.L., M.J. Rosek, J.C. Gardner, D.L. Allan, M.J. Alms, D.F. Bezdicek, M. Flock, D.R. Huggins, B.S. Miller, and M.L. Staben. (1999). Conservation reserve program effects on soil quality indicators. *Journal of Soil and Water Conservation* 54: 439-444.

Kauffman, S.A. (1993). *The Origins of Order*. New York: Oxford University Press.

Kindiger, B. and J.B. Beckett. (1992). Popcorn germplasm as a parental source for maize × *Tripsacum dactyloides* hybridization. *Maydica* 37:245-249.

Knapp, A.K., J.M. Blair, J.M. Briggs, S.L. Collins, D.C. Hartnett, L.C. Johnson, and E.G. Towne. (1999). The keystone role of bison in North American tallgrass prairie. *Bioscience* 49:39-50.

Knapp, A.K and M.D. Smith. (2001). Variation among biomes in temporal dynamics of aboveground primary production. *Science* 291: 481-484.

Kulakow, P.A., L.L. Benson, and J.G. Vail. (1990). Prospects for domesticating Illinois bundleflower. In *Advances in New Crops*, eds. J. Janick and J.E. Simon, Portland, OR: Timber Press, pp. 168-171.

Kulakow, P.A. (1999). Variation in Illinois bundleflower (*Desmanthus illinoensis* (Michaux) MacMillan): A potential perennial grain legume. *Euphytica* 110:7-20.

Ladizinsky, G. (1985). Founder effect in crop-plant evolution. *Economic Botany* 39:191-199.

Levins, R. (1998). Dialectics and systems theory. *Science and Society* 62:375-399.

Lowdermilk, W.C. (1953). Conquest of the land through seven thousand years. *Agriculture Information Bulletin* 99:1-30.

Mangelsdorf, P.C. and R.G. Reeves. (1931). Hybridization of maize, *Tripsacum*, and *Euchlaena*. *Journal of Heredity* 22:329-343.

Mitchell, C.E., D. Tilman, and J.V. Groth. (2002). Effects of plant species diversity, abundance, and composition on foliar fungal disease. *Ecology* 83:1713-1726.

Newell, C.A. and J.M.J. deWet. (1974). Morphological and cytological variability in *Tripsacum dactyloides* (Gramineae). *American Journal of Botany* 61:652-664.

Noll, M.G., C.J. Sorenson, and C.W. Rice. (1995). Biological condition of an agricultural soil six years after conservation reserve. *Transactions of the Kansas Academy of Sciences* 98: 102-112.

Nyahoza, F., C. Marshall, and G.R. Sagar. (1973). The interrelationship between tillers and rhizomes of *Poa pretensis* L.–autoradiographic study. *Weed Research* 13:304-309.

Oram, R.N. (1996). *Secale montanum*: A wider role in Australasia? *New Zealand Journal of Agricultural Research* 39:629-633.

Pimentel, D., C. Harvey, P. Resosudarmo, K. Sinclair, D. Kurz, M. McNair, S. Crist, L. Shpritz, L. Fitton, R. Saffouri, and R. Blair. (1995). Environmental and economic costs of soil erosion and conservation benefits. *Science* 267:1117-1123.

Pimm, S. (1991). *The Balance of Nature? : Ecological Issues in the Conservation of Species and Communities*. Chicago, IL: University of Chicago Press.

Pimm, S. (1997). In search of perennial solutions. *Nature* 389:126-127.

Piper, J.K. (1992). Size structure and seed yield over 4 years in an experimental *Cassia marilandica* (Leguminosae) population. *Canadian Journal of Botany* 70:1324-1330.

Piper, J.K. (1993). A grain agriculture fashioned in nature's image: The work of The Land Institute. *Great Plains Research* 3:249-272.

Piper, J. (1994). Composition of prairie plant communities on productive versus unproductive sites in wet and dry years. *Canadian Journal of Botany* 73: 1635-1644.

Piper, J.K. (1996). Density of *Anomoea flavokansiensis* on *Desmanthus illinoensis* in monoculture and polyculture. *Entomologia Experimentalis et Applicata* 81:105-111.

Piper, J.K. (1998). Growth and seed yield of three perennial grains within monocultures and mixed stands. *Agriculture, Ecosystems and Environment* 68:1-11.

Piper, J.K. (1999). Natural Systems Agriculture. In *Biodiversity in Agroecosystems*, eds. W.W. Collins and C.O. Qualset, Boca Raton, FL: CRC Press, pp. 167-195.

Piper, J.K., M.K. Handley, and P.A. Kulakow. (1996). Incidence and severity of viral disease on eastern gamagrass within monoculture and polycultures. *Agriculture, Ecosystems and Environment* 59:139-147.

Piper, J.K. and P.A. Kulakow. (1994). Seed yield and biomass allocation in *Sorghum bicolor* and F_1 and backcross generations of *S. bicolor* × *S. halapense* hybrids. *Canadian Journal of Botany* 72:468-474.

Reimann-Philipp, R. (1995). Breeding perennial rye. *Plant Breeding Reviews* 13:265-292.

Runnels, C.N. (1995). Environmental degradation in ancient Greece. *Scientific American* 272:96-99.

Scheinost, P.L., D.L. Lammer, X. Cai, T.D. Murray, and S.S. Jones. (2001). Perennial wheat: the development of a sustainable cropping system for the U.S. Pacific Northwest. *American Journal of Alternative Agriculture* 16:147-151.

Sharma, H.C. (1995). How wide can a wide cross be? *Euphytica* 82:43-64.

Slafer, G.A. and R. Savin. (1994). Source-sink relationships and grain mass at different positions within the spike in wheat. *Field Crops Research* 37:39-49.

Smil, V. (2001). *Enriching the Earth*. Cambridge, MA: MIT Press.

Smith, S.E. and D.J. Read. (1997). *Mycorrhizal Symbiosis*. New York, NY: Academic Press.

Snaydon, R.W. (1984). Plant demography in an agricultural context. In *Perspectives on Plant Population Ecology*, eds. R. Dirzo and J. Sarukhan, Sunderland, MA: Sinauer Assoc., pp. 389-407.

Soule, J.D. and J.K. Piper. (1992). *Farming in Nature's Image*. Washington, DC: Island Press.

Steingraber, S. (1998). *Living Downstream: A Scientist's Personal Investigation of Cancer and the Environment.* New York, NY: Vintage Books.

Stutz, H.C. (1957). A cytogenetic analysis of the hybrid *Secale cereale* L. × *S. montanum* Guss. and its progeny. *Genetics* 42:199-221.

Sukno, S., J. Ruso, C.C. Jan, J.M. Melero-Vara, and J.M. Fernandez-Martinez. (1999). Interspecific hybridization between sunflower and wild perennial *Helianthus* species via embryo rescue. *Euphytica* 106:69-78.

Suneson, C.A., A. El Sharkawy, and W.E. Hall. (1963). Progress in 25 years of perennial wheat development. *Crop Science* 3:437-438.

Tilman, D., J. Fargione, B. Wolff, C. D'Antonio, A. Dobson, R. Howarth, D. Schindler, W. Schlesinger, D. Simberloff, and D. Swackhamer. (2001). Forecasting agriculturally driven global environmental change. *Science* 292:281-284.

Tsitsin, N.V. (1965). Remote hybridisation as a method of creating new species and varieties of plants. *Euphytica* 14:326-330.

United States Department of Agriculture (USDA). (1999). *1997 National Resources Inventory Summary Report.* Washington, DC: Natural Resources Conservation Service, USDA.

Wagoner, P. (1990). Perennial grain development: Past efforts and potential for the future. *Critical Reviews in Plant Sciences* 9:381-408.

Wagoner, P. (1995). Intermediate wheatgrass (*Thinopyrum intermedium*): Development of a perennial grain crop. In *Cereals and Pseudocereals*, ed. J.T. Williams, London, UK: Chapman and Hall, pp. 248-259.

Wagoner, P., M. van der Grinten, and L.E. Drinkwater. (1996). Breeding intermediate wheatgrass (*Thinopyrum intermedium*) for use as a perennial grain. *Agronomy Abstracts* 1996:93.

Weaver, J.E. (1954). *North American Prairie.* Lincoln, NE: Johnsen Publishing Co.

Whyte, R.O. (1977). The botanical Neolithic revolution. *Human Ecology* 5:209-222.

Wright, L.S., C.M. Taliaferro, and F.P. Horn (1983). Variability of morphological and agronomic traits in eastern gamagrass accessions. *Crop Science* 23:135-137.

Index

Abawi, G.S., 113-114
Abiotic stresses, 475-479
Above-ground biomass (AGB), 345
Abrams, M., 295
Acid(s)
  fatty, phospholipid, 259
  organic, and mineral dissolution, 218-221,219t
Actinomycete(s), 121
Action learning, described, 37
Addy, H.D., 184
Aebischer, N.J., 374
AGB. *See* Above-ground biomass (AGB)
Agricultural chemicals, 271-273
Agricultural ecosystems, *vs.* natural ecosystems, 84-87,85t
Agricultural landscapes, 365-404. *See also* Field margin habitats, interactions with crop production
Agricultural practices, microbial diversity and, 268-276,268f
Agricultural production
  ecological effects on, 463-466, 469-472
    community-related, 465-466, 471-472,474-475
    cultivation practices, 463
    gene flow, 463-465,470,474
    insecticide use, 469-470
    resistance, 465,470-471,474
  trends in, natural resources and, 320-321
Agricultural science, expanding frontiers of, 3-13,5t. *See also* Agroecology, dimensions of
Agricultural soils
  C storage in, SOM preservation and, 302-303
  improved management of, biotic-abiotic interactions in, 209-248. *See also* Nutrient dynamics, biotic-abiotic interactions in, in agricultural soils
Agricultural Systems, 36
Agricultural universities, in agroecology, 24-26,24f
Agriculture
  contemporary, described, 406
  livestock-based, synergies of, capturing of, 418-426
  livestock-dominated, polarity in
    economic effects of, 411-412
    environmental effects of, 412-414
    health effects of, 414
    problems associated with, 411-414
  management of field margins for, 385-391,386t,391f
  monoculture, 99
  natural systems, research priorities in, 511-531. *See also* Natural systems agriculture (NSA), research priorities in
  philosophy of, agroecology as, 4-6,5t
  sustainable, diversity through, 327-331,328f
  traditional building on, 87-89

http://www.haworthpress.com/web/JCRIP

Agrobiodiversity
as basis for sustainable crop production, 317-320
decline of, 320-326,322t-324t
defined, 319
enhancement of, 326-335,328f
habitat conservation and, merging of, 333-334
losses due to, causes of, 326
Agroecological Index, 425
Agroecological processes
conversion to, 73-78,75t
integration into cropping systems research, 61-80
approach to, 63-73,64f,68t,69t
dynamic equilibrium in, 72
ecological concepts and principles in, 64-67
ecological design and management of, 67-70,68t, 69t
energy flow in, 70
future perspectives on, 78-79
introduction to, 62-63
moving toward sustainability in, 72-73
nutrient cycling in, 70-71
population regulating mechanisms in, 71-72
Agroecology
in biological approach to crop production, 1-20
defined, 3,23,62,63
described, xiii,1-2
in designing diversified cropping systems in Tropics, 81-103
diversification strategies, 94-98
introduction to, 82-84
dimensions of, 1,3-13,5t
closing materials cycle, 5t,10-11
crop autecology, 5t,9-10
encompassing agricultural landscape, 5t,10
human ecology, 5t,12-13
local adaptation, 5t,7-8
natural dimension, 5t,13
new philosophy of agriculture, 4-6,5t
non-crop biota, 5t,8-9
systems thinking, 5t,6-7
technology and ecology, 5t,11-12
as ecology of food systems, 22-24
education in, 21-43
agricultural universities in, 24-26,24f
courses addressing ecological topics, 26-30,27t
introduction to, 2-3
Agroecology: Ecological Processes in Sustainable Agriculture, 35
Agroecology and Farming Systems, 37
Agroecology and Food Systems, 37
Agroecology courses, 35-38
design of, 30-35,31f,33f
Agroecosystem(s)
biodiversity of, 315-337
introduction to, 316-317
cropping systems as, 70-72
derivation of, 2-3
industrial
development of, 47-50,47f-49f
redesigning of, 45-60
Georgia case study, 50-55, 51f,52f
introduction to, 46-47
management of, implications of, 300-304
succession analog, designing of, 89-91,91t
temperate, nutrient dynamics in, soil organisms effects on, 175-207
future directions in, 198-199
introduction to, 176-177
weed communities in, 137-151. *See also* Crop-weed interactions, ecology of
Agroforestry, described, 341
Agroforestry systems, research on, in southern Ontario, Canada, 341,341t

Agrolandscape, 55-56
Ahnstedt, B., 413
Ahohuendo, B.C., 380
Algae, eukaryotic, biology and ecology of, 178
Altieri, M.A., xiv,53,68,81
Andow, D.A., 68,96,457
Andrews, P.K., 52
Animal Production Systems, 445
Annual crops systems, disturbance in, and generalist predators, 159-160
Ant(s), 124
Anthelmintic properties, 428-429
Apse, M.P., 478
Archea, biology and ecology of, 177
Armesto, J.J., 7
Arthropod(s), 123
  abundance of, resource predictability and, 156-158,157t
Assembly Rule Theory, 148
Autecological study of cropping systems, 65
Autecology, crop, 5t,9-10

Bacon, F., 6
Bacteria
  biology and ecology of, 177
  N_2-fixing, nutrient transformations by, 184-185
  introduced, effects of, 266-267
Baermann funnel method, 126
Baker, F.H., 411,430
Baldock, J.O., 422,425
Balesdent, J., 230
Bamforth, S.S., 196
Barbash, J.E., 415
Barker, K., xv
Barley yellow dwarf virus (BYDV), 380
Barrett, G.W., xiv,xv,45
Bashkin, M.A., 299
Basra, A., xiv
Bastiaans, L., xv
Bawden, R.J., 36
Beare, M.H., 196,197
Beauchamp, E.G., 426
Beetle(s), carabid, 161-164,162f-164f
Beetle banks, 388
Belowground processes, altered C and N availability and temperature effects on, 295-300
Benbrook, C.M., 419
Beneficial insects, in field margin habitats, 376-380,379f
Berlese funnel, 126
Bethlenfalvay, G.J., 183
"Beyond Bookchin," 493
Bezemer, T.M., 301
Bilbro, J.D., 429
Biodiversity, of agroecosystems, 315-337
BIOLOG system, 269
Biological diversity, field margins and, 372-373
Biological Farming Systems, 445
Biological processes, in soil, global change effects on, 289-314
  introduction to, 290-291
Biomarker(s), FAME, 120
Biomass plantations, 429
BioScience, xiv
Biotechnology, for transgenic crops in production systems, 462
Biotic-abiotic interactions, in nutrient dynamics, in agricultural soils, 209-248. *See also* Nutrient dynamics, biotic-abiotic interactions in, in agricultural soils
Bird(s), in temperate tree-based intercropping systems, diversity of, 350-351
Bird, E.A.R., 411
Bookchin, M., 493
Booth, B., xv
Bosch-Haber process, 184
Boström, S., 179
Boutin, C., xv,372

Bouwman, L.A., 196
Bowman, W.D., 292
Breland, T.A., 423,424
Bridge, P., 252
Brinfield, R.B., 421
Briones, M.J.I., 296
British Trust for Ornithology (BTO), 373
Bromfield, L., 405,407,408
Brown, P.D., xv
BTO. *See* British Trust for Ornithology (BTO)
Bugg, R.L., 505
Bullied, W.J., 419
Bulson, H.A.J., 425
Bultena, G.L., 411
Bunce, J.A., 301
Bureau of Land Management, 166
Butler Flora, C., xv
Butterfield, J., 502
BYDV. *See* Barley yellow dwarf virus (BYDV)

"C" value, 427
Cady, A.B., 181
Cambardella, C.A., 113
Cambouris, A.N., 235
Cannon, R.J.C., 302
Carabid beetles, 161-164,162f-164f
Carbon sequestration, 426-427
Carlson, I.H., 414
Carreck, N.L., 389
Casler, M.D., 425
Cavigelli, M.A., 7
CBD. *See* Convention on Biological Diversity (CBD)
Centipede(s), 123
Chang, C., 413
Checkland, P.B., 36
Chemical(s), agricultural, 271-273
Cheng, W.X., 303
Chenu, C., 230
CIAT (International Center for Tropical Agriculture), 332
Clark, E.A., 405
Clark, F.E., 212
Clark, J.S., 142
Clayton, G.W., 121
Clements, D.R., 1
Climate change mitigation, in temperate tree-based intercropping systems, interactions related to, 356-358
CLPPs. *See* Community level physiological profiles (CLPPs)
CMS. *See* Cytoplasmic male sterility (CMS)
CO_2 elevation, effects on plant primary production, 291-293,292f
Cobbett, W., 388
Cole, C.V., 357
Coleman, D.C., xv
Collier, R.H., 301
Collins, A., xv
Collins, S.L., 7
Commission on Plant Genetic Resources, 333
Common Agricultural Policy, 441-442
Commoner, B., 480
Community(ies)
- agricultural production and, 465-466,471-472,474-475
- plant
  - in field boundaries, 370
  - structure of, theories of, 138-147. *See also* Plant community structure, theories of
- weed, in agroecosystems, 137-151. *See also* Crop-weed interactions, ecology of

Community Assembly Rules, 148
Community level physiological profiles (CLPPs), 264
Competition
- among transgenic crops in production systems, 461-467
- in temperate tree-based intercropping systems, interactions related to, 352-355,354t,355f

Competitive Exclusion Principle, 146-147
Connor, D.J., 28
Conservation, habitat, agrobiodiversity and, merging of, 333-334
Conservation Agriculture, 444
Conservation headlands, 385-387
Conservation Reserve Program, 429,514
Consultative Group on International Agricultural Research, 333
Contemporary agriculture, described, 406
Convention on Biodiversity, 251
Convention on Biological Diversity (CBD), 327,333
Conventional tillage (CT) systems, 53
Conway, G.R., 36
Cooperative Extension, 25
"Cordon sanitaire," 389
Costanza, R., 50
Countryside Survey 1990, 372,373
Coutinho, H.L.C., 269
Cox, T.A., 6
Cox, T.S., 511,516,518
Crafoord Prize, of Royal Swedish Academy, xiv
Crop(s), 414-418
  erosion of, 321-324,322t,323t
  genetically modified, 53
  grain, breeding of, for domestic prairie, 515-521
  transgenic. *See* Transgenic crops
Crop autecology, 5t,9-10
Crop community structure, and crop-weed interactions, 300-301
Crop diseases, in field margin habitats, 380
Crop management, margin effects of, 382-385,383f
Crop plants, nutrient uptake by, seasonal variation in, 232,233f
Crop production
  biological approach to, agroecology in, 1-20. *See also* Agroecology, in biological approach to crop production
  field margin habitats interactions with, 365-404. *See also* Field margin habitats, interactions with crop production
  introduction to, 2-3
  sustainable, agroecosystems as basis for, 317-320
Crop production systems, re-integration of livestock and forages in
  benefits of, 405-436
  historical context of, 407-411,410f
  introduction to, 406
Crop species, effect on soil ecosystem changes during transition to no-till cropping, 110-111
Crop systems, annual, disturbance in, and generalist predators, 159-160
Crop yield effects, in field margin habitats, 380-381,381f
Cropping systems
  as agroecosystems, 70-72
  autecological study of, 65
  diversified, in Tropics, agroecological basis for designing of, 81-103. *See also* Agroecology, in designing diversified cropping systems in Tropics
  no-till, soil ecosystem changes during transition to, 105-135. *See also* No-till cropping, soil ecosystem changes during transition to
Crop-weed interactions
  crop community structure and, 300-301
  ecology of, 137-151
    introduction to, 138

plant community structure in, theories of, 138-147,139t,143f. *See also* Plant community structure, theories of
Crossing systems research, agroecological processes integrated into, 61-80. *See also* Agroecological processes, integration into cropping systems research
CT systems. *See* Conventional tillage (CT) systems
Cultivar(s), plant, effect of, 264-265
Cultivation practices, agricultural production effects of, 463
Cultural controls, in pest control, 502-505
Cunningham, J.E., 228
Cyanobacteria, biology and ecology of, 178
Cyclic colonization, applicability to IPM, 157t
Cytoplasmic male sterility (CMS), 472-473

Dail, B., xv
Database(s), LOMBRI-ASSESS, 181
Davis, J.G., 413
de Bary, 480
de Boer, I.J.M., 442
De Marke, 445-446
De Minderhoudhoeve, 445
de Ruter, P.C., 214
de Snoo, G.R., 370
de Wit, J., 443,443t
"Dead hedge," 367
Decanting and sieving, 126
Decomposition, organic matter, 185-190
den Biggelaar, C., xv
Denaturing gradient gel electrophoresis (DGGE), 125,269,273
Deng, S.P., 423
Department of Soil Science, 53
Developed countries, mixed farming systems in, renaissance of, 439-440
DGGE. *See* Denaturing gradient gel electrophoresis (DGGE)
Dhillion, S.S., 295
Dick, W.A., 249
Didden, W.A.M., 181
Dindal, D.L., 181
Disease(s), agriculture-related, 472-475
  biotechnology applications, 473-474
  epidemics of, 472-475
  magnitude of problem, 472-473
Dissolved form (DOM), 53
Disturbance, applicability to IPM, 157t
Diversity-stability hypothesis, 141-143,143f
DNA reassociation (diversity), 125
DNA:DNA hybridization (similarity and relative diversity), 125
DOM. *See* Dissolved form (DOM)
Dormaar, J.F., 187
Douds, D.D., Jr., 252
Dow, B.W., 235
Drijber, R.A., 110,252
Drinkwater, L.E., 419
Duvick, D.N., 408
Dyer, L.E., 153
Dynamic equilibrium, in agroecosystems, 72

Earthworm(s), 124
Ecological interactions, theoretical models of, 143-146
Ecosystem(s)
  described, xiii,xiv
  disruption of succession in, 86
  inefficient nutrient cycling in, 87
  landscape and field simplification in, 85-86
  lowering of plant defenses in, 86-87
  natural *vs.* agricultural, 84-87,85t
  soil, during transition to no-till cropping, 105-135. *See also* No-till cropping, soil ecosystem changes during transition to

Ecosystem C balance, decomposition of organic matter and, 297
Education, in agroecology and integrated systems, 21-43. *See also* Agroecology, education in; Integrated systems, education in
Einstein, A., 14
Ekschmitt, V., 181
Elmi, A., 209
Eltun, R., 423,424
Elutriation, 126
Enemy impact hypothesis, applicability to IPM, 157t
Energy flow, in agroecosystems, 70
5-Enol-pyruvyl-shikimate-3-phosphate synthase (EPSPS), 462
Entz, M.H., 419
Environment(s)
 enhancement of, forages and livestock for, 426-431
 soil
  favorable, maintenance of, 230-231
  microbial and genetic diversity in, 249-287
EPSPS. *See* 5-Enol-pyruvyl-shikimate-3-phosphate synthase (EPSPS)
Eriksen, J., 422
Erosion, soil, 427
"Eugene P. Odum: Pioneer in Ecosystem Ecology," xiv
Eukaryotic algae, biology and ecology of, 178
European Union, 451
 Nitrate Directive of, 451,453
Evans, L.T., 512
Ewel, J.J., 100
EXPAGR. *See* Expansion Agriculture (EXPAGR)
Expansion Agriculture (EXPAGR), 444
Extension, in support of ecologically based IPM systems, 153-174
Fabreguettes, J., 297
Fairbairn, S., 377
Falconer, K.E., 384
FAME analysis. *See* Fatty acid methyl ester (FAME) analysis
FAME profile. *See* Fatty acid methyl ester (FAME) profile
Farming systems
 classification of, 443-446,443t
 mixed, nitrogen efficiency in, 437-455. *See also* Mixed farming systems, nitrogen efficiency in
Farmyard manure (FYM), 425
Farvar, M.T., 480
Fatty acid(s), phospholipid, 259
Fatty acid methyl ester (FAME) analysis, 264
Fatty acid methyl ester (FAME) biomarker, 120
Fatty acid methyl ester (FAME) profile, 110
Faulkner, S., 407
Fauna, soil, dynamics of, 296-297
Fertilizer(s), nitrogen, use efficiency of, 303-304
Fertilizer movement, 382-384,383f
Field boundaries
 habitat types in, 370
 plant communities in, 370
Field edge management
 combined approaches to, 390
 modified, benefits of, 390-391,391f
Field level diversity, temporal and tillage effects on, 268-269,268f
Field margin(s). *See also* Field margin habitats
biological diversity and, 372-373
 components of, 369f
 crop management effects on, 382-385,383f
 described, 368-372,369f,371t
 management of, for wildlife and agriculture, 385-391,386t,391f

role of, 370-372,371t
semi-natural, functions of, 370,371t
structure of, 368-370,369f
Field margin habitats. *See also* Field margin(s)
interactions with crop production, 365-404
beneficial insects, 376-380,379f
crop diseases, 380
crop yield effects, 380-381,381f
introduction to, 366-368
margin-related, 374-381,375t, 376t,379f,381f
pests, 376-380,379f
weed ingress, 374-376,375t,376t
Fillery, I.R.P., 234
Filtration, 126
N_2-Fixing bacteria, nutrient transformations by, 184-185
Flora, J.L., xv
Flotation/centrifugation, 126
Flower margins, and strips across fields, 389
Food and Agricultural Organization, 330,333
Food systems, ecology of, agroecology as, 22-24
Food web interactions, 296-297
Forage(s)
for environmental enhancement, 426-431
re-integration into crop production systems, benefits of, 405-436. *See also* Crop production systems, re-integration of livestock and forages in, benefits of
Foster, R.C., 187
Foster, R.F., 430
Fox, C.A., 124
Francis, C.A., xv,3,21,26
Frank, T., 377
Franson, R.L., 183
Fraser, T.J., 428
Frey, K.J., 518
Fründ, H.-C., 181
Fryrear, D.W., 429
"Fundamentals of Ecology," xiv
Fungus(i)
mycorrhizal, nutrient transformations by, 182-184
soil, biology and ecology of, 177-178
FYM. *See* Farmyard manure (FYM)

Game Conservancy Trust, 388
Gardner, J.C., 411,515
Garrett, H.E., 352
Gause, G.F., 146,147
Gene flow, agricultural production effects of, 463-465,470,474
General Agreement on Tariffs and Trade, 333
Genetic diversity, in soil environments, 249-287. *See also* Soil environments, microbial and genetic diversity in
Genetically modified (GM) crops, 53
Georgia Institute of Ecology, 50
Gerzabek, M.H., 270
Ghersa, G.M., xv
Giller, K., xv
Giroux, M., 236,237
Gliessman, S.R., xv,3,26,35,36,61,90
Global change, effect on biological processes in soil, 289-314
Global warming, effects on plant primary production, 291-293,293f
Glover, J.D., 52
Goddard College, 493
Goh, K.M., 439
Goodell, P.B., 8
Gordon, A.M., 339,351
GPA. *See* Green peach aphid (GPA)
Graefe, U., 181
Grain crops, breeding of, for domestic prairie, 515-521
Grant, A.J., 377

Grass, sown, 388-389
Gray, G.R.A., 353
Grayston, S.J., 263,296
Greaves, M.P., 369
Green peach aphid (GPA), 481
Green Revolution, 321,322
Greer, J., 50
Gregson, M.E., 408
Groot, J.C.J., 442
Groth, J.V., 522
Gupta, V.V.S.R., 196

Habitat(s)
in field boundaries, 370
field margin, interactions with crop production, 365-404. *See also* Field margin habitats, interactions with crop production
losses in, 325-326
Habitat conservation, agrobiodiversity and, merging of, 333-334
Habitat management, applicability to IPM, 157t,160-165,162f-164f
Habitat simplification and fragmentation, 158-159
Halaj, J., 181
Hamel, C., 175,209
Harmon, J.P., 457
Harper, W.S., xv
Haynes, R.J., 424,439
He, Z.L., 186,235
Headland(s), conservation, 385-387
Health, soil, conserving/improving of, 422-426
HEIA. *See* High-External-Input Agriculture (HEIA)
Herbivory, 468-472
biotechnology applications, 468-469
magnitude of problem, 468
High-External-Input Agriculture (HEIA), 444
reaction in Netherlands to, 444-446
Hill, S.B., xv,13,491,505
Holling, C.S., 166
Hopkins, A., 421
HorseShoe Bend (HSB) experimental site, at University of Georgia, xiv,53
Host-parasite interactions, 301-302
House, L.R., 518
Hoxey, A.M., 421
HSB experimental site. *See* HorseShoe Bend (HSB) experimental site
Hu, S., 262,289
Hubbell, S.P., 140
Hulsmann, A., 123
Hungate, B.A., 298,304
Hungria, M., 299
Huxley, J., 493
Huxley, P., 352

ICIPE, 96
Idle, A.A., 421
IFOAM. *See* International Federation of Organic Agriculture Movements (IFOAM)
Ikerd, J., 412
Industrial agroecosystems, redesigning of, 45-60. *See also* Agroecosystem(s), industrial, redesigning of
Ineson, P., 296
Ingham, R.E., 179,187,235
Inorganic nutrients, microbial transformations of, 215-218,216t
Insect(s), beneficial, in field margin habitats, 376-380,379f
Insect control, cultural methods of, 502-505
Insect diseases, vulnerability to, increased, 324-325,324t
Insect pests, vulnerability to, increased, 324-325,324t
Insecticide use, agricultural production effects of, 469-470

Institute of Ecology, xiv,53
Integrated pest management (IPM), 28,503
  dimensions of, 8
  ecological concepts and, 156-160, 157t
Integrated pest management (IPM) methods, 329,330
Integrated pest management (IPM) systems
  ecological concepts related to, 157t,160-165,162f-164f
  ecologically based, research and extension supporting, 153-174
    introduction to, 154-156
  history of, 154-156
Integrated systems, education in, 21-43
  agricultural universities in, 24-26,24f
  courses addressing ecological topics, 26-30,27t
Intercropping systems, temperate tree-based
  biophysical and ecological interactions in, 339-363
    avian diversity due to, 350-351
    climate change mitigation–related, 356-358
    competition-related, 352-355, 354t,355f
    introduction to, 341-344,341t,343f
    microclimatic modifications–related, 351-352
    nutrient additions through hydrologic pathways, 347-349,348t
    nutrient and soil conservation–related, 356
    soil fertility–related, 344-347, 344t,346f,347f
  earthworm populations under, dynamics of, 349-350,350t
International Center for Tropical Agriculture (CIAT), 332
International Federation of Organic Agriculture Movements (IFOAM), 420
International Plant Genetic Resources Institute, 333
IPM. *See* Integrated pest management (IPM)
IPM methods. *See* Integrated pest management (IPM) methods
IPM systems. *see* Integrated pest management (IPM) systems
Island biogeography, applicability to IPM, 157t
Ison, R.L., 421
iWUE. *See* Water use efficiencies (iWUE)

Jackson, W., 6,511
Jaeger, J.W., 414
Jan, C.C., 520
Jensen, L.S., 422
Jeranyama, P., xv
Joergensen, R.G., 234
Johnson, D.W., 303
Jones, J.H., 301
Jones, S., xv
Jones, T.H., 296
Jordan, N., xv
Journal of Crop Improvement, xiv

Kabir, Z., 183,194,232,233f
Kaiser, O., 263
Karr, J.D., 417
Karsten, H., xv
Katepa-Mupondwa, F., 419
Keeney, D., 439
Kegley, S., 415,416
Kennedy, A.C., 105,121
Kerridge, P.C., xv
Kirschbaum, M.U.F., 295
Kladivko, E.J., 197

Kleijn, D., 384
Knight, K.J., 301
Knight, T.L., 428
Knowledge, losses in, 325-326
Koblents, H., 351
Kogan, M., 154
Kovacic, D.A., 428
Kuiack, C., 228
Kulakow, P.A., 517,518

Labile nutrient pool, microorganisms as, 229-230
Landis, D.A., 153,159
Landry, C., 209
Landscape(s), agricultural, 365-404. *See also* Field margin habitats, interactions with crop production
Langer, V., 421-422
Lantinga, E.A., 437,442
Latour, X., 264
Learning, action, described, 37
Lee, J., xiv
Lehman, C.L., 141,142
LEIA. *See* Low-External-Input Agriculture (LEIA)
Leitch, A., 409,418
Letourneau, D.K., xv
Levins, R., 158
Libra, R.D., 415
Lilley, J.M., 299
LISA. *See* Low-Input Sustainable Agriculture (LISA)
Liu, A., 209,228
Livestock, 417-418
  in agriculture, polarity in, problems associated with, 411-414
  for environmental enhancement, 426-431
  re-integration into crop production systems, benefits of, 405-436. *See also* Crop production systems, re-integration of livestock and forages in, benefits of
Livestock diversity, erosion of, 321-324,322t,323t
Livestockbased agriculture, synergies of, capturing of, 418-426
Logsdon, S.D., 113
Loiseau, P., 298
LOMBRI-ASSESS database, 181
Londsdale, W.M., xv
Loomis, R.S., 28
Lotka, A.J., 145,146
Lotka-Volterra equations, 145
Lotka-Volterra model, 146
Lowell, K.A., 424
Low-External-Input Agriculture (LEIA), 444
Low-Input Sustainable Agriculture (LISA), 49,50f
Luna, J.M., xv
Lupwayi, N.Z., 121,194
Luschei, E., 137
Lynch, J.M., 234

MacArthur, R.H., 158
Macrofauna, 123-125
  soil, biology and ecology of, 180-181
"Macuna" system, 89
Maize, in Africa, case study, 466-467
Malabar Farm, 407,408
Maloney, P.E., 262
Manglesdorf, P.C., 519
Margin(s)
  set-aside, 389
  wild flower, 388-389
Marsh, T.L., xv
Marshall, E.J.P., 365,369,377,382, 387,391
Marvier, M., xv
Matter, organic, decomposition of, and ecosystem C balance, 297
Maurer, B.A., 145
Maxwell, B.D., 137,141
McDonough, W., 431
McGonigle, T.P., 184

McLaughlin, A., 385
McLendon, T., 292
MeBr. *See* Methyl bromide (MeBr)
Menalled, F.D., 153,159
Mesofauna, 122-123
  soil, biology and ecology of, 178-180
Metapopulation(s), applicability to IPM, 157t
Methyl bromide (MeBr), 272
Meyer, B., 234
"Michigan Field Crop Ecology," 165
"Michigan Field Crop Pest Ecology and Management," 165
Microbial diversity
  and agricultural practices, 268-276,268f
  described, 253-255,254f
  measurement of, techniques for, 255-261,257t-258t
  in rhizosphere, 261-267
  in rice field soil, 273-276
  in soil environments, 249-287. *See also* Soil environments, microbial and genetic diversity in
Microbial transformations, of inorganic nutrients, 215-218,216t
Microclimatic modifications, in temperate tree-based intercropping systems, interactions related to, 351-352
Microfauna, 122-123
  soil, biology and ecology of, 178-180
Microorganism(s), 120-122
  as labile nutrient pool, 229-230
  soil, tillage effects on, 194-195
Miller, D.N., 259
Miller, M.H., 184
Millipede(s), 123
Millner, P.D., 252
Milton, J.P., 480
Minderhoudhoeve project, 446-452, 448t,449t,451f,452t
  farm profiles in, 447,448t
  nitrate leaching in, 451,451f,452t
  nitrogen balance in, 447-450,449t
  pests in, 452
  results of, discussion of, 447-452, 448t,449t,451f,452t
Mineau, P., 385
Mineral(s), soil, nutrient solubilization from, 190-191
Mineral dissolution, organic acids and, 218-221,219t
Mitchell, C.E., 522
Mite(s), 122-123
  biology and ecology of, 180
Mixed farming systems
  in developed countries, renaissance of, 439-440
  nitrogen efficiency in, 437-455
    introduction to, 438-440
  research on, 446-452,448t,449t, 451f,452t
Moënne-Loccoz, Y., 267
Molina, J.A.E., xv
Monoculture agriculture, 99
Montrealegre, C.M., 299
Moonen, A.C., 391
Mooney, P.F., xv
Moore, J.C., 214
Moore, J.M., 423
Morrison, J., 421
Mueller, J.P., xv
Mueller, T., 234
Murphy, D.V., 234
Murphy, S.D., xv
Murray, P., xv
Murray-Darling Basin, Southeastern Australia, 478,479
Musgrave, R.B., 422
Mycorrhizal fungi, nutrient transformations by, 182-184
Mycorrhizosphere
  nature of, 224-226
  nutrient cycling in, 231-232
  nutrient mobilization in, 226-229

N deposition, effects on plant primary production, 291-293
N efficiency, defined, 450
Nair, P.K.R., 342,352
National Academy of Sciences, xiv
Natural ecosystems, *vs.* agricultural ecosystems, 84-87,85t
Natural resources, agricultural production trends in relation to, 320-321
Natural succession mimicry, management options for, 91-94
Natural systems agriculture (NSA), 13
  prognosis for, 526-527
  research priorities in, 511-531
    introduction to, 512-513
    prairie as model of, 513-515. *See also* Prairie(s)
NCA. *See* New-Conservation Agriculture (NCA)
Neave, P., 124
Neetespm, J.J., 418
Nematode(s), 122
  biology and ecology of, 179-180
Neolithic Near East crop assembly, 441
Net Primary production (NPP), 291
Netherlands
  HEIA in, reaction to, 444-446
  nitrogen problems in, historical background of, 440-443
Neumeister, L., 415, 416
Neutel, A.M., 214
New-Conservation Agriculture (NCA), 437,444
Nguyen, M.L., 439
Niche-Assembly Theory, 140,147
Nicholls, C.I., 81
Nijs, I., 297
Nitrate Directive of the European Union, 451,453
Nitrate leaching, in Minderhoudhoeve project, 451, 451f,452t
Nitrogen, in Netherlands, problems associated with, historical background of, 440-443
Nitrogen balance, in Minderhoudhoeve project, 447-450,449t
Nitrogen efficiency, in mixed farming systems, 437-455. *See also* Mixed farming systems, nitrogen efficiency in
Nitrogen fertilizers, use efficiency of, 303-304
Nitrogen mineralization, 298
Nobel Prize, xiv
Nolin, M.C., 235
Nordic Forestry, Veterinary and Agricultural (NOVA) University, 35,37
No-till cropping, soil ecosystem changes during transition to, 105-135
  crop selection, 110-111
  introduction to, 106
  macrofauna in, 123-125
  measurement of, methods for, 125-126
  mesofauna in, 122-123
  microfauna in, 122-123
  microorganisms in, 120-122
  organisms in, 115-125,116f, 117t-119t
  physical and chemical soil characteristics, 111-115,112t
  residue, 110-111
  rotation sequence, 110-111
  tillage practices, 107-108,108f
  transition period in, 109-110,109f
No-tillage (NT) systems, 53,54
NOVA University, 35,37
NPP. *See* Net primary production (NPP)
NSA. *See* Natural systems agriculture (NSA)
NT systems. *See* No-tillage (NT) systems
Nutrient(s)
  additions to, via hydrologic pathways, in temperate tree-based intercropping systems, 347-349,348t

dynamics of, soil organisms effects on, in temperate agroecosystems, 175-207. *See also* Agroecosystem(s), temperate, nutrient dynamics in, soil organisms effects on
inorganic, microbial transformations of, 215-218,216t
management of, by plants, 221-231,222f
recycling of, inefficient, 87
in temperate tree-based intercropping systems, interactions related to, 356
Nutrient cycling
in agroecosystems, 70-71
in mycorrhizosphere, 231-232
Nutrient dynamics, biotic-abiotic interactions in, in agricultural soils, 209-248
introduction to, 210-211
soil processes in, 211-221,216t,219t
Nutrient losses, soil organisms and, 192-193
Nutrient mobilization, in mycorrhizosphere, 226-229
Nutrient pool, labile, microorganisms as, 229-230
Nutrient release
from soil, dynamics of, 232,234-237,236f-238f
spatial and temporal variation in, 231-237,233f,236f-238f
Nutrient solubilization, from soil minerals, 190-191
Nutrient transformations, soil organisms in, 182-193
Nutrient uptake
by crop plants, seasonal variation in, 232,233f
by plants, 222-224,222f
Nutrition, losses in, 325-326

Odum, E.P., xiii,xiv,xv,45
Odum, H.T., xiv
Olsson, P.A., 186,229
OM. *See* Organic matter (OM)
Ong, C.K., 342,352
Oomen, G.J.M., 437
Opio, C., xv
Oren, R., 297
Organic acids, and mineral dissolution, 218-221,219t
Organic matter (OM), 422
decomposition of, 212-215
and ecosystem C balance, 297
nutrient transformations by, 185-190
Orme, S., 415,416

Packham, R.G., 36
PAHs. *See* Polycyclic aromatic hydrocarbons (PAHs)
Panting, L.M., 234
PAR. *See* Photosynthetically-active radiation (PAR)
Particulate form (POM), 53
Paschke, M.W., 292
Patra, D.D., 235
Paul, E.A., 212
PCB–polluted soil. *See* Polychlorinated biphenyl (PCB)–polluted soil
Pearson, C.J., 421
Percent G+C composition of community DNA (species composition), 125
Perrott, K.W., 235
Pest(s)
in field margin habitats, 376-380,379f
insect, vulnerability to, increased, 324-325,324t
in Minderhoudhoeve project, 452
outbreaks of, vegetation diversity and, 95-97
producer-level 'redesign' initiative for, 499-500
Pest control, cultural methods of, 502-505
Pest management

redesigning of, 491-510. *See also* Pest management, social ecology approach to
social ecology approach to, 491-510
ecologically based 'redesign' approaches to
implementation of, 505-507
legislation related to, 505-506
policy of, 505
public education in, 507
regulations in, 505-506
research related to, 506
services for, 507
training and education in, 507
introduction to, 492-493
producer-level 'redesign' initiative, example of, 499-500
redesign approaches at pest manager level, 500-502,501f
Pesticide drift, 384-385
Pesticide Management Zones (PMZs), 416
Phospholipid fatty acid (PLFA), 259
Phospholipid fatty acid (PLFA) analysis, 272
Photosynthetic photon flux density (PPFD), 351
Photosynthetically-active radiation (PAR), 353
Pickett, C.H., 505
Pickett, S.T.A., 7
Picone, C., 511
Picone, J.C., 6
Piearce, T.G., 296
Piper, J.K., 517,518
Plant(s)
nutrient management by, 221-231, 222f
nutrient uptake by, 222-224,222f
transgenic, effect of, 265-266
Plant Agriculture, University of Guelph, xiv
Plant communities, in field boundaries, 370
Plant community structure, theories of, 138-147,139t,143f
Competitive Exclusion Principle, 146-147
diversity-stability hypothesis, 141-143,143f
Lotka-Volterra model, 146
Plant cultivars, effect of, 264-265
Plant growth, global change–induced alteration in
effects on soil C availability, 294-295
effects on soil N availability, 294-295
Plant primary production
CO_2 elevation effects on, 291-293, 292f
global warming effects on, 291-293, 293f
N deposition effects on, 291-293
Plant Production Systems, 445
Plantation(s), biomass, 429
Plant-microbial symbiotic interactions, 298-300
PLFA. *See* Phospholipid fatty acids (PLFA)
PLFA analysis. *See* Phospholipid fatty acid (PLFA) analysis
Plowman's Folly, 407
PLRV. *See* Potato leafroll virus (PLRV)
PMZs. *See* Pesticide Management Zones (PMZs)
Polychlorinated biphenyl (PCB)–polluted soil, 271-272
Polycyclic aromatic hydrocarbones (PAHs), 272
POM. *See* Particulate form (POM)
Population regulating mechanisms, in agroecosystems, 71-72
Porter, W.P., 414
Posner, J.L., 425
Postgraduate School in Chapingo, 35
Potato leafroll virus (PLRV), 480
Potvin, C., 300

PPFD. *See* Photosynthetic photon flux density (PPFD)
Prairie(s)
domestic
breeding grain crops for, 515-521
building of, agroecological principles for, 521-524
designing of, 525,526f
NSA in, 513-515. *See also* Natural systems agriculture (NSA), research priorities in, prairie as model of
Predator(s), generalists, disturbance in annual crop systems and, 159-160
Pregitzer, K.S., 295
Pretty, J.N., 417
Price, G., 339
Production systems, transgenic crops in, examining effects of, ecological context for, 457-489. *See also* Transgenic crops, in production systems, examining effects of, ecological context for
Protozoa, 122
biology and ecology of, 178-179
Puget, P., 230
Püler, A., 263

Random Amplified Polymorphic DNA (RAPD) analysis, 271
Rao, M.R., 342,352
RAPD analysis. *See* Random Amplified Polymorphic DNA (RAPD) analysis
Raun, N.S., 411,430
Rayburn, E.B., 427
Redente, E.F., 292
Reeves, R.G., 519
Reganold, J.P., 52
Reid, F., 430
Rekolainen, S., 414
Remediation, 427-428
Research
on agroforestry systems, in southern Ontario, Canada, 341,341t
on cropping systems, agroecological processes integrated into, 61-80. *See also* Agroecological processes, integration into cropping systems research
on mixed farming systems, 446-452,448t,449t,451f,452t
NSA-related, 511-531. *See also* Natural systems agriculture (NSA), research priorities in
in support of ecologically based IPM systems, 153-174. *See also* Integrated pest management (IPM) systems, ecologically based, research and extension supporting
Residue, effect on soil ecosystem changes during transition to no-till cropping, 110-111
Resistance
agricultural production effects of, 465,470-471,474
viral, 473
Resource concentration hypothesis, applicability to IPM, 157t
Resource predictability
applicability to IPM, 157t
arthropod abundance and, 156-158,157t
Reyes, I., 220,228
Reynolds, P.E., 339
Rhizoctonia root rot, 114
Rhizosphere, microbial diversity in, 261-267
Rhizotonia solani (Kühn) AG8, 114
Rice, C.W., 296
Rice field soil, microbial diversity in, 273-276
Rice, W.A., 121
Richter, D.D., 299
Rickerl, D.H., 413

Rillig, M.C., xv
RISA. *See* rRNA intergenic spacer analysis (RISA)
Rodale Institute, 516
Roper, M.M., 196
Rosemeyer, M.E., 68
Rotation sequence, effect on soil ecosystem changes during transition to no-till cropping, 110-111
"Roundup Ready," transgenic, 462
Roundup Ready (RR) soybeans, 419
Rowarth, J.S., 428
Roy, J., 295,297
Royal Swedish Academy, Crafoord Prize of, xiv
RR soybeans. *See* Roundup Ready (RR) soybeans
rRNA intergenic spacer analysis (RISA), 270
Ruess, L., 296
"Rural Rides," 388
Russelle, M.P., 427
Rustad, L.E., 294
Rydberg, T., 113

Saïd, F., 272
Salager, J.L., 297
Salt-affected lands, in Australia, 477-479
Sanderson, M., xv
Sapelo Island Marine Biology Laboratory, xiv
Sarathchandra, S.U., 235
Sarkar, S., 380
Sauerborn, J., xv
Savannah River Ecology Laboratory, xiv
Saviozzi, A., 425
Savory, A., 502
Schiere, J.B., 437,443,443t
Schilling, K.E., 415
Schillinger, W.F., 105
Schoeneberger, M.M., xv
Schreinemachers, D.M., 416
Seasonal variation, in nutrient uptake by crop plants, 232,233f
Selbitschka, W., 263
Sessitsch, A., 270
Set-aside margins, 389
Sewage sludge (SS), 425
Shade, H.M., 424
Shannon-Weiner diversity indices, 261,271,272-273
Sharpley, A.N., 414
Shrestha, A., 1
Simard, R.R., 235
Simpson Diversity Index, 261
Simpson, J.A., 339
Sing, S., 144
Singh, S., xiv
Single strand conformation polymorphism (SSCP), 125
Single strand conformation polymorphism (SSCP) profiles, 267
"Slash and burn," 89
Smit, E., 252
Smith, B.D., 387
Smith, K.L., 121
Snoeijing, G.I.J., 384
Social ecology
  described, 493
  as framework for improving pest management, 493-499, 495t,497t
  in pest management, 491-510. *See also* Pest management, social ecology approach to
"Social Ecology After Bookchin," 493
Sohlenius, B., 179
Soil
  biological processes in, global change effects on, 289-314
  formation of, 250
  nutrient release from, dynamics of, 232, 234-237,236f-238f
  PCB–polluted, 271-272
  rice field, microbial diversity in, 273-276

Soil(s)
agricultural
C storage in, SOM preservation and, 302-303
improved management of, biotic-abiotic interactions in nutrient dynamics in, 209-248. *See also* Nutrient dynamics, biotic-abiotic interactions in, in agricultural soils
biological processes in, global change effects on. *See also* Biological processes, in soil, global change effects on
"Soil Biology Guide," 181
Soil C
availability of
altered, belowground processes effects of, 295-300
effects of global change–induced alteration in plant growth on, 294-295
storage of, in agricultural soils, SOM preservation and, 302-303
Soil conservation, in temperate tree-based intercropping systems, 356
Soil environments
favorable, maintenance of, 230-231
microbial and genetic diversity in, 249-287
introduction to, 250-253
Soil erosion, 427
Soil fauna, dynamics of, 296-297
Soil fertility, in temperate tree-based intercropping systems, interactions related to, 344-347,344t,346f,347f
Soil food web, 116f
Soil health, conserving/improving of, 422-426
Soil macrofauna, biology and ecology of, 180-181
Soil management, integrating effects of, 97-98
Soil mesofauna, biology and ecology of, 178-180
Soil microbial biomass, activity, and community structure, 295-296
Soil microfauna, biology and ecology of, 178-180
Soil microorganisms, tillage effects on, 194-195
Soil minerals, nutrient solubilization from, 190-191
Soil N, availability of
altered, belowground processes effects of, 295-300
effects of global change–induced alteration in plant growth on, 294-295
Soil organic matter (SOM), 419
preservation of, and C storage in agricultural soils, 302-303
Soil organic matter (SOM) accumulation, 87
Soil organisms
biology and ecology of, 177-181
effects on nutrient dynamics in temperate agroecosystems, 175-207. *See also* Agroecosystem(s), temperate, nutrient dynamics in, soil organisms effects on
nutrient losses due to, 192-193
nutrient transformations by, 182-193
tillage effects on, 193-197
microorganisms, 194-195
Soil processes, 211-221,216,219t
Soil properties, 270-271
SOM. *See* Soil organic matter (SOM)
SOM accumulation. *See* Soil organic matter (SOM) accumulation
Sotherton, N.W., 377
Soussana, J.F., 298
Sown grass, 388-389

Sown grass margins, 387-388
Soybean(s), Roundup Ready, 419
Sparling, G.P., 234
Spatial variation, in nutrient release, 231-237,233f,236f-238f
Spedding, C.R.W., 421
Spedding, T., 209
Spooner, B., 252
Springtail(s), biology and ecology of, 180
Sriskandarajah, N., 36
SS. *See* Sewage sludge (SS)
SSCP. *See* Single strand conformation polymorphism (SSCP)
SSCP profiles. *See* Single strand conformation polymorphism (SSCP) profiles
St. John's wort, 470
Stahl, P., xv
Staver, C.P., 167
Staver, K.W., 421
Steltzer, H., 292
Stenberg, B., 113
Stenberg, M., 113
Steppuhn, H., 429
Sterile strips, 389-390
Sterility, cytoplasmic male, 472-473
Stress(es), abiotic, 475-479
Strips across fields, flower margins and, 389
Stubbs, T.L., 105,114
Sukno, S., 520
Swanton, C., xiv
Switchgrass, 429
Synecology, 65-66
Systems courses, 35-38

Tabatabai, M.A., 423
Tansley, A.C., xiv
Taylor, D.C., 413
Teasdale, J.R., 301
Temperate tree-based intercropping systems, biophysical and ecological interactions in, 339-363. *See also* Intercropping systems, temperate tree-based, biophysical and ecological interactions in
Temperature gradient gel electrophoresis (TGGE), 264
Temporal variation, in nutrient release, 231-237,233f,236f-238f
TGGE. *See* Temperature gradient gel electrophoresis (TGGE)
The Land Institute, xiv,12,29,513,514, 517,518,520,521
Thevathasan, N.V., 339
Thomas, R.G.B., 299
Thrupp, L.A., 315
Tillage
  effects on field level diversity, 268-269,268f
  effects on soil organisms, 193-197
Tillage practices, during transition to no-till cropping, 107-108,108f
Tillage systems, 123-124
Tilman, D., 141,142,146,522
Tomlin, A.D., xv
Tracy, B., xv
Tran, T.S., 236,237
Transgenic crops, in production systems, examining effects of
  agricultural production effects of, 463-466
  biotechnology applications of, 462
  ecological context for, 457-489
    case study, 466-467
    competition in, 461-467
    diseases, 472-475
    herbivory, 468-472. *See also* Herbivory
    introduction to, 458-461
  magnitude of problem, 461-462
Transgenic plants, effect of, 265-266
Transgenic "Roundup Ready," 462
Tree-based intercropping systems, temperate, biophysical and ecological interactions in, 339-363. *See also* Intercropping systems, temperate tree-based, biophysical and ecological interactions in
Troiano, J., 415,418

Tropics, diversified cropping systems in, agroecological basis for designing of, 81-103. *See also* Agroecology, in designing diversified cropping systems in Tropics
Tsiouris, S., 382
Tullgren apparatus, 126
Tyler Prize for Experimental Achievement, xiv

UCSC. *See* University of California at Santa Cruz (UCSC)
Udo de Haes, H.A., 370
Uetz, G.W., 181
UK Countryside Stewardship scheme, 387
UK Environmentally Sensitive Areas scheme, 387
Unified Neutral Theory, 140
Universal Soil Loss Equation, 427
University(ies)
  agricultural, in agroecology, 24-26,24f
  of future, designing of, 38-39,39f
University of California at Santa Cruz (UCSC), 35
University of Georgia, xiv
  at HorseShoe Bend (HSB) experimental site, xiv,53
University of Guelph, 340
  Plant Agriculture of, xiv
University of Guelph Agroforestry Research Station, 342,349,355,357
University of Nebraska, 36
University of the Yucutan, 35
University of Western Sydney, 494
  at Hawkesbury, New South Wales, Australia, 36
U.S. Fisheries and Wildlife, 166
U.S. Forest Service, 166
US National Water Quality Assessment program, 415
USDA-ARS, 516

Value(s), "c," 427
van Bruggen, A.H.C., 262
Van der Meer, H.G., 442
van Emden, H.F., 377
Van Keulen, H., 442
Van Wychen, L., 144
Vargas, M.A.T., 299
Vasseur, L., 300
Vegetation diversity, pest outbreaks due to, 95-97
Vegetation management, 430-431
"Veritable pleasure grounds," 388
Viral resistance, 473
Vogel, K.P., 429
Voisin, A., 494
Volterra, 146
Voroney, R.P., 426

Waddington, J., 429
Wageningen Agricultural University, 445,446
Wagoner, P., 518
Walsingham, J.M., 421
Walters, C.J., 166
Wang, P., 249
Wardle, D.A., 196
Water use efficiencies (iWUE), 353
Weed communities, in agroecosystems, 137-151. *See also* Crop-weed interactions, ecology of
Weed ingress, in field margin habitats, 374-376,375t,376t
Weil, R.R., 424
Weilharter, A., 270
Western Region Plant Introduction Station, 519
Whalen, J.K., 175,413
White, J.A., 457
Whitelaw, M.A., 220
Widmer, T.L., 113-114
Wild flower margins, 388-389
Wildlife, management of field margins for, 385-391,386t,391f

Wildlife mixtures, 389
Wildlife strips, uncropped, 387
Williams, I.H., 389
Williams, P.A., 351
Willis, A.J., 382
Wilson, E.O., 158
Wilson, P.J., 374
Windbreak(s), 429
Wolters, V., 123,181
World Intellectual Properties Organization, 333
World War II, 372,408,430,441
Wurtz, T.L., 430

Young, M., 413

Zalom, F.G., 8
Zhang, P., 339,348
Zhang, W., 289
Ziska, L.H., 301
Zwart, K.B., 196